Morningside Lectures in Mathematics

www.intlpress.com/mlm

MLM 4 Lectures on the Analysis of Nonlinear Partial Differential Equations: Part 4

MLM 3 Lectures on the Analysis of Nonlinear Partial Differential Equations: Part 3

MLM 2 Lectures on the Analysis of Nonlinear Partial Differential Equations: Part 2

MLM 1 Lectures on the Analysis of Nonlinear Partial Differential Equations: Part 1

MORNINGSIDE LECTURES IN MATHEMATICS

Series Editors

MORNINGSIDE LECTURES IN MATHEMATICS
VOLUME IV

Lectures on the Analysis of Nonlinear Partial Differential Equations: Part 4

edited by

Jean-Yves Chemin
Fanghua Lin
Ping Zhang

International Press
www.intlpress.com

高等教育出版社
HIGHER EDUCATION PRESS

Morningside Lectures in Mathematics, Volume IV
Lectures on the Analysis of Nonlinear Partial Differential Equations: Part 4

Editors:

Jean-Yves Chemin
Laboratoire J.-L. Lions, Université Pierre et Marie, Paris

Fanghua Lin
Courant Institute of Mathematical Sciences, New York University

Ping Zhang
Academy of Mathematics and Systems Sciences, Chinese Academy of Sciences

2010 Mathematics Subject Classification. 00Bxx, 35-XX.

ISBN: 978-1-57146-317-3

Printed in the United States of America.

20 19 18 17 16   1 2 3 4 5 6 7 8 9

Preface to the First Volume

In summer of 2001, we initiated a summer school program on the harmonic analysis and its applications in nonlinear partial differential equations, with special emphases on nonlinear Schrödinger equations, kinetic equations of Bolzmann type and classical fluid equations. Over the years, there have been many distinguished mathematicians working in these fields who have come to help our program and to give series of special lectures. The program has been shown to be particularly helpful to young researchers and students. The lectures involved have gradually turned into more formal and regular seminars on *Analysis in Partial Differential Equations* at the Morningside Center of Mathematics, Academy of Mathematics and Systems Science, Chinese Academy of Sciences.

From June 2007 to January 2008, we held a special semester PDE–program. We invited many mathematicians and experts in mathematical theory of fluid mechanics and quantum mechanics. The visitors during that period include: Bresch Didier, Carles Rémi, Jean-Yves Chemin, Desvillettes Laurent, Lopes Filho Milton C, Nussenzveig Lopes Helena J., Novotny Antonin, Chao-Jiang Xu, Chongchun Zeng, Ping Zhang and Yuxi Zheng, who gave a series of lectures and provided excellent lecture notes. It is no doubt that these lecture notes would be very useful for many researchers and students. In this volume, we have collected lecture notes by M. C. Lopes concerning the boundary layers of incompressible fluid flow; by C. J. Xu on the micro-local analysis and its applications to the regularities of kinetic equations; by Y. X. Zheng on the weak solutions of variational wave equation from liquid crystals; and by P. Zhang and Z. F. Zhang on the free boundary problem of Euler equations. In addition, we also included the notes by F. Nier on the hypoellipticity of Fokker-Planck operator and Witten-Laplace operator that were given earlier in the summer of 2006.

We have planned to publish in the forthcoming volumes the other lecture notes. Some are from past lectures at our program and some will be collected from the newly scheduled seminars. We hope that the publication of these lecture notes may provide valuable references and up-to-date descriptions of current developments of various and related research topics, that will benefit many young researchers or graduate students. We wish to take this opportunity to thank the Morningside Center of Mathematics, the Institute of Mathematics of AMSS that provides all necessary supports. We are particularly grateful to professor Lo Yang for his constant help, supports and encouragements to our program. We also would like to thank Guilong Gui for his careful preparations of the Latex file of the entire book. We finally appreciate for the financial support from the Chinese Academy of Sciences.

Fanghua Lin in New York
Xueping Wang in Nantes
Ping Zhang in Beijing
November 3, 2008

Preface to the Fourth Volume

This book is a sequel to the previous volumes *Lectures on the Analysis of Nonlinear Partial Differential Equations*, Vol. I, Vol. II, and Vol. III. In this fourth volume we have collected those lecture notes by Jean-Yves Chemin, Hongjie Dong, Xiaochun Li, Fanghua Lin, Alexis F. Vasseur, Jiahong Wu, and Xiaoyi Zhang. We appreciate very much their time and efforts to provide us with these excellent notes. We believe these lecture notes will serve as valuable references on current developments in various research topics in nonlinear partial differential equations (PDE).

The notes collected here came initially from the seminars on Analysis in Partial Differential Equations. It was a part of the PDE program (from April through October 2012) at the Academy of Mathematics and Systems Science, Chinese Academy of Sciences (CAS). We would like to take this opportunity to thank the National Center for Mathematics and Interdisciplinary Sciences, Academy of Mathematics and Systems Science as well as the Morningside Center of Mathematics of CAS for the financial support. We thank also several other distinguished visitors who had delivered excellent lectures at the program: Hammadi Abidi, Albert Cohen, Marius Paicu, Benoit Perthame and Jean-Claude Saut. We wish their notes would be included in our future volumes.

Finally, we wish to thank Prof. Lo Yang, Prof. Nanhua Xi, and Prof. Yuefei Wang, for their continuing support over all these years which was essential to the success of this PDE program. We also thank Dr. Guilong Gui for the careful preparation of the Latex files that make up this entire book.

Jean-Yves Chemin in Paris
Fanghua Lin in New York
Ping Zhang in Beijing
June, 2015

Contents

Lectures on the Analysis of Nonlinear Partial Differential Equations Vol. 4
MLM 4, pp. 1–53

Profile Decomposition and Its Applications to Navier-Stokes System

Jean-Yves Chemin*

Abstract

The purpose of theses lectures is the description, up to an extraction, of lack of compactness of the Sobolev embeddings as achieved in the work [18] of P. Gérard. We apply this result to the description of the possible blow up in the three dimensional incompressible Navier-Stokes.

Introduction

This text is notes of a series of lectures given first in the Morningside Center of Mathematics in February and March 2014 in Beijing and then at Jacques-Louis Lions laboratory in May and June 2014. I thank very warmly the audience of this two series of lectures for their interest remarks. I want to thank especially Professor Ping Zhang for his rereading of the notes and a lot of suggestions of improvements.

The purpose of theses lectures was the description, up to an extraction, of lack of compactness of the Sobolev embeddings as achieved in the work [18] of P. Gérard. We apply this result to the description of the possible blow up in the three dimensional incompressible Navier-Stokes.

More precisely, the first section will be devoted to the proof of a refined Sobolev inequality which is in particular invariant under translation in Fourier space (i.e. multiplication by an oscillating function). This involves some particular classes of Besov spaces we shall define and study.

The second section is devoted to the statement and the detailed proof of the P. Gérard's result which provides a precise description, up an extraction, of a sequence of function which is bounded in the homogeneous Sobolev $\dot{H}^s(\mathbb{R}^d)$. It claims that in some sense, it is the sum of dilation and translation of some

*Laboratoire J.-L. Lions, Case 187, Université Pierre et Marie Curie, 75 230 Paris Cedex 05, France. E-mail: chemin@ann.jussieu.fr

given function in $\dot{H}^s(\mathbb{R}^d)$. For a given sequence $(\lambda_n)_{n\in\mathbb{N}}$, the concept of $(\lambda_n)_{n\in\mathbb{N}}$-oscillating sequence is defined. Together with some particular class of Besov spaces, this turns out to be the crucial tool of the proof.

In the third section, we first recall some basic results about incompressible Navier-Stokes system. Then we prove some result about bounded sequences of initial data in the spirit of the work by I. Gallagher (see [14]). And we apply this result to prove the celebrated result by L. Escauriaza, G. Serëgin, V. Sverak (see [12]) which claims that if a solution u to the incompressible three dimensional Navier-Stokes equation develops a singularity at time $T^\star$, then we have

$$\limsup_{t\to T^\star} \|u(t)\|_{\dot{H}^{\frac12}} = \infty.$$

We follow the approach developed by C. Kenig and G. Koch in [20].

1 Sobolev embeddings revisited

1.1 Sobolev spaces and Sobolev embedding

Definition 1.1. *Let s be a real number. The homogeneous Sobolev space $\dot{H}^s(\mathbb{R}^d)$ is the set of tempered distributions u the Fourier transform of which $\widehat{u}$ belongs to $L^1_{loc}(\mathbb{R}^d)$ and satisfies*

$$\|u\|^2_{\dot{H}^s} \stackrel{\text{def}}{=} \int_{\mathbb{R}^d} |\xi|^{2s}|\widehat{u}(\xi)|^2 d\xi < \infty.$$

The following interpolation inequality is of constant use. We left as an exercise to the reader the proof that $\dot{H}^s(\mathbb{R}^d)$ is a Hilbert space in the case when s is less than $d/2$.

Proposition 1.1. *If u belongs to $\dot{H}^{s_1} \cap \dot{H}^{s_2}$, for any s between s_1 and s_2, u belongs on $\dot{H}^s$ and*

$$\|u\|_{\dot{H}^s} \leq \|u\|^{\theta}_{\dot{H}^{s_1}} \|u\|^{1-\theta}_{\dot{H}^{s_2}} \quad \textit{with} \quad s = \theta s_1 + (1-\theta)s_2.$$

Proof. Let us simply apply Hölder inequality with $p = \frac{1}{\theta}$ and $q = \frac{1}{1-\theta}$ to the functions $\xi \mapsto |\xi|^{2\theta s_1}$, $\xi \mapsto |\xi|^{2(1-\theta)s_2}$ and the Borel measure $|\widehat{u}(\xi)|^2 d\xi$ and this gives the result. □

Let us state the classical Sobolev inequality and its dual version.

Theorem 1.1. *If s belongs to $[0, \frac{d}{2}[$, then the space $\dot{H}^s(\mathbb{R}^d)$ is continuously embedded in $L^{\frac{2d}{d-2s}}(\mathbb{R}^d)$. If p belongs to $]1,2]$, then the space $L^p(\mathbb{R}^d)$ is continuously included in the space $\dot{H}^{d(\frac12-\frac1p)}$.*

Proof. The second part is easily deduced from the first one proceeding by duality. Let us write that

$$\|a\|_{\dot{H}^s} = \sup_{\|\varphi\|_{\dot{H}^{-s}(\mathbb{R}^d)}\leq 1} \langle a, \varphi\rangle.$$

As $s = d\left(\frac{1}{2} - \frac{1}{p}\right) = d\left(1 - \frac{1}{p} - \frac{1}{2}\right)$, we have by the first part

$$\|\varphi\|_{L^{\overline{p}}} \leq C\|\varphi\|_{\dot{H}^{-s}}$$

where $\overline{p}$ is the conjugate of p defined by $\frac{1}{p} + \frac{1}{\overline{p}} = 1$ and thus

$$\begin{aligned}\|a\|_{\dot{H}^s} &\leq C \sup_{\|\varphi\|_{L^{\overline{p}}}\leq 1} \langle a, \varphi\rangle \\ &\leq C\|a\|_{L^p}.\end{aligned}$$

This concludes the proof of the second part. □

There are many different proofs of the first part. We shall use a frequency cut off argument which gives for free a refined version of this inequality which will be crucial in the second chapter. Let us introduce the following definition.

Definition 1.2. *Let θ be a function of $\mathcal{S}(\mathbb{R}^d)$ such that $\widehat{\theta}$ be compactly supported, has value 1 near 0 and satisfies $0 \leq \widehat{\theta} \leq 1$. For u in $\mathcal{S}'(\mathbb{R}^d)$ and $\sigma > 0$, we set*

$$\|u\|_{\dot{B}^{-\sigma}} \stackrel{\text{def}}{=} \sup_{A>0} A^{d-\sigma}\|\theta(A\cdot) \star u\|_{L^\infty}.$$

The fact that $B^{-\sigma}$ is a Banach space is an exercise left to the reader. We shall see later on that the space is independent of the choice of the function θ. Let us observe that if u belongs to $\dot{H}^s$, then $\widehat{u}$ is locally in L^1 and the function $\widehat{\theta}(A^{-1}\cdot)\widehat{u}$ is in L^1. The inverse Fourier theorem implies that

$$\begin{aligned}\|A^d\theta(A\cdot) \star u\|_{L^\infty} &\leq (2\pi)^{-d}\|\widehat{\theta}(A^{-1}\cdot)\widehat{u}\|_{L^1} \\ &\leq (2\pi)^{-d}\int_{\mathbb{R}^d} \widehat{\theta}(A^{-1}\xi)|\xi|^{-s}|\xi|^s|\widehat{u}(\xi)|\,d\xi.\end{aligned}$$

Using the fact that $\widehat{\theta}$ is compactly supported, Cauchy-Schwarz inequality implies that

$$\|A^d\theta(A\cdot) \star u\|_{L^\infty} \leq \frac{C}{\left(\frac{d}{2} - s\right)^{\frac{1}{2}}} A^{\frac{d}{2}-s}\|u\|_{\dot{H}^s}.$$

This means exactly that the space $\dot{H}^s$ is continuously included in $\dot{B}^{s-\frac{d}{2}}$. By multiplication, we can assume that $\|u\|_{\dot{B}^{s-\frac{d}{2}}} = 1$. Then let us estimate $\|u\|_{L^p}$. We decompose the function u in low and high frequencies. More precisely, let us write

$$u = u_{\ell,A} + u_{h,A} \quad \text{with} \quad u_{\ell,A} = \mathcal{F}^{-1}(\widehat{\theta}(A^{-1}\cdot)\widehat{u}) \tag{1.1}$$

where θ is the function of Definition 1.2. The triangle inequality implies that

$$\big(|u| > \lambda\big) \subset \big(|u_{\ell,A}| > \lambda/2\big) \cup \big(|u_{h,A}| > \lambda/2\big)\cdot$$

By definition of $\|\cdot\|_{\dot{B}^{s-\frac{d}{2}}}$, we have $\|u_{\ell,A}\|_{L^\infty} \leq A^{\frac{d}{2}-s}$. From this, we deduce that

$$A = A_\lambda \stackrel{\text{def}}{=} \left(\frac{\lambda}{2}\right)^{\frac{p}{d}} \Longrightarrow \mu\big(|u_{\ell,A}| > \lambda/2\big) = 0.$$

We deduce that

$$\|u\|_{L^p}^p \leq p \int_0^\infty \lambda^{p-1} \mu\big(|u_{h,A_\lambda}| > \lambda/2\big)\, d\lambda.$$

Using that

$$\mu\big(|u_{h,A_\lambda}| > \lambda/2\big) \leq 4\frac{\|u_{h,A_\lambda}\|_{L^2}^2}{\lambda^2},$$

we get

$$\|u\|_{L^p}^p \leq 4p \int_0^\infty \lambda^{p-3} \|u_{h,A_\lambda}\|_{L^2}^2\, d\lambda.$$

Because the Fourier transform is (up to a constant) an isometry on $L^2(\mathbb{R}^d)$ and the function $\widehat{\theta}$ has value 1 near 0, we thus get for some $c > 0$ depending only on $\widehat{\theta}$,

$$\|u\|_{L^p}^p \leq 4p\,(2\pi)^{-d} \int_0^\infty \lambda^{p-3} \int_{(|\xi| \geq cA_\lambda)} |\widehat{u}(\xi)|^2\, d\xi\, d\lambda \tag{1.2}$$

for some positive constant c. Now, by definition of A_λ, we have

$$|\xi| \geq cA_\lambda \Longleftrightarrow \lambda \leq C_\xi \stackrel{\text{def}}{=} 2\Big(\frac{|\xi|}{c}\Big)^{\frac{d}{p}}.$$

Fubini's theorem thus implies that

$$\begin{aligned}\|u\|_{L^p}^p &\leq 4p\,(2\pi)^{-d} \int_{\mathbb{R}^d} \bigg(\int_0^{C_\xi} \lambda^{p-3} d\lambda\bigg) |\widehat{u}(\xi)|^2\, d\xi\\ &\leq (2\pi)^{-d} \frac{p2^p}{p-2} \int_{\mathbb{R}^d} \Big(\frac{|\xi|}{c}\Big)^{\frac{d(p-2)}{p}} |\widehat{u}(\xi)|^2\, d\xi.\end{aligned}$$

As $s = d\left(\frac{1}{2} - \frac{1}{p}\right)$, the theorem is proved. □

1.2 Interpretation in terms of Besov spaces and oscillations

In fact the above proof tells more than the classical Sobolev theorem, namely the following theorem.

Theorem 1.2. *Let s be in $]0, d/2[$. There exists a constant C depending only on d and $\widehat{\theta}$ such that*

$$\|u\|_{L^p} \leq \frac{C}{(p-2)^{\frac{1}{p}}} \|u\|_{\dot{B}^{s-\frac{d}{2}}}^{1-\frac{2}{p}} \|u\|_{\dot{H}^s}^{\frac{2}{p}} \quad \textit{with} \quad p = \frac{2d}{d-2s}\cdot$$

Let us see what type of improvement it is compared with the classical inequality. Let φ be a given function in the Schwartz class $\mathcal{S}(\mathbb{R}^d)$ and ω a unit vector of $\mathbb{R}^d$. Let us consider the family of functions defined by

$$u_\varepsilon(x) = e^{i\frac{(x|\omega)}{\varepsilon}} \varphi(x).$$

Let us prove that for any σ in $]0, d]$ we have

$$\|u_\varepsilon\|_{\dot{B}^{-\sigma}} \lesssim \varepsilon^\sigma. \tag{1.3}$$

By Hölder's inequality, we have

$$A^d \|\theta(A\cdot) \star \phi_\varepsilon\|_{L^\infty} \leq \|\theta\|_{L^1} \|\phi\|_{L^\infty}.$$

From this we deduce that, if $A\varepsilon \geq 1$ then we have

$$A^{d-\sigma} \|\theta(A\cdot) \star \phi_\varepsilon\|_{L^\infty} \leq \varepsilon^\sigma \|\theta\|_{L^1} \|\phi\|_{L^\infty}. \tag{1.4}$$

If $A\varepsilon \leq 1$, we perform integrations by parts. More precisely, using that

$$(-i\varepsilon\partial_1)^d e^{i\frac{x_1}{\varepsilon}} = e^{i\frac{x_1}{\varepsilon}}$$

and Leibniz formula, we get

$$\begin{aligned} A^d(\theta(A\cdot) \star \phi_\varepsilon)(x) &= (iA\varepsilon)^d \int_{\mathbb{R}^d} \partial_{y_1}^d \left(\theta(A(x-y))\phi(y)\right) e^{i\frac{y_1}{\varepsilon}}\, dy \\ &= (iA\varepsilon)^d \sum_{k\leq d} \binom{d}{k} A^k((-\partial_1)^k\theta)(A\cdot) \star (e^{i\frac{y_1}{\varepsilon}} \partial_1^{d-k}\phi)(x). \end{aligned}$$

Using Hölder's inequalities, we get that

$$A^k \left\|((-\partial_1)^k\theta)(A\cdot) \star (e^{i\frac{y_1}{\varepsilon}} \partial_1^{d-k}\phi)\right\|_{L^\infty} \leq \|\partial_1^k\theta\|_{L^{\frac{d}{k}}} \|\partial_1^{d-k}\phi\|_{L^{(\frac{d}{k})'}}.$$

Thus, we get $A^d\|\theta(A\cdot) \star \phi_\varepsilon\|_{L^\infty} \leq C(A\varepsilon)^d$. As we are in the case when $A\varepsilon \leq 1$, we get, for any $\sigma \leq d$,

$$A^d \|\theta(A\cdot) \star \phi_\varepsilon\|_{L^\infty} \leq C(A\varepsilon)^\sigma.$$

Together with (1.4), this concludes the proof of Inequality (1.3).

Considering that $\|u_\varepsilon\|_{\dot{H}^s} \lesssim \varepsilon^{-s}$, then we can check that

$$\|u_\varepsilon\|_{\dot{B}^{s-\frac{d}{2}}}^{1-\frac{2}{p}} \|u_\varepsilon\|_{\dot{H}^s}^{\frac{2}{p}} \lesssim \varepsilon^{(\frac{d}{2}-s)(1-\frac{2}{p})-\frac{2s}{p}} \sim 1.$$

This shows that the refined inequality of Theorem 1.2 is invariant under translation in Fourier spaces (i.e. multiplication by oscillating functions).

The spaces defined in Definition 1.2 have a universal property: they are the biggest normed spaces which are translation invariant and which have the same scaling. More precisely we have the following proposition.

Proposition 1.2. *Let E be a norm space continuously included in the space of tempered distribution. Let us assume that the space E is globally invariant under dilations and translations and that a constant C and a positive real number σ exists such that*

$$\|u(\lambda \cdot - \vec{a})\|_E \leq C\lambda^{-\sigma}\|u\|_E.$$

Then the space E is continuously embedded in $B^{-\sigma}$.

Proof. As E is continuously included in $\mathcal{S}'$, then we have

$$|\langle u, \theta \rangle| \leq C\|u\|_E.$$

Because of the hypothesis on E, we get

$$|\langle u(A^{-1} \cdot +x), \theta \rangle| \leq \|u(A^{-1} \cdot +x)\|_E \leq CA^\sigma \|u\|_E.$$

As we have

$$\langle u(A^{-1} \cdot +x), \theta \rangle = A^d(\theta(A\cdot) \star u)(x),$$

we get the result. □

1.3 The link with Besov norms

The following proposition is an exercise about partition of unity.

Proposition 1.3. *Let $\mathcal{C}$ be the annulus $\{\xi \in \mathbb{R}^d : 3/4 \leq |\xi| \leq 8/3\}$. There exist two radial functions χ and φ valued in the interval $[0,1]$, belonging respectively to $\mathcal{D}(B(0,4/3))$ and to $\mathcal{D}(\mathcal{C})$, and such that*

$$\forall \xi \in \mathbb{R}^d, \ \chi(\xi) + \sum_{j \geq 0} \varphi(2^{-j}\xi) = 1, \tag{1.5}$$

$$\forall \xi \in \mathbb{R}^d \backslash \{0\}, \ \sum_{j \in \mathbb{Z}} \varphi(2^{-j}\xi) = 1, \tag{1.6}$$

$$|j - j'| \geq 2 \Rightarrow \mathrm{Supp}\ \varphi(2^{-j}\cdot) \cap \mathrm{Supp}\ \varphi(2^{-j'}\cdot) = \emptyset, \tag{1.7}$$

$$j \geq 1 \Rightarrow \mathrm{Supp}\ \chi \cap \mathrm{Supp}\ \varphi(2^{-j}\cdot) = \emptyset, \tag{1.8}$$

the set $\widetilde{\mathcal{C}} \stackrel{\text{def}}{=} B(0,2/3) + \mathcal{C}$ is an annulus and we have

$$|j - j'| \geq 5 \Rightarrow 2^{j'}\widetilde{\mathcal{C}} \cap 2^j \mathcal{C} = \emptyset. \tag{1.9}$$

Besides, we have

$$\forall \xi \in \mathbb{R}^d, \ \frac{1}{2} \leq \chi^2(\xi) + \sum_{j \geq 0} \varphi^2(2^{-j}\xi) \leq 1, \tag{1.10}$$

$$\forall \xi \in \mathbb{R}^d \backslash \{0\}, \ \frac{1}{2} \leq \sum_{j \in \mathbb{Z}} \varphi^2(2^{-j}\xi) \leq 1. \tag{1.11}$$

Let us state the following definition.

Definition 1.3. *Let s be a real number, and (p,r) be in $[1,\infty]^2$. The homogeneous Besov space $\dot{B}^s_{p,r}$ is the subset of distributions u of $\mathcal{S}'_h$ such that*

$$\|u\|_{\dot{B}^s_{p,r}} \stackrel{\text{def}}{=} \Big(\sum_{j \in \mathbb{Z}} 2^{rjs} \|\dot{\Delta}_j u\|^r_{L^p}\Big)^{\frac{1}{r}} < \infty.$$

Let us point out that in the case when $p = r = 2$, this is homogeneous Sobolev spaces and in the case when $r = \infty$, this definition coincides with Definition 1.2.

Proposition 1.4. *The $\dot{H}^s$ norm and the $\dot{H}^s$ norm are equivalent. For any positive σ, the two norms $\dot{B}^{-\sigma}$ and $\dot{B}^{-\sigma}_{\infty,\infty}$ norm are equivalent. Moreover, we have, for s less than $d/2$,*

$$\|a\|_{\dot{B}^s_{2,\infty}} \lesssim \|a\|_{\dot{B}^{s-\frac{d}{2}}} \sim \|a\|_{\dot{B}^{s-\frac{d}{2}}_{\infty,\infty}}.$$

Proof. It is possible to use Proposition 1.2. Let us give here a direct proof of this inequality. Let us write that

$$A^d\|\theta(A\cdot) \star u\|_{L^\infty} \leq (2\pi)^{-d}\|\widehat{\theta}(A^{-1}\cdot)\widehat{u}\|_{L^1}.$$

Because the Fourier transform of θ is compactly supported, we have

$$\widehat{\theta}(A^{-1}\cdot)\widehat{u} = \sum_{2^j \lesssim A} \widehat{\theta}(A^{-1}\cdot)\mathcal{F}(\Delta_j u).$$

Thus, we get

$$\begin{aligned}
A^d\|\theta(A\cdot) \star u\|_{L^\infty} &\lesssim \sum_{2^j \lesssim A} \|\theta(A^{-1}\cdot)\mathcal{F}(\Delta_j u)\|_{L^1} \\
&\lesssim \sum_{2^j \lesssim A} 2^{j\frac{d}{2}}\|\Delta_j u\|_{L^2} \\
&\lesssim \Big(\sum_{2^j \lesssim A} 2^{j(\frac{d}{2}-s)}\Big)\|u\|_{\dot{B}^s_{2,\infty}} \\
&\lesssim A^{\frac{d}{2}-s}\|u\|_{\dot{B}^s_{2,\infty}}.
\end{aligned}$$

This proves the proposition. □

In order to figure out the difference between the norm $\dot{H}^s = \dot{B}^s_{2,2}$ and the norm $\dot{B}^s_{2,\infty}$, let us consider the following example which based on the idea of lacunar series. Let us consider a function χ in $\mathcal{S}$ such that its Fourier transform is supported in a (small) ball of center 0 and radius ε_0. Let us consider, for a vector ω of $\mathbb{R}^d$ of Euclidian norm 3/2, the sequence of functions $(f_n)_{n\in\mathbb{N}}$ defined by

$$f_n(x) = \sqrt{n}\sum_{j\geq n} 2^{-js}\frac{1}{j+1}e^{i2^j(x|\omega)}\chi(x).$$

As the support of the Fourier transform of the function $e^{i2^j(\cdot|\omega)}\chi$ is included in $2^j\omega + B(0,\varepsilon_0)$, we get

$$\|f_n\|^2_{H^s} \sim n\sum_{j\geq n}\frac{1}{(j+1)^2} \sim 1 \quad \text{and} \quad \|f_n\|_{\dot{B}^s_{2,\infty}} \lesssim \frac{1}{\sqrt{n}}.$$

2 The theory of profiles

2.1 The fundamental theorem about bounded sequences

Let us state the following definition.

Definition 2.1. *We say that a sequence* $(\lambda_n^j, x_n^j)_{(j,n)\in\mathbb{N}^2}$ of $]0,\infty[\times\mathbb{R}^3$ *is a sequence of scales and cores if it satisfies*

$$j\neq k \Longrightarrow \text{ either } \lim_{n\to\infty}\left|\log\Bigl(\frac{\lambda_n^j}{\lambda_n^k}\Bigr)\right| = +\infty \quad \text{or} \quad \lambda_n^j=\lambda_n^k \text{ and } \lim_{n\to\infty}\frac{|x_n^j-x_n^k|}{\lambda_n^j}=\infty. \tag{2.1}$$

The following theorem has been proved by P. Gérard in [18] and describes, up to extraction, the defect of compactness of Sobolev embeddings.

Theorem 2.1. *Let* $(u_n)_{n\in\mathbb{N}}$ *be a bounded sequence of* $\dot H^s$ *with* s *less than* $d/2$. *Then there exists a sequence of scales and cores* $(\lambda_n^j, x_n^j)_{(j,n)\in\mathbb{N}^2}$ *in the sense of Definition* 2.1, *a sequence* $(\varphi^j)_{j\in\mathbb{N}}$ *in* $\dot H^s$ *and a sequence* $(r_n^j)_{(j,n)\in\mathbb{N}^2}$ *of functions which satisfies, up to an extraction on* $(u_n)_{n\in\mathbb{N}}$, *the following properties:*

$$\forall J\in\mathbb{N},\ u_{\phi(n)}(x)=\sum_{j=0}^{J}\frac{1}{(\lambda_n^j)^{\frac d2-s}}\varphi^j\Bigl(\frac{x-x_n^j}{\lambda_n^j}\Bigr)+r_n^J(x),$$

$$\lim_{J\to\infty}\limsup_{n\to\infty}\|r_n^J\|_{\dot B^{s-\frac d2}_{\infty,\infty}}=0 \quad \text{and}$$

$$\forall J\in\mathbb{N},\ \lim_{n\to\infty}\Bigl(\|u_n\|_{\dot H^s}^2-\sum_{j=0}^{J}\|\varphi^j\|_{\dot H^s}^2-\|r_n^J\|_{\dot H^s}^2\Bigr)=0.$$

Remarks. The functions φ^j are called the profiles and they satisfies

$$\lim_{n\to\infty}(\lambda_n^j)^{\frac d2-s}u_n(\lambda_n^j\cdot+x_n^j)\rightharpoonup\varphi^j$$

in the sense of distributions. In particular φ^0 is the weak limit of $(u_n)_{n\in\mathbb{N}}$.

Moreover, the refined Sobolev embeddings proved in Theorem 1.2 implies that

$$\lim_{J\to\infty}\limsup_{n\to\infty}\|r_n^J\|_{L^p}=0 \quad \text{with} \quad p=\frac{2d}{d-2s}\cdot$$

The above theorem has the following corollary.

Corollary 2.1. *Let us consider a bounded sequence* $(u_n)_{n\in\mathbb{N}}$ *of* $\dot H^s$ *such that for any sequence* $(\lambda_n, x_n)_{n\in\mathbb{N}}$ *of* $]0,\infty[\times\mathbb{R}^d$, *we have*

$$\lambda_n^{\frac d2-s}u_n(\lambda_n\cdot+x_n)\rightharpoonup 0.$$

Then the infimum limit of $\|u_n\|_{L^p}$ *is* 0 *for* $p=\frac{2d}{d-2s}\cdot$

2.2 The extraction of the scales

Let us state some definitions.

Definition 2.2. *A scale is simply a sequence of positive real numbers. We say that two scales* $(\lambda_n)_{n\in\mathbb{N}}$ *and* $(\lambda'_n)_{n\in\mathbb{N}}$ *are orthogonal (denotes by* $(\lambda_n)_{n\in\mathbb{N}} \perp (\lambda'_n)_{n\in\mathbb{N}}$*) if*

$$\lim_{n\to\infty}\left|\log\Big(\frac{\lambda_n}{\lambda'_n}\Big)\right| = +\infty.$$

The following proposition describes some effects of orthogonality or not orthogonality of scales.

Proposition 2.1. *Let* $(\lambda_n)_{n\in\mathbb{N}}$ *and* $(\lambda'_n)_{n\in\mathbb{N}}$ *be two scales. If they are orthogonal, then*

$$\forall R>1\,,\ \exists n_R:\ n\geq n_R \Longrightarrow (\lambda'_n)^{-1}\mathcal{C}_R \subset \lambda_n^{-1}\mathcal{C}_R^c$$

where, as in all that follows, $\mathcal{C}_R \stackrel{\text{def}}{=} \{\xi\in\mathbb{R}^d : |\xi|\in[R^{-1},R]\}$.

If $(\lambda_n)_{n\in\mathbb{N}}$ *and* $(\lambda'_n)_{n\in\mathbb{N}}$ *are not orthogonal, an extraction* ϕ *exists such that*

$$\frac{1}{C} \leq \frac{\lambda_{\phi(n)}}{\lambda'_{\phi(n)}} \leq C.$$

Proof. Let us observe, for any R greater than 1 and any n,

$$\frac{1}{R} \leq \lambda'_n|\xi| \leq R \Longleftrightarrow \frac{\lambda_n}{\lambda'_n R} \leq \lambda_n|\xi| \leq \frac{\lambda_n R}{\lambda'_n}\cdot \tag{2.2}$$

As the scales $(\lambda_n)_{n\in\mathbb{N}}$ and $(\lambda'_n)_{n\in\mathbb{N}}$ are assumed to be orthogonal, it exists n_R such that

$$n\geq n_R \Longrightarrow \max\Big\{\frac{\lambda_n}{\lambda'_n},\frac{\lambda'_n}{\lambda_n}\Big\} \geq R^2.$$

Using (2.2), we observe that $\frac{\lambda_n}{\lambda'_n} \geq R^2$, then

$$\frac{1}{R} \leq \lambda'_n|\xi| \Longrightarrow \lambda_n|\xi| \geq R$$

and if $\frac{\lambda_n}{\lambda'_n} \geq R^2$, then

$$\lambda'_n|\xi| \leq R \Longrightarrow \lambda_n|\xi| \leq \frac{1}{R}$$

and the first assertion is proved.

The fact that the scales $(\lambda_n)_{n\in\mathbb{N}}$ and $(\lambda'_n)_{n\in\mathbb{N}}$ are not orthogonal exactly means that

$$\liminf_{n\to\infty}\left|\log\Big(\frac{\lambda_n}{\lambda'_n}\Big)\right| = C < \infty.$$

Thus an extraction ϕ, exists such that, for any n large enough

$$C^{-\frac{1}{2}} \leq \frac{\lambda_{\phi(n)}}{\lambda'_{\phi(n)}} \leq C^{\frac{1}{2}}.$$

The proposition is proved. □

Let us define the concept of sequence which is oscillatory or unrelated with respect to a scale.

Definition 2.3. *Let $(f_n)_{n\in\mathbb{N}}$ be a bounded sequence of functions in L^2 and $(\lambda_n)_{n\in\mathbb{N}}$ a scale. The sequence $(f_n)_{n\in\mathbb{N}}$ is said to be $(\lambda_n)_{n\in\mathbb{N}}$-oscillatory if*

$$\lim_{R\to\infty}\limsup_{n\to\infty}\int_{\lambda_n^{-1}\mathcal{C}_R^c}|\widehat{f}_n(\xi)|^2d\xi=0.$$

The sequence $(f_n)_{n\in\mathbb{N}}$ is said $(\lambda_n)_{n\in\mathbb{N}}$-unrelated if

$$\forall R>1\,,\ \lim_{n\to\infty}\int_{\lambda_n^{-1}\mathcal{C}_R}|\widehat{f}_n(\xi)|^2d\xi=0.$$

A first remark is that, if $(f_n)_{n\in\mathbb{N}}$ is $(\lambda_n)_{n\in\mathbb{N}}$-oscillatory, then the sequence $(g_n)_{n\in\mathbb{N}}$ defined by

$$g_n(x)\stackrel{\text{def}}{=}\lambda_n^{\frac{d}{2}}f_n(\lambda_nx)$$

is 1-oscillating. Let us notice that a typical example of a 1-oscillating sequence in a convergent sequence in $L^2(\mathbb{R}^d)$.

Now the following proposition translates in terms of orthogonality the properties of being oscillating or unrelated.

Proposition 2.2. *If $(f_n)_{n\in\mathbb{N}}$ is a $(\lambda_n)_{n\in\mathbb{N}}$ oscillating sequence and $(g_n)_{n\in\mathbb{N}}$ a $(\lambda_n)_{n\in\mathbb{N}}$ unrelated sequence, then*

$$\lim_{n\to\infty}(f_n|g_n)_{L^2}=0.$$

Proof. Let us write that

$$\begin{aligned}(2\pi)^d\big|(f_n|g_n)_{L^2}\big| &= \Big|\int_{\mathbb{R}^d}\widehat{f}_n(\xi)\overline{\widehat{g}}_n(\xi)d\xi\Big|\\ &\leq \Big|\int_{\lambda_n^{-1}\mathcal{C}_R^c}\widehat{f}_n(\xi)\overline{\widehat{g}}_n(\xi)d\xi\Big|+\Big|\int_{\lambda_n^{-1}\mathcal{C}_R}\widehat{f}_n(\xi)\overline{\widehat{g}}_n(\xi)d\xi\Big|\\ &\leq \|\widehat{f}_n\|_{L^2(\lambda_n^{-1}\mathcal{C}_R^c)}\|\widehat{g}_n\|_{L^2}+\|\widehat{f}_n\|_{L^2}\|\widehat{g}_n\|_{L^2(\lambda_n^{-1}\mathcal{C}_R)}.\end{aligned}$$

Let ε be a positive real number; because $(f_n)_{n\in\mathbb{N}}$ is $(\lambda_n)_{n\in\mathbb{N}}$-oscillating, a radius R_ε exists such that

$$\limsup_{n\to\infty}\int_{\lambda_n^{-1}\mathcal{C}_{R_\varepsilon}^c}|\widehat{f}_n(\xi)|^2d\xi\leq\frac{\varepsilon}{2}\cdot$$

As $(g_n)_{n\in\mathbb{N}}$ is $(\lambda_n)_{n\in\mathbb{N}}$-unrelated we have $\lim_{n\to\infty}\|\widehat{g}_n\|_{L^2(\lambda_n^{-1}\mathcal{C}_{R_\varepsilon})}=0$. Thus

$$\forall\varepsilon>0\,,\ \limsup_{n\to\infty}|(f_n|g_n)_{L^2}|\leq\varepsilon$$

and the result is proved. □

The following corollary will be useful.

Corollary 2.2. *Let (f_n^j, λ_n^j) be an element of $(L^2 \times]0,\infty[)^{\mathbb{N}^2}$ such that for any j, the sequence $(f_n^j)_{n\in\mathbb{N}}$ is $(\lambda_n^j)_{n\in\mathbb{N}}$-oscillating and for any j different from j' two scales $(\lambda_n^j)_{n\in\mathbb{N}}$ and $(\lambda_n^{j'})_{n\in\mathbb{N}}$ are orthogonal. Then, for any J,*

$$\lim_{n\to\infty}\left(\Big\|\sum_{j=0}^{J} f_n^j\Big\|_{L^2}^2 - \sum_{j=0}^{J}\|f_n^j\|_{L^2}^2\right) = 0.$$

Proposition 2.3. *Let $(f_n)_{n\in\mathbb{N}}$ be a bounded sequence of L^2 and $(\lambda_n)_{n\in\mathbb{N}}$ and $(\lambda'_n)_{n\in\mathbb{N}}$ two scales. Let us assume that $(f_n)_{n\in\mathbb{N}}$ is $(\lambda_n)_{n\in\mathbb{N}}$ oscillating.*

If the scales $(\lambda_n)_{n\in\mathbb{N}}$ and $(\lambda'_n)_{n\in\mathbb{N}}$ are orthogonal then $(f_n)_{n\in\mathbb{N}}$ is $(\lambda'_n)_{n\in\mathbb{N}}$ unrelated.

Conversely, if $\liminf_{n\to\infty}\|f_n\|_{L^2}$ is positive and the sequence $(f_n)_{n\in\mathbb{N}}$ is $(\lambda'_n)_{n\in\mathbb{N}}$-unrelated, then the scales $(\lambda_n)_{n\in\mathbb{N}}$ and $(\lambda'_n)_{n\in\mathbb{N}}$ are orthogonal.

Proof. Let R be a real number greater than 1 and a positive real number ε. Proposition 2.1 implies that

$$\forall R>1\,,\ \exists n_R : n\geq n_R \Longrightarrow \int_{(\lambda'_n)^{-1}\mathcal{C}_R} |\widehat{f}_n(\xi)|^2 d\xi \leq \int_{\lambda_n^{-1}\mathcal{C}_R^c} |\widehat{f}_n(\xi)|^2 d\xi. \tag{2.3}$$

As $(f_n)_{n\in\mathbb{N}}$ is $(\lambda_n)_{n\in\mathbb{N}}$-oscillating, a real number R_ε (which can be chosen greater than R) exists such that

$$\limsup_{n\to\infty}\int_{\lambda_n^{-1}\mathcal{C}_R^c} |\widehat{f}_n(\xi)|^2 d\xi \leq \varepsilon.$$

This means that an integer $\widetilde{n}_\varepsilon$ (which can be chosen greater than n_R) exists such that

$$\forall n\geq \widetilde{n}_\varepsilon\,,\ \int_{\lambda_n^{-1}\mathcal{C}_{R_\varepsilon}^c} |\widehat{f}_n(\xi)|^2 d\xi \leq \varepsilon.$$

Using (2.3) allows to conclude that $(f_n)_{n\in\mathbb{N}}$ is $(\lambda'_n)_{n\in\mathbb{N}}$ -unrelated.

Now let us argue by contraposition. If $(\lambda_n)_{n\in\mathbb{N}}$ and $(\lambda'_n)_{n\in\mathbb{N}}$ are not orthogonal, Proposition 2.1 claims that an extraction ϕ exists such that $\lambda_{\phi(n)} \equiv \lambda'_{\phi(n)}$. For the scale $(\lambda_{\phi(n)})_{n\in\mathbb{N}}$, the sequence $(f_{\phi(n)})_{n\in\mathbb{N}}$ is oscillating and unrelated. Proposition 2.2 implies that $\|f_{\phi(n)}\|_{L^2}$ tends to 0 when n tends to infinity. □

The following proposition describes the relation between scales and Besov spaces.

Proposition 2.4. *Let $(f_n)_{n\in\mathbb{N}}$ be a bounded sequence of L^2. Then the sequence $(f_n)_{n\in\mathbb{N}}$ tends to 0 in $\dot{B}^0_{2,\infty}$ if and only if the sequence $(f_n)_{n\in\mathbb{N}}$ is unrelated to any scale $(\lambda_n)_{n\in\mathbb{N}}$.*

Proof. Let us assume that the sequence $(f_n)_{n\in\mathbb{N}}$ is unrelated to any scale $(\lambda_n)_{n\in\mathbb{N}}$. Let us observe that

$$\begin{aligned}\limsup_{n\to\infty} \|f_n\|_{\dot{B}^0_{2,\infty}} &= \limsup_{n\to\infty} \sup_{k\in\mathbb{Z}} \|\Delta_k f_n\|_{L^2}\\ &= \sup_{(k_n)\in\mathbb{Z}^{\mathbb{N}}} \lim_{n\to\infty} \|\Delta_{k_n} f_n\|_{L^2}.\end{aligned}$$

Let us consider any sequence $(k_n)_{n\in\mathbb{N}}$ of elements of $\mathbb{Z}$. Let us define $\lambda_n = 2^{-k_n}$. By definition of the operator Δ_k, we have, for some R large enough,

$$\|\Delta_{k_n} f\|^2_{L^2} \lesssim \int_{2^{k_n}\mathcal{C}_R} |\widehat{f}_n(\xi)|^2 d\xi.$$

As $(f_n)_{n\in\mathbb{N}}$ is $(\lambda_n)_{n\in\mathbb{N}}$-unrelated then $\|\Delta_{k_n} f_n\|_{L^2}$ tends to 0.

Now let us assume that $(f_n)_{n\in\mathbb{N}}$ is not $(\lambda_n)_{n\in\mathbb{N}}$-unrelated for some scale $(\lambda_n)_{n\in\mathbb{N}}$. By rescaling, we can assume that $(f_n)_{n\in\mathbb{N}}$ is not 1-unrelated. This means that

$$\liminf_{n\to\infty} \int_{\mathcal{C}_R} |\widehat{f}_n(\xi)|^2 d\xi$$

is positive. As we have

$$\int_{\mathcal{C}_R} |\widehat{f}_n(\xi)|^2 d\xi \leq C_R \sup_{j'_R\leq j\leq j_R} \|\Delta_j f_n\|_{L^2}$$

we have $\liminf_{n\to\infty} \|f_n\|_{\dot{B}^0_{2,\infty}} > 0$ and the proposition is proved. □

The main theorem of this section of the following one.

Theorem 2.2. *Let $(f_n)_{n\in\mathbb{N}}$ be a bounded sequence of L^2. It exists an extraction ϕ, a sequence $(g^j_n, r^j_n, \lambda^j_n)_{(j,n)\in\mathbb{N}^2}$ of $L^2\times L^2\times]0,\infty[$ such that we have the following decomposition of $f_{\phi(n)}$. For any integer J, we have*

$$\forall n\geq J\,,\ f_{\phi(n)} - \sum_{j=0}^{J} g^j_n = r^J_n$$

with the following properties:

- *for any couple (j,j') with j different from j', the two scales $(\lambda^j_n)_{n\in\mathbb{N}}$ and $(\lambda^{j'}_n)_{n\in\mathbb{N}}$ are orthogonal,*
- *for any j, the sequence $(g^j_n)_{n\in\mathbb{N}}$ is $(\lambda^j_n)_{n\in\mathbb{N}}$-oscillatory,*
- *The sequence $(r^j_n)_{(j,n)\in\mathbb{N}^2}$ tends to 0 in the sense that*

$$\lim_{j\to\infty} \limsup_{n\to\infty} \|r^j_n\|_{\dot{B}^0_{2,\infty}} = 0\,, \tag{2.4}$$

- *for any couple (j',j) such that j' is less than or equal to j, the sequence $(r^j_n)_{n\in\mathbb{N}}$ is $(\lambda^{j'}_n)_{n\in\mathbb{N}}$-unrelated which implies that*

$$\forall J\in\mathbb{N}\,,\ \lim_{n\to\infty}\Big(\|f_{\phi(n)}\|^2_{L^2} - \sum_{j=0}^{J} \|g^j_n\|^2_{L^2} - \|r^J_n\|^2_{L^2}\Big) = 0.$$

Proof. The proof is based on the repeated application of the following lemma.

Lemma 2.1. *Let us consider $(f_n)_{n\in\mathbb{N}}$ a bounded sequence of L^2 such that*

$$\limsup_{n\to\infty} \|f_n\|_{\dot{B}^0_{2,\infty}} = L > 0.$$

Then a scale $(\lambda_n)_{n\in\mathbb{N}}$, a sequence $(g_n)_{n\in\mathbb{N}}$ and an extraction ϕ exist such that

- *the sequence $(g_n)_{n\in\mathbb{N}}$ is $(\lambda_n)_{n\in\mathbb{N}}$-oscillating,*
- *the sequence $(f_{\phi(n)} - g_n)_{n\in\mathbb{N}}$ is $(\lambda_n)_{n\in\mathbb{N}}$-unrelated and for any scale $(\lambda'_n)_{n\in\mathbb{N}}$ such that the sequence $(f_n)_{n\in\mathbb{N}}$ is $(\lambda'_n)_{n\in\mathbb{N}}$-unrelated, the sequence $(f_{\phi(n)} - g_n)_{n\in\mathbb{N}}$ is also $({\lambda'}_{\phi(n)})_{n\in\mathbb{N}}$-unrelated,*
- *we have $\displaystyle\lim_{n\to\infty} \|g_n\|_{L^2} \geq \frac{L}{2}\cdot$*

Proof. By definition of the Besov norm $\|\cdot\|_{\dot{B}^0_{2,\infty}}$, the hypothesis implies that a sequence of integers $(k_n)_{n\in\mathbb{N}}$ exists such that, for large enough n,

$$\|\Delta_{k_n} f_n\|_{L^2} \geq \frac{L}{2}\cdot \tag{2.5}$$

Now let us consider the scale $\lambda_n \stackrel{\text{def}}{=} 2^{-k_n}$ and let us consider the sequence of functions $(F_n)_{n\in\mathbb{N}}$ defined by

$$F_n \begin{cases}]1,\infty[\longrightarrow [0,\infty[\\ \quad R \quad \longmapsto \|\mathbb{1}_{\lambda_n^{-1}\mathcal{C}_R} \widehat{f}_n\|^2_{L^2}. \end{cases}$$

For any n, the function F_n is non decreasing and, for any R, $F_n(R)$ is less than or equal to the supremum of $\|f_n\|_{L^2}$. Helly's lemma implies that an extraction ϕ_1 and a (of course non decreasing) function exist such that

$$\forall R > 1\,, \ \lim_{n\to\infty} F_{\phi_1(n)}(R) = F(R).$$

This implies that

$$\forall n\,,\ \exists m_n : \forall m \geq m_n\,,\ |F_{\phi_1(m)}(n) - F(n)| \leq \frac{1}{n+1}\cdot$$

Let us define by induction by $\phi_2(n) \stackrel{\text{def}}{=} \max\{\phi_1(n-1)+1, m_n\}$ and state $\phi \stackrel{\text{def}}{=} \phi_1 \circ \phi_2$. Now we can define the sequence $(g_n)_{n\in\mathbb{N}}$ by

$$\widehat{g}_n \stackrel{\text{def}}{=} \mathbb{1}_{\lambda^{-1}_{\phi(n)}\mathcal{C}_n} \widehat{f}_{\phi(n)}.$$

Let us check that $(g_n)_{n\in\mathbb{N}}$ satisfies the conclusions of the lemma. Let us observe that

$$n \geq R \Longrightarrow \mathbb{1}_{\mathcal{C}^c_R} = \mathbb{1}_{\mathcal{C}_n} - \mathbb{1}_{\mathcal{C}_R}.$$

Thus by definition of ϕ and F_n, we get, for n greater than or equal to R,

$$\begin{aligned} \int_{\lambda_n^{-1}\mathcal{C}^c_R} |\widehat{g}_n(\xi)|^2 d\xi &= \int_{\lambda_n^{-1}\mathcal{C}_n} |\widehat{g}_n(\xi)|^2 d\xi - \int_{\lambda_n^{-1}\mathcal{C}_R} |\widehat{g}_n(\xi)|^2 d\xi \\ &= F_{\phi(n)}(n) - F_{\phi(n)}(R). \end{aligned} \tag{2.6}$$

The function F is bounded and non decreasing. Let us denote by F_∞ its limits at infinity and let us write

$$|F_{\phi(n)}(n) - F_\infty| \leq |F_{\phi(n)}(n) - F(n)| + F_\infty - F(n) \leq \frac{1}{n+1} + F_\infty - F(n),$$

thus we have $\lim_{n\to\infty} F_{\phi(n)}(n) = F_\infty$. Then we deduce from (2.6) that

$$\lim_{n\to\infty} \int_{\lambda_n^{-1}\mathcal{C}_R^c} |\widehat{g}_n(\xi)|^2 d\xi = F_\infty - F(R).$$

Thus $(g_n)_{n\in\mathbb{N}}$ is $(\lambda_{\phi(n)})_{n\in\mathbb{N}}$-oscillating.

Moreover, if n is greater or equal to R, the ring $\mathcal{C}_R$ is included in the ring $\mathcal{C}_n$. Thus we have

$$\mathbb{1}_{\lambda_{\phi(n)}^{-1}\mathcal{C}_R}(\widehat{f}_{\phi(n)} - \widehat{g}_n) = \mathbb{1}_{\lambda_{\phi(n)}^{-1}\mathcal{C}_R}\big(\mathbb{1}_{\lambda_{\phi(n)}^{-1}\mathcal{C}_n}\widehat{f}_{\phi(n)} - \widehat{g}_n\big) = 0.$$

Thus $(f_{\phi(n)} - g_n)_{n\in\mathbb{N}}$ is $(\lambda_{\phi(n)})_{n\in\mathbb{N}}$-unrelated.

By definition of g_n and λ_n, we have

$$\begin{aligned}\|\widehat{g}_n\|_{L^2}^2 &= \int_{\lambda_{\phi(n)}^{-1}\mathcal{C}_n} |\widehat{f}_{\phi(n)}(\xi)|^2 d\xi \\ &= \int_{2^{k_{\phi(n)}}\mathcal{C}_n} |\widehat{f}_{\phi(n)}(\xi)|^2 d\xi.\end{aligned}$$

If n is large enough, the ring $\mathcal{C}_n$ contains the support of the smooth cut-off function of the Littlewood-Paley theory. Thus, because of (2.5), we have, for large enough n,

$$\|g_n\|_{L^2} \geq \frac{L}{2}\cdot$$

Up the an extraction, we can assume that $\|g_n\|_{L^2}$ converges and the lemma is proved. □

Conclusion of the proof of Theorem 2.2. We proceed by induction. For an integer J let us define the property $\mathcal{P}_J$ as for any j, it exists a family $(\phi_j)_{0\leq j\leq J}$ of extractions, families of sequences of functions $(g_n^j)_{n\in\mathbb{N}}$ and $(r_n^j)_{n\in\mathbb{N}}$ such that

- up to extraction, the sequence $(f_n)_{n\in\mathbb{N}}$ can be decompose as

$$f_{\phi_0\circ\phi_1\circ\cdots\circ\phi_J(n)} = \sum_{j=0}^{J-1} g^j_{\phi_{j+1}\circ\cdots\circ\phi_J(n)} + g_n^J + r_n^J,$$

- for any couple (j',j) such that j' is different from j, the two scales $(\lambda_n^j)_{n\in\mathbb{N}}$ and $(\lambda_n^{j'})_{n\in\mathbb{N}}$ are orthogonal,
- for any couple (j',j) such that j' is less than or equal to j, the sequence $(r_n^j)_{n\in\mathbb{N}}$ is $({\lambda^{j'}}_{\phi_{j'}\circ\cdots\circ\phi_j(n)})_{n\in\mathbb{N}}$-unrelated,
- for any j, the sequence $(g_n^j)_{n\in\mathbb{N}}$ is $(\lambda_n^j)_{n\in\mathbb{N}}$-oscillatory,

$$\lim_{n\to\infty} \|g_n^j\|_{L^2} \geq \frac{1}{2}\limsup_{n\to\infty} \|r_n^{j-1}\|_{\dot{B}^0_{2,\infty}}\cdot \tag{2.7}$$

Let us prove $\mathcal{P}_{J+1}$. If $\liminf_{n\to\infty} \|r_n^J\|_{\dot{B}^0_{2,\infty}} = 0$, then we choose

$$g_n^{J+1} = r_n^{J+1} = 0 \quad \text{and} \quad \phi_{J+1}(n) = n.$$

If $\liminf_{n\to\infty} \|r_n^J\|_{\dot{B}^0_{2,\infty}}$ is positive, then we apply Lemma 2.1 to the sequence $(r_n^J)_{n\in\mathbb{N}}$ which provides the existence of an extraction ϕ_{J+1}, and sequences $(g_n^{J+1})_{n\in\mathbb{N}}$ and $(r_n^{J+1})_{n\in\mathbb{N}}$, and a scale $(\lambda_n^{J+1})_{n\in\mathbb{N}}$ which satisfies $\mathcal{P}_{J+1}$.

Now let us prove Assertion (2.4). Because of Proposition 2.2, we have, for any J,

$$\lim_{n\to\infty}\Big(\|f_{\phi_0\circ\phi_1\circ\cdots\circ\phi_J(n)}\|^2_{L^2} - \sum_{j=0}^{J-1}\|g^j_{\phi_{j+1}\circ\cdots\circ\phi_J(n)}\|^2_{L^2} - \|g_n^J\|^2_{L^2} - \|r_n^J\|^2_{L^2}\Big) = 0.$$

This implies that

$$\sup_n \|f_n\|^2_{L^2} \geq \sum_{j=0}^{J-1}\|g^j_{\phi_{j+1}\circ\cdots\circ\phi_J(n)}\|^2_{L^2} + \|g_n^J\|^2_{L^2} + o_n^J(1)$$

with $\lim_{n\to\infty} o_n^J(1) = 0$. Passing to the limit when n tends to infinity gives thanks to Inequality (2.7)

$$\frac{1}{4}\sum_{j=0}^{J-1}\limsup_{n\to\infty}\|r^{j-1}_{\phi_{j+1}\circ\cdots\circ\phi_J(n)}\|^2_{\dot{B}^0_{2,\infty}} \leq \sup_n \|f_n\|_{L^2}$$

which obviously implies that

$$\lim_{J\to\infty}\limsup_{n\to\infty}\|r^{j-1}_{\phi_{j+1}\circ\cdots\circ\phi_J(n)}\|_{\dot{B}^0_{2,\infty}} = 0.$$

Now we use the diagonal process and define

$$\begin{aligned}
\phi(n) &\stackrel{\text{def}}{=} \phi_0\circ\cdots\circ\phi_n(n)\\
\widetilde{g}_n^j &\stackrel{\text{def}}{=} g^j_{\phi_{j+1}\circ\cdots\circ\phi_n(n)} \quad \text{for } n\geq j \quad\text{and}\quad 0 \text{ if } n<j\,,\\
\widetilde{r}_n^j &\stackrel{\text{def}}{=} r^j_{\phi_{j+1}\circ\cdots\circ\phi_n(n)} \quad \text{for } n\geq j \quad\text{and}\quad 0 \text{ if } n<j\,.
\end{aligned}$$

Then, for any n greater than J, we have

$$\forall n\geq J\,,\ f_{\phi(n)} = \sum_{j=0}^{J}\widetilde{g}_n^j + r_n^J$$

and the theorem is proved. □

2.3 The extraction of the cores for 1-oscillating sequences

The purpose of this section is the proof of the following theorem which describes, up to an extraction, the structure of the 1-oscillating sequences.

Theorem 2.3. *Let $(g_n)_{n\in\mathbb{N}}$ be a 1-oscillating sequence. It exists an extraction ϕ, two sequences $(\psi^k)_{k\in\mathbb{N}}$ and $(r_n^k)_{n\in\mathbb{N}}$ of L^2 functions and a sequence $(x_n^k)_{n\in\mathbb{N}}$ of points of $\mathbb{R}^d$ such that*

- *for any integer K, $g_{\phi(n)} = \displaystyle\sum_{k=0}^{K} \psi^k(x - x_n^k) + R_n^K$,*
- *for any couple (k', k) such that k and k' are different, $|x_n^k - x_n^{k'}|$ tends to ∞ when n tends to infinity,*
- *we have*

$$\lim_{n\to\infty}\Big(\|g_{\phi(n)}\| - \sum_{k=0}^{K}\|\psi^k\|_{L^2}^2 - \|R_n^K\|_{L^2}^2\Big) = 0$$

and

$$\lim_{K\to\infty}\limsup_{n\to\infty}\|R_n^K\|_{\dot B^{-\frac{d}{2}}_{\infty,\infty}} = 0.$$

Proof. It relies on the repeated application of the following lemma.

Lemma 2.2. *A positive constant c_0 exists such that for any 1-oscillating sequence $(g_n)_{n\in\mathbb{N}}$, it exists an extraction ϕ, a function ψ and a sequence $(x_n)_{n\in\mathbb{N}}$ such that $x_n \equiv 0$ or $|x_n|$ tends to infinity when n tends to infinity which satisifies*

$$g_{\phi(n)}(x_n + \cdot) \rightharpoonup \psi \quad\text{and}\quad \|\psi\|_{L^2} \geq c_0 \limsup_{n\to\infty}\|g_n\|_{\dot B^{-\frac{d}{2}}_{\infty,\infty}}.$$

Proof. If $(g_n)_{n\in\mathbb{N}}$ tends to 0 in the space $\dot B^{-\frac{d}{2}}_{\infty,\infty}$, there is nothing to prove because any element ψ of the weak closure of $(g_n)_{n\in\mathbb{N}}$ works with $x_n \equiv 0$. Let us assume that $(g_n)_{n\in\mathbb{N}}$ does not tend to 0 in the space $\dot B^{-\frac{d}{2}}_{\infty,\infty}$. Up to an extraction we omit to note, we can assume that

$$\lim_{n\to\infty}\|g_n\|_{\dot B^{-\frac{d}{2}}_{\infty,\infty}} = L > 0.$$

By definition of the Besov norm $\dot B^{-\frac{d}{2}}_{\infty,\infty}$, it implies that a sequence $(\ell_n, \widetilde{x}_n)_{n\in\mathbb{N}}$ of elements of $\mathbb{Z}\times\mathbb{R}^d$ exists such that

$$2^{-\ell_n\frac{d}{2}}|\Delta_{\ell_n} g_n(\widetilde{x}_n)| \geq \frac{L}{2} \tag{2.8}$$

if n is large enough. The fact that $(g_n)_{n\in\mathbb{N}}$ is 1-oscillating implies that the sequence $(\ell_n)_{n\in\mathbb{N}}$ takes only a finite number of values. Indeed, Bernstein inequality implies that

$$2^{-\ell_n\frac{d}{2}}\|\Delta_{\ell_n} g_n\|_{L^\infty} \leq C_0\|\Delta_{\ell_n} g_n\|_{L^2}. \tag{2.9}$$

As $(g_n)_{n\in\mathbb{N}}$ is 1-oscillatory, a radius R_0 greater than 1 exists such that

$$\limsup_{n\to\infty}\int_{\mathcal{C}^c_{R_0}} |\widehat{g}_n(\xi)|^2 d\xi \leq \Big(\frac{L}{4C_0}\Big)^2.$$

Moreover, a constant C_1 exists such that

$$|\ell| \geq C_1 \log R_0 \Longrightarrow \int_{\mathcal{C}_{R_0}} |\mathcal{F}(\Delta_\ell g_n)(\xi)|^2 d\xi = 0.$$

Thus, an integer n_R exists such that

$$\Big(n \geq n_R \quad \text{and} \quad |\ell_n| \geq C_1 \log R_0\Big) \Longrightarrow 2^{-\ell_n \frac{d}{2}} \|\Delta_{\ell_n} g_n\|_{L^\infty} \leq \frac{L}{4} \cdot$$

Inequality (2.9) implies the sequence $(\ell_n)_{n\in\mathbb{N}}$ defining by (2.8) takes only a finite number of values. Thus, up to an extraction we omit to note, we can assume that the sequence $(\ell_n)_{n\in\mathbb{N}}$ is constant and equal to some integer k. By definition of the operator Δ_k, we have

$$2^{-k\frac{d}{2}} \Delta_k g_n(\widetilde{x}_n) = 2^{-k\frac{d}{2}} 2^{kd} \int_{\mathbb{R}^d} h(2^k(\widetilde{x}_n - y)) g_n(y) dy.$$

As h is even, let us change variable $z = -2^k(\widetilde{x}_n - y)$ in the above integral. This gives

$$2^{-k\frac{d}{2}} \Delta_k g_n(\widetilde{x}_n) = 2^{-k\frac{d}{2}} \int_{\mathbb{R}^d} h(z) g_n(\widetilde{x}_n + 2^{-k} z) dz.$$

The sequence $\big(2^{-k\frac{d}{2}} g_n(\widetilde{x}_n + 2^{-k}\cdot)\big)_{n\in\mathbb{N}}$ is bounded in L^2. Thus an extraction ϕ exists and a function $\widetilde{\psi}$ such that the sequence $\big(2^{-k\frac{d}{2}} g_{\phi(n)}(\widetilde{x}_{\phi(n)} + 2^{-k}\cdot)\big)_{n\in\mathbb{N}}$ converges weakly to $\widetilde{\psi}$. Then we deduce that

$$g_{\phi(n)}(\widetilde{x}_{\phi(n)} + \cdot) \rightharpoonup \psi \quad \text{with} \quad \psi(x) \stackrel{\text{def}}{=} 2^{k\frac{d}{2}} \widetilde{\psi}(2^k x).$$

Now let us remark that by definition of k and of the sequence $(\widetilde{x}_n)_{n\in\mathbb{N}}$, we have

$$\begin{aligned} \frac{L}{2} &\leq 2^{-k\frac{d}{2}} \big|\Delta_k g_{\phi(n)}(\widetilde{x}_{\phi(n)})\big| \\ &\leq \big|\langle 2^{-k\frac{d}{2}} g_{\phi(n)}(\widetilde{x}_{\phi(n)} + 2^{-k}\cdot), h\rangle\big|. \end{aligned}$$

Passing to the limit when n tends to infinity gives

$$\frac{L}{2} \leq \big|\langle \widetilde{\psi}, h\rangle\big| \leq \|h\|_{L^2} \|\widetilde{\psi}\|_{L^2}.$$

Once observed that $\|\widetilde{\psi}\|_{L^2} = \|\psi\|_{L^2}$, we define $x_n = \widetilde{x}_{\phi(n)}$ and then get the lemma with the constant c_0 equal to $\frac{1}{2\|h\|_{L^2}}$ up to the fact that $x_n \equiv 0$ or $|x_n|$ tends to infinity when n tends to infinity.

If $|x_n|$ does not tend to infinity, then, up to an extraction we omit to note, we can assume that the sequence $(x_n)_{n\in\mathbb{N}}$ converges to some point x_∞ of $\mathbb{R}^d$. Let us write that, for any test function θ in L^2, we have

$$\begin{aligned} &\int_{\mathbb{R}^d} \big(g_{\phi(n)}(x_n + x) - g_{\phi(n)}(x_\infty + x)\big)\theta(x) dx \\ &= \int_{\mathbb{R}^d} g_{\phi(n)}(x_\infty + x)\big(\theta(x + x_\infty - x_n) - \theta(x)\big) dx. \end{aligned}$$

Because $(g_n)_{n\in\mathbb{N}}$ is bounded on L^2, the fact that translation are continuous on L^2 ensures that the sequence $\big(g_{\phi(n)}(x_n+\cdot)-g_{\phi(n)}(x_\infty+\cdot)\big)_{n\in\mathbb{N}}$ tends weakly to 0. This concludes the proof of the lemma. □

Continuation of the proof of Theorem 2.3. By repeated application of the above lemma, let us define a sequence $(\phi_k)_{k\in\mathbb{N}}$, a sequence $(\psi^k)_{k\in\mathbb{N}}$, a sequence $(x_n^k)_{(k,n)\in\mathbb{N}^2}$ of points of $\mathbb{R}^d$, a sequence $(R_n^k)_{(k,n)\in\mathbb{N}^2}$ which satisfies $R_0=g_n$, $\psi^0=0$, $x_n^0=0$ and $\phi^0=\mathrm{Id}$ and

$$R_n^k=R^{k-1}_{\phi_k(n)}-\psi^k(\cdot-x_n^k)\,,\ \|\psi^k\|_{L^2}\geq c_0\limsup_{n\to\infty}\|R_n^{k-1}\|_{\dot B^{-\frac{d}{2}}_{\infty,\infty}}\,,\ R_n^k(x_n^k+\cdot)\rightharpoonup 0. \tag{2.10}$$

Let us prove that

$$\forall k'<k\,,\ \lim_{n\to\infty}\big|x_n^k-x^{k'}_{\phi_{k'+1}\circ\cdots\circ\phi_k(n)}\big|=\infty. \tag{2.11}$$

Let us first prove it for $k'=k-1$. By Definition (2.10) of the sequences, we have

$$R^{k-1}_{\phi_k(n)}(x_n^k+\cdot)\rightharpoonup\psi^k\quad\text{and}\quad R_n^{k-1}(x_n^{k-1}+\cdot)\rightharpoonup 0.$$

Property (2.11) for $k'=k-1$ will follow from the following easy lemma, the proof of which is omitted.

Lemma 2.3. *Let $(f_n)_{n\in\mathbb{N}}$ be a bounded sequence of L^2 which converges weakly to some function f which is not the zero function. Let $(y_n)_{n\in\mathbb{N}}$ be any sequence of $\mathbb{R}^d$. Then the sequence $(f_n(y_n+\cdot))_{n\in\mathbb{N}}$ tends weakly to 0 if and only if $|y_n|$ tends to infinity.*

Now let us proceed by induction assuming that

$$\forall k':k-p\leq k'<k\,,\ \lim_{n\to\infty}\big|x_n^k-x^{k'}_{\phi_{k'+1}\circ\cdots\circ\phi_k(n)}\big|=\infty. \tag{2.12}$$

By Definition (2.10) of the sequences, we have

$$R^{k-p-1}_{\phi_{k-p}\circ\cdots\circ\phi_k(n)}=R_n^k+\sum_{p'=0}^{p-1}\psi^{k-p'}\big(\cdot-x^{k-p'}_{\phi_{k-p'}\circ\cdots\circ\phi_k(n)}\big). \tag{2.13}$$

Using that $(R_n^k(x_n^k+\cdot))_{n\in\mathbb{N}}$ tends weakly to 0, and the induction hypothesis (2.12), we get

$$R^{k-p-1}_{\phi_{k-p}\circ\cdots\circ\phi_k(n)}(x_n^k+\cdot)\rightharpoonup\psi^k\quad\text{and}\quad R^{k-p-1}_{\phi_{k-p}\circ\cdots\circ\phi_k(n)}(x_n^{k-p-1}+\cdot)\rightharpoonup 0.$$

Lemma 2.3 ensures (2.11).

Moreover, we have

$$\|R^{k-1}_{\phi_k(n)}\|^2_{L^2}=\|\psi^k\|^2_{L^2}+\|R_n^k\|^2_{L^2}+2\big(\psi^k|R_n^k(x_n^k+\cdot)\big)_{L^2}.$$

Using that $(R_n^k(x_n^k+\cdot))_{n\in\mathbb{N}}$ tends weakly to 0, we get that

$$\lim_{n\to\infty}\big(\|R^{k-1}_{\phi_k(n)}\|^2_{L^2}-\|\psi^k\|^2_{L^2}-\|R_n^k\|^2_{L^2}\big)=0.$$

Using Formula (2.13), we get by iteration that, for any k,

$$\lim_{n\to\infty}\Big(\|g_{\phi_1\circ\cdots\circ\phi_k(n)}\|_{L^2}^2-\sum_{k'=1}^{k}\|\psi^k\|_{L^2}^2-\|R_n^k\|_{L^2}^2\Big)=0.$$

As the sequence $(g_n)_{n\in\mathbb{N}}$ is bounded in L^2, the series $(\|\psi^k\|_{L^2}^2)_{k\in\mathbb{N}}$ is convergent. Using that

$$\|\psi^k\|_{L^2}\geq c_0\limsup_{n\to\infty}\|R^{k-1}_{\phi_k(n)}\|_{\dot B^{-\frac{d}{2}}_{\infty,\infty}}.$$

As a term of a square convergent series, we have that

$$\lim_{k\to\infty}\limsup_{n\to\infty}\|R^{k-1}_{\phi_k(n)}\|_{\dot B^{-\frac{d}{2}}_{\infty,\infty}}=0.$$

Now let us argue with a diagonal argument by defining $\phi(n)\stackrel{\text{def}}{=}\phi_1\circ\cdots\circ\phi_n(n)$. For any n greater than k, we have

$$g_{\phi(n)}(x)=\sum_{k=0}^{K}\psi^k(x-x^k_{\phi_{k+1}\circ\cdots\circ\phi_n(n)})+R^K_{\phi_{k+1}\circ\cdots\circ\phi_n(n)}.$$

Defining $\widetilde{x}_n^k=x^k_{\phi_{k+1}\circ\cdots\circ\phi_n(n)}$ and $\widetilde{R}_n^K=R^K_{\phi_{k+1}\circ\cdots\circ\phi_n(n)}$ allows to conclude the proof of Theorem 2.3. □

Conclusion of the proof of Theorem 2.1. First, let us apply Theorem 2.2 to the sequence $(\Lambda^s f_n)_{n\in\mathbb{N}}$ where $\Lambda^s f\stackrel{\text{def}}{=}\mathcal{F}^{-1}(|\cdot|^s\widehat{f})$. This provides the existence of a sequence of an extraction ϕ_0, two by two orthogonal scales $(\widetilde{\lambda}_n^j)_{(j,n)\in\mathbb{N}^2}$, a sequence $(g_n^j)_{(j,n)\in\mathbb{N}}$ of functions in L^2 such that, for any j, the sequence $(g_n^j)_{n\in\mathbb{N}}$ is $(\widetilde{\lambda}_n^j)_{n\in\mathbb{N}}$-oscillating, a sequence $(r_n^j)_{(j,n)\in\mathbb{N}^2}$ such that

$$\Lambda^s f_{\phi_0(n)}=\sum_{j=0}^{J}g_n^j+r_n^J\quad\text{with}$$

$$\lim_{n\to\infty}\Big(\Lambda^s f_{\phi_0(n)}-\sum_{j=0}^{J}\|g_n^j\|_{L^2}^2-\|r_n^J\|_{L^2}^2\Big)=0\quad\text{and}\quad\lim_{J\to\infty}\limsup_{n\to\infty}\|r_n^J\|_{\dot B^0_{2,\infty}}=0.\tag{2.14}$$

Theorem 2.3 implies that a sequence $(\phi_j)_{j\in\mathbb{N}}$ of extraction, two sequences of functions of L^2 $(\widetilde{\psi}^{j,k})_{(j,k)\in\mathbb{N}^2}$ and $(\widetilde{R}_n^k)_{(k,n)\in\mathbb{N}^2}$, a sequence $(\widetilde{x}_n^k)_{(k,n)\in\mathbb{N}^2}$ of points of $\mathbb{R}^d$ such that, for any j, we have

$$g^j_{\phi_1\circ\cdots\circ\phi_j(n)}(x)=\sum_{k=1}^{K}\frac{1}{(\widetilde{\lambda}_n^j)^{\frac{d}{2}}}\widetilde{\psi}^{j,k}\Big(\frac{x-\widetilde{x}_n^{j,k}}{\widetilde{\lambda}_n^j}\Big)+\frac{1}{(\widetilde{\lambda}_n^j)^{\frac{d}{2}}}\widetilde{R}_n^{j,K}\Big(\frac{x}{\widetilde{\lambda}_n^j}\Big)\quad\text{with}$$

$$\lim_{n\to\infty}\Big(\|g^j_{\phi_1\circ\cdots\circ\phi_j(n)}\|_{L^2}^2-\sum_{k=0}^{K}\|\widetilde{\psi}^{j,k}\|_{L^2}^2-\|\widetilde{R}_n^{j,K}\|_{L^2}^2\Big)=0\quad\text{and}\tag{2.15}$$

$$\lim_{K\to\infty}\limsup_{n\to\infty}\|\widetilde{R}_n^{j,K}\|_{\dot B^{-\frac{d}{2}}_{\infty,\infty}}=0.$$

Now let us define

$$
\begin{aligned}
\phi(n) &\stackrel{\text{def}}{=} \phi_0 \circ \phi_1 \circ \cdots \circ \phi_n(n)\,,\\
\lambda_n^j &\stackrel{\text{def}}{=} \widetilde{\lambda}^j_{\phi_{j+1}\circ\cdots\circ\phi_n(n)} \quad \text{if} \quad n \geq j+1 \quad \text{and} \quad 0 \quad \text{if not},\\
x_n^{j,k} &\stackrel{\text{def}}{=} \widetilde{x}^{j,k}_{\phi_{j+1}\circ\cdots\circ\phi_n(n)} \quad \text{if} \quad n \geq j+1 \quad \text{and} \quad 0 \quad \text{if not},\\
\psi^{j,k} &= \Lambda^{-s}\widetilde{\psi}^{j,k} \quad \text{and}\\
\rho_n^{J,K} &\stackrel{\text{def}}{=} \Lambda^{-s} r_n^J + \sum_{j=0}^{J} \frac{1}{(\lambda_n^j)^{s-\frac{d}{2}}} (\Lambda^{-s}\widetilde{R}^{j,k}_{\phi_{j+1}\circ\cdots\circ\phi_n(n)}) \Big(\frac{\cdot}{\lambda_n^j}\Big)\cdot
\end{aligned}
$$

Using (2.14) and (2.15), we have, for n greater than j

$$
f_{\phi(n)} = \sum_{j=0}^{J}\sum_{k=1}^{K} \frac{1}{(\lambda_n^j)^{\frac{d}{2}-s}} \psi^{j,k}\Big(\frac{\cdot - x_n^{j,k}}{\lambda_n^j}\Big) + \rho_n^{J,K}.
$$

We get the theorem by reordering the sequences. □

3 Some basic facts about the Navier-Stokes equation

3.1 The concept of solutions and some historical results

Let us define the concept of (weak) solutions of the incompressible Navier-Stokes system. Let us first recall what the incompressible Navier-Stokes system is. We consider as unknown the speed $u = (u^1, u^2, u^3)$ a time dependant divergence free vector field on $\mathbb{R}^3$ and the pressure p. We consider the system

$$
(NS) \quad \begin{cases} \partial_t u + u\cdot\nabla u - \Delta u = -\nabla p + f \quad \text{in} \quad \mathbb{R}^+\times\mathbb{R}^3,\\ \operatorname{div} u = 0,\\ u|_{t=0} = u_0. \end{cases}
$$

Let us immediately notice that this system has two fundamental properties. The first one is the energy inequality. Formally, and in the case when $f = 0$, let us take the L^2 scalar product with u in the equation. We get

$$
\frac{1}{2}\frac{d}{dt}\|u(t)\|^2_{L^2} + \|\nabla u(t)\|^2_{L^2} = -\int_{\mathbb{R}^3} (u(t)\cdot\nabla u(t)|u(t))_{L^2} - \int_{\mathbb{R}^3} (\nabla p(t)|u(t))_{L^2}.
$$

Thanks to the divergence free condition, obvious integration by parts implies that, any function vector field a

$$
(u\cdot\nabla a|a)_{L^2} = 0 = (\nabla p|u)_{L^2} = 0. \tag{3.1}
$$

This gives

$$
\frac{1}{2}\frac{d}{dt}\|u(t)\|^2_{L^2} + \|\nabla u(t)\|^2_{L^2} = 0. \tag{3.2}
$$

The second one is the scaling invariance. It is easy to see that if u is a (smooth) solution of (NS) on $[0,T]\times\mathbb{R}^3$ with pressure p associated with the initial data u_0, then, for any positive λ, the vector field and the pressure

$$u_\lambda(t,x)\stackrel{\text{def}}{=}\lambda u(\lambda^2 t,\lambda x)\quad\text{and}\quad p_\lambda(t,x)=\lambda^2 p(\lambda^2 t,\lambda x)$$

is a solution of (NS) on the interval $[0,\lambda^{-2}T]\times\mathbb{R}^3$ associated with the initial data $\lambda u_0(\lambda x)$.

The notion of C^2 solution (i.e. classical solution) is not efficient because singularity can appear here and also we can be interested in rough initial data. It has been pointed out by C. Oseen in the beginning of the 20th century (see [27] and [28]) that another concept of solution must be used. This has been formalized by J. Leray in 1934 in his seminal work [24]. Let us define the notion of weak solution (that we shall denote simply solution in all that follows).

Definition 3.1. *A time-dependent vector field u with components in $L^2_{loc}([0,T]\times\mathbb{R}^d)$ is a weak solution (simply a solution in these notes) of (NS) if for any smooth compactly supported divergence free vector field Ψ,*

$$\begin{aligned}&\sum_{j=1}^{3}\int_{\mathbb{R}^3}u^j(t,x)\Psi^j(t,x)dx\\&=\sum_{j=1}^{3}\int_{\mathbb{R}^3}u_0^j(x)\Psi^j(0,x)dx+\sum_{j=1}^{3}\int_0^t\int_{\mathbb{R}^3}u^j(t',x)\big(\partial_t\Psi^j(t',x)\\&\quad+\Delta\Psi^j(t',x)\big)dt'dx+\sum_{j,k}\int_0^t\int_{\mathbb{R}^3}(u^ju^k)(t',x)\partial_j\Psi^k(t',x)\,dt'dx\\&\quad+\sum_{j=1}^{3}\int_0^t\langle f^j(t'),u^j(t')\rangle dt'.\end{aligned}$$

This definition is too weak in the sense there is not enough constraints on the solution. In particular it ignores the fundamental concept of energy. J. Leray introduced in his seminal paper [24] the concept of turbulent solution.

Definition 3.2. *A turbulent solution of (NS) is a divergence free vector field u which is a weak solution, has component in $L^\infty_T(L^2)\cap L^2_T(H^1)$ and satisfies in addition the energy inequality*

$$\frac12\|u(t)\|_{L^2}^2+\int_0^t\|\nabla u(t')\|_{L^2}^2dt'\le\frac12\|u_0\|_{L^2}^2+\int_0^t\int_{\mathbb{R}^3}f(t',x)\cdot u(t',x)dt'dx.\quad(3.3)$$

Remark. For a turbulent solution, Definition 3.1 of a weak solution becomes

$$\begin{aligned}&\sum_{j=1}^{3}\int_{\mathbb{R}^3}u^j(t,x)\Psi^j(t,x)dx\\&=\sum_{j=1}^{3}\int_{\mathbb{R}^3}u_0^j(x)\Psi^j(0,x)dx+\sum_{j=1}^{3}\int_0^t\int_{\mathbb{R}^3}u^j(t',x)\partial_t\Psi^j(t',x)dt'dx\end{aligned}$$

$$
-\sum_{j,k}\int_0^t\int_{\mathbb{R}^3}\partial_k u^j(t',x)\partial_k\Psi^j(t',x))\,dt'dx
$$

$$
+\sum_{j,k}\int_0^t\int_{\mathbb{R}^3}(u^ju^k)(t',x)\partial_j\Psi^k(t',x)\,dt'dx+\sum_{j=1}^3\int_0^t\langle f^j(t'),u^j(t')\rangle dt'.
$$

In [24], J. Leray proved the following theorem.

Theorem 3.1. *Let u_0 be a divergence free vector field in $L^2(\mathbb{R}^d)$. Then a turbulent solution u exists on $\mathbb{R}^+\times\mathbb{R}^3$.*

The proof of this theorem relies on compactness methods and thus no uniqueness is proved.

In this text, we are going to focus on solution which are regular enough to be unique in their own class. For this type of solutions let us state a theorem of existence of solutions by J. Leray which he called semi-regular solutions.

Theorem 3.2. *Let u_0 be a divergence free vector field in $L^2(\mathbb{R}^3)$ such that ∇u_0 belongs to $L^2(\mathbb{R}^3)$. Then a positive time, which can be chosen greater or equal to $\rho_0\|\nabla u_0\|_{L^2}^{-4}$ for some constant ρ_0, exists such that a unique solution u exists in $C([0,T];\dot H^1)\cap L^2([0,T];\dot H^2)$. Moreover a constant c_1 exists such that if*

$$
\|u_0\|_{L^2}\|\nabla u_0\|_{L^2}\le c_1,
$$

then T can be chosen equal to ∞.

Proof. We simply prove a formal control on the $\|\nabla u(t)\|_{L^2}$ norm. By differentiation of the equation, we get

$$
\partial_t\partial_j u+u\cdot\nabla\partial_j u-\Delta\partial_j u=-\nabla p_j-\partial_j u\cdot\nabla u.
$$

Taking the L^2 scalar product of the equation and summing in j gives

$$
\frac{1}{2}\frac{d}{dt}\|\nabla u(t)\|_{L^2}^2+\|\nabla^2u(t)\|_{L^2}^2=\sum_{j=1}\int_{\mathbb{R}^3}\partial_j u(t,x)\cdot\nabla u(t,x)\partial_j u(t,x)dx.
$$

Sobolev embeddings (see Theorem 1.1) implies that

$$
\begin{aligned}
\int_{\mathbb{R}^3}\partial_j u(t,x)\cdot\nabla u(t,x)\partial_j u(t,x)dx &\le \|\nabla u(t)\|_{L^3}^3\\
&\lesssim \|\nabla u(t)\|_{\dot H^{\frac12}}^3.
\end{aligned}
$$

The interpolation inequality between Sobolev spaces (see Proposition 1.1) implies that

$$
\int_{\mathbb{R}^3}\partial_j u(t,x)\cdot\nabla u(t,x)\partial_j u(t,x)dx\lesssim\|\nabla u(t)\|_{L^2}^{\frac32}\|\nabla^2u(t)\|_{L^2}^{\frac32}.
$$

We shall very often use the familiar convexity inequality

$$
ab\le\theta a^{\frac1\theta}+(1-\theta)b^{\frac{1}{1-\theta}}.
$$

Used with $\theta = 1/4$ this gives

$$\int_{\mathbb{R}^3} \partial_j u(t,x) \cdot \nabla u(t,x) \partial_j u(t,x) dx \leq \frac{1}{2} \|\nabla^2 u(t)\|_{L^2}^2 + C\|\nabla u(t)\|_{L^2}^6.$$

This gives

$$\frac{d}{dt}\|\nabla u(t)\|_{L^2}^2 + \|\nabla^2 u(t)\|_{L^2}^2 \lesssim \|\nabla u(t)\|_{L^2}^6.$$

Thus, as long as $\|\nabla u(t)\|_{L^2} \leq 2\|\nabla u_0\|_{L^2}$, we have

$$\|\nabla u(t)\|_{L^2}^2 \leq \|\nabla u_0\|_{L^2}^2 + Ct\|\nabla u_0\|_{L^2}^6.$$

Thus for $t \leq \frac{1}{2C}\|\nabla u_0\|_{L^2}^4$, the quantity $\|\nabla u(t)\|_{L^2}^2$ remains less than or equal to $2\|\nabla u_0\|_{L^2}^2$. This proves the local part of the theorem. In order to treat the case of small initial data, we estimate the term $(\partial_j u \cdot u|\partial_j u)_{L^2}$ in a different way. By integration by parts, let us write that

$$\begin{aligned}\int_{\mathbb{R}^3} \partial_j u(t,x) \cdot \nabla u(t,x) \partial_j u(t,x) dx &= -\int_{\mathbb{R}^3} u(t,x) \cdot \nabla \partial_j u(t,x) \partial_j u(t,x) dx \\ &\quad - \int_{\mathbb{R}^3} u(t,x) \cdot \nabla u(t,x) \partial_j^2 u(t,x) dx.\end{aligned}$$

As u is divergence free, we get that

$$\int_{\mathbb{R}^3} u(t,x) \cdot \nabla \partial_j u(t,x) \partial_j u(t,x) dx = 0.$$

Using Hölder inequality and Sobolev embeddings, we deduce that

$$\begin{aligned}\int_{\mathbb{R}^3} \partial_j u(t,x) \cdot \nabla u(t,x) \partial_j u(t,x) dx &\leq \|u(t)\|_{L^3} \|\nabla u(t)\|_{L^6} \|\nabla^2 u(t)\|_{L^2} \\ &\lesssim \|u(t)\|_{L^3} \|\nabla^2 u(t)\|_{L^2}^2.\end{aligned}$$

Using again Sobolev embeddings and interpolation inequality, we infer that

$$\|u(t)\|_{L^3} \lesssim \|u(t)\|_{L^2}^{\frac{1}{2}} \|\nabla u(t)\|_{L^2}^{\frac{1}{2}}.$$

The energy inequality implies that

$$\|u(t)\|_{L^3} \leq C_0 \|u_0\|_{L^2}^{\frac{1}{2}} \|\nabla u(t)\|_{L^2}^{\frac{1}{2}}.$$

Now, as long as

$$C_0 \|u_0\|_{L^2}^{\frac{1}{2}} \|\nabla u(t)\|_{L^2}^{\frac{1}{2}} \leq \frac{1}{2},$$

we have

$$\frac{d}{dt}\|\nabla u(t)\|_{L^2}^2 + \|\nabla^2 u(t)\|_{L^2}^2 \leq 0.$$

In particular $\|\nabla u(t)\|_{L^2}^2$ is a decreasing function. Thus, if it is small enough at initial time, it remains small and the theorem is proved. □

Corollary 3.1. *A constant c_0 exists such that if u_0 is a divergence free vector field in $L^2(\mathbb{R}^3)$ such that if ∇u_0 belongs to $L^2(\mathbb{R}^3)$ and if the maximal time of existence $T^\star$ of the solution associated with u_0 is finite, then we have*

$$\forall t \in [0, T^\star[, \ \|\nabla u(t)\|_{L^2}^4 \geq c_0 (T^\star - t)^{-1} \quad \text{and} \quad T^\star \leq c_0^{-1} \|u_0\|_{L^2}^4.$$

Proof. Applying this bound on the life span with $u(t)$ as an initial data gives the first inequality. Then the energy estimate implies that

$$\begin{aligned}(T^\star)^{\frac{1}{2}} &= 2 \int_0^{T^\star} \frac{1}{(T^\star - t)^{\frac{1}{2}}} dt \\ &\leq \frac{2}{\sqrt{c_0}} \int_0^{T^\star} \|\nabla u(t)\|_{L^2}^2 dt \\ &\leq \frac{1}{\sqrt{c_0}} \|u_0\|_{L^2}^2.\end{aligned}$$

The corollary is proved. □

3.2 The Kato method; the case of $\dot{H}^{\frac{1}{2}}$ initial data

Let us first define operators which we are going to use in this chapter.

Definition 3.3. *Let τ be a non negative real number. We denote by $L_0^{(\tau)}$ the operator defined by the fact that $L_0^{(\tau)} f$ is the solution of*

$$\begin{cases} \partial_t L_0^{(\tau)} f - \Delta L_0^{(\tau)} f = f - \nabla p, \\ \operatorname{div} L_0^{(\tau)} f = 0 \quad \text{and} \quad L_0^{(\tau)} f_{|t=\tau} = 0. \end{cases}$$

Let us also define for k in $\{1,2,3\}$, the operator $L_k^{(\tau)}$ by

$$\begin{cases} \partial_t L_k^{(\tau)} f - \Delta L_k^{(\tau)} f = \partial_k f - \nabla p, \\ \operatorname{div} L_k^{(\tau)} f = 0 \quad \text{and} \quad L_k^{(\tau)} f_{|t=\tau} = 0. \end{cases}$$

In the case when $\tau = 0$, we simply note $L_j^{(0)} = L_j$. Let us also define the bilinear operator B by

$$B(u,v) = -\frac{1}{2} L_0 \Big(\sum_{k=1}^3 \partial_k (v^k u + u^k v) \Big) = -\frac{1}{2} \sum_{k=1}^3 L_k (v^k u + u^k v).$$

Let us remark that if u and v are divergence free, we also have

$$B(u,v) = -\frac{1}{2} L_0 (v \cdot \nabla u + u \cdot \nabla v). \tag{3.4}$$

It is obvious that u a solution of (NS) if and only if u satisfies

$$u = e^{t\Delta} u_0 + B(u,u).$$

Solving (NS) is equivalent to find a fixed point for the map

$$u \longmapsto e^{t\Delta}u_0 + B(u,u).$$

Now let us assume that we have a Banach space X of functions locally in L^2 on $\mathbb{R}^+ \times \mathbb{R}^3$ such that B is a bilinear map from $X \times X$ into X. Then Picard's fixed point theorem implies the existence of a unique solution. Such a space X will be called "adapted".

Let us remark there is a strong constrain on X due to the scaling property. If X adapted, it must be scaling invariant (and also translation invariant) in the sense that

$$\begin{aligned}&\forall \lambda > 0\,,\ \forall \vec{a} \in \mathbb{R}^3\,,\ u \in X\\ \Longleftrightarrow\ & u\big(\lambda^2 t, \lambda(\cdot - \vec{a})\big) \in X \quad \text{and} \quad \|u\|_X \sim \lambda \big\|u\big(\lambda^2 t, \lambda(\cdot - \vec{a})\big)\big\|_X.\end{aligned}$$

Let us give a first example of an adapted spaces: the space $L^4(\mathbb{R}^+; \dot{H}^1)$.

The wellposedness of (NS) for initial data in the space $\dot{H}^{\frac{1}{2}}$ is described by the following theorem.

Theorem 3.3. *Let u_0 be a divergence free vector field in $\dot{H}^{\frac{1}{2}}$. A positive time T exists such that the system (NS) has a unique solution u in $L^4([0,T]; \dot{H}^1)$ which also belongs to*

$$C([0,T]; \dot{H}^{\frac{1}{2}}) \cap L^2([0,T]; \dot{H}^{\frac{3}{2}}).$$

A constant ρ_0 exists if $\|u_0\|_{\dot{H}^{\frac{1}{2}}} \leq \rho_0$, then the solution is global and satisfies

$$\|u\|^2_{L^\infty(\mathbb{R}^+; \dot{H}^{\frac{1}{2}})} + \|u\|^2_{L^2(\mathbb{R}^+; \dot{H}^{\frac{3}{2}})} \leq \|u_0\|^2_{\dot{H}^{\frac{1}{2}}}.$$

Proof. It relies on the following lemma.

Lemma 3.1. *The operator L_0 maps continuously the space $L^2([0,T]; \dot{H}^{-\frac{1}{2}})$ into the space*

$$C([0,T]; \dot{H}^{\frac{1}{2}}) \cap L^2([0,T]; \dot{H}^{\frac{3}{2}}).$$

Proof. In Fourier space, we can write that

$$\mathcal{F}L_0 f(t,\xi) = \int_0^t e^{-(t-t')|\xi|^2}\, \mathbb{P}(\xi) \widehat{f}(t',\xi)dt'$$

where $\mathbb{P}(\xi)$ is the orthogonal projection in $\mathbb{R}^3$ on the orthogonal of ξ. Thus, we get

$$|\mathcal{F}L_0 f(t,\xi)| \leq \int_0^t e^{-(t-t')|\xi|^2} |\xi|^{\frac{1}{2}} \theta(t',\xi) \|f(t',\cdot)\|_{\dot{H}^{-\frac{1}{2}}} dt'$$

with $\|\theta(t',\cdot)\|_{L^2} = 1$ for any t' of $\mathbb{R}^+$. Convolution inequality gives

$$|\xi|\, \|\mathcal{F}L_0 f(\cdot,\xi)\|^2_{L^\infty([0,T])} + |\xi|^3 \|\mathcal{F}L_0 f(\cdot,\xi)\|^2_{L^2([0,T])} \lesssim \int_0^T \theta^2(t',\xi) \|f(t',\cdot)\|^2_{\dot{H}^{-\frac{1}{2}}} dt'.$$

Taking the L^2 in ξ norm in the above inequality gives the result except the continuity. Let us assume that t_1 and t_2 are two points of $[0,T]$ such that $t_1 \leq t_2$. Thus, we have

$$L_0 f(t_2) = e^{(t_2-t_1)\Delta} L_0 f(t_1) + L_k^{(t_1)} f(t_2 - t_1). \tag{3.5}$$

Thus if t_2 is such that $\|f\|_{L^2([t_1,t_2];\dot{H}^{-\frac{1}{2}})} \leq \varepsilon$ is small, then

$$\|L_0 f(t_2) - e^{(t_2-t_1)\Delta} L_0 f(t_1)\|_{\dot{H}^{\frac{1}{2}}} \lesssim \varepsilon.$$

The continuity of the heat flow on $\dot{H}^{\frac{1}{2}}$ allows to conclude the proof. □

Conclusion of the proof of Theorem 3.3. Let us observe that dual Sobolev embedding and Sobolev embedding imply that

$$\begin{aligned}
\|u\cdot\nabla v + v\cdot\nabla u\|_{\dot{H}^{-\frac{1}{2}}} &\lesssim \|u\cdot\nabla v + v\cdot\nabla u\|_{L^{\frac{3}{2}}}\\
&\lesssim \|u\|_{L^6}\|\nabla v\|_{L^2} + \|v\|_{L^6}\|\nabla u\|_{L^2}\\
&\lesssim \|\nabla u\|_{L^2}\|\nabla v\|_{L^2}.
\end{aligned} \tag{3.6}$$

We infer that the bilinear operator B maps the space $L^4([0,T];\dot{H}^1)\times L^4([0,T];\dot{H}^1)$ into the space $C([0,T];\dot{H}^{\frac{1}{2}}) \cap L^2([0,T];\dot{H}^{\frac{3}{2}})$. As we have

$$\mathcal{F}(e^{t\Delta}u_0)(\xi) = e^{-t|\xi|^2}\widehat{u}_0(\xi),$$

we infer that

$$\|\mathcal{F}(e^{t\Delta}u_0)(\xi)\|_{L^4(\mathbb{R}^+)} \leq |\xi|^{-\frac{1}{2}}|\widehat{u}_0(\xi)|.$$

Because of Minkowski inequality, we get

$$\|e^{t\Delta}u_0\|_{L^4(\mathbb{R}^+;\dot{H}^1)} \lesssim \|u_0\|_{\dot{H}^{\frac{1}{2}}}.$$

Thus if $\|u_0\|_{\dot{H}^{\frac{1}{2}}}$ is small enough, then a unique global solution exists in $L^4(\mathbb{R}^+;\dot{H}^1)$. If $\|u_0\|_{\dot{H}^{\frac{1}{2}}}$ is not small the result is only local. Indeed let us define

$$u_0 = u_{0,\flat} + u_{0,\sharp} \quad \text{with} \quad u_{0,\flat} \stackrel{\text{def}}{=} \mathcal{F}^{-1}(\mathbf{1}_{B(0,\lambda_0)}\widehat{u}_0). \tag{3.7}$$

We have

$$\|e^{t\Delta}u_0\|_{L^4([0,T];\dot{H}^1)} \lesssim \|u_{0,\flat}\|_{\dot{H}^{\frac{1}{2}}} + T^{\frac{1}{4}}\lambda_0^{\frac{1}{2}}\|u_0\|_{\dot{H}^{\frac{1}{2}}}.$$

We can choose λ_0 such that $\|u_{0,\flat}\|_{\dot{H}^{\frac{1}{2}}}$ small which proves the theorem up to the last inequality. In order to prove it, let us perform an $\dot{H}^{\frac{1}{2}}$ energy estimate which gives

$$\frac{1}{2}\frac{d}{dt}\|u(t)\|^2_{\dot{H}^{\frac{1}{2}}} + \|\nabla u(t)\|^2_{\dot{H}^{\frac{1}{2}}} = (u\cdot\nabla u|u)_{\dot{H}^{\frac{1}{2}}}.$$

As we have

$$\begin{aligned}
(a|b)_{\dot{H}^{\frac{1}{2}}} &= \int_{\mathbb{R}^3} |\xi|^{-\frac{1}{2}}\widehat{a}(\xi)|\xi|^{\frac{3}{2}}\widehat{b}(-\xi)d\xi\\
&\leq \|a\|_{\dot{H}^{-\frac{1}{2}}}\|\nabla b\|_{\dot{H}^{\frac{1}{2}}},
\end{aligned}$$

Inequality (3.6) together with an interpolation argument gives that

$$\begin{aligned}(u\cdot\nabla u|u)_{\dot H^{\frac12}} &\le C\|\nabla u\|_{L^2}^2\|\nabla u\|_{\dot H^{\frac12}}\\ &\le C\|u\|_{\dot H^{\frac12}}\|\nabla u\|^2_{\dot H^{\frac12}}.\end{aligned}$$

Thus as long as $C\|u\|_{\dot H^{\frac12}} \le 1/2$, we have

$$\frac{d}{dt}\|u(t)\|^2_{\dot H^{\frac12}} + \|\nabla u(t)\|^2_{\dot H^{\frac12}} \le 0.$$

Thus if $\|u_0\|_{\dot H^{\frac12}} \le (2C_0)^{-1}$, we get the required inequality and the theorem is proved. □

Remark. Because the $\dot H^{\frac12}$ norm is scaling invariant, the life span cannot be bounded from below by a function of the norm. Let us notice that even if we assume that u_0 belongs to $L^2\cap \dot H^1$ the best known life span is the one coming from the arguments we give here in the proof of Theorem 3.2.

Now let us establish a criteria for blow up of the regularity. In other words, it is a necessary condition for the appearance of singularities.

Proposition 3.1. *If u_0 is in $\dot H^{\frac12}$ and if $T^\star$ the maximal time of existence of a solution in the space $L^\infty_{loc}[0,T^\star[;\dot H^{\frac12})\cap L^2_{loc}(T^\star;\dot H^{\frac32})$ is finite. Then we have*

$$\forall (p,q)\in[2,\infty[\times]3,\infty] : \frac2p+\frac3q = 1\,,\quad \int_0^{T^\star}\|u(t)\|^p_{L^q}dt = \infty. \tag{3.8}$$

Proof. Let us write that

$$\frac12\frac{d}{dt}\|u(t)\|^2_{\dot H^{\frac12}} + \|\nabla u(t)\|^2_{\dot H^{\frac12}} = -(u\cdot\nabla u|u)_{\dot H^{\frac12}}.$$

Using that, if a and b are real valued, we have

$$\begin{aligned}(a|b)_{\dot H^{\frac12}} &= \int_{\mathbb R^3}\widehat a(\xi)|\xi|\widehat b(-\xi)d\xi\\ &= \big(a\,\big|\,|D|b\big)_{L^2}\quad\text{with}\quad |D|b \stackrel{\text{def}}{=} \mathcal F^{-1}(|\cdot|\widehat b).\end{aligned}$$

Thus we have

$$(u\cdot\nabla u|u)_{\dot H^{\frac12}} = \int_{\mathbb R^3} u(x)\cdot\nabla u(x)(|D|u)(x)dx.$$

Let $\widetilde q$ defined by

$$\frac1q+\frac2{\widetilde q} = 1.$$

Hölder inequality, Sobolev embedding and interpolation inequality give using the relation between p and q,

$$\begin{aligned}(u\cdot\nabla u|u)_{\dot H^{\frac12}} &\lesssim \|u\|_{L^q}\|\nabla u\|_{L^{\widetilde q}}\||D|u\|_{L^{\widetilde q}}\\ &\lesssim \|u\|_{L^q}\|\nabla u\|^2_{\dot H^{\frac{3}{2q}}}\\ &\lesssim \|u\|_{L^q}\|\nabla u\|^2_{\dot H^{\frac12-\frac1p}}\\ &\lesssim \|u\|_{L^q}\|u\|^{\frac2p}_{\dot H^{\frac12}}\|\nabla u\|^{\frac{2}{p'}}_{\dot H^{\frac12}}.\end{aligned}$$

Then using the convexity inequality and Relation (3.8), we infer that

$$(u\cdot\nabla u|u)_{\dot H^{\frac12}}\leq\frac12\|\nabla u(t)\|^2_{\dot H^{\frac12}}+C\|u(t)\|^p_{L^q}\|u(t)\|^2_{\dot H^{\frac12}}.$$

By Gronwall lemma, we deduce that

$$\|u(t)\|^2_{\dot H^{\frac12}}+\int_0^t\|\nabla u(t')\|^2_{\dot H^{\frac12}}dt'\leq\|u_0\|^2_{\dot H^{\frac12}}\exp\Bigl(C\int_0^t\|u(t')\|^p_{L^q}dt'\Bigr).$$

The theorem is proved. □

As a conclusion of this introduction, let us introduce the set of initial data which gives birth to global regular solution.

Definition 3.4. *Let us denote by $\mathcal{G}$ the set of initial data u_0 such that the solution u given by Theorem 3.3 is global i.e. belongs to $L^4_{\rm loc}(\mathbb{R}^+;\dot H^1)$.*

We denote as $\rho_{\rm c}$ the supremum of the positive real number ρ such that if $\|u_0\|_{\dot H^{\frac12}}<\rho$ implies that u_0 gives birth to a global solution in the space in $L^4_{loc}(\mathbb{R}^+;\dot H^1)$.

3.3 Global stability results in $\dot H^{\frac12}$

The basic theorem in this section is the following.

Theorem 3.4. *Let u be a global solution of (NS) in $L^\infty_{loc}(\mathbb{R}^+;\dot H^{\frac12})\cap L^4_{loc}(\mathbb{R}^+;\dot H^1)$. Then we have*

$$\lim_{t\to\infty}\|u(t)\|_{\dot H^{\frac12}}=0\quad\textit{and}\quad\int_0^\infty\|u(t)\|^4_{\dot H^1}dt<\infty.$$

Proof. We shall decompose the initial data u_0 as a sum of a low and a high frequency part. A positive real number ρ being given, let us state

$$u_0=u_{0,\sharp}+u_{0,\flat}\quad\text{with}\quad u_{0,\flat}\stackrel{\rm def}{=}\mathcal{F}^{-1}(\mathbf{1}_{B(0,r)}(\xi)\widehat{u}_0(\xi)).$$

Let ε be any positive real number. We can choose r such that

$$\|u_{0,\flat}\|_{\dot H^{\frac12}}\leq\min\{\rho_0,\varepsilon/2\},$$

where ρ_0 is the constant of Theorem 3.3. Let us denote by $u_\flat$ the global solution of (NS) given by Theorem 3.3 for the initial data $u_{0,\flat}$. We have

$$\|u_\flat\|^2_{L^\infty(\mathbb{R}^+;\dot H^{\frac12})}+\|u_\flat(t)\|^2_{L^2(\mathbb{R}^+;\dot H^{\frac32})}\leq\|u_{0,\flat}\|^2_{\dot H^{\frac12}}.\tag{3.9}$$

Let us define $u_\sharp\stackrel{\rm def}{=}u-u_\flat$. Let us notice that $u_\sharp$ is globally defined because so are u and $u_\flat$. It satisfies

$$\begin{cases}\partial_t u_\sharp-\Delta u_\sharp+(u_\sharp+u_\flat)\cdot\nabla u_\sharp+u_\sharp\cdot\nabla u_\flat=-\nabla p,\\ u_{\sharp|t=0}=u_{0,\sharp}.\end{cases}$$

By energy estimate, we infer

$$\frac{1}{2}\|u_\sharp(t)\|_{L^2}^2 + \int_0^t \|\nabla u_\sharp(t')\|_{L^2}^2 dt' = \frac{1}{2}\|u_{0,\sharp}\|_{L^2}^2 + \int_0^t (u_\sharp \cdot \nabla u_\flat | u_\sharp) dt'.$$

Using Sobolev embedding, we claim that

$$\begin{aligned}\big|\langle u_\sharp(t)\cdot\nabla u_\flat(t), u_\sharp(t)\rangle\big| &\leq \|u_\sharp(t)\cdot\nabla u_\flat(t)\|_{L^2}\|\nabla u(t)\|_{L^2}\\ &\leq \|u_\sharp(t)\|_{L^6}\|u_\flat(t)\|_{L^3}\|\nabla u_\sharp(t)\|_{L^2}\\ &\leq C\|u_\flat(t)\|_{\dot H^{\frac12}}\|\nabla u_\sharp(t)\|_{L^2}^2.\end{aligned}$$

Then we deduce that

$$\frac{1}{2}\|u_\sharp(t)\|_{L^2}^2 + \int_0^t \|\nabla u_\sharp(t')\|_{L^2}^2 dt' \leq \frac{1}{2}\|u_{0,\sharp}\|_{L^2}^2 + C\|u_{0,\flat}\|_{\dot H^{\frac12}} \int_0^t \|\nabla u_\sharp(t')\|_{L^2}^2 dt'.$$

Choosing ε small enough ensures that

$$\|u_\sharp(t)\|_{L^2}^2 + \int_0^t \|\nabla u_\sharp(t')\|_{L^2}^2 dt' \leq \|u_{0,\sharp}\|_{L^2}^2.$$

This implies that a positive time t_ε exists such that $\|u_\sharp(t_\varepsilon)\|_{\dot H^{\frac12}} < \varepsilon/2$. Thus $\|u(t_\varepsilon)\|_{\dot H^{\frac12}}$ is less than ε. Then Theorem 3.3 allows to conclude the proof. □

Let us remark that the set $\mathcal{G}$ contains the open ball of radius ρ_c and centered at origin. Let us state the following corollary of Theorem 3.4.

Theorem 3.5. *The set $\mathcal{G}$ is an open connected subset of $\dot H^{\frac12}$.*

Proof. Let us consider u_0 in $\dot H^{\frac12}$ such that the associated solution is global. Let us consider w_0 in $\dot H^{\frac12}$ and the (a priori) local solution v associated with the initial data $v_0 \stackrel{\text{def}}{=} u_0 + w_0$. The function $w \stackrel{\text{def}}{=} v - u$ is solution of

$$\begin{cases}\partial_t w - \Delta w + u\cdot\nabla w + w\cdot\nabla u + w\cdot\nabla w = -\nabla p,\\ w_{|t=0} = w_0.\end{cases}$$

Sobolev embeddings together with interpolation inequality gives

$$\begin{aligned}\big|(u\cdot\nabla w + w\cdot\nabla u|w)_{\frac12}\big| &\leq C\|u\|_{\dot H^1}\|w\|_{\dot H^{\frac12}}^{\frac12}\|\nabla w\|_{\dot H^{\frac12}}^{\frac32} \quad \text{and}\\ \big|(w\cdot\nabla w|w)_{\frac12}\big| &\leq C\|w\|_{\dot H^{\frac12}}\|\nabla w\|_{\dot H^{\frac12}}^2.\end{aligned}$$

Let us assume that $\|w_0\|_{\dot H^{\frac12}} \leq \frac{1}{8C}$ and define

$$T_{w_0} \stackrel{\text{def}}{=} \sup\Big\{t \,/ \max_{0\leq t'\leq t}\|w(t')\|_{\dot H^{\frac12}} \leq \frac{1}{4C}\Big\}.$$

Thanks to the convexity inequality we infer that, for any $t < T_{w_0}$,

$$\|w(t)\|_{\dot H^{\frac12}}^2 + \int_0^t \|\nabla w(t')\|_{\dot H^{\frac12}}^2 dt' \leq \|w_0\|_{\dot H^{\frac12}}^2 + C\int_0^t \|u(t')\|_{\dot H^1}^4 \|w(t')\|_{\dot H^{\frac12}}^2 dt'.$$

Gronwall's Lemma and Theorem 3.4 imply that, for any $t < T_{w_0}$,

$$\|w(t)\|^2_{\dot H^{\frac12}} + \int_0^t \|\nabla w(t')\|^2_{\dot H^{\frac12}} dt' \leq \|w_0\|^2_{\dot H^{\frac12}} \exp\Big(C\int_0^\infty \|u(t)\|^4_{\dot H^1} dt\Big).$$

If the smallness condition

$$\|w_0\|^2_{\dot H^{\frac12}} \exp\Big(C\int_0^\infty \|u(t)\|^4_{\dot H^1} dt\Big) \leq \frac{1}{16C^2} \tag{3.10}$$

is satisfied, the blow up condition for v is never satisfied. Thus $\mathcal{G}$ is open.

The fact that $\mathcal{G}$ is connected is due to the fact that as $\lim\limits_{t\to\infty} u(t) = 0$ in $\dot H^{\frac12}$, any u_0 in $\mathcal{G}$ is connected to 0. Thus the corollary is proved. □

3.4 The Kato theory in the L^p framework

The purpose of this section is the proof of the following theorem.

Theorem 3.6. *Let u_0 be a divergence free vector field in L^3. A positive time T exists such that the system (NS) has a unique solution u in $L^5([0,T]\times\mathbb{R}^3)$ which also belongs to $C([0,T];L^3)$. Moreover, a constant ρ_1 exists if $\|u_0\|_{L^3} \leq \rho_1$, then the solution is global and satisfies*

$$\|u\|_{L^\infty(\mathbb{R}^+;L^3)} + \|u\|_{L^5(\mathbb{R}^+\times\mathbb{R}^3)} \lesssim \|u_0\|_{L^3}.$$

Proof. It relies on the following lemma.

Lemma 3.2. *For j in $\{1,2,3\}$, we have*

$$L_j f(x) = \sum_{k=1}^3 \int_0^t \int_{\mathbb{R}^3} \Gamma^j_k(t-t', x-y) f(t',y)\,dy\,dt',$$

where the functions $\Gamma^j_{k,\ell}$ belongs to the space $L^{\frac54}_w$ and satisfies

$$|\Gamma^j_{k,\ell}(\tau,z)| \lesssim \frac{1}{(\sqrt{\tau}+|z|)^4}\cdot$$

Proof. In Fourier space, we have

$$\mathcal{F}L_j f(t,\xi) = i\int_0^t e^{-(t-t')|\xi|^2} \sum_{k,\ell} \alpha_{j,k\ell}\xi_j\xi_k\xi_\ell|\xi|^{-2}\widehat{f}(t',\xi)\,dt'.$$

In order to write this operator as a convolution operator, it is enough to compute the inverse Fourier transform of $\xi_j\xi_k\xi_\ell|\xi|^{-2}e^{-t|\xi|^2}$. Using the fact that

$$e^{-t|\xi|^2}|\xi|^{-2} = \int_t^\infty e^{-t'|\xi|^2}dt',$$

we get that

$$\begin{aligned}\Gamma^j_{k,\ell}(t,x) &= i\int_t^\infty\int_{\mathbb{R}^3}\xi_j\xi_k\xi_\ell e^{i(x|\xi)-t'|\xi|^2}dt'd\xi\\ &= \partial_j\partial_k\partial_\ell\int_t^\infty\int_{\mathbb{R}^3}e^{i(x|\xi)-t'|\xi|^2}dt'd\xi.\end{aligned}$$

Using the formula about the Fourier transform of the Gaussian functions, we get

$$\begin{aligned}\Gamma^j_{k,\ell}(t,x) &= \partial_j\partial_k\partial_\ell\int_t^\infty\frac{1}{(4\pi t')^{\frac32}}e^{-\frac{|x|^2}{4t'}}dt'\\ &= \frac{1}{\pi^{\frac32}}\int_t^\infty\frac{1}{(4t')^3}\Psi^j_{k,\ell}\Big(\frac{x}{\sqrt{4t'}}\Big)dt' \quad\text{with}\quad \Psi^j_{k,\ell}(z)\overset{\text{def}}{=}\partial_j\partial_k\partial_\ell e^{-|z|^2}.\end{aligned}$$

Changing variable $r=(4t')^{-1}|x|^2$ gives

$$|\Gamma^j_{k,\ell}(t,x)|\leq\frac{1}{\pi^{\frac32}}\frac{1}{|x|^4}\int_0^{\frac{|x|^2}{4t}}r\Psi^j_{k,\ell}\Big(\frac{x}{|x|}r\Big)dr.$$

This implies that

$$|\Gamma^j_{k,\ell}(t,x)|\lesssim\min\Big\{\frac{1}{t^2},\frac{1}{|x|^4}\Big\}\lesssim\frac{1}{(\sqrt{t}+|x|)^4}\cdot$$

The fact that $(\sqrt{t}+|x|)^{-4}$ belongs to $L^{\frac54}_w$ from the fact that the function is homogeneous of order -4 in the space of dimension 5 because the homogeneity is with respect of the dilation $(t,x)\mapsto(\lambda^2t,\lambda x)$. This proves the lemma. □

Corollary 3.2. *The operators L_k maps continuously from the space $L^{\frac52}([0,T]\times\mathbb{R}^3)$ into the space $C([0,T];L^3)\cap L^5([0,T];L^5)$.*

Proof. Using Lemma 3.2, we immediately infer that Γ^j_k belongs to $L^{\frac54}_w$. As we have

$$1+\frac15=\frac45+\frac25,$$

then using Hardy-Littlewood-Sobolev inequality (see for instance Theorem 1.7 on page 10 of [2]) we infer that

$$\|L_kf\|_{L^5([0,T]\times\mathbb{R}^3)}\lesssim\|f\|_{L^{\frac52}([0,T]\times\mathbb{R}^3)}\cdot \tag{3.11}$$

Now let us observe that Lemma 3.2 implies that

$$\|\Gamma^j_k(\cdot,z)\|_{L^{\frac53}(\mathbb{R}^+)}\lesssim\frac{1}{|z|^{\frac{14}{5}}}\cdot$$

Hölder inequality in time implies that

$$\begin{aligned}\|L_kf(\cdot,x)\|_{L^\infty([0,T])} &\lesssim\int_{\mathbb{R}^3}\|\Gamma^j_k(\cdot,x-y)\|_{L^{\frac53}(\mathbb{R}^+)}\|f(\cdot,y)\|_{L^{\frac52}([0,T])}dy\\ &\lesssim\int_{\mathbb{R}^3}\frac{1}{|x-y|^{\frac{14}{5}}}\|f(\cdot,y)\|_{L^{\frac52}([0,T])}dy.\end{aligned}$$

As

$$1+\frac{1}{3}=\frac{14}{15}+\frac{2}{5},$$

Hardy-Littlewood-Sobolev inequality implies that

$$\|L_k f\|_{L^3(\mathbb{R}^3;L^\infty([0,T]))} \lesssim \|f\|_{L^{\frac{5}{2}}([0,T]\times\mathbb{R}^3)}.$$

As $\|g\|_{L^\infty([0,T];L^3(\mathbb{R}^3))} \leq \|g\|_{L^3(\mathbb{R}^3;L^\infty([0,T]))}$ this gives the result. In order to prove the continuity, we use (3.8) and then proceed exactly as in the proof of Lemma 3.1. □

Conclusion of the proof of theorem 3.6. The fact that the bilinear operator B maps continuously $L^5([0,T]\times\mathbb{R}^3)\times L^5([0,T]\times\mathbb{R}^3)$ into $L^\infty(\mathbb{R}_+;L^3)\cap L^5([0,T]\times\mathbb{R}^3)$ is an obvious consequence of Corollary 3.2. Then we simply have to prove that

$$\|e^{t\Delta}u_0\|_{L^5(\mathbb{R}^+\times\mathbb{R}^3)} \lesssim \|u_0\|_{L^3}. \tag{3.12}$$

As we have

$$|e^{t\Delta}u_0(x)| \leq \frac{1}{(4\pi t)^{\frac{3}{2}}}\int_{\mathbb{R}^3} e^{-\frac{|x-y|^2}{4t}}|u_0(y)|dy,$$

we get

$$\begin{aligned}\|e^{\cdot\Delta}u_0(x)\|_{L^5([0,T])} &\leq \int_{\mathbb{R}^3}\Big\|\frac{1}{(4\pi\cdot)^{\frac{3}{2}}}e^{-\frac{|x-y|^2}{4\cdot}}\Big\|_{L^5([0,T])}|u_0(y)|dy\\ &\lesssim \int_{\mathbb{R}^3}\frac{1}{|x-y|^{\frac{13}{5}}}|u_0(y)|dy.\end{aligned}$$

As

$$1+\frac{1}{5}=\frac{13}{15}+\frac{1}{3},$$

Hardy-Littlewood-Sobolev inequality give the result in the case when u_0 is small. For the local version, let us use that the space $L^3\cap L^5$ is dense in L^3. Thus for any positive ε, a function $u_{0,\varepsilon}$ exists in $L^3\cap L^5$ such that $\|u_0-u_{0,\varepsilon}\|_{L^3}\leq\varepsilon$. Using (3.12) gives that

$$\|e^{t\Delta}u_0\|_{L^5([0,T]\times\mathbb{R}^3)} \lesssim \varepsilon + T^{\frac{1}{5}}\|u_{0,\varepsilon}\|_{L^5}.$$

This concludes the proof of the theorem. □

3.5 A stability result in the L^p framework

The following theorem refines Theorem 3.5. It will be useful in the proof of forthcoming Theorem 4.1.

Theorem 3.7. *A constant C_0 exists such that for any couple (u_0, v_0) of divergence free vector fields in $\dot H^{\frac{1}{2}}$ and an external force f in $L^2(\mathbb{R}^+;\dot H^{\frac{1}{2}})\cap L^{\frac{5}{2}}(\mathbb{R}^+\times\mathbb{R}^3)$. Let us assume that u belongs to $L^\infty(\mathbb{R}^+;\dot H^{\frac{1}{2}})\cap L^4(\mathbb{R}^+;\dot H^1)$ and that*

$$\|v_0\|_{L^3}+\|f\|_{L^{\frac{5}{2}}(\mathbb{R}^+\times\mathbb{R}^3)} \leq \frac{1}{C_0}\exp\Big(-C_0\|u\|^5_{L^5(\mathbb{R}^+\times\mathbb{R}^3)}\Big).$$

Then the system

$$(NS_u)\quad \begin{cases} \partial_t v - \Delta v + v\cdot\nabla v + \operatorname{div}\big(u\otimes v + v\otimes u\big) = -\nabla p + \displaystyle\sum_{\ell=1}^{3}\partial_\ell f^\ell, \\ \operatorname{div} v = 0 \quad \textit{and} \quad v_{|t=0} = v_0 \end{cases}$$

has a global solution which satisfies

$$\begin{aligned} &\|v\|_{L^\infty(\mathbb{R}^+;L^3)} + \|v\|_{L^5(\mathbb{R}^+\times\mathbb{R}^3)} \\ &\leq \frac{1}{C_0}\big(\|v_0\|_{L^3} + \|f\|_{L^{\frac52}(\mathbb{R}^+\times\mathbb{R}^3)}\big)\exp\big(C_0\|u\|^5_{L^5(\mathbb{R}^+\times\mathbb{R}^3)}\big). \end{aligned}$$

Proof. Let us first notice that for any subinterval $I = [a, b)$, the bilinear operator B_I defined by

$$\begin{cases} \partial_t B_I - \Delta B_I = -\dfrac{1}{2}\operatorname{div}\big(u\otimes v + v\otimes u\big) - \nabla p, \\ \operatorname{div} B_I = 0 \quad \text{and} \quad B_{I|t=a} = 0 \end{cases}$$

maps continuously $L^5(I\times\mathbb{R}^3)\times L^5(I\times\mathbb{R}^3)$ into $L^\infty(I;L^3(\mathbb{R}^3))\cap L^5(I\times\mathbb{R}^3)$ with a constant of continuity which does not depend on I; let us denote by $\|B\|$ this constant. Let us also denote by C_1 the best constant such that

$$\|e^{t\Delta}a\|_{L^5(\mathbb{R}^+\times\mathbb{R}^3)} \leq C_1\|a\|_{L^3}.$$

Let us remark that Theorem 3.4 implies that u belongs to $L^\infty(\mathbb{R}^+;\dot H^{\frac12})\cap L^2(\mathbb{R}^+;\dot H^{\frac32})$. Thus Sobolev embedding and interpolation inequalities between Sobolev spaces implies that u belongs to the space $L^5(\mathbb{R}^+\times\mathbb{R}^3)$. The local theory of wellposedness for initial data in $\dot H^{\frac12}$ claims in particular that a unique maximal solution w exists in $C([0,T^\star[;\dot H^{\frac12})\cap L^2_{loc}([0,T^\star[;\dot H^{\frac32})$ and if $T^\star$ is finite, then

$$\int_0^{T^\star}\|w(t,\cdot)\|^5_{L^5}dt = \infty.$$

We want to control the $\|w\|_{L^5([0,T]\times\mathbb{R}^3)}$ to prevent blow up. In order to do so, let us write the solution $w = u + v$. Then v is the solution of

$$(NS_u)\quad \begin{cases} \partial_t v - \Delta v + v\cdot\nabla v + \operatorname{div}\big(u\otimes v + v\otimes u\big) = -\nabla p + \displaystyle\sum_{\ell=1}^{3}\partial_\ell f^\ell, \\ \operatorname{div} v = 0 \quad \text{and} \quad v_{|t=0} = v_0. \end{cases}$$

Let us decompose $\mathbb{R}^+$ as a disjoint union of intervals $(I_j)_{1\leq j\leq N}$ such that if $I_j = [I_j^-, I_j^+]$, then $I_j^- = 0$, $I_j^+ = I_{j+1}^-$ and $I_N^+ = \infty$ and

$$\forall j\in\{1,\cdots,N\},\ \int_{I_j}\|u(t,\cdot)\|^5_{L^5}dt = \Big(\frac{1}{8\|B\|}\Big)^5$$

and

$$\int_{I_N} \|u(t,\cdot)\|_{L^5}^5 dt \leq \Big(\frac{1}{8\|B\|}\Big)^5.$$

Let us notice that, by addition, we get

$$(N-1)\Big(\frac{1}{8\|B\|}\Big)^5 \leq \int_{\mathbb{R}^+} \|u(t,\cdot)\|_{L^5}^5 dt \leq N\Big(\frac{1}{8\|B\|}\Big)^5.$$

We shall prove by induction that if

$$\|v_0\|_{L^3} + \|f\|_{L^{\frac{5}{2}}(\mathbb{R}^+\times\mathbb{R}^3)} \leq \frac{1}{8C_1\|B\|}(2C_1)^{-N-1}, \tag{3.13}$$

then, for any j, we have $\widehat{I}_j \stackrel{\text{def}}{=} \bigcup_{j'=1}^{j} I'_j \subset [0, T^\star[$ and

$$T^\star > I_j^+ \quad \text{and} \quad \|v(I_j^+)\|_{L^3} \leq \big(\|v_0\|_{L^3} + \|f\|_{L^{\frac{5}{2}}(I_1\times\mathbb{R}^3)}\big)(2C_1)^j. \tag{3.14}$$

Let us point out that it prevents blow up and thus $u_0 + v_0$ belongs to $\mathcal{G}$. Let us prove the induction hypothesis for $j = 1$. Let us define

$$\begin{aligned}\overline{T} \stackrel{\text{def}}{=} \sup\Big\{T < \min\{T^\star, I_1^+\} \,/\, \max\big\{\|v\|_{L^\infty([0,T];L^3(\mathbb{R}^3))}, \|v\|_{L^5([0,T]\times\mathbb{R}^3)}\big\} \\ \leq \big(\|v_0\|_{L^3} + \|f\|_{L^5(I_1\times\mathbb{R}^3)}\big)(2C_1)\Big\}.\end{aligned}$$

The system (NS_u) writes

$$v = e^{t\Delta}v_0 + 2B(u,v) + B(v,v) + \sum_{k=1}^{3} L_k f^k.$$

For any $T < \overline{T}$, we have

$$\begin{aligned}&\max\big\{\|v\|_{L^\infty([0,T];L^3(\mathbb{R}^3))}, \|v\|_{L^5([0,T]\times\mathbb{R}^3)}\big\} \\ &\leq C_1\big(\|v_0\|_{L^3} + \|f\|_{L^{\frac{5}{2}}(I_1\times\mathbb{R}^3)}\big) \\ &\quad +2\|B\|\|u\|_{L^5(I_1\times\mathbb{R}^3)}\|v\|_{L^5([0,T]\times\mathbb{R}^3)} + \|B\|\|v\|_{L^5([0,T]\times\mathbb{R}^3)}^2.\end{aligned}$$

By definition of T and the intervals I_j, we get, thanks to Hypothesis (3.13)

$$\begin{aligned}&\max\big\{\|v\|_{L^\infty([0,T];L^3(\mathbb{R}^3))}, \|v\|_{L^5([0,T]\times\mathbb{R}^3)}\big\} \\ &\leq C_1\big(\|v_0\|_{L^3} + \|f\|_{L^{\frac{5}{2}}(I_1\times\mathbb{R}^3)}\big) + \frac{1}{2}\|v\|_{L^5([0,T]\times\mathbb{R}^3)}\end{aligned}$$

which gives

$$\forall T < \overline{T},\ \max\big\{\|v\|_{L^\infty([0,T];L^3(\mathbb{R}^3))}, \|v\|_{L^5([0,T]\times\mathbb{R}^3)}\big\} \leq 2C_1\big(\|v_0\|_{L^3} + \|f\|_{L^{\frac{5}{2}}(I_1\times\mathbb{R}^3)}\big).$$

This implies that $T^\star > I_1^+$ and $\overline{T} = I_1^+$ which is exactly the induction hypothesis for $j = 1$.

Let us assume the induction hypothesis for $j < N$. The system (NS_u) can be written

$$v = e^{(t-I_{j+1}^-)\Delta}v(I_{j+1}^-) + 2B_{I_{j+1}}(u,v) + B(v,v) + \sum_{k=1}^{3} L_k f^k.$$

Let us define

$$\overline{T}_{j+1} \stackrel{\text{def}}{=} \sup\Big\{T < \min\{T^\star, I_{j+1}^+\} \;/\; \max\{\|v\|_{L^\infty([I_{j+1}^-,T];L^3(\mathbb{R}^3))}, \|v\|_{L^5([I_{j+1}^-,T]\times\mathbb{R}^3)}\} \leq (2C_1)^{j+1}\big(\|v_0\|_{L^3} + \|f\|_{L^{\frac{5}{2}}(I_{j+1}\times\mathbb{R}^3)}\big)\Big\}.$$

For any $I_{j+1}^- \leq T < \overline{T}_{j+1}$, we have

$$\begin{aligned}
&\max\{\|v\|_{L^\infty([I_{j+1}^-,T];L^3(\mathbb{R}^3))}, \|v\|_{L^5([I_{j+1}^-,T]\times\mathbb{R}^3)}\}\\
&\leq C_1\big(\|v(I_{j+1}^-)\|_{L^3} + \|f\|_{L^{\frac{5}{2}}(I_{j+1}\times\mathbb{R}^3)}\big)\\
&\quad +2\|B\|\|u\|_{L^5(I_1\times\mathbb{R}^3)}\|v\|_{L^5([I_{j+1}^-,T]\times\mathbb{R}^3)} + \|B\|\|v\|^2_{L^5([I_{j+1}^-,T]\times\mathbb{R}^3)}.
\end{aligned}$$

By definition of T and of the intervals I_j, we get that

$$\begin{aligned}
&\max\{\|v\|_{L^\infty([I_{j+1}^-,T];L^3(\mathbb{R}^3))}, \|v\|_{L^5([I_{j+1}^-,T]\times\mathbb{R}^3)}\}\\
&\leq C_1\big(\|v(I_{j+1}^-)\|_{L^3} + \|f\|_{L^{\frac{5}{2}}(I_{j+1}\times\mathbb{R}^3)}\big) + \frac{1}{2}\|v\|_{L^5([I_{j+1}^-,T]\times\mathbb{R}^3)}
\end{aligned}$$

which gives

$$\begin{aligned}
\forall T < \overline{T},\ &\max\{\|v\|_{L^\infty([I_{j+1}^-,T];L^3(\mathbb{R}^3))}, \|v\|_{L^5([I_{j+1}^-,T]\times\mathbb{R}^3)}\}\\
&\leq 2C_1\big(\|v(I_{j+1}^-)\|_{L^3} + \|f\|_{L^{\frac{5}{2}}(I_{j+1}\times\mathbb{R}^3)}\big).
\end{aligned}$$

This implies that $T^\star > I_{j+1}^+$ and $\overline{T} = I_{j+1}^+$ which is exactly the induction hypothesis for $j+1$ and the theorem is proved. □

4 Applications of profile decomposition to the Navier-Stokes equations

The following theorem is the basis of application of profiles theory to incompressible Navier-Stokes system.

4.1 Bounded sequences of initial data

Theorem 4.1. *Let $(u_{0,n})_{n\in\mathbb{N}}$ be a sequence of initial data which is bounded in $\dot{H}^{\frac{1}{2}}$. Let us consider an extraction ϕ, a sequence of profiles $(\varphi^j)_{j\in\mathbb{N}}$, a sequence of scales* *and cores* $(\lambda_n^j, x_n^j)_{(j,n)\in\mathbb{N}^2}$ *in the sense of Definition 2.1, a sequence* $(r_n^j)_{(j,n)\in\mathbb{N}^2}$ *of functions given by Theorem 2.1.*

Let us define $\mathcal{J}_f$ *as the set of indices* j *such that* φ^j *does not belongs to* $\mathcal{G}$ *i.e.* φ^j *does not give birth to a global solution in* $L^4(\mathbb{R}^+; \dot H^1)$*. This set is finite.*

If $\mathcal{J}_f$ *is empty, then, for* n *large enough,* $u_{\phi(n)}$ *belongs to* $\mathcal{G}$*.*

If $\mathcal{J}_f$ *is non empty, an index* j_0 *exists in* $\mathcal{J}_f$ *such that*

$$\forall j \in \mathcal{J}_f\,,\ \exists n_j : n \geq n_j \Longrightarrow (\lambda_n^{j_0})^2 T^\star(\varphi^{j_0}) \leq (\lambda_n^j)^2 T^\star(\varphi^j).$$

For any positive real number ε*, let us define* $\tau_n^\varepsilon \stackrel{\text{def}}{=} (\lambda_n^{j_0})^2 (T^\star(\varphi^{j_0}) - \varepsilon)$*. Then*

$$\exists n_\varepsilon : n \geq n_\varepsilon \Longrightarrow T^\star(u_{0,\phi(n)}) \geq \tau_n^\varepsilon.$$

Proof. Let us search the solution u_n^J associated with $u_{0,\phi(n)}$ of the form

$$\begin{aligned} u_n^J &= u_{n,\text{app}}^J + R_n^J \quad \text{with} \quad u_{n,\text{app}}^J \stackrel{\text{def}}{=} \sum_{j=0}^J \Phi_n^j + e^{t\Delta} r_n^J \\ &\text{and} \quad \Phi_n^j(t,x) = \frac{1}{\lambda_n^j} \Phi^j\Big(\frac{t}{(\lambda_n^j)^2}, \frac{x - x_n^j}{\lambda_n^j}\Big) \end{aligned} \tag{4.1}$$

where Φ^j denotes the solution of (NS) associated with φ^j. The vector field R_n^J satisfies

$$\begin{aligned} &\partial_t R_n^J - \Delta R_n^J + R_n^J \cdot \nabla R_n^J + u_{n,\text{app}}^J \cdot \nabla R_n^J + R_n^J \cdot \nabla u_{n,\text{app}}^J = -\nabla p - \operatorname{div} \sum_{\ell=1}^3 F_n^{J,\ell} \\ &\quad \text{with} \quad F_n^{J,1} \stackrel{\text{def}}{=} \sum_{\substack{0\leq j,k\leq J\\ j\neq k}} F_n^{j,k} \quad \text{with} \quad F_n^{j,k} \stackrel{\text{def}}{=} \Phi_n^j \otimes \Phi_n^k\,, \\ &F_n^{J,2} \stackrel{\text{def}}{=} \Big(\sum_{j=0}^J \Phi_n^j\Big) \otimes e^{t\Delta} r_n^J + e^{t\Delta} r_n^J \otimes \Big(\sum_{j=0}^J \Phi_n^j\Big) \quad \text{and} \\ &F_n^{J,3} \stackrel{\text{def}}{=} e^{t\Delta} r_n^J \otimes e^{t\Delta} r_n^J. \end{aligned} \tag{4.2}$$

Now let us prove the following lemma.

Lemma 4.1. *A real number* M *exists such that, for* J*,*

$$\limsup_{n\to\infty} \Big\| \sum_{j=0}^J \Phi_n^j \Big\|_{L^5([0,\tau_n^\varepsilon]\times\mathbb{R}^3)} \leq M$$

with the agreement that τ_n^ε *equal to infinity if* $\mathcal{J}_f$ *is empty.*

Proof. Using the scaling invariance of the space-time L^5 norm, let us write that

$$\left| \Big\| \sum_{j=0}^{J} \Phi_n^j \Big\|_{L^5([0,\tau_n^\varepsilon]\times\mathbb{R}^3)}^5 - \sum_{j=0}^{J} \|\Phi^j\|_{L^5([0,\tau_n^\varepsilon]\times\mathbb{R}^3)}^5 \right|$$
$$\leq \sum_{(j_1,j_2,j_3,j_4,j_5)\in\{0,\cdot,J\}^5\setminus\Delta} \int_{[0,\tau_n^\varepsilon]\times\mathbb{R}^3} \prod_{\ell=1}^{5} |\Phi_n^{j_\ell}(t,x)|dxdt$$

where $\Delta \stackrel{\text{def}}{=} \{(j,j,j,j,j) : j \in \{0,\cdots,J\}^5\}$. Up to a permutation of indices, we can assume that j_1 is different from j_2. Using Hölder inequality and the scaling invariance of the space-time L^5 norm we get

$$\int_{[0,\tau_n^\varepsilon]\times\mathbb{R}^3} \prod_{\ell=1}^{5} |\Phi_n^{j_\ell}(t,x)|dxdt \leq \Big(\prod_{\ell=3}^{5} \|\Phi^{j_\ell}\|_{L^5([0,T^\star(\varphi^j)-\varepsilon]\times\mathbb{R}^3)}\Big)$$
$$\times\Big(\int_{[0,\tau_n^\varepsilon]\times\mathbb{R}^3} |\Phi_n^{j_1}(t,x)|^{\frac{5}{2}} |\Phi_n^{j_2}(t,x)|^{\frac{5}{2}} dxdt\Big)^{\frac{2}{5}}.$$

Let us assume that $\lim_{n\to\infty} \left|\log\Big(\frac{\lambda_n^{j_1}}{\lambda_n^{j_2}}\Big)\right| = +\infty$. Using the scaling, we get

$$\mathcal{J}_n^{j_1,j_2} \stackrel{\text{def}}{=} \int_{[0,\tau_n^\varepsilon]\times\mathbb{R}^3} |\Phi_n^{j_1}(t,x)|^{\frac{5}{2}} |\Phi_n^{j_2}(t,x)|^{\frac{5}{2}} dxdt$$
$$\leq \int_0^{\frac{\tau_n^\varepsilon}{(\lambda_n^{j_2})^2}} \frac{\lambda_n^{j_2}}{\lambda_n^{j_1}} \Big\|\Phi^{j_1}\Big(\Big(\frac{\lambda_n^{j_2}}{\lambda_n^{j_1}}\Big)^2 t,\cdot\Big)\Big\|_{L^5(\mathbb{R}^3)}^{\frac{5}{2}} \|\Phi^{j_2}(t,\cdot)\|_{L^5(\mathbb{R}^3)}^{\frac{5}{2}} dt.$$

Using that the weak converge in L^p namely the fact that, if p is in $]1,\infty[$, for any (f,g) in $L^p(\mathbb{R}^d)\times L^{p'}(\mathbb{R}^d)$, we have

$$\lim_{\Lambda\to\infty} \int_{\mathbb{R}^d} \Lambda^{-\frac{d}{p}} f(\Lambda^{-1}x)g(x)dx = 0. \tag{4.3}$$

Thus we have

$$\lim_{n\to\infty} \left|\log\Big(\frac{\lambda_n^{j_1}}{\lambda_n^{j_2}}\Big)\right| = +\infty \Longrightarrow \lim_{n\to\infty} \mathcal{J}_n^{j_1,j_2} = 0. \tag{4.4}$$

If the case when $\lambda_n^{j_1} = \lambda_n^{j_2}$, let us write that

$$\mathcal{J}_n^{j_1,j_2} \leq \int_0^{\frac{\tau_n^\varepsilon}{(\lambda_n^{j_2})^2}} \int_{\mathbb{R}^3} \left|\Phi^j\Big(t, x - \frac{x_n^j - x_n^k}{\lambda_n^j}\Big)\right|^{\frac{5}{2}} |\Phi^k(t,x)|^{\frac{5}{2}} dxdt.$$

Using the fact that $\lim_{n\to\infty} \frac{|x_n^j - x_n^k|}{\lambda_n^j}$ tends to infinity this implies that together with (4.4)

$$\forall (j,k) \in \{0,\cdots,J\}^2 : j \neq k\,,\ \lim_{n\to\infty} \mathcal{J}_n^{j_1,j_2} = 0. \tag{4.5}$$

Together with (4.4), this implies that

$$\forall J\,,\ \limsup_{n\to\infty}\Big\|\sum_{j=0}^{J}\Phi_n^j\Big\|^5_{L^5([0,\tau_n^\varepsilon]\times\mathbb{R}^3)} \leq \sum_{j=0}^{J}\|\Phi^j\|^5_{L^5([0,T^\star(\varphi^j)-\varepsilon]\times\mathbb{R}^3)}. \tag{4.6}$$

Now we have to estimate the right hand side term of the above inequality independently of J. Let us denote by $\mathcal{J}_{\rho_0}$ the set of indices such that $\|\varphi^j\|_{\dot H^{\frac12}}$ is greater than or equal to than $\rho_0/2$ (where ρ_0 is the radius given by Theorem 3.3). As the series $(\|\varphi^j\|^2_{\dot H^{\frac12}})_{j\in\mathbb{N}}$ is summable, this set is finite. If j does not belongs to $\mathcal{J}_{\rho_0}$, Theorem 3.3 claims that

$$\|\Phi^j\|^2_{L^\infty(\mathbb{R}^+;\dot H^{\frac12})}+\|\Phi^j\|^2_{L^2(\mathbb{R}^+;\dot H^{\frac32})}\leq\|\varphi^j\|^2_{\dot H^{\frac12}}.$$

Sobolev embeddings and interpolation inequality between Sobolev spaces imply that

$$\begin{aligned}\|\Phi^j\|^2_{L^5(\mathbb{R}^+;L^5(\mathbb{R}^3))} &\leq C\|\Phi^j\|^2_{L^5(\mathbb{R}^+;\dot H^{\frac12+\frac25}(\mathbb{R}^3))}\\ &\leq C\|\varphi^j\|^2_{\dot H^{\frac12}}.\end{aligned}$$

Thus, as the series $(\|\varphi^j\|^2_{\dot H^{\frac12}})_{j\in\mathbb{N}}$ is summable and as $\ell^2(\mathbb{Z})$ is continuously included in $\ell^5(\mathbb{Z})$, we get that

$$\sum_{j=0}^{J}\|\Phi^j\|^5_{L^5([0,T^\star(\varphi^j)-\varepsilon]\times\mathbb{R}^3)}\leq C\sup_n\|u_{0,n}\|^5_{\dot H^{\frac12}}+\sum_{j\in\mathcal{J}_{\rho_0}}\|\Phi^j\|^5_{L^5([0,T^\star(\varphi^j)-\varepsilon]\times\mathbb{R}^3)}.$$

Using Theorem 3.4 and Inequality (4.6), we get the lemma. □

Continuation of the proof of Theorem 4.1. Let η be a positive real number which will be chosen later on. Two integers J_η and n_η such that

$$n\geq n_\eta\Longrightarrow\|r_n^{J_\eta}\|_{L^3}\leq\eta. \tag{4.7}$$

Lemma 4.1 and Inequality (3.12) imply that, if n large enough (depending on η)

$$\begin{aligned}\|u^{J_\eta}_{n,\mathrm{app}}\|_{L^5([0,T^\star(\varphi^j)-\varepsilon]\times\mathbb{R}^3)} &\leq M+\|e^{t\Delta}r_n^{J_\eta}\|_{L^5(\mathbb{R}^+\times\mathbb{R}^3)}\\ &\leq M+C\eta.\end{aligned} \tag{4.8}$$

Moreover, Inequality (3.12) implies that, if n large enough,

$$\|F_n^{J_\eta,3}\|_{L^{\frac52}([0,\tau_n^\varepsilon]\times\mathbb{R}^3)}\leq C\eta^2. \tag{4.9}$$

Hölder inequality and Lemma 4.1 imply that, if n is large enough,

$$\begin{aligned}\|F_n^{J\eta,2}\|_{L^{\frac52}([0,\tau_n^\varepsilon]\times\mathbb{R}^3)} &\leq\Big\|\sum_{j=0}^{J_\eta}\Phi_n^j\Big\|_{L^5([0,\tau_n^\varepsilon]\times\mathbb{R}^3)}\|e^{t\Delta}r_{n,J}\|_{L^5(\mathbb{R}^+\times\mathbb{R}^3)}\\ &\leq CM\eta.\end{aligned} \tag{4.10}$$

Inequality (4.5) implies that

$$\lim_{n\to\infty} \|F_n^{J_\eta,1}\|_{L^{\frac{5}{2}}([0,\tau_n^\varepsilon]\times\mathbb{R}^3)} = 0.$$

Together with Inequalities (4.8)–(4.10), this implies that, choosing η small enough, we have, for n large enough,

$$\|F_n^{J_\eta}\|_{L^{\frac{5}{2}}([0,\tau_n^\varepsilon]\times\mathbb{R}^3)} \leq C_0^{-1}\exp\bigl(-C_0\|u_{n,\mathrm{app}}^{J_\eta}\|^5_{L^5([0,\tau_n^\varepsilon]\times\mathbb{R}^3)}\bigr).$$

In the case when τ_n^ε equals to infinity, then the theorem is proved. In the case when τ_n^ε in finite, the proof of Theorem 3.7 can be repeated words for words to ensures that $T^\star(u_{\phi(n)})$ is greater than or equal to τ_n^ε. □

4.2 A first application to the structure of the set $\mathcal{G}$

Theorem 4.2. *Let us assume that $\mathcal{G} \neq \dot{H}^{\frac{1}{2}}$. Let us define*

$$\rho_c \stackrel{\text{def}}{=} \sup\{\rho \in]0,\infty[: B(0,\rho) \subset \mathcal{G}\}.$$

Then if $S(0,\rho)$ denotes the sphere on radius ρ and center 0 in $\dot{H}^{\frac{1}{2}}$, we have that $S(0,\rho_c)\cap\mathcal{G}^c$ is non empty. It is compact up to translation and dilation in the following sense. If $(u_n)_{n\in\mathbb{N}}$ is a sequence of $S(0,\rho_c)\cap\mathcal{G}^c$, then a sequence $(\lambda_n, x_n)_{n\in\mathbb{N}}$ of $(]0,\infty[\times\mathbb{R}^3)^{\mathbb{N}}$ and a function v in $S(0,\rho_c)\cap\mathcal{G}^c$ exists such that, up to an extraction, we have

$$\lim_{n\to\infty} \|\lambda_n u_n(\lambda_n(\cdot + x_n)) - v\|_{\dot{H}^{\frac{1}{2}}} = 0.$$

Proof. By definition of ρ_c, a sequence $(u_n)_{n\in\mathbb{N}}$ of elements of $\dot{H}^{\frac{1}{2}}$ exists such that

$$T^\star(u_n) < \infty \quad \text{and} \quad \lim_{n\to\infty}\|u_n\|_{\dot{H}^{\frac{1}{2}}} = \rho_c. \tag{4.11}$$

Theorem 2.1 claims that a sequence of scales and cores $(\lambda_n^j, x_n^j)_{(j,n)\in\mathbb{N}^2}$, a sequence $(\varphi^j)_{j\in\mathbb{N}}$ in $\dot{H}^{\frac{1}{2}}$, a sequence $(r_n^j)_{(j,n)\in\mathbb{N}^2}$ which satisfies the following, up to an extraction on $(u_n)_{n\in\mathbb{N}}$,

$$\forall J\in\mathbb{N},\ u_n(x) = \sum_{j=0}^{J}\frac{1}{\lambda_n^j}\varphi^j\Bigl(\frac{x-x_n^j}{\lambda_n^j}\Bigr) + r_n^J(x) \quad \text{with}$$

$$\lim_{J\to\infty}\limsup_{n\to\infty}\|r_n^J\|_{L^3} = 0 \quad \text{and}$$

$$\forall J\in\mathbb{N},\ \lim_{n\to\infty}\Bigl(\|u_n\|^2_{\dot{H}^{\frac{1}{2}}} - \|r_n^J\|^2_{\dot{H}^{\frac{1}{2}}}\Bigr) = \sum_{j=0}^{J}\|\varphi^j\|^2_{\dot{H}^{\frac{1}{2}}}.$$

The last relation implies in particular that

$$\sum_{j=0}^{\infty}\|\varphi^j\|^2_{\dot{H}^{\frac{1}{2}}} \leq \rho_c^2. \tag{4.12}$$

Theorem 4.1 implies that an integer j_0 exists such that φ^{j_0} does not belongs to $\mathcal{G}$. As the (open) ball of center 0 and radius ρ_c is included in $\mathcal{G}$, then an integer j_0 exists such that $\|\varphi^{j_0}\|_{\dot{H}^{\frac{1}{2}}} \geq \rho_c$. Inequality (4.12) implies that

$$\|\varphi^{j_0}\|_{\dot{H}^{\frac{1}{2}}} = \rho_c \quad \text{and} \quad j \neq j_0 \Longrightarrow \varphi^j = 0.$$

Thus φ^{j_0} is an element of $S(0,\rho_c) \cap \mathcal{G}^c$ which proves the first part of the theorem. For the second part, let us consider a sequence $(u_n)_{n\in\mathbb{N}}$ of $S(0,\rho_c) \cap \mathcal{G}$. Arguing exactly as above, we deduce that, up to an extraction

$$u_n = \frac{1}{\lambda_n^{j_0}} \varphi^{j_0}\Big(\frac{\cdot - x_n^{j_0}}{\lambda_n^{j_0}}\Big) + r_n \quad \text{with} \quad \lim_{n\to\infty} \|r_n\|_{\dot{H}^{\frac{1}{2}}} = 0.$$

As the $\dot{H}^{\frac{1}{2}}$ norm is scaling invariant, this gives the whole theorem. □

4.3 Description of bounded sequences of solutions

The purpose of this section is to prove the following theorem.

Theorem 4.3. *Let $(u_{0,n})_{n\in\mathbb{N}}$ be a sequence of initial data bounded in $\dot{H}^{\frac{1}{2}}$. With notations of Theorem 4.1, we have*

$$\lim_{J\to\infty} \limsup_{n\to\infty} \big(\|R_n^J\|_{L^\infty([0,\tau_n^\varepsilon];\dot{H}^{\frac{1}{2}})} + \|R_n^J\|_{L^2([0,\tau_n^\varepsilon];\dot{H}^{\frac{3}{2}})}\big) = 0. \tag{4.13}$$

Moreover, for any J, we have

$$\lim_{n\to\infty} \Big\| \|u_{\phi(n)}(t)\|^2_{\dot{H}^{\frac{1}{2}}} - \sum_{j=0}^{J} \|\Phi_n^j(t)\|^2_{\dot{H}^{\frac{1}{2}}} - \|e^{t\Delta} r_n^J\|^2_{\dot{H}^{\frac{1}{2}}} \Big\|_{L^\infty([0,\tau_n^\varepsilon])} = 0. \tag{4.14}$$

Proof. The first step is the proof of the following lemma which is the analog of Lemma 4.1.

Lemma 4.2. *A real number M exists such that, for J,*

$$\limsup_{n\to\infty} \Big\| \sum_{j=0}^{J} \Phi_n^j \Big\|_{L^4([0,\tau_n^\varepsilon];\dot{H}^1)} \leq M$$

with the agreement that τ_n^ε equal to infinity if $\mathcal{J}_f$ is empty.

Proof. Let us write that for any t in $[0,\tau_n^\varepsilon]$, we have

$$\Big\| \sum_{j=0}^{J} \Phi_n^j(t) \Big\|^2_{\dot{H}^1} = \sum_{j=0}^{J} \|\Phi_n^j(t)\|^2_{\dot{H}^1} + \sum_{\substack{0\leq j,k\leq J \\ j\neq k}} \big(\Phi_n^j(t)|\Phi_n^k(t)\big)_{\dot{H}^1}. \tag{4.15}$$

Let us prove that for any (j,k) in $\{0,\cdots,J\}^2$ such that j and k are different, we have

$$\lim_{n\to\infty} \big\| \big(\Phi_n^j(t)|\Phi_n^k(t)\big)_{\dot{H}^1} \big\|_{L^2([0,\tau_n^\varepsilon[)} = 0. \tag{4.16}$$

Let us first consider the case when the two scales $(\lambda_n^j)_{n\in\mathbb{N}}$ and $(\lambda_n^k)_{n\in\mathbb{N}}$ are different.

Using the scaling invariance of the space $L_t^4(\dot H^1)$ norm, let us write that

$$\begin{aligned}\big\|\big(\Phi_n^j(t)|\Phi_n^k(t)\big)_{\dot H^1}\big\|^2_{L^2([0,\tau_n^\varepsilon[)} &\leq \int_0^\infty \frac{1}{\lambda_n^j\lambda_n^k}\mathbb{1}_{[0,T^\star(\varphi^j)-\varepsilon]}\Big(\frac{t}{(\lambda_n^j)^2}\Big)\Big\|\Phi^j\Big(\frac{t}{(\lambda_n^j)^2}\Big)\Big\|^2_{\dot H^1}\\
&\qquad\times\mathbb{1}_{[0,T^\star(\varphi^k)-\varepsilon]}\Big(\frac{t}{(\lambda_n^k)^2}\Big)\Big\|\Phi^k\Big(\frac{t}{(\lambda_n^k)^2}\Big)\Big\|^2_{\dot H^1}dt\\
&\leq \int_0^\infty \frac{\lambda_n^k}{\lambda_n^j}\mathbb{1}_{[0,T^\star(\varphi^j)-\varepsilon]}\Big(\Big(\frac{\lambda_n^k}{\lambda_n^j}\Big)^2t\Big)\Big\|\Phi^j\Big(\Big(\frac{\lambda_n^k}{\lambda_n^j}\Big)^2t\Big)\Big\|^2_{\dot H^1}\\
&\qquad\times\mathbb{1}_{[0,T^\star(\varphi^k)-\varepsilon]}(t)\|\Phi^k(t)\|^2_{\dot H^1}dt.\end{aligned}$$

Because of the hypothesis of orthogonality on the scales, this implies (4.16) in the case when the two scales $(\lambda_n^j)_{n\in\mathbb{N}}$ and $(\lambda_n^k)_{n\in\mathbb{N}}$ are different.

In the case when the two scales $(\lambda_n^j)_{n\in\mathbb{N}}$ and $(\lambda_n^k)_{n\in\mathbb{N}}$ are equal, let us observe

$$\big\|\big(\Phi_n^j(t)|\Phi_n^k(t)\big)_{\dot H^1}\big\|^2_{L^2([0,\tau_n^\varepsilon[)} \leq \int_0^{T^\varepsilon_{j,k}}\Big|\Big(\Phi^j(t,\cdot)\Big|\Phi^k\Big(t,\cdot-\frac{x_n^j-x_n^k}{\lambda_n^j}\Big)\Big)_{\dot H^1}\Big|^2dt$$

with $T^\varepsilon_{j,k}\stackrel{\text{def}}{=}\min\{T^\star(\varphi^j),T^\star(\varphi^k)\}-\varepsilon$. Lebesgue's theorem implies that for any positive η, a positive real number α_J exists such that, for any j in $\{0,\cdots,J\}$, we have

$$\begin{aligned}&\|\Phi^j-\Phi^j_{\alpha_J}\|_{L^4([0,T^\star(\varphi^j)-\varepsilon];\dot H^1)}\leq\frac{\eta}{\sup_{j\in J}\|\Phi^j\|_{L^4([0,T^\star(\varphi^j)-\varepsilon];\dot H^1)}}\quad\text{with}\\
&\qquad\Phi^j_{\alpha_J}\stackrel{\text{def}}{=}\Big(1-\chi\Big(\frac{\cdot}{\alpha_J}\Big)+\chi(\cdot\,\alpha_J)\Big)\Phi^j\end{aligned}\tag{4.17}$$

where χ if a compactly supported function in $\mathbb{R}^3$ with value 1 near the origin. Thus we have

$$\big\|\big(\Phi_n^j(t)|\Phi_n^k(t)\big)_{\dot H^1}\big\|^2_{L^2([0,\tau_n^\varepsilon[)} \leq 2\eta+\int_0^{T^\varepsilon_{j,k}}\Big|\Big(\Phi^j_{\alpha_J}(t,\cdot)\Big|\Phi^k_{\alpha_J}\Big(t,\cdot-\frac{x_n^j-x_n^k}{\lambda_n^j}\Big)\Big)_{\dot H^1}\Big|^2dt.$$

As $\lim\limits_{n\to\infty}\frac{|x_n^j-x_n^k|}{\lambda_n^j}=\infty$, for n large enough, the support of the two functions

$$\Phi^j_{\alpha_J}(t,\cdot)\quad\text{and}\quad\Phi^k_{\alpha_J}\Big(t,\cdot-\frac{x_n^j-x_n^k}{\lambda_n^j}\Big)$$

are disjoint. Thus we get (4.16) in all cases. Taking the L^2 norm and passing to the limit when n tends to infinity in Relation (4.15), we get

$$\limsup_{n\to\infty}\Big\|\sum_{j=0}^J\Phi_n^j(t)\Big\|^2_{L^4([0,\tau_n^\varepsilon];\dot H^1)}\leq\sum_{j=0}^J\|\Phi^j\|^2_{L^4([0,T^\star(\varphi^j)-\varepsilon];\dot H^1)}.$$

With the notations of Theorem 4.1, let us write that

$$\sum_{j=0}^{J} \|\Phi^j\|^2_{L^4([0,T^\star(\varphi^j)-\varepsilon];\dot H^1)} \leq \sum_{j\in\mathcal{J}_{\rho_0}} \|\Phi^j\|^2_{L^4([0,T^\star(\varphi^j)-\varepsilon];\dot H^1)} + \sum_{\substack{j\notin\mathcal{J}_{\rho_0}\\ j\leq J}} \|\Phi^j\|^2_{L^4(\mathbb{R}^+;\dot H^1)}.$$

Using Theorem 3.3, we get that

$$\sum_{j=0}^{J} \|\Phi^j\|^2_{L^4([0,T^\star(\varphi^j)-\varepsilon];\dot H^1)} \leq \sum_{j\in\mathcal{J}_{\rho_0}} \|\Phi^j\|^2_{L^4([0,T^\star(\varphi^j)-\varepsilon];\dot H^1)} + \limsup_{n\in\mathbb{N}} \|u_n\|^2_{\dot H^{\frac12}}$$

and the lemma is proved. □

Continuation of the proof of Theorem 4.3. Thanks to Lemma 4.2, it is enough to prove that

$$\lim_{J\to\infty}\lim_{n\to\infty} \|F_n^J\|_{L^2([0,\tau_n^\varepsilon];\dot H^{\frac12})} = 0. \tag{4.18}$$

We use the notations of Definition (4.2) of R_n^J and F_n^J. Let us first study $F_n^{J,1}$. If the two scales $(\lambda_n^j)_{n\in\mathbb{N}}$ and $(\lambda_n^k)_{n\in\mathbb{N}}$ are different, we have, because of law of product in Sobolev spaces

$$\begin{aligned}
\|F_n^{j,k}\|^2_{L^2([0,\tau_n^\varepsilon];\dot H^{\frac12})} &\stackrel{\text{def}}{=} \|\Phi^j\otimes\Phi^k\|^2_{L^2([0,\tau_n^\varepsilon];\dot H^{\frac12})}\\
&\leq \int_0^\infty \frac{1}{\lambda_n^j\lambda_n^k}\mathbb{1}_{[0,T^\star(\varphi^j)-\varepsilon]}\left(\frac{t}{(\lambda_n^j)^2}\right)\left\|\Phi^j\left(\frac{t}{(\lambda_n^j)^2}\right)\right\|^2_{\dot H^1}\\
&\qquad\times\mathbb{1}_{[0,T^\star(\varphi^k)-\varepsilon]}\left(\frac{t}{(\lambda_n^k)^2}\right)\left\|\Phi^k\left(\frac{t}{(\lambda_n^k)^2}\right)\right\|^2_{\dot H^1}dt\\
&\leq \int_0^\infty \frac{\lambda_n^k}{\lambda_n^j}\mathbb{1}_{[0,T^\star(\varphi^j)-\varepsilon]}\left(\frac{(\lambda_n^k)^2 t}{(\lambda_n^j)^2}\right)\left\|\Phi^j\left(\frac{(\lambda_n^k)^2 t}{(\lambda_n^j)^2}\right)\right\|^2_{\dot H^1}\\
&\qquad\times\mathbb{1}_{[0,T^\star(\varphi^k)-\varepsilon]}(t)\|\Phi^k(t)\|^2_{\dot H^1}dt.
\end{aligned}$$

Thus we infer that if the two scales $(\lambda_n^j)_{n\in\mathbb{N}}$ and $(\lambda_n^k)_{n\in\mathbb{N}}$ are different, then

$$\lim_{n\to\infty} \|F_n^{j,k}\|_{L^2([0,\tau_n^\varepsilon];\dot H^{\frac12})} = 0. \tag{4.19}$$

In the case when the two scales $(\lambda_n^j)_{n\in\mathbb{N}}$ and $(\lambda_n^k)_{n\in\mathbb{N}}$ are equal, let us observe, using (4.17), we have, for any appropriated choice of α_J,

$$\begin{aligned}
\|F_n^{j,k}\|^2_{L^2([0,\tau_n^\varepsilon];\dot H^{\frac12})} &\leq \int_0^{T^\varepsilon_{j,k}} \left\|\Phi^j(t,\cdot)\otimes\Phi^k\left(t,\cdot-\frac{x_n^j-x_n^k}{\lambda_n^j}\right)\right\|^2_{\dot H^1}dt\\
&\leq 2\eta + \int_0^{T^\varepsilon_{j,k}} \left\|\Phi^j_{\alpha_J}(t,\cdot)\otimes\Phi^k_{\alpha_J}\left(t,\cdot-\frac{x_n^j-x_n^k}{\lambda_n^j}\right)\right\|^2_{\dot H^1}dt
\end{aligned}$$

with $T^\varepsilon_{j,k} \stackrel{\text{def}}{=} \min\{T^\star(\varphi^j),T^\star(\varphi^k)\}-\varepsilon$. As $\lim_{n\to\infty}\frac{|x_n^j-x_n^k|}{\lambda_n^j} = \infty$, for n large enough, the support of the two functions

$$\Phi^j_{\alpha_J}(t,\cdot) \quad\text{and}\quad \Phi^k_{\alpha_J}\left(t,\cdot-\frac{x_n^j-x_n^k}{\lambda_n^j}\right)$$

are disjoint and thus, thanks to (4.19), we have

$$\forall J\,,\ \lim_{n\to\infty} \|F_n^{J,1}\|_{L^2([0,\tau_n^\varepsilon];\dot H^1)} = 0. \tag{4.20}$$

In order to treat the term $F_n^{J,2}$, let us observe that interpolation inequalities imply that

$$\|a\|_{\dot B^{\frac32\theta-1}_{\frac2\theta,\frac2\theta}} \leq \|a\|^\theta_{\dot H^{\frac12}} \|a\|^{1-\theta}_{\dot B^{-1}_{\infty,\infty}}\,.$$

Using Lemma 2.4 of [2] and the Minkowski inequality, we have

$$\|e^{t\Delta} r_n^J\|_{L^4(\mathbb{R}^+;\dot B^{\frac32\theta-\frac12}_{\frac2\theta,\frac2\theta})} \leq \|r_n^J\|^\theta_{\dot H^{\frac12}} \|r_n^J\|^{1-\theta}_{\dot B^{-1}_{\infty,\infty}}\,.$$

Properties of the sequence $(r_n^j)_{(j,n)\in\mathbb{N}^2}$ implies that

$$\forall \theta\in[0,1[\,,\ \lim_{J\to\infty}\limsup_{n\to\infty} \|e^{t\Delta} r_n^J\|_{L^4(\mathbb{R}^+;\dot B^{\frac32\theta-\frac12}_{\frac2\theta,\frac2\theta})} = 0.$$

Law of product in Besov spaces (see for instance Chapter 2 of [2]) say that, for $\theta > 2/3$, we have

$$\dot B^{\frac32\theta-\frac12}_{\frac2\theta,\frac2\theta}\cdot \dot B^{\frac32\theta-\frac12}_{\frac2\theta,\frac2\theta} \subset \dot H^{\frac12}.$$

It immediately ensures that

$$\lim_{J\to\infty}\limsup_{n\to\infty} \|F_n^{J,3}\|_{L^4(\mathbb{R}^+;\dot H^{\frac12})} = 0. \tag{4.21}$$

Moreover, as $\dot H^1$ is continuously embedded in $\dot B^{\frac32\theta-\frac12}_{\frac2\theta,\frac2\theta}$, then Lemma 4.2 ensures that

$$\lim_{J\to\infty}\limsup_{n\to\infty} \|F_n^{J,2}\|_{L^4(\mathbb{R}^+;\dot H^{\frac12})} = 0.$$

Then Inequalitites (4.20) and (4.21) imply (4.18) and the first part of Theorem 4.3 is proved.

In order to prove the second part of Theorem 4.3, let us first write that

$$\begin{aligned}
\|u_{\phi(n)}(t)\|^2_{\dot H^{\frac12}} &= \sum_{j=0}^J \|\Phi_n^j(t)\|^2_{\dot H^{\frac12}} + \|e^{t\Delta} r_n^J\|^2_{\dot H^{\frac12}} + \|R_n^J(t)\|^2_{\dot H^{\frac12}} + 2\sum_{\ell=1}^4 E_n^{J,\ell}(t) \quad \text{with}\\
E_n^{J,1}(t) &\stackrel{\text{def}}{=} \Big(\sum_{j=0}^J \Phi_n^j(t)\big|R_n^J(t)\Big)_{\dot H^{\frac12}}\,,\\
E_n^{J,2}(t) &\stackrel{\text{def}}{=} \big(e^{t\Delta} r_n^J\big|R_n^J(t)\big)_{\dot H^{\frac12}}\,,\\
E_n^{J,3}(t) &\stackrel{\text{def}}{=} \Big(\sum_{j=0}^J \Phi_n^j(t)\Big|e^{t\Delta} r_n^J\Big)_{\dot H^{\frac12}}\,,\\
E_n^{J,4}(t) &\stackrel{\text{def}}{=} \sum_{\substack{0\leq j,k\leq J\\ j\neq k}} \big(\Phi_n^j(t)\big|\Phi_n^k(t)\big)_{\dot H^{\frac12}}\,.
\end{aligned} \tag{4.22}$$

We shall make a very frequent use of the fact that

$$\forall (j,n)\in\mathbb{N}^2\,,\ t\in[0,\tau_n^\varepsilon]\Longrightarrow \frac{t}{(\lambda_n^j)^2}\in[0,T^\star(\varphi^j)-\varepsilon].$$

The first step is the proof of the following lemma.

Lemma 4.3. *For any (j,k) in $\mathbb{N}^2$ such that j and k are different, we have*

$$\lim_{n\to\infty}\big\|\big(\Phi_n^j(t)|\Phi_n^k(t)\big)_{\dot H^{\frac12}}\big\|_{L^\infty([0,\tau_n^\varepsilon])}=0.$$

Proof. Let us consider a positive real number η. We first study the case when $(\lambda_n^j)_{n\in\mathbb{N}}$ and $(\lambda_n^k)_{n\in\mathbb{N}}$ are different. For any j, the set $\big\{\Phi^j(t)\,,t\in[0,T^\star(\varphi^j)-\varepsilon]\big\}$ is a compact subset of $\dot H^{\frac12}$ as the range of the compact interval $[0,T^\star(\varphi^j)-\varepsilon]$ by the continuous ap Φ^j. Thus it can be covered by a finite number of balls centered at functions the Fourier transform of which is included in a ring of $\mathbb{R}^3$ and of radius

$$\frac{\eta}{2}\max\{\|\Phi^j\|^{-1}_{L^\infty([0,T^\star(\varphi^j)-\varepsilon];\dot H^{\frac12})},\|\Phi^k\|^{-1}_{L^\infty([0,T^\star(\varphi^k)-\varepsilon];\dot H^{\frac12})}\}.$$

Then we deduce that for any t in $[0,\tau_n^\varepsilon]$, it exists two functions f_j and g_k which are chosen a finite family of functions depending on η, on (j,k) and on ε such that

$$\begin{aligned}&\Big|\big(\Phi_n^j(t)|\Phi_n^k(t)\big)_{\dot H^{\frac12}}-\frac{1}{\lambda_n^j\lambda_n^k}\Big(f_j\Big(\frac{\cdot-x_n^j}{\lambda_n^j}\Big)\Big|g_k\Big(\frac{\cdot-x_n^k}{\lambda_n^k}\Big)\Big)_{\dot H^{\frac12}}\Big|\\&=\Big|\big(\Phi^j((\lambda_n^j)^2t)|\Phi^k((\lambda_n^k)^2t)\big)_{\dot H^{\frac12}}-(f_j|g_k)_{\dot H^{\frac12}}\Big|\leq\eta.\end{aligned}$$

When $\Big|\log\frac{\lambda_n^j}{\lambda_n^k}\Big|$ large enough, the support of the Fourier transform of two functions

$$f_j\Big(\frac{\cdot-x_n^j}{\lambda_n^j}\Big)\quad\text{and}\quad g_k\Big(\frac{\cdot-x_n^k}{\lambda_n^k}\Big)$$

are disjoint; thus the two functions are orthogonal in $\dot H^{\frac12}$ and we have, for $\Big|\log\frac{\lambda_n^j}{\lambda_n^k}\Big|$ large enough

$$\big\|\big(\Phi_n^j(t)|\Phi_n^k(t)\big)_{\dot H^{\frac12}}\big\|_{L^\infty([0,\tau_n^\varepsilon])}\leq\eta. \tag{4.23}$$

In the case when the two scales $(\lambda_n^j)_{n\in\mathbb{N}}$ and $(\lambda_n^k)_{n\in\mathbb{N}}$ are equal, let us write that because of the scaling invariance of the space $\dot H^{\frac12}$ we have

$$\big(\Phi_n^j(t)|\Phi_n^k(t)\big)_{\dot H^{\frac12}}=\Big(\Phi^j\Big(\frac{t}{(\lambda_n^j)^2},\cdot\Big)\Big|\Phi^k\Big(\frac{t}{(\lambda_n^j)^2},\cdot-y_n^{j,k}\Big)\Big)_{\dot H^{\frac12}}$$

$$\text{with}\quad y_n^{j,k}\stackrel{\text{def}}{=}\frac{x_n^k-x_n^j}{\lambda_n^k}\cdot$$

By definition of the $\dot H^{\frac12}$ scalar product, we get

$$\big(\Phi_n^j(t)|\Phi_n^k(t)\big)_{\dot H^{\frac12}}=\Big\langle(|D|\Phi^j\Big(\frac{t}{(\lambda_n^j)^2},\cdot\Big),\Phi^k\Big(\frac{t}{(\lambda_n^j)^2},\cdot-y_n^{j,k}\Big)\Big\rangle.$$

Using that the set

$$\{|D|\Phi^j(t)\,, t\in[0,T^\star(\varphi^j)-\varepsilon]\}$$

is a compact subset of $\dot H^{-\frac12}$, it can be covered with a finite number of ball of radius less than

$$\frac{\eta}{2}\big(\|\Phi^j\|^{-1}_{L^\infty([0,T^\star(\varphi^j)-\varepsilon])}+1\big)^{-1}$$

and centered on functions of $\mathcal{D}(\mathbb{R}^3\setminus\{0\})$ for the $\dot H^{-\frac12}$ topology. The same is true for the set

$$\{\Phi^k(t)\,, t\in[0,T^\star(\varphi^j)-\varepsilon]\}$$

in the $\dot H^{\frac12}$ topology. Thus two functions f_j and g_k exists in $\mathcal{D}(\mathbb{R}^3\setminus\{0\})$ such that

$$\big|(\Phi^j_n(t)|\Phi^k_n(t))_{\dot H^{\frac12}}\big|\leq\eta+\Big|\int_{\mathbb{R}^3}f_j(x)g_k(x-y^{j,k}_n)dx\Big|.$$

As $\lim_{n\to\infty}|y^{j,k}_n|=\infty$, the above integral is 0. With (4.23), the lemma is proved. □

Continuation of the proof of Theorem 4.3. Lemma 4.3 implies that, for any J

$$\lim_{n\to\infty}\bigg\|\Big\|\sum_{j=0}^J\Phi^j_n(t)\Big\|^2_{\dot H^{\frac12}}-\sum_{j=0}^J\|\Phi^j_n(t)\|^2_{\dot H^{\frac12}}\bigg\|_{L^\infty([0,\tau^\varepsilon_n])}=0.$$

Let us introduce $\mathcal{J}_{\rho_0}$ the set of indices such that the associated profile φ^j does not belong to the open ball centered at the origin and of radius ρ_0 where ρ_0 is given by Theorem 3.3. It is a finite set. Moreover thanks to Theorem 3.3, we have

$$\forall j\notin\mathcal{J}_{\rho_0}\,,\ \|\Phi^j_n\|_{L^\infty(\mathbb{R}^+;\dot H^{\frac12})}=\|\Phi^j\|_{L^\infty(\mathbb{R}^+;\dot H^{\frac12})}\leq\|\varphi^j\|_{\dot H^{\frac12}}. \tag{4.24}$$

Thus we infer that

$$\limsup_{n\to\infty}\Big\|\sum_{j=0}^J\Phi^j(t)\Big\|^2_{L^\infty([0,\tau^\varepsilon_n];\dot H^{\frac12})}\leq M_\varepsilon\stackrel{\rm def}{=}\sum_{j\in\mathcal{J}_{\rho_0}}\|\Phi^J\|^2_{L^\infty([0,\tau^\varepsilon_n];\dot H^{\frac12})}+\sum_j\|\varphi^j\|^2_{\dot H^{\frac12}}.$$

Using Cauchy-Schwarz inequality, we get, with the notations of (4.22),

$$\begin{aligned}&\limsup_{n\to\infty}\|E^{J,1}_n\|_{L^\infty([0,\tau^\varepsilon_n])}\\ &\leq\limsup_{n\to\infty}\Big\|\sum_{j=0}^J\Phi^j(t)\Big\|_{L^\infty([0,\tau^\varepsilon_n])}\limsup_{n\to\infty}\|R^J_n\|_{L^\infty([0,\tau^\varepsilon_n];\dot H^{\frac12})}\\ &\leq M_\varepsilon\|R^J_n\|_{L^\infty([0,\tau^\varepsilon_n];\dot H^{\frac12})}.\end{aligned}$$

Thus, using Inequality (4.13) (which is now proven), we get

$$\lim_{J\to\infty}\limsup_{n\to\infty}\|E^{J,1}_n\|_{L^\infty([0,\tau^\varepsilon_n])}=0. \tag{4.25}$$

The estimate of $E_n^{J,2}$ is easy. Indeed, Cauchy-Schwarz inequality implies that

$$\big\|\big(e^{t\Delta}r_n^J|R_n^J(t)\big)_{\dot H^{\frac12}}\big\|_{L^\infty([0,\tau_n^\varepsilon])}\leq \|r_n^J\|_{\dot H^{\frac12}}\|R_n^J\|_{L^\infty([0,\tau_n^\varepsilon];\dot H^{\frac12})}.$$

Because $\limsup_{n\to\infty}\|r_n^J\|_{\dot H^{\frac12}}$ is less than or equal to $M\stackrel{\rm def}{=}\sup_n\|u_{0,n}\|_{\dot H^{\frac12}}$, we get

$$\limsup_{n\to\infty}\big\|\big(e^{t\Delta}r_n^J|R_n^J(t)\big)_{\dot H^{\frac12}}\big\|_{L^\infty([0,\tau_n^\varepsilon])}\leq M\limsup_{n\to\infty}\|R_n^J\|_{L^\infty([0,\tau_n^\varepsilon];\dot H^{\frac12})}.$$

Using Lemma 4.13, we get

$$\lim_{J\to\infty}\limsup_{n\to\infty}\|E_n^{J,2}\|_{L^\infty([0,\tau_n^\varepsilon])}=0. \tag{4.26}$$

The term $E_n^{J,3}$ requires more care. Let us consider an integer J' greater than the maximum of $\mathcal{J}_{\rho_0}$. For any J greater than J', we have, because of Lemma 4.3 and Assertion (4.24),

$$\begin{aligned}\limsup_{n\to\infty}\Big\|\sum_{j=J'}^J\Phi_n^j(t)\Big\|^2_{L^\infty([0,\tau_n^\varepsilon];\dot H^{\frac12})}&\leq\limsup_{n\to\infty}\sum_{j=J'}^J\|\Phi_n^j\|^2_{L^\infty([0,\tau_n^\varepsilon];\dot H^{\frac12})}\\&\leq\sum_{j=J'}^J\|\Phi^j\|^2_{L^\infty(\mathbb{R}^+;\dot H^{\frac12})}\\&\leq\sum_{j=J'}^J\|\varphi^j\|^2_{\dot H^{\frac12}}.\end{aligned}$$

Let us define J_η such that

$$\sum_{j\geq J_\eta}\|\varphi^j\|^2_{\dot H^{\frac12}}\leq\frac{\eta}{2\sup_{n,J}\|r_n^J\|_{\dot H^{\frac12}}}\cdot$$

Now let us consider J such that J_η. By definition of $E_n^{J,3}(t)$, we have

$$\limsup_{n\to\infty}\|E_n^{J,3}(t)\|_{L^\infty([0,\tau_n^\varepsilon])}\leq\frac{\eta}{2}+\sum_{j=0}^{J_\eta}\limsup_{n\to\infty}\big\|\big(\Phi_n^j(t)|e^{t\Delta}r_n^J\big)_{\dot H^{\frac12}}\big\|_{L^\infty([0,\tau_n^\varepsilon])}.$$

Using the scaling properties of the space $\dot H^{\frac12}$, let us write that, by definition of the scalar product of $\dot H^{\frac12}$,

$$\big(\Phi_n^j(t)|e^{t\Delta}r_n^J\big)_{\dot H^{\frac12}}=\Big\langle|D|\Phi^j\Big(\frac{t}{(\lambda_n^j)^2},\cdot\Big),\lambda_n^j(e^{(\lambda_n^j)^2t\Delta})r_n^j(\lambda_n^j\cdot+x_n)\Big\rangle.$$

Using the compactness of $\big\{\Phi^j(t)\,,t\in[0,T^\star(\varphi^j)-\varepsilon]\big\}$, we can recover the set

$$\big\{|D|\Phi^j(t)\,,t\in[0,T^\star(\varphi^j)-\varepsilon]\big\}$$

by a finite number of balls (for the $\dot{H}^{-\frac{1}{2}}$-topology) of radius

$$\frac{\eta}{4J_\eta \sup_{n,J} \|r_n^J\|_{\dot{H}^{\frac{1}{2}}}}$$

and centered at function of $\mathcal{D}(\mathbb{R}^3)$. Thus we get, for any t in $[0,\tau_n^\varepsilon]$,

$$\begin{aligned}\left|(\Phi_n^j(t)|e^{t\Delta}r_n^J)_{\dot{H}^{\frac{1}{2}}}\right| &\leq \frac{\eta}{4J_\eta} + \langle f_\ell, \lambda_n^j(e^{(\lambda_n^j)^2 t\Delta}) r_n^j(\lambda_n^j \cdot + x_n)\rangle \\ &\leq \frac{\eta}{4J_\eta} + \|f_\ell\|_{\dot{B}^1_{1,1}} \|r_n^J\|_{\dot{B}^{-1}_{\infty,\infty}}.\end{aligned}$$

Thus we get

$$\|E_n^{J,3}\|_{L^\infty([0,\tau_n^\varepsilon])} \leq \frac{3}{4\eta} + C\eta\|r_n^J\|_{\dot{B}^{-1}_{\infty,\infty}}.$$

Using (4.25), (4.26) and Lemma 4.3, we conclude the proof of the Theorem. □

4.4 A blow up theorem

The purpose of this section is to prove the following theorem.

Theorem 4.4. *Let us consider a maximal solution u to (NS) which belongs to the space*

$$C([0,T^\star[;\dot{H}^{\frac{1}{2}}) \cap L^2_{\mathrm{loc}}([0,T^\star[;\dot{H}^{\frac{3}{2}}).$$

We have

$$T^\star < \infty \Longrightarrow \limsup_{t\to T^\star} \|u(t)\|_{\dot{H}^{\frac{1}{2}}} = \infty.$$

Proof. We argue by contradiction. Let us denote by $I_{\mathcal{B}}$ the interval of real numbers A such that it exists a solution u to (NS) which blows up for some finite $T^\star$ and is such that

$$\limsup_{t\to T^\star} \|u(t)\|_{\dot{H}^{\frac{1}{2}}} \leq A.$$

The theorem claims that $I_{\mathcal{B}}$ is empty. Let us assume that $I_{\mathcal{B}}$ is not empty; this will lead to some contradiction.

Because of Theorem 3.3, $I_{\mathcal{B}}$ does not intersect the ball centered at the origin and of radius ρ_0 defined in this theorem. Thus if we denote by $\mathcal{A}_c$ the infimum of $I_{\mathcal{B}}$, we have

$$\mathcal{A}_c \geq \rho_0. \tag{4.27}$$

Now let us consider a sequence $(u_n)_{n\in\mathbb{N}}$ of solutions to (NS) which blow up for finite time $(T_n^\star)_{n\in\mathbb{N}}$ such that

$$\mathcal{A}_c \leq \limsup_{t\to T_n^\star} \|u_n(t)\|_{\dot{H}^{\frac{1}{2}}} \leq \mathcal{A}_c + \frac{1}{2n}\cdot \tag{4.28}$$

By definition of the upper limit, a sequence $(t_n)_{n\in\mathbb{N}}$ exists such that

$$\|u_n(t_n)\|_{\dot{H}^{\frac{1}{2}}} \geq \mathcal{A}_c - \frac{1}{n} \quad\text{and}\quad \|u_n\|_{L^\infty([t_n,T_n^\star[;\dot{H}^{\frac{1}{2}})} \leq \mathcal{A}_c + \frac{1}{n}\cdot \tag{4.29}$$

Let us consider the sequence

$$v_{0,n} \stackrel{\text{def}}{=} \sqrt{T_n^\star - t_n}\, u_n\Big(t_n, \sqrt{T_n^\star - t_n}\, x\Big)\cdot$$

By definition of $(u_n)_{n\in\mathbb{N}}$, the sequence $(v_{0,n})_{n\in\mathbb{N}}$ gives birth to a family of solutions $(v_n)_{n\in\mathbb{N}}$ to (NS) the life span of which is 1; because of (4.29), we have

$$\forall n \in \mathbb{N}\,,\ \|v_{0,n}\|_{\dot H^{\frac12}} \geq \mathcal{A}_c - \frac{1}{n} \quad\text{and}\quad \|v_n\|_{L^\infty([0,1[;\dot H^{\frac12})} \leq \mathcal{A}_c + \frac{1}{n}\cdot \tag{4.30}$$

Let us prove the following proposition, which is the one of the main step of the proof.

Proposition 4.1. *Let $(v_{0,n})_{n\in\mathbb{N}}$ be a family of initial data such that for any n, the life span of the solution associated with $v_{0,n}$ is 1 and which satisfies (4.30). Let us consider an extraction, a sequence of scales and cores, and a sequence of profiles given by Theorem 2.1. For any j, we have that λ_n^j is less than or equal to 1.*

Proof. Let us apply Theorem 2.1. We have

$$v_{0,\phi(n)} = \sum_{j=0}^{J} \frac{1}{\lambda_n^j}\varphi^j\Big(\frac{\cdot - x_n^j}{\lambda_n^j}\Big) + r_n^J. \tag{4.31}$$

A consequence of (4.30) is that

$$\lim_{n\to\infty} \|v_n\|_{L^\infty([0,\tau_n^\varepsilon];\dot H^{\frac12})} = \mathcal{A}_c.$$

Inequality (4.14) of Theorem 4.3 allows us to write that, for any t in $[0,\tau_n^\varepsilon]$,

$$\|v_n(t)\|^2_{\dot H^{\frac12}} \geq \Big\|\frac{1}{\lambda_n^0}\Phi^0\Big(\frac{t}{(\lambda_n^0)^2},\frac{x-x_n^0}{\lambda_n^0}\Big)\Big\|^2_{\dot H^{\frac12}} + \sum_{j=1}^{J}\Big\|\frac{1}{\lambda_n^j}\Phi^j\Big(\frac{t}{(\lambda_n^j)^2},\frac{x-x_n^j}{\lambda_n^j}\Big)\Big\|^2_{\dot H^{\frac12}} - o_n^J$$

where o_n^J is non negative and $\displaystyle\lim_{J\to\infty}\limsup_{n\to\infty} o_n^J = 0$. We deduce that, for any t in $[0,\tau_n^\varepsilon]$ we get

$$\begin{aligned}\|v_n(t)\|^2_{\dot H^{\frac12}} \geq{}& \Big\|\frac{1}{\lambda_n^0}\Phi^0\Big(\frac{t}{(\lambda_n^0)^2},\frac{x-x_n^0}{\lambda_n^0}\Big)\Big\|^2_{\dot H^{\frac12}}\\ &+\sum_{j=1}^{J}\inf\Big\{\|\Phi^j(t)\|^2_{\dot H^{\frac12}}\,, t\in\Big[0,\Big(\frac{\lambda_n^0}{\lambda_n^j}\Big)^2(T^\star(\varphi_0)-\varepsilon)\Big]\Big\} - o_n^J.\end{aligned}$$

Passing to the supremum in time gives

$$\begin{aligned}\|v_n\|^2_{L^\infty([0,\tau_n^\varepsilon];\dot H^{\frac12})} \geq{}& \|\Phi^0\|^2_{L^\infty([0,T^\star(\varphi^0)-\varepsilon];\dot H^{\frac12})}\\ &+\sum_{j=1}^{J}\inf\Big\{\|\Phi^j(t)\|^2_{\dot H^{\frac12}}\,, t\in\Big[0,\Big(\frac{\lambda_n^0}{\lambda_n^j}\Big)^2(T^\star(\varphi_0)-\varepsilon)\Big]\Big\} - o_n^J.\end{aligned} \tag{4.32}$$

Let us analyze this inequality. First of all, we get

$$\|v_n\|^2_{L^\infty([0,\tau_n^\varepsilon];\dot H^{\frac12})} + o_n^J \geq \|\Phi^0\|^2_{L^\infty([0,T^\star(\varphi^0)-\varepsilon];\dot H^{\frac12})}.$$

Passing to the limit in n and J gives that

$$\|\Phi^0\|^2_{L^\infty([0,T^\star(\varphi^0)-\varepsilon];\dot H^{\frac12})} \leq \mathcal{A}_c^2.$$

By definition of $\mathcal{A}_c$, this implies the following proposition.

Proposition 4.2. *A solution Φ to (NS) exists such that $\|\Phi\|_{L^\infty([0,T^\star[;\dot H^{\frac12})}$ is equal to $\mathcal{A}_c$ and $\limsup_{t\to T^\star} \|\Phi(t)\|_{\dot H^{\frac12}} = \mathcal{A}_c$.*

Let us continue the proof of Proposition 4.1. Let us consider in Inequality (4.32) the indices j such that φ^j does not belongs to $\mathcal{G}$. Because of Theorem 3.3, we get

$$\inf\{\|\Phi^j(t)\|_{\dot H^{\frac12}},\ t\in[0,T^\star(\varphi^j)[\} \geq \rho_0.$$

Inequality (4.32) implies that this set is empty. Let us consider that the set of j such that

$$\varphi^j \in \mathcal{G} \quad \text{and} \quad \lim_{n\to\infty} \frac{\lambda_n^0}{\lambda_n^j} = 0.$$

As the solution is continuous in time with value in $\dot H^{\frac12}$, we have

$$\lim_{n\to\infty} \inf\left\{\|\Phi^j(t)\|^2_{\dot H^{\frac12}}, t\in\left[0, \left(\frac{\lambda_n^0}{\lambda_n^j}\right)^2 (T^\star(\varphi_0)-\varepsilon)\right]\right\} = \|\varphi^j\|^2_{\dot H^{\frac12}}.$$

Thus for such indices j, $\varphi^j \equiv 0$ and the proposition is proved. □

Continuation of the proof of Theorem 4.4. It reduces to a backward uniqueness argument once we have proved the following proposition.

Proposition 4.3. *Let Φ be a maximal solution of (NS) given by Proposition* 4.2. *Then, for all s in $[-3/2, 1/2[$,*

$$\Phi \in C([0,T^\star[;\dot H^s) \quad \text{and} \quad \lim_{t\to T^\star} \|\Phi(t)\|_{\dot H^s} = 0.$$

Proof. Now let us consider a critical blow up solution given by Proposition 4.2. Let us consider a sequence $(t_n)_{n\in\mathbb{N}}$ which tends to $T^\star$ such that $\|\Phi(t_n)\|_{\dot H^{\frac12}}$ tends to $\mathcal{A}_c$. Let us apply Proposition 4.1 to the sequence

$$v_{0,n}(x) \stackrel{\text{def}}{=} \sqrt{T^\star - t_n}\,\Phi(t_n, \sqrt{T^\star - t_n}\, x).$$

Profiles decomposition of $v_{0,n}$ contains only scales less than or equal to 1. By rescaling, we deduce that

$$\Phi(t_n, x) = \sum_{j=0}^J \frac{1}{\lambda_n^j} \varphi^j\left(\frac{x - x_n^j}{\lambda_n^j}\right) + r_n^J \quad \text{with} \quad \forall j\in\mathbb{N}, \lambda_n^j \leq \sqrt{T^\star - t_n}.$$

This implies that the sequence $(\Phi(t_n))_{n\in\mathbb{N}}$ tends weakly to 0.

Thanks to law of product in Sobolev spaces, the fact that Φ is a solution of Navier-Stokes equation gives that

$$\begin{aligned}\|\partial_t\Phi\|_{L^\infty([0,T^\star[;\dot H^{-\frac{3}{2}})} &\leq \|\Delta\Phi\|_{L^\infty([0,T^\star[;\dot H^{-\frac{3}{2}})} + \big\|\mathrm{div}(\Phi\otimes\Phi)\big\|^2_{L^\infty([0,T^\star[;\dot H^{-\frac{3}{2}})}\\ &\lesssim \|\Phi\|_{L^\infty([0,T^\star[;\dot H^{\frac{1}{2}})} + \|\Phi\|^2_{L^\infty([0,T^\star[;\dot H^{\frac{1}{2}})}. \end{aligned}\tag{4.33}$$

This implies that the function $\widetilde\Phi(t) \stackrel{\text{def}}{=} \Phi(t)-\Phi(0)$ in time with value in $\dot H^{-\frac{3}{2}}$ and satisfies a Cauchy condition when t tends to $T^\star$ which means that

$$\forall\varepsilon>0\,,\ \exists t_\varepsilon>0\,/\ \forall(t,t')\in]T^\star-t_\varepsilon,T^\star[\,,\ \|\widetilde\Phi(t)-\widetilde\Phi(t')\|_{\dot H^{-\frac{3}{2}}}<\varepsilon.$$

This implies that the function $\widetilde\Phi$ has a limit in the space $\dot H^{-\frac{3}{2}}$ when t tend to $T^\star$. As the sequence $(\Phi(t_n))_{n\in\mathbb{N}}$ tends weakly to 0 when n tends to infinity, we deduce that $\Phi(0)$ belongs to $\dot H^{-\frac{3}{2}}$ and then (4.33) implies the result after interpolation. □

Continuation of the proof of Theorem 4.4. Now, we are going to explain why such a solution must be 0 which gives a contradiction.

Let us introduce the concept of suitable solution of Navier-Stokes equations on a domain of $\mathbb{R}^+\times\mathbb{R}^3$.

Definition 4.1. *Let ω be an open set in $\mathbb{R}^3$. We say that a pair (u,p) is a suitable weak solution of the Navier-Stokes equations on the set $\omega\times]-T_1,T[$ if the following conditions hold:*

$$u\in L^\infty(]-T_1,T[;L^2(\Omega))\cap L^2(-T_1,T[;\dot H^1(\omega))\quad\text{and}\quad p\in L^{\frac{3}{2}}(]-T_1,T[\times\omega)$$

and (u,p) satisfy the Navier-Stokes equations in the distribution sense and in addition the following local energy inequality

$$\begin{aligned}&\int_\omega\varphi(t,x)|u(x,t)|^2dx+2\int_{-T_1}^T\int_\omega\varphi(t,x)|\nabla u(t,x)|^2dxdt\\ &\leq\int_{-T_1}^T\int_\omega\Big(|u(t,x)|^2(\Delta\varphi(t,x)+\partial_t\varphi(t,x))\\ &\qquad+u(t,x)\cdot\nabla\varphi(t,x)(|u(t,x)|^2+2p(t,x))\Big)dxdt\end{aligned}$$

for almost all t in $]-T_1,T[$ and for any non negative function φ in $\mathcal{D}(\mathbb{R}^+\times\mathbb{R}^3)$ in a neighbourhood of the parabolic boundary $\partial'Q\stackrel{\text{def}}{=}\omega\times\{t=-T_1\}\cup\partial\omega\times[-T_1,T]$.

The following statement is a local regularity criterion proved in [11].

Lemma 4.4. *There exist positive absolute constants ε_0 and $(c_k)_{k\in\mathbb{N}}$ with the following property: If a suitable weak solution (u,p) of (NSE) on Q_1 where $Q_r\stackrel{\text{def}}{=}]-r^2,0[\times B(0,r)$ satisfies the condition*

$$\int_{Q_1}(|u(t,x)|^3+|p(t,x)|^{\frac{3}{2}})dxdt<\varepsilon_0,$$

then u is smooth on $\overline{Q}_{\frac{1}{2}}$ *and satisfies the estimate*

$$\sup_{(t,x)\in\overline{Q}_{\frac{1}{2}}} |\nabla^{k-1}u(t,x)| \leq c_k.$$

Let us observe that because of Sobolev embeddings, Φ belongs to $L^\infty([0,T^\star];L^3)$. Because Riesz operators maps $L^{\frac{3}{2}}(\mathbb{R}^3)$ into itself, we have the pressure p belongs to $L^\infty([0,T^\star];L^{\frac{3}{2}})$. Thus

$$\int_0^{T^\star}\int_{\mathbb{R}^3}\big(|\Phi(t,x)|^3+|p(t,x)|^{\frac{3}{2}}dxdt\big)<\infty.$$

It implies that, for any positive ε, a radius R_ε exists such that, for any x_0 such that $|x_0|$ greater than or equal to R,

$$\int_0^{T^\star}\int_{B(x_0,\sqrt{T^\star})}\big(|\Phi(t,x)|^3+|p(t,x)|^{\frac{3}{2}}\big)dxdt\leq\varepsilon.$$

Then applying Lemma 4.4 after a rescaling and time translation we get that Φ is smooth on $[(3/4)T^\star,T^\star]\times(\mathbb{R}^3\setminus B(0,R_\varepsilon))$. Let us write the vorticity equation

$$\partial_t\Omega-\Delta\Omega+\Phi\cdot\nabla\Omega+\Omega\cdot\nabla\Phi=0$$

on the set $[(3/4)T^\star,T^\star]\times(\mathbb{R}^3\setminus B(0,R_\varepsilon))$. Because Φ is smooth up to $T^\star$, the vorticity $\Omega(T^\star)$ is identically equal to 0 on $\mathbb{R}^3\setminus B(0,R_\varepsilon))$. As Φ in bounded in $[(3/4)T^\star,T^\star]\times(\mathbb{R}^3\setminus B(0,R_\varepsilon))$, we have the pointwise inequality

$$\big|\partial_t\Omega(t,x)-\Delta\Omega(t,x)\big|\leq C\big(|\Omega(t,x)|+|\nabla\Omega(t,x)|\big).$$

Then we can apply the following backward uniqueness result proved in [12].

Theorem 4.5 (Backwards uniqueness). *Let us consider a vector valued distribution* Ω *in* $\dot H^2_{\rm loc}((\mathbb{R}^3\setminus B(0,R_\varepsilon))\times]-\delta,0])$ *and satisfies*

$$\big|\partial_t\Omega(t,x)-\Delta\Omega(t,x)\big|\leq C\big(|\Omega(t,x)|+|\nabla\Omega(t,x)|\big)$$

on $(\mathbb{R}^3\setminus B(0,R_\varepsilon))\times]-\delta,0]$. *If* $\Omega(0,x)=0$ *on* $B(\mathbb{R}^3\setminus B(0,R_\varepsilon))$, *then*

$$\Omega\equiv0 \text{ on the set } (\mathbb{R}^3\setminus B(0,R_\varepsilon))\times]-\delta,0].$$

In order to conclude the proof of Theorem 4.4, let us notice that for any time t in $]0,T^\star[$, the vector field u has components which are analytic in the variable x on the whole space $\mathbb{R}^3$ (see for instance [6] and [23]) . Thus for any t in $[3/4T^\star,T^\star]$, $\Omega(t)\equiv0$ and thus $\Phi(t)\equiv0$ which is obviously in contradiction with the definition of Φ. □

References

[1] P. Auscher, S. Dubois and P. Tchamitchiann, On the stability of global solutions to Navier-Stokes equations in the space, *Journal de Mathématiqeus Pures et Appliquées*, **83**, 2004, pages 673–697 .

[2] H. Bahouri, J.-Y. Chemin and R. Danchin, *Fourier Analysis and Nonlinear Partial Differential Equations*, Grundlehren der mathematischen Wissenschaften, **343**, Springer-Verlag Berlin Heidelberg, 2011.

[3] H. Bahouri and P. Gérard, High frequency approximation of solutions to critical nonlinear wave equations, *American Journal of Math*, **121**, 1999, pages 131–175.

[4] L. Cafarelli, R. Kohn and L. Nirenberg, Partial regularity of suitable weal solutions of Navier-Stokes equations, *Communications in Pure and Applied Mathematics*, **35**, 1982, pages 771–831.

[5] M. Cannone, Y. Meyer and F. Planchon, Solutions autosimilaires des équations de Navier-Stokes, Séminaire "Équations aux Dérivées Partielles de l'École polytechnique", Exposé VIII, 1993–1994.

[6] J.-Y. Chemin, Localization in Fourier space and Navier-Stokes system, *Phase Space Analysis of Partial Differential Equations*, Proceedings 2004, CRM series, pages 53–136.

[7] J.-Y. Chemin and I. Gallagher, Large, global solutions to the Navier-Stokes equations, slowly varying in one direction, *Transactions of the Americal Mathematical Society*, **362**, 2010, pages 2859–2873.

[8] J.-Y. Chemin, I. Gallagher and P. Zhang, On large perturbation of global solutions of the 3D incompressible Navier-Stokes equations, *Journal für reine und angewandte Mathematik*, **681**, 2013, pages 65–82.

[9] J.-Y. Chemin, I. Gallagher and M. Paicu, Global regularity for some classes of large solutions to the Navier-Stokes equations, *Annals of Mathematics*, **173**, 2011, pages 986–1012.

[10] J.-Y. Chemin and N. Lerner, Flot de champs de vecteurs non-lipschitziens et équations de Navier-Stokes, *Journal of Differential Equations*, **121**, 1995, pages 314–328.

[11] L. Escauriaza, G. A. Serëgin and V. Sverak, $L^{3,\infty}$-solutions of Navier-Stokes equations and backward uniqueness, (Russian) *Uspekhi Mat. Nauk*, **58**, 2003, no. 2 (350), 3–44; translation in *Russian Math. Surveys*, **58**, 2003, no. 2, 211–250.

[12] L. Escauriaza, G. Serëgin and V. Sverak, Backward uniqueness for parabolic equations, *Archiv for Rational Mechanic and Analysis*, **169**, 2003, pages 147–157.

[13] H. Fujita and T. Kato, On the Navier-Stokes initial value problem I, *Archiv for Rational Mechanic Analysis*, **16**, 1964, pages 269–315.

[14] I. Gallagher, Profile decomposition for solutions of the Navier-Stokes equations, *Bulletin de la Société Mathématique de France*, **129**, 2001, pages 285–316.

[15] I. Gallagher, D. Iftimie and F. Planchon, Non-explosion en temps grand et stabilité de solutions globales des équations de Navier-Stokes, *Notes aux Comptes*

Rendus de l'Académie des Sciences, **334**, Série I, 2002, pages 289–292.

[16] I. Gallagher, D. Iftimie and F. Planchon, Asymptotics and stability for global solutions to the Navier-Stokes equations, *Annales de l'Institut Fourier*, **53**, 2003, pages 1387–1424.

[17] I. Gallagher, G. Koch and F. Planchon, A profile decomposition approach to the $L^\infty_t(L^3_x)$ Navier-Stokes regularity criterion, *Mathematische Annalen*, **355**, 2013, pages 1527–1559.

[18] P. Gérard, Description des défauts de compacité de l'injection de Sobolev, *ESAIM, Control Optimal et Calcul des variations*, **3**, 1998, pages 213–233.

[19] T. Kato, Nonstationary flows of viscous and ideal fluids in $\mathbb{R}^3$, *Journal of functionnal Analysis*, **9**, 1972, pages 296–305.

[20] C. Kenig and G. Koch, An alternative approach to regularity for the Navier-Stokes equations in critical spaces, *Annales de l'Institut Henri Poincaré Analyse non Linéaire*, **28**, 2011, pages 15–187.

[21] H. Koch and D. Tataru, Well-posedness for the Navier-Stokes equations, *Advances in Mathematics*, **157**, 2001, pages 22–35.

[22] P.-G. Lemarié-Rieusset, *Recent Developments in the Navier-Stokes Problem*, Research Notes in Mathematics, **431**, 2002.

[23] P.-G. Lemarié-Rieusset, Nouvelles remarques sur l'analyticité des solutions milds des équations de Navier-Stokes dans $\mathbb{R}^3$, (French) [Further remarks on the analyticity of mild solutions for the Navier-Stokes equations in R^3], *Compte Rendus de l'Académie des Sciences de Paris, Série I*, **338 (6)**, 2004, pages 443–446.

[24] J. Leray, Essai sur le mouvement d'un liquide visqueux emplissant l'espace, *Acta Matematica*, **63**, 1933, pages 193–248.

[25] F. Lin, A new proof of the Caffarelli-Kohn-Nirenberg theorem, *Communication in Pure and Applied Mathematics*, **51**, 1998, pages 241–257.

[26] Y. Meyer, *Wavelets, Paraproducts and Navier-Stokes*, Current Developments in Mathematics, International Press, Cambridge, Massachussets, 1996.

[27] C. Oseen, Sur les formules de Green généralisées qui se présentent dans l'hydrodynamique et sur quelques unes de leurs applications, (première partie) *Acta Mathematica*, **34**, 1911, pages 205–284.

[28] C. Oseen, Sur les formules de Green généralisées qui se présentent dans l'hydrodynamique et sur quelques unes de leurs applications, (seconde partie) *Acta Mathematica*, **35**, 1912, pages 97–192.

[29] T. Runst and W. Sickel, *Sobolev Spaces of Fractional Order, Nemytskij Operators, and Nonlinear Partial Differential Equations*, Nonlinear Analysis and Applications, **3**, Walter de Gruyter & Co., Berlin, 1996.

[30] W. Rusin and V. Sverák, Minimal initial data for potential Navier-Stokes singularities, *Journal of Functional Analysis*, **260**, 2011, pages 879–891.

[31] F. Weissler, The Navier-Stokes initial value problem in L^p, *Archiv for Rational Mechanics and Analysis*, **74**, 1980, pages 219–230.

Lectures on the Analysis of Nonlinear Partial Differential Equations Vol. 4
MLM 4, pp. 55–91

L_p Estimates for Parabolic Equations

Hongjie Dong*†

Abstract

In this mini course, we present a self-contained approach of the L_p estimates for second-order parabolic equations with irregular coefficients. Both interior and boundary estimates are covered, and most of the results can be extended to higher-order equations and systems. We focus on non-divergence form equations, while only illustrate the ideas of how to extend these results to divergence form equations.

1 L_2 estimates

In the first section, we derive L_2 estimates for non-divergence form parabolic equations. Consider the following linear parabolic partial differential equation:

$$u_t - a_{ij}D_{ij}u + \lambda u = f \quad \text{in} \quad \mathbb{R}^{d+1}_T, \tag{1.1}$$

where $T \in (-\infty, +\infty]$ and $\mathbb{R}^{d+1}_T = (-\infty, T) \times \mathbb{R}^d$. Here the solution space is

$$W^{1,2}_2(\mathbb{R}^{d+1}_T) := \{u \in L_2(\mathbb{R}^{d+1}_T) \,:\, \partial_t u, Du, D^2u \in L_2(\mathbb{R}^{d+1}_T)\}.$$

We assume that $u \in W^{1,2}_2(\mathbb{R}^{d+1}_T)$, and our goal is to give an a priori bound of $\|u\|_{W^{1,2}_2(\mathbb{R}^{d+1}_T)}$ in terms of $\|f\|_{L_2(\mathbb{R}^{d+1}_T)}$.

The L_p-theory of non-divergence form second-order elliptic and parabolic equations in the whole space or smooth domains with discontinuous coefficients was studied extensively in the last fifty years. According to the well-known counterexamples of Ural'tseva, Talenti, and Nadirashvili, in general there does not exist a solvability theory for uniformly elliptic or parabolic operators with bounded and measurable coefficients. For the solvability and regularity theory, we need to impose certain regularity assumptions on the coefficients.

Our starting point is the following type of coefficients, which depend only on the time variable and satisfy the uniform ellipticity condition:

*Division of Applied Mathematics, Brown University, 182 George Street, Providence, RI 02912, USA, E-mail: Hongjie_Dong@brown.edu

†H. Dong was partially supported by the NSF under agreement DMS-1056737.

(a) $a_{ij} = a_{ji}$, where a_{ij} are measurable functions of t only.
(b) $\kappa|\xi|^2 \le a_{ij}(t)\xi_i\xi_j \le |\xi|^2/\kappa$ for some $\kappa \in (0,1)$.

Lemma 1.1. *Let $T \in (-\infty, +\infty]$ and $\lambda \ge 0$. Suppose that $u \in W_2^{1,2}(\mathbb{R}_T^{d+1})$, $f \in L_2(\mathbb{R}_T^{d+1})$, and u satisfies (1.1). Then we have the following estimate:*

$$\|u_t\|_{L_2(\mathbb{R}_T^{d+1})} + \|D^2u\|_{L_2(\mathbb{R}_T^{d+1})} + \sqrt{\lambda}\|Du\|_{L_2(\mathbb{R}_T^{d+1})} + \lambda\|u\|_{L_2(\mathbb{R}_T^{d+1})}$$
$$\le N(d,\kappa)\|f\|_{L_2(R_T^{d+1})}.$$

Proof. We prove the lemma by using the Fourier transform. Set $\hat{u}(t,\xi)$ to be the Fourier transform of $u(t,x)$ with respect to x, and $\hat{f}(t,\xi)$ to be the Fourier transform of $f(t,x)$ with respect to x. Then we have $\widehat{D_j u} = i\xi_j\hat{u}$. Take Fourier transform of the equation, we have

$$\hat{u}_t(t,\xi) = (-a_{ij}(t)\xi_i\xi_j - \lambda)\hat{u} + \hat{f}(t,\xi). \tag{1.2}$$

Solving the ODE above, we have

$$\hat{u}(t,\xi) = \int_{-\infty}^t e^{-\int_s^t a_{ij}(r)dr\xi_i\xi_j - \lambda(t-s)}\hat{f}(s,\xi)\,ds.$$

Taking the absolute value of both sides, we get

$$\begin{aligned}|\hat{u}(t,\xi)| &\le \int_{-\infty}^t e^{-\int_s^t a_{ij}(r)dr\xi_i\xi_j - \lambda(t-s)}|\hat{f}(s,\xi)|\,ds\\ &\le \int_{-\infty}^t e^{-(\kappa|\xi|^2+\lambda)(t-s)}|\hat{f}(s,\xi)|\,ds.\end{aligned}$$

Therefore,

$$|\hat{u}(t,\xi)| \le I_{t\ge 0}e^{-(\kappa|\xi|^2+\lambda)t} * |\hat{f}(t,\xi)|.$$

Taking the L_2 norm with respect to t, by using the Minkowski inequality for convolutions, we get

$$\int_{-\infty}^T |\hat{u}(t,\xi)|^2\,dt \le \int_{-\infty}^T |\hat{f}(t,\xi)|^2\,dt\left(\int_0^\infty e^{-(\kappa|\xi|^2+\lambda)t}\,dt\right)^2$$

and

$$(\kappa|\xi|^2+\lambda)^2\int_{-\infty}^T |\hat{u}(t,\xi)|^2\,dt \le \int_0^T |\hat{f}(t,\xi)|^2\,dt.$$

Integrating in ξ and by using Parseval's identity, we obtain

$$\kappa^2\sum_{i,j=1}^d \|D_{ij}u\|_{L_2}^2 + 2\kappa\lambda\sum_{i=1}^d \|\partial_i u\|_{L_2}^2 + \lambda^2\|u\|_{L_2}^2 \le \|f\|_{L_2}^2.$$

The estimate of u_t follows from the equation and the estimate above. The lemma is proved. ☐

Remark 1.2. From the proof above, we also get an explicit representation of the solution. For $t > s$, let

$$A_{s,t}^{i,j} = \int_s^t a_{ij}(r)\, dr.$$

Then $(A_{s,t}^{i,j})$ is non-degenerate. We set $[B_{s,t}^{i,j}] = [A_{s,t}^{i,j}]^{-1}$. Taking the inverse Fourier transform of (1.2), we get

$$u(t,x) = \int_{-\infty}^t G(t,s,x) * f(s,x) e^{-\lambda(t-s)}\, ds,$$

where

$$G(t,s,x) = I_{t>s} \left(\frac{1}{4\pi}\right)^{d/2} (\det B_{s,t})^{1/2} e^{\frac{-(B_{s,t}x,x)}{4}}.$$

Note that if $a_{ij} = \delta_{ij}$, $G(t,s,x)$ is reduced to the standard heat kernel. As an exercise, the reader can check that $\int_{\mathbb{R}^d} G(t,s,x)\, dx = 1$ for any $t > s$, which implies that

$$\lambda \|u\|_{L_p(\mathbb{R}_T^{d+1})} \le \|f\|_{L_p(\mathbb{R}_T^{d+1})}$$

for any $p \in [1, \infty]$.

Now we present another proof of Lemma 1.1 by using integration by parts.

Another proof of Lemma 1.1. We may assume $u \in C_0^\infty((-\infty, T] \times \mathbb{R}^d)$. In fact, since $C_0^\infty((-\infty, T] \times \mathbb{R}^d)$ is dense in $W_2^{1,2}(\mathbb{R}_T^{d+1})$, we can find $u_k \in C_0^\infty((-\infty, T] \times \mathbb{R}^d)$ such that $u_k \to u$ in $W_2^{1,2}(\mathbb{R}_T^{d+1})$. Then we have

$$(u_k)_t - a_{ij}(t) D_{ij} u_k + \lambda u_k = (u_k - u)_t + a_{ij}(t) D_{ij}(u - u_k) + \lambda(u_k - u) + f.$$

Note that the right-hand side above tends to f in L_2. Once we prove the lemma for $u \in C_0^\infty((-\infty, T] \times \mathbb{R}^d)$, by taking the limit as $k \to \infty$, we then get the result for general $u \in W_2^{1,2}(\mathbb{R}_T^{d+1})$.

Now we multiply the equation by λu and integrate over the time-space. We have

$$\int_{\mathbb{R}_T^{d+1}} \lambda a_{ij}(t) D_i u D_j u + \lambda^2 u^2\, dx\, dt \le \int_{\mathbb{R}_T^{d+1}} \lambda |fu|\, dx\, dt.$$

Using Young's inequality on the right-hand side and the ellipticity condition, we have

$$\lambda\kappa \int_{\mathbb{R}_T^{d+1}} |Du|^2\, dx\, dt + \lambda^2 \int_{\mathbb{R}_T^{d+1}} u^2\, dx\, dt \le N \int_{\mathbb{R}_T^{d+1}} f^2\, dx\, dt.$$

Next we multiply the equation by Δu and integrate by parts. We have the following estimate:

$$\int_{\mathbb{R}_T^{d+1}} \kappa |D^2 u|^2 + \lambda |Du|^2\, dx\, dt \le N \int_{\mathbb{R}_T^{d+1}} f^2\, dx\, dt.$$

Combining the above two estimate yields the desired estimate. □

Next let us move onto the case when the coefficients depends on both the space and the time variables.

We consider the 1-D case first. In this case, the equation becomes

$$u_t - a(t,x)u_{xx} + \lambda u = f \quad \text{in} \quad \mathbb{R}^2_T, \tag{1.3}$$

where for some fixed constant $0 < \kappa \le 1$, we have $\kappa \le a \le 1/\kappa$ and a is a measurable function.

Lemma 1.3. *Assume that $f \in L_2(\mathbb{R}^2_T)$ and $u \in W_2^{1,2}(\mathbb{R}^2_T)$ satisfies (1.3). Then we have*

$$\|u_t\|_{L_2(\mathbb{R}^2_T)} + \|u_{xx}\|_{L_2(\mathbb{R}^2_T)} + \sqrt{\lambda}\|u_x\|_{L_2(\mathbb{R}^2_T)} + \lambda\|u\|_{L_2(\mathbb{R}^2_T)} \le \frac{2}{\kappa}\|f\|_{L_2(\mathbb{R}^2_T)}.$$

Proof. By using a density argument as above, without loss of generality we may assume that $u \in C_0^\infty((-\infty,T]\times\mathbb{R})$. Since $a(x,t)$ depends on x now, we cannot use the Fourier transform any longer. Divide the equation by $\sqrt{a}$, we get

$$\frac{1}{\sqrt{a}}(u_t + \lambda u) - \sqrt{a}u_{xx} = \frac{1}{\sqrt{a}}f.$$

Take the square of both sides and integrate over the whole time-space. We then get

$$\int_{\mathbb{R}^2_T} \frac{1}{a}(u_t^2 + 2\lambda u_t u + \lambda^2 u^2) - 2(u_t + \lambda u)(u_{xx}) + a(u_{xx})^2\,dx\,dt = \int_{\mathbb{R}^2_T} \frac{f^2}{a}\,dx\,dt.$$

Using the ellipticity condition and integrating by parts, we have

$$\int_{\mathbb{R}^2_T} \kappa(u_t^2 + \lambda^2 u^2) + 2\lambda u_x^2 + \kappa(u_{xx})^2\,dx\,dt \le \int_{\mathbb{R}^2_T} \frac{1}{\kappa}f^2\,dx\,dt,$$

where we used

$$\int_{\mathbb{R}^2_T} u_t u\,dx\,dt = \frac{1}{2}\int_{\mathbb{R}} u^2(T,x)\,dx \ge 0$$

and

$$-\int_{\mathbb{R}^2_T} u_t u_{xx}\,dx\,dt = \int_{\mathbb{R}^2_T} u_{tx}u_x\,dx\,dt = \frac{1}{2}\int_{\mathbb{R}}(u_x)^2(T,x)\,dx \ge 0.$$

Hence the lemma is proved. □

Now we turn to the multi-dimensional case. First, we consider the case that $a_{ij}(t,x) = a_{ij}(t,x_1)$ are bounded measurable functions depending only on t and x_1. We use the notation

$$x = (x_1, x_2, ..., x_n) = (x_1, x'), \quad x' = (x_2, ..., x_d).$$

Theorem 1.4. *Let $\lambda \geq 0$ be a constant, and $a_{ij}(t,x) = a_{ij}(t,x_1)$ be bounded and measurable functions depending on t and x_1 only. Assume that $u \in W_2^{1,2}(\mathbb{R}_T^{d+1})$ and $f \in L_2(\mathbb{R}_T^{d+1})$, and u satisfies*

$$u_t - a_{ij}D_{ij}u + \lambda u = f \quad in \quad \mathbb{R}_T^{d+1}.$$

Then the following estimate holds:

$$\begin{aligned}
&\|u_t\|_{L_2(\mathbb{R}_T^{d+1})} + \|D^2u\|_{L_2(\mathbb{R}_T^{d+1})} + \sqrt{\lambda}\|Du\|_{L_2(\mathbb{R}_T^{d+1})} + \lambda\|u\|_{L_2(\mathbb{R}_T^{d+1})} \\
&\leq N(d,\kappa)\|f\|_{L_2(R_T^{d+1})}.
\end{aligned}$$

Unlike in the 1-D case, where we can divide the equation by $\sqrt{a}$, for the multi-dimensional case we need a new idea which uses the Fourier transform in x'.

For $\xi \in \mathbb{R}^{d-1}$, we denote by $v(t,x,\xi)$ the Fourier transform of $u(t,x,x')$ with respect to x'. Take the Fourier transform of the equation and denote by $g(t,x,\xi)$ the Fourier transform of f in the x' variable. We have an equation for v:

$$v_t - a_{11}D_{11}v - i\,2bD_1v + (c+\lambda)v = g, \tag{1.4}$$

where

$$b = \sum_{j=2}^{d} a_{1j}\xi_j, \quad c = \sum_{j,k=2}^{d} a_{jk}\xi_j\xi_k.$$

We will use the following lemma regarding a pointwise estimate of v.

Lemma 1.5. *Assume that $u, f \in C_0^\infty((-\infty,T] \times \mathbb{R}^d)$, $|\xi|^2 + \lambda > 0$, and $a_{ij}(t,x_1)$ are smooth functions. Then for any $\xi \in \mathbb{R}^{d-1}$, we have*

$$|v(t,x_1,\xi)| \leq u^*(t,x_1,\xi),$$

where u^* is the solution of

$$u_t^* - a_{11}D_{11}u^* + (\kappa|\xi|^2 + \lambda)u^* = |g|$$

in $\mathbb{R}_T^2$.

Proof. First of all, by Lemma 1.3 and the method of continuity, $u^* \in W_2^{1,2}(\mathbb{R}_T^2)$ exists and is unique. Since $|g|$ is a Lipschitz function, using interior Schauder's estimates we have

$$u^* \in C^{1+\alpha/2,2+\alpha}$$

for any $\alpha \in (0,1)$. Furthermore, by the maximum principle, we know $u^* \geq 0$ on $\mathbb{R}_T^{d+1}$. Next, we derive the equation for $|v|$ on $\Omega = \{(t,x_1) \in \mathbb{R}_T^2 : v(t,x_1,\xi) \neq 0\}$. Since $u \in C_0^\infty((-\infty,T] \times \mathbb{R}^d)$, it is easily seen that Ω is an open and bounded set in $\mathbb{R}^2$.

By simple calculations, we have

$$\begin{aligned}
\partial_t|v| &= \frac{1}{2|v|}(\overline{v}\partial_t v + v\partial_t\overline{v}) = \frac{1}{|v|}\Re(\overline{v}\partial_t v), \\
D_1|v| &= \frac{1}{2|v|}(\overline{v}D_1 v + vD_1\overline{v}) = \frac{1}{|v|}\Re(\overline{v}\partial_1 v),
\end{aligned}$$

and

$$
\begin{aligned}
D_{11}|v| &= \frac{1}{2|v|}(\bar{v}D_{11}v + vD_{11}\bar{v} + 2D_1v\cdot D_1\bar{v}) - \frac{1}{4|v|^3}(\bar{v}D_1v + vD_1\bar{v})^2 \\
&= \frac{1}{2|v|}(\bar{v}D_{11}v + vD_{11}\bar{v}) - \frac{1}{4|v|^3}(\bar{v}D_1v - vD_1\bar{v})^2 \\
&= \frac{1}{|v|}\Re(\bar{v}D_{11}v) + \frac{1}{|v|^3}\big(\Im(\bar{v}D_1v)\big)^2.
\end{aligned}
$$

Multiplying the both sides of (1.4) by $\bar{v}/|v|$ and taking the real parts, we see that $|v|$ satisfies

$$
\partial_t|v| - a_{11}D_{11}|v| = \frac{1}{|v|}\Re(\bar{v}g) - \frac{a_{11}}{|v|^3}\big[\Im(\bar{v}D_1v)\big]^2 - \frac{2b}{|v|}\Im(\bar{v}D_1v) - (c+\lambda)|v|
$$

in Ω. Denote $M = \Im(\bar{v}D_1v)$. By the uniform ellipticity condition with ξ_1 replaced by M and ξ_j replaced by $\xi_j|v|^2$ for $j = 2,\dots,d$, we obtain

$$
\begin{aligned}
&\frac{a_{11}}{|v|^3}\big[\Im(\bar{v}D_1v)\big]^2 + \frac{2b}{|v|}\Im(\bar{v}D_1v) + (c+\lambda)|v| \\
&= \frac{1}{|v|^3}\Big(a_{11}M^2 + 2\sum_{j=2}^{d} a_{1j}\xi_jM|v|^2 + \sum_{j,k=2}^{d} a_{jk}\xi_j\xi_k|v|^4\Big) + \lambda|v| \\
&\geq (\lambda + \kappa|\xi|^2)|v|.
\end{aligned}
$$

Therefore, we reach

$$
\partial_t|v| - a_{11}D_{11}|v| + (\kappa|\xi|^2 + \lambda)|v| \leq |g| \quad \text{in} \quad \Omega,
$$

which together with the comparison principle implies that $|v| \leq u^*$. □

Corollary 1.6. *Under the conditions of Lemma 1.5, for any $\xi \in \mathbb{R}^{d-1}$,*

$$
(|\xi|^2 + \lambda)\|v(\cdot,\cdot,\xi)\|_{L_2(\mathbb{R}^2_T)} \leq N\|g(\cdot,\cdot,\xi)\|_{L_2(\mathbb{R}^2_T)}. \tag{1.5}
$$

Proof. For $|\xi|^2 + \lambda > 0$, it follows immediately from Lemmas 1.3 and 1.5. When $|\xi|^2 + \lambda = 0$, the inequality is obvious. □

Proof of Theorem 1.4. By mollification, we can find a sequence of smooth coefficients $a^k_{ij} \to a_{ij}$ (*a.e.*), which satisfy the same ellipticity condition. By the density argument, we can also find $u_k \in C_0^\infty((-\infty,T]\times\mathbb{R}^d)$ such that $u_k \to u \in W_2^{1,2}(\mathbb{R}^{d+1}_T)$. Thus for any k, u_k satisfies

$$
\begin{aligned}
&\partial_t u_k - a^k_{ij}D_{ij}u_k + \lambda u_k \\
&= (\partial_t u_k - \partial_t u) + (a_{ij} - a^k_{ij})D_{ij}u + a^k_{ij}(D_{ij}u - D_{ij}u_k) + \lambda(u_k - u) + f.
\end{aligned}
$$

Denote the right-hand side above by f_k. Then by the dominated convergence theorem, we have

$$
f_k \to f \quad \text{in} \quad L_2(\mathbb{R}^{d+1}_T).
$$

If we can prove the estimate with u_k and f_k in place of u and f, then we can get the general result by letting $k \to +\infty$. Therefore, in the sequel we may assume that $u, f \in C_0^\infty((-\infty, T] \times \mathbb{R}^d)$ and a_{ij} are infinitely differentiable.

Take the square of (1.5), and then integrate with respect to $\xi \in \mathbb{R}^{d-1}$. By Parseval's identity, we have

$$\int_{\mathbb{R}_T^{d+1}} |D_{x'}^2 u|^2 + \lambda |D_{x'} u|^2 + \lambda^2 u^2 \, dx \, dt \le N \|f\|_{L_2(\mathbb{R}_T^2)}^2. \tag{1.6}$$

Now we rewrite the equation into

$$u_t - a_{11} D_{11} u + \lambda u = \sum_{i \,\mathrm{or}\, j > 1} a_{ij} D_{ij} u + f.$$

We fix x' and use Lemma 1.3 to get

$$\begin{aligned} &\|u_t\|_{L_2(\mathbb{R}_T^{d+1})} + \|D_{11} u\|_{L_2(\mathbb{R}_T^{d+1})} + \sqrt{\lambda} \|D_1 u\|_{L_2(\mathbb{R}_T^{d+1})} \\ &\le N \|f\|_{L_2(\mathbb{R}_T^{d+1})} + \|D D_{x'} u\|_{L_2(\mathbb{R}_T^{d+1})}. \end{aligned} \tag{1.7}$$

Next we use a scaling argument to control $\|D_{1x'} u\|_{L_2}$. Integrating by parts and using Young's inequality, we have

$$\|D_{1x'} u\|_{L_2} \le N(\|D_{11} u\|_{L_2} + \|D_{x'x'} u\|_{L_2}),$$

where N is a constant depending on the dimension only. By applying the inequality to $v(x, x') := u(\epsilon x_1, x')$, we get

$$\|D_{1x'} u\|_{L_2} \le \epsilon \|D_{11} u\|_{L_2} + \frac{N(d)}{\epsilon} \|D_{x'x'} u\|_{L_2}. \tag{1.8}$$

By substituting (1.8) with a sufficiently small ϵ into (1.7) and using (1.6), we complete the proof of the theorem. □

As an extension, we now consider a 2-D problem:

$$D_t u - a_{ij}(t, x) D_{ij} u + \lambda u = f \quad \text{in} \quad (-\infty, T) \times \mathbb{R}^2. \tag{1.9}$$

We have the following result.

Theorem 1.7. (i) *Suppose that $a_{11} + a_{22} = 1$. Then there exists $\theta_0 = \theta_0(\kappa) < 1$ such that the following holds true. For any $p \in (2 - \theta_0, 2 + \theta_0)$, $\lambda \ge 0$, $f \in L_p(\mathbb{R}_T^3)$, and $u \in W_p^{1,2}(\mathbb{R}_T^3)$ satisfying (1.9), we have*

$$\begin{aligned} &\|D_t u\|_{L_p(\mathbb{R}_T^3)} + \|D^2 u\|_{L_p(\mathbb{R}_T^3)} + \sqrt{\lambda} \|D u\|_{L_p(\mathbb{R}_T^3)} + \lambda \|u\|_{L_p(\mathbb{R}_T^3)} \\ &\le N \|f\|_{L_p(\mathbb{R}_T^3)}. \end{aligned}$$

(ii) *The same result holds if $a_{11} + a_{22}$ is a bounded functions of t only.*

Proof of Theorem 1.7 (ii) *by using* (i). As before, we may assume that a_{ij} are smooth functions and $u \in C_0^\infty((-\infty, T] \times \mathbb{R}^2)$. We make a change of variables in t:

$$t \to \int_0^t (a_{11} + a_{22})(r)\,dr := s(t),$$

which is a bi-Lipschitz transformation by the ellipticity condition. The equation can be rewritten as

$$\partial_s u - \frac{a_{ij}}{a_{11}+a_{22}} D_{ij} u + \frac{\lambda}{a_{11}+a_{22}} u = \frac{f}{a_{11}+a_{22}}.$$

To apply Assertion (i), we further rewrite the equation as following

$$\partial_s u - \frac{a_{ij}}{a_{11}+a_{22}} D_{ij} u + \lambda u = \frac{f}{a_{11}+a_{22}} + \left(\lambda - \frac{\lambda}{a_{11}+a_{22}}\right) u.$$

Then it follows from Assertion (i) under the trace 1 condition, that

$$\begin{aligned} &\|\partial_t u\|_{L_p} + \|D^2 u\|_{L_p} + \sqrt{\lambda}\|Du\|_{L_p} + \lambda\|u\|_{L_p} \\ &\le N\|f\|_{L_p} + N\lambda\|u\|_{L_p}. \end{aligned}$$

Note that $N = N(p, \kappa)$ is independent of λ.

Next we use show how to bound $\lambda\|u\|_{L_p}$ by $\|f\|_{L_p}$. By Assertion (i) and the method of continuity, we can find a unique solution $v \in W_2^{1,2}(\mathbb{R}^3_{s(T)})$ to the equation

$$\partial_s v - \frac{a_{ij}}{a_{11}+a_{22}} D_{ij} v + \frac{\kappa\lambda}{2} v = \left|\frac{f}{a_{11}+a_{22}}\right|$$

in $\mathbb{R}^3_{s(T)}$. By the maximum principle, we infer that $v \ge 0$ and thus

$$\partial_s v - \frac{a_{ij}}{a_{11}+a_{22}} D_{ij} v + \frac{\lambda}{a_{11}+a_{22}} v \ge \left|\frac{f}{a_{11}+a_{22}}\right|$$

in $\mathbb{R}^3_{s(T)}$. By the comparison principle, we get $v \ge |u|$. Hence it follows that

$$\lambda\|u\|_{L_p} \le \lambda\|v\|_{L_p} \le N\|f\|_{L_p}. \qquad \square$$

It remains to prove Assertion (i). Before its proof, let us show a consequence of it. Consider the 2D elliptic equation

$$-a_{ij}(x) D_{ij} u + \lambda u = f, \tag{1.10}$$

where a_{ij} are bounded measure functions and satisfy the same ellipticity condition.

Corollary 1.8. *There exists $\theta_0 = \theta_0(\kappa) < 1$ such that the following holds true. For any $p \in (2 - \theta_0, 2 + \theta_0)$, $\lambda \ge 0$, $f \in L_p(\mathbb{R}^2)$, and $u \in W_p^{1,2}(\mathbb{R}^2)$ satisfying* (1.10), *we have*

$$\|D^2 u\|_{L_p(\mathbb{R}^2)} + \sqrt{\lambda}\|Du\|_{L_p(\mathbb{R}^2)} + \lambda\|u\|_{L_p(\mathbb{R}^2)} \le N\|f\|_{L_p(\mathbb{R}^2)}.$$

Proof. We divide the equation by $h = a_{11} + a_{22}$ and denote $b_{ij} = a_{ij}/h$. Then we have the following equation:

$$-b_{ij}(x)D_{ij}u + \frac{\lambda}{h}u = \frac{f}{h}.$$

As before, by the maximum principle, if w is the W_2^2-solution of the following equation:

$$-b_{ij}(x)D_{ij}w + \frac{\lambda\kappa}{2}w = \left|\frac{f}{h}\right|,$$

then $|u| \le w$. This enables us to control the L_p norm of u by that of w, and further by f as follows.

We choose a smooth cutoff function η on $\mathbb{R}$ such that $\eta(t) = 1$ in $[-1, 1]$, $\eta(t) = 0$ for $|t| > 2$, and $0 \le \eta \le 1$. Define $\eta_R = \eta(\cdot/R)$ and $v(x,t) = w(x)\eta_R(t)$, which satisfies the following parabolic equation:

$$v_t - b_{ij}(x)D_{ij}v + \frac{\lambda\kappa}{2}v = \left|\frac{f}{h}\right|\eta_R + w(x)\eta_R'.$$

By Theorem 1.7 (i), we have

$$\begin{aligned}&\|\eta_R\|_{L_p}(\|D^2w\|_{L_p} + \sqrt{\lambda}\|Dw\|_{L_p} + \lambda\|w\|_{L_p})\\ &\le N\big(\|\eta_R\|_{L_p}\|f\|_{L_p} + \|w\|_{L_p}\|\eta_R'\|_{L_p}\big).\end{aligned} \tag{1.11}$$

It is easily seen that

$$\frac{\|\eta_R'\|_{L_p}}{\|\eta_R\|_{L_p}} = \frac{1}{R}\frac{\|\eta'\|_{L_p}}{\|\eta\|_{L_p}} \to 0$$

as $R \to \infty$. Therefore, by dividing both sides of (1.11) by $\|\eta_R\|_{L_p}$ and sending $R \to \infty$, we obtain

$$\|D^2w\|_{L_p} + \sqrt{\lambda}\|Dw\|_{L_p} + \lambda\|w\|_{L_p} \le N\|f\|_{L_p}, \tag{1.12}$$

which together with $|u| \le w$ implies that

$$\lambda\|u\|_{L_p} \le N\|f\|_{L_p}. \tag{1.13}$$

Now we rewrite the equation of u as

$$-b_{ij}(x)D_{ij}u + \frac{\lambda\kappa}{2}u = \frac{f}{h} + \frac{\lambda\kappa}{2}u - \frac{\lambda}{h}u := g.$$

By the proof above and (1.13), we have

$$\|D^2u\|_{L_p} + \sqrt{\lambda}\|Du\|_{L_p} + \lambda\|u\|_{L_p} \le N\|g\|_{L_p} \le N\|f\|_{L_p}.$$

The corollary is proved. □

For the proof of Theorem 1.7 (i), we need a couple of lemmas.

Lemma 1.9. *For any $\lambda \geq 0$, we denote*

$$u_{ij,\lambda} = D_{ij}u - (u_t + \lambda u)\delta_{ij}.$$

Then there exists a continuous positive function $N(p)$, such that $N(2) = 1$ and

$$\left\| \sqrt{\sum u_{ij,\lambda}^2} \right\|_{L_p(\mathbb{R}_T^3)} \leq N(p)\|u_{11,\lambda} + u_{22,\lambda}\|_{L_p(\mathbb{R}_T^3)}.$$

Proof. First we note that

$$u_{11,\lambda} + u_{22,\lambda} = \Delta u - 2(u_t + \lambda u).$$

Recall the Riesz-Thorin interpolation theorem for linear bounded operators: if T is both a (p,p) and (q,q)-type linear bounded operator for $1 < p < q < \infty$, then the operator norm

$$\|T\|_{L_{1/\alpha} \to L_{1/\alpha}} = \left\{ \frac{\|Tg\|_{L_{1/\alpha}}}{\|g\|_{L_{1/\alpha}}} : g \in L_{1/\alpha}, g \neq 0 \right\}$$

is a log-convex function of $\alpha \in (1/q, 1/p)$.

Next we claim the following existence result, which will be proved in the next section.

Claim. Let $1 < p < \infty$. For any $\lambda > 0$ and $g \in L_p(\mathbb{R}_T^3)$, there is a unique solution $u \in W_p^{1,2}(\mathbb{R}_T^3)$ to the equation

$$\Delta u - 2(u_t + \lambda u) = g$$

and u satisfies

$$\|u_t\|_{L_p} + \|D^2 u\|_{L_p} + \sqrt{\lambda}\|Du\|_{L_p} + \lambda\|u\|_{L_p} \leq N\|f\|_{L_p},$$

where N is a constant independent of λ and u.

Now we define

$$T_\sigma^\lambda(g) = \sigma_{ij} u_{ij,\lambda},$$

where for $i, j = 1, 2$, $\sigma_{ij}(t,x)$ is a bounded measurable function such that $\|\sigma(t,x)\|_{l^2} = 1$ for each $(t,x) \in \mathbb{R}_T^3$. Then it is easily seen that T_σ^λ is a (p,p)-type operator for any $p \in (1, \infty)$. Clearly,

$$\sqrt{\sum u_{ij,\lambda}^2} = \sup\{\sigma_{ij} u_{ij,\lambda} : \|\sigma_{ij}\|_{l^2} = 1\}.$$

Since the supremum of log-convex function is also log-convex, by the Riesz-Thorin theorem, we know that

$$N^*(\alpha) = \sup \left\{ \frac{\|T_\sigma^\lambda g\|_{1/\alpha}}{\|g\|_{1/\alpha}} : g \in L_{1/\alpha}, g \neq 0, \|\sigma(t,x)\|_{l_2} = 1, \forall\, (t,x) \in \mathbb{R}_T^3 \right\}$$

is log-convex with respect to α in $(0,1)$, which is certainly continuous. Therefore, $N(p) = N^*(1/p)$ is also a continuous function on $(1, \infty)$. Finally, it is easy to check that $N(2) = 1$ by using integration by parts. We leave it to the reader as an exercise. □

Since $N(2) = 1$ and $N(p)$ is continuous, from the previous lemma we immediately get the following

Corollary 1.10. *Let $\lambda \geq 0$. Then there exists $\delta_0 \in (0, 1/2)$ depending on κ such that for any $p \in (2 - \delta_0, 2 + \delta_0)$ and $u \in W_p^{1,2}(\mathbb{R}_T^3)$, we have*

$$\left\| \sqrt{\sum u_{ij,\lambda}^2} \right\|_{L_p(\mathbb{R}_T^3)} \leq (1 + \kappa^2/2)^{1/p} \|u_{11,\lambda} + u_{22,\lambda}\|_{L_p(\mathbb{R}_T^3)}.$$

Lemma 1.11. *Suppose that $d = 2$, a_{ij} satisfy the ellipticity condition, $a_{11} + a_{22} = 1$. Then for any symmetric matrix (σ_{ij}) and $p \geq 1$, we have*

$$|\mathrm{tr}\,\sigma|^p \leq (1 - \kappa^2/2)(\sum \sigma_{ij}^2)^{p/2} + (2/\kappa)^p |\mathrm{tr}\,(a\sigma)|^p.$$

Proof. Noting the qualities in the equality above are invariant under orthogonal transformation of matrices. Therefore, without loss of generality we may assume that $(\sigma_{ij}) = \mathrm{diag}[t, 1-t]$, where $t \in \mathbb{R}$. We only need to show

$$1 \leq (1 - \kappa^2/2)(t^2 + (1-t)^2)^{p/2} + (2/\kappa)^p |a_{11}t + a_{22}(1-t)|^p := \mathrm{I} + \mathrm{II}.$$

Of course, by symmetry we may also assume that $t \geq 1/2$. We consider three cases: (a) $t \in [1/2, 1]$; (b) $t \in [1, 2/(2-\kappa^2)]$; (c) $t \in [2/(2-\kappa^2), \infty]$.

For case (a), since $a_{11}, a_{22} \geq \kappa$, we have $\mathrm{II} \geq 2^p \geq 1$.

For case (c), we have $\mathrm{I} \geq 1$ because

$$(t^2 + (1-t)^2)^{p/2} > t^p \geq [2/(2-\kappa^2)]^p.$$

Finally, for case (b), we want to show that $\mathrm{II} \geq 1$. This is done by solving a minimization problem:

$$A := \inf\left\{ |a_{11}t + a_{22}(1-t)| \;:\; t \in [1, 2/(2-\kappa^2)], a_{11}, a_{22} \in [\kappa, 1/\kappa] \right\}.$$

We define

$$\phi(t) = \inf\left\{ |a_{11}t + a_{22}(1-t)| \;:\; a_{11}, a_{22} \in [\kappa, 1/\kappa] \right\}.$$

It is easily seen that for $t \in [1, 2/(2-\kappa^2)]$, $\phi(t) = \kappa t + (1-t)/\kappa$ and it is a decreasing function. Thus

$$A \geq \phi(2/(2-\kappa^2)) \geq \kappa/2,$$

and $\mathrm{II} \geq 1$. The lemma is proved. □

Now we are ready to give

Proof of Theorem 1.7 (i). By Corollary 1.10, we have

$$\left\| \sqrt{\sum u_{ij,\lambda}^2} \right\|_{L_p}^p \leq (1 + \kappa^2/2) \|u_{11,\lambda} + u_{22,\lambda}\|_{L_p}^p.$$

It follows from Lemma 1.11 with $\sigma_{ij} = u_{ij,\lambda}$ that

$$\begin{aligned}&\|u_{11,\lambda} + u_{22,\lambda}\|_{L_p}^p\\ &\le (1-\kappa^2/2)\left\|\sqrt{\sum u_{ij,\lambda}^2}\right\|_{L_p}^p + (2/\kappa)^p\|a_{ij}u_{ij} - u_t - \lambda u\|_{L_p}^p.\end{aligned}$$

Hence, we get

$$\begin{aligned}&\left\|\sqrt{\sum u_{ij,\lambda}^2}\right\|_{L_p}^p\\ &\le (1-\kappa^4/4)\left\|\sqrt{\sum u_{ij,\lambda}^2}\right\|_{L_p}^p + (1+\kappa^2/2)(2/\kappa)^p\|a_{ij}u_{ij} - u_t - \lambda u\|_{L_p}^p,\end{aligned}$$

which implies

$$\left\|\sqrt{\sum u_{ij,\lambda}^2}\right\|_{L_p} \le N\|f\|_{L_p}.$$

In particular, we obtain

$$\|u_{11} - u_t - \lambda u\|_{L_p} \le N\|f\|_{L_p}$$

and

$$\|u_{22} - u_t - \lambda u\|_{L_p} \le N\|f\|_{L_p}.$$

Therefore,

$$\|\Delta u - 2u_t - 2\lambda u\|_{L_p} \le N\|f\|_{L_p},$$

and the theorem follows by using the classical L_p-estimates for parabolic equations with constant coefficients (the heat equation); see the next section. □

Remark 1.12. (1) We can also treat parabolic equations with lower-order terms. Suppose that $u \in W_2^{1,2}$ satisfies the following equation:

$$u_t - a_{ij}D_{ij}u - (b_iu_i + cu) + \lambda u = f.$$

We move the lower-order terms to the right and rewrite the equation as

$$u_t - a_{ij}D_{ij}u + \lambda u = f + (b_iu_i + cu).$$

By the estimate for equations considered above, we have

$$\begin{aligned}&\|u_t\|_{L_2} + \|D^2u\|_{L_2} + \sqrt{\lambda}\|Du\|_{L_2} + \lambda\|u\|_{L_2}\\ &\le N(\|f\|_{L_2} + \|Du\|_{L_2} + \|u\|_{L_2}).\end{aligned}$$

Upon taking λ sufficiently large, we reach

$$\|u_t\|_{L_2} + \|D^2u\|_{L_2} + \sqrt{\lambda}\|Du\|_{L_2} + \lambda\|u\|_{L_2} \le N\|f\|_{L_2}.$$

(2) For the existence and uniqueness theorem, we can apply the a priori estimates and the method of continuity. Indeed, by using the Hahn–Banach theorem

and the Riesz representation theorem, it is easy to show that $(\partial_t - \Delta + \lambda)C_0^\infty$ is dense in L_p for any $p \in (1, \infty)$ and $\lambda \geq 0$. This together with the available a priori estimates gives the unique $W_p^{1,2}$ solvability for the equation

$$\partial_t u - \Delta u + \lambda u = f$$

for any $\lambda > 0$ and $f \in L_p$. For the general operators, we can use the method of continuity by considering $\partial_t u - ((1-\mu)\Delta + \mu a_{ij} D_{ij})u + \lambda u = f$.

(3) To deal with the Cauchy problem

$$\partial_t u - a_{ij} D_{ij} u + \lambda u = f \quad \text{in} \quad \mathbb{R}^d \times (0, T)$$

with the initial condition $u(0, x) = 0$, one can extend $f = 0$ for $t < 0$ and solve the equation in $\mathbb{R}_T^{d+1}$. By the method of continuity and the property of the heat equation, we have $u = 0$ for $t \leq 0$. Therefore, the initial condition is satisfied automatically. It is also worth noting that if T is finite, we can get rid of λ by considering the equation of $v = ue^{-\lambda t}$. Indeed, if u satisfies

$$\partial_t u - a_{ij} D_{ij} u = f \quad \text{in} \quad \mathbb{R}^d \times (0, T),$$

then v satisfies

$$\partial_t v - a_{ij} D_{ij} v + \lambda v = f e^{-\lambda t} \quad \text{in} \quad \mathbb{R}^d \times (0, T).$$

We finish this section by extending Theorems 1.7 and 1.4. We denote $x = (y, z)$, where $y = (x_1, x_2)$ and $z = (x_3, ..., x_d)$.

Theorem 1.13. *Let $\lambda \geq 0$ be a constant, and $a_{ij}(t, x) = a_{ij}(t, y)$ are bounded measurable functions depending on t and y only. Assume $u \in W_2^{1,2}(\mathbb{R}_T^{d+1})$ and $f \in L_2(\mathbb{R}_T^{d+1})$, and u satisfies*

$$u_t - a_{ij} D_{ij} u + \lambda u = f \quad in \quad \mathbb{R}_T^{d+1}.$$

Then the following estimate holds

$$\|u_t\|_{L_2(\mathbb{R}_T^{d+1})} + \|D^2 u\|_{L_2(\mathbb{R}_T^{d+1})} + \sqrt{\lambda}\|Du\|_{L_2(\mathbb{R}_T^{d+1})} + \lambda\|u\|_{L_2(\mathbb{R}_T^{d+1})}$$
$$\leq N(d, \kappa)\|f\|_{L_2(R_T^{d+1})}.$$

Proof. We give a sketched proof and leave the details to the reader. Take the Fourier transform of u in the z-variable and denote it by $v(t, y, \xi)$. Following the proof of Theorem 1.4, using the pointwise estimate of $|v|$ achieved by the maximum principle, we have the estimate below for the derivatives in z:

$$\|u_{zz}\|_{L_2} + \sqrt{\lambda}\|u_z\|_{L_2} + \lambda\|u\|_{L_2} \leq N\|f\|_{L_2}. \tag{1.14}$$

Then we rewrite the equation as following:

$$\partial_t u - \sum_{i,j=1}^{2} a_{ij} D_{ij} u + \lambda u = f + \sum_{i \, or \, j > 2} D_{ij} u.$$

Applying Theorem 1.7, we have

$$\begin{aligned}&\|\partial_t u\|_{L_2} + \|u_{yy}\|_{L_2} + \sqrt{\lambda}\|u_y\|_{L_2} + \lambda\|u\|_{L_2}\\ &\le N\|f\|_{L_2} + N\|u_{yz}\|_{L_2} + N\|u_{zz}\|_{L_2}.\end{aligned} \tag{1.15}$$

Similar to (1.8), we have

$$\|u_{yz}\|_{L_2} \le \epsilon\|u_{yy}\|_{L_2} + N(\epsilon)\|u_{zz}\|_{L_2} \tag{1.16}$$

for any $\epsilon \in (0,1)$. Combining (1.15), (1.16), and (1.14), we reach the desired estimate. □

Historical Remarks. Lemma 1.3 and Theorem 1.4 were first established in [16], in which the proof of Lemma 1.5 is slightly different. Here we follow an argument in [7]. Theorem 1.7 is originally due to Krylov [17], and Corollary 1.8 was proved earlier by Campanato [5]. In the case when $d = 2$ and $p = 2$, the result is dated back to Bernstein [1]. For the case when $d \ge 3$ and $p = 2$, there is a counterexample due to Talenti [24]. The proof of Theorem 1.13 can be found in [7].

2 Interior L_p estimates

2.1 Equations with coefficients depending only on t

Consider the following linear parabolic equation:

$$\partial_t u - a_{ij}(t)D_{ij}u + \lambda u = f \tag{2.1}$$

in $(-\infty, T) \times \mathbb{R}^d$. In this section, we shall prove the L_p estimates for this equation, where $p \in (1, +\infty)$. We use some tools from harmonic analysis such as the maximal function theorem and the Fefferman–Stein theorem on sharp functions. As in the previous section, the assumptions imposed on the coefficients are

(a) $a_{ij} = a_{ji}$, where a_{ij} are measurable functions of t only.

(b) $\kappa|\xi|^2 \le a_{ij}(t)\xi_i\xi_j \le |\xi|^2/\kappa$ for some $\kappa \in (0,1)$.

The following example shows that the L_∞ estimate for second derivatives is in general not possible.

Example. Let

$$u(x_1, x_2) = x_1 x_2(-\log|x|)^{1/3}\phi(x),$$

where $\phi(x)$ is smooth a cutoff function, which takes the value 1 in $B_{1/3}$ and 0 outside $B_{2/3}$. Then it is easily seen that u and Δu are bounded and continuous. However, $D_{12}u$ is unbounded near the origin.

On the other hand, it is known that if $f \in L_\infty$ (or BMO), then D^2u is in BMO.

Theorem 2.1. (i) *Let $p \in (1,\infty)$, $\lambda \ge 0$, and $f \in L_p(\mathbb{R}^{d+1}_T)$. Assume that $u \in W^{1,2}_p(\mathbb{R}^{d+1}_T)$ satisfies* (2.1). *Then we have*

$$\begin{aligned}&\|u_t\|_{L_p(\mathbb{R}^{d+1}_T)} + \|D^2u\|_{L_p(\mathbb{R}^{d+1}_T)} + \sqrt{\lambda}\|Du\|_{L_p(\mathbb{R}^{d+1}_T)} + \lambda\|u\|_{L_p(\mathbb{R}^{d+1}_T)}\\ &\le N(d,\kappa,p)\|f\|_{L_p(\mathbb{R}^{d+1}_T)}.\end{aligned}$$

(ii) *For any* $f \in L_p(\mathbb{R}^{d+1}_T)$ *and* $\lambda > 0$, *there is a unique solution* $u \in W^{1,2}_p(\mathbb{R}^{d+1}_T)$ *to the equation* (2.1).

We note that the solution u in Theorem 2.1 is called a strong solution. It satisfies the equation (2.1) almost everywhere.

Let us introduce some notation and function spaces. Denote

$$Q_r = (-r^2, 0) \times B_r(0)$$

to be the standard parabolic cylinder. For any function u defined in $\Omega \subset \mathbb{R}^{d+1}$, we denote its Hölder semi-norm by

$$[u]_{C^{\alpha/2,\alpha}(\Omega)} = \sup\left\{\frac{|u(t,x)-u(s,y)|}{|t-s|^{\alpha/2}+|x-y|^{\alpha}} : \forall\,(t,x) \neq (s,y) \in \Omega\right\}.$$

Lemma 2.2. *Suppose that* $u \in C^\infty_b(Q_1)$ *and satisfies*

$$u_t - a_{ij}(t)D_{ij}u + \lambda u = 0 \quad in \quad Q_1.$$

Then we have

$$[u]_{C^{1/2,1}(Q_{1/2})} \le N(d,\kappa)\|u\|_{L_2(Q_1)}. \tag{2.2}$$

Proof. For any $1/2 \le r < R \le 1$, choose a smooth cutoff function η satisfying $0 \le \eta \le 1$, $\eta = 1$ in Q_r, and $\eta = 0$ outside $(-R^2, R^2) \times B_R$. Multiplying the equation by $u\eta^2$ and integrating by parts, we get for any $s \in (-R^2, 0)$,

$$\begin{aligned}
&\frac{1}{2}\int_{B_R} u^2(s,x)\eta^2\,dx \\
&+\int_{(-R^2,s)\times B_R} \left(a_{ij}D_iuD_ju\eta^2 + 2a_{ij}D_iu\eta\cdot D_j\eta u + \lambda u^2\eta^2\right)dx\,dt \\
\le &\int_{(-R^2,s)\times B_R} u^2\eta\eta_t\,dx\,dt.
\end{aligned}$$

Using the ellipticity condition and Young's inequality, we have

$$\begin{aligned}
&\frac{1}{2}\int_{B_R} u^2(s,x)\eta^2\,dx + \int_{(-R^2,s)\times B_R}\left(\kappa|Du|^2\eta^2 + \lambda u^2\eta^2\right)dx\,dt \\
\le &\int_{(-R^2,s)\times B_R} (u^2\eta\eta_t - 2a_{ij}D_iu\eta\cdot D_j\eta u)\,dx\,dt \\
\le &\int_{(-R^2,s)\times B_R} (Nu^2 + \frac{\kappa}{2}|Du|^2\eta^2 + N|D\eta|^2u^2)\,dx\,dt.
\end{aligned}$$

Taking the supremum with respect to $s \in (-r^2, 0)$, we obtain the following Caccioppoli's inequality

$$\sup_{s\in(-r^2,0)}\int_{B_r} u^2\,dx + \int_{Q_r}(|Du|^2 + \lambda u^2)\,dx\,dt \le N(d,\kappa,r,R)\int_{Q_R} u^2\,dx\,dt.$$

Because Du satisfies the same equation, we have

$$\sup_{s\in(-r^2,0)} \int_{B_r} |Du|^2\,dx + \int_{Q_r} (|D^2u|^2 + \lambda|Du|^2)\,dx\,dt$$
$$\le N(d,\kappa,r,R) \int_{Q_R} |Du|^2\,dx\,dt.$$

By iteration,

$$\sup_{s\in(-r^2,0)} \int_{B_r} |D^k u|^2\,dx + \int_{Q_r} (|D^{k+1}u|^2 + \lambda|D^k u|^2)\,dx\,dt$$
$$\le N(d,\kappa,r,R,k) \int_{Q_R} |u|^2\,dx\,dt,$$

which further implies that

$$(1+\lambda^2) \sup_{s\in(-r^2,0)} \int_{B_r} |D^k u|^2\,dx \le N(d,\kappa,r,R,k) \int_{Q_R} |u|^2\,dx\,dt.$$

Now we take $k > d/2$. By the Sobolev embedding theorem, we get

$$(1+\lambda) \sup_{Q_{1/2}} |u| \le N \left(\int_{Q_R} |u|^2\,dx\,dt \right)^{1/2}.$$

Again, because $D^k u$ satisfies the same equation, it follows from the two inequalities above that, for any nonnegative integer k,

$$(1+\lambda) \sup_{Q_{1/2}} |D^k u| \le N \left(\int_{Q_1} |u|^2\,dx\,dt \right)^{1/2}.$$

Since $u_t = a_{ij}(t) D_{ij} u - \lambda u$, we also have

$$\sup_{Q_{1/2}} |\partial_t D^k u| \le N(d,\kappa,k) \left(\int_{Q_1} |u|^2\,dx\,dt \right)^{1/2}.$$

The lemma is proved. □

Remark 2.3. From the proof above, actually we have

$$\sup_{Q_{1/2}} (1+\lambda^l)(|\partial_t D^k u| + |D^k u|) \le N(r,\kappa,k,l) \left(\int_{Q_1} |u|^2\,dx\,dt \right)^{1/2}$$

for any nonnegative integer l and multi-index k. Heuristically, $\sqrt{\lambda}$ is corresponding to one differentiation in x.

In the sequel, we denote $(f)_{\mathcal{D}}$ to be the average of f in $\mathcal{D}$:

$$(f)_{\mathcal{D}} = \frac{1}{|\mathcal{D}|} \int_{\mathcal{D}} f(t,x)\,dx\,dt = \fint_{\mathcal{D}} f(t,x)\,dx\,dt,$$

where $|\mathcal{D}|$ is the $d+1$-dimensional Lebesgue measure of $\mathcal{D}$.

Corollary 2.4. *Let $r \in (0,\infty)$, $\mu \in (0,1/2)$, and $\lambda \ge 0$. Suppose that $u \in C_b^\infty(Q_r)$ satisfies*

$$u_t - a_{ij}(t)D_{ij}u + \lambda u = 0$$

in Q_r. Then for any $\mu \in (0,1/2)$ we have

$$\fint_{Q_{\mu r}} |D^2u - (D^2u)_{Q_{\mu r}}|\,dx\,dt \le N\mu \left(\fint_{Q_r} |D^2u|^2\,dx\,dt\right)^{1/2},$$

$$\fint_{Q_{\mu r}} |Du - (Du)_{Q_{\mu r}}|\,dx\,dt \le N\mu \left(\fint_{Q_r} |Du|^2\,dx\,dt\right)^{1/2},$$

$$\fint_{Q_{\mu r}} |u - (u)_{Q_{\mu r}}|\,dx\,dt \le N\mu \left(\fint_{Q_r} |u|^2\,dx\,dt\right)^{1/2}.$$

Proof. By scaling, we may assume that $r = 1$. Then by Lemma 2.2,

$$\fint_{Q_\mu} |u - (u)_{Q_{\mu r}}|\,dx\,dt \le N\mu[u]_{C^{1/2,1}(Q_{1/2})} \le N\mu \left(\fint_{Q_1} |u|^2\,dx\,dt\right)^{1/2}.$$

The other two inequalities can be proved in the same way because Du and D^2u satisfy the same equation as u. □

In the sequel, we define a vector-valued function $U = (D^2u, \sqrt{\lambda}Du, \lambda u)$. Similarly, for functions w and v, we define

$$W = (D^2w, \sqrt{\lambda}Dw, \lambda w), \quad V = (D^2v, \sqrt{\lambda}Dv, \lambda v).$$

Next we prove a mean oscillation estimate for non-divergence form equations.

Lemma 2.5. *Let $r \in (0,\infty)$, $\mu \in (0,1/4)$, $\lambda \ge 0$, $X_0 = (t_0,x_0) \in \mathbb{R}^{d+1}$, and $f \in L_{2,loc}(\mathbb{R}^{d+1})$. Assume that $u \in C_{loc}^\infty$ satisfies (2.1) in $Q_r(X_0)$. Then we have*

$$\left(|U - (U)_{Q_{\mu r}(X_0)}|\right)_{Q_{\mu r}(X_0)} \le N\mu(|U|^2)^{\frac{1}{2}}_{Q_r(X_0)} + N\mu^{-1-\frac{d}{2}}(|f|^2)^{\frac{1}{2}}_{Q_r(X_0)}, \tag{2.3}$$

where $N = N(d,\kappa) > 0$.

Proof. By a translation of the coordinates, we may assume that $X_0 = (0,0)$. We can certainly assume that f and a_{ij} are infinitely differentiable. If not, we take the standard mollifications and prove the estimate for the mollifications. Then we take the limit because the concerned constants are independent of the smoothness of the functions involved.

Take an infinitely differentiable function ζ defined on $\mathbb{R}^{d+1}$ such that

$$\zeta = 1 \quad \text{on} \quad Q_{r/2}, \quad \zeta = 0 \quad \text{outside} \quad (-r^2, r^2) \times B_r.$$

By Lemma 1.1 and Remark 1.12 (3), for any $\lambda \ge 0$, there exists a unique solution $w \in W_2^{1,2}((-r^2,0) \times \mathbb{R}^d)$ to the equation

$$w_t - a_{ij}(t)D_{ij}w + \lambda w = \zeta f$$

with the initial condition $w(-r^2,0)=0$. Since all functions and coefficients involved are infinitely differentiable, by the classical parabolic theory, w is infinitely differentiable. Let $v:=u-w$. Then the function v is also infinitely differentiable and satisfies

$$v_t - a_{ij}(t)D_{ij}v + \lambda v = 0 \quad \text{in} \quad Q_{r/2}.$$

By Corollary 2.4 applied to v, we have

$$\big(|V-(V)_{Q_{\mu r}}|\big)_{Q_{\mu r}} \le N\mu(|V|^2)^{\frac{1}{2}}_{Q_{r/2}}. \tag{2.4}$$

Next we estimate w. By Lemma 1.1, we have

$$\|W\|_{L_2(\mathbb{R}^{d+1}_0)} \le N\|\zeta f_\alpha\|_{L_2(\mathbb{R}^{d+1}_0)}.$$

In particular,

$$\big(|W|^2\big)^{\frac{1}{2}}_{Q_{\mu r}} \le N\mu^{-1-\frac{d}{2}}(|f|^2)^{\frac{1}{2}}_{Q_r}, \tag{2.5}$$

$$\big(|W|^2\big)^{\frac{1}{2}}_{Q_r} \le N(|f|^2)^{\frac{1}{2}}_{Q_r}. \tag{2.6}$$

Now we are ready to prove (2.3). Since

$$(|U-(U)_{Q_{\mu r}}|)_{Q_{\mu r}} \le 2(|U-c|)_{Q_{\mu r}}$$

for any constant vector c, by taking $c=(V)_{Q_{\mu r}}$ we bound the left-hand side of (2.3) by a constant times

$$(|U-(V)_{Q_{\mu r}}|)_{Q_{\mu r}} \le \big(|V-(V)_{Q_{\mu r}}|\big)_{Q_{\mu r}} + \big(|W|^2\big)^{\frac{1}{2}}_{Q_{\mu r}}.$$

Using (2.4) and (2.5), we see that the above is less than

$$N\mu(|V|^2)^{\frac{1}{2}}_{Q_{r/2}} + N\mu^{-1-\frac{d}{2}}(|f|^2)^{\frac{1}{2}}_{Q_r}.$$

Finally, we use the fact that $v=u-w$ and (2.6) to prove that the terms above are not greater than the right-hand side of (2.3). □

We recall the maximal function theorem and the Fefferman–Stein theorem. Let

$$\mathcal{Q}=\{Q_r(z): z=(t,x)\in\mathbb{R}^{d+1}, r\in(0,\infty)\}.$$

For a function g defined in $\mathbb{R}^{d+1}$, the (parabolic) maximal and sharp function of $g\in L_{1,\text{loc}}$ are given by

$$\mathcal{M}g(t,x) = \sup_{Q\in\mathcal{Q},(t,x)\in Q} \fint_Q |g(s,y)|\,dy\,ds,$$

$$g^{\#}(t,x) = \sup_{Q\in\mathcal{Q},(t,x)\in Q} \fint_Q |g(s,y)-(g)_Q|\,dy\,ds.$$

Then for any $p \in (1, \infty)$ and $g \in L_p(\mathbb{R}^{d+1})$, we have

$$\|g\|_{L_p(\mathbb{R}^{d+1})} \le N \|g^{\#}\|_{L_p(\mathbb{R}^{d+1})}, \quad \|\mathcal{M} g\|_{L_p(\mathbb{R}^{d+1})} \le N \|g\|_{L_p(\mathbb{R}^{d+1})},$$

where $N = N(d, p)$. Indeed, the first of the inequalities above is due to the Fefferman–Stein theorem on sharp functions and the second one to the Hardy–Littlewood maximal function theorem; see, for instance, [20, Chapter 3]. Note that by the triangle inequality, we have

$$g^{\#}(t, x) \le 2 \mathcal{M} g(t, x), \quad \forall \ (t, x) \in \mathbb{R}^{d+1},$$

which implies that

$$\|g^{\#}\|_{L_p(\mathbb{R}^{d+1})} \le 2 \|\mathcal{M} g\|_{L_p(\mathbb{R}^{d+1})}.$$

We are now ready to complete the proof of Theorem 2.1.

Proof of Theorem 2.1. First we consider the case when $T = \infty$ and $p \in (2, \infty)$. Since C_0^∞ is dense in $W_p^{1,2}$, by mollifications we may assume that $u \in C_0^\infty$ and a_{ij} are infinitely differentiable. Then it follows from Lemma 2.5 that for any $X_0 \in \mathbb{R}^{d+1}$,

$$U^{\#}(X_0) \le N \mu (\mathcal{M} U^2)^{1/2}(X_0) + N \mu^{-1-d/2} (\mathcal{M} f^2)^{1/2}(X_0).$$

By using the Fefferman–Stein theorem on sharp functions and the Hardy–Littlewood maximal function theorem,

$$\begin{aligned}
\|U\|_{L_p} \le \|U^{\#}\|_{L_p} &\le N\mu \|(\mathcal{M} U^2)^{1/2}\|_{L_p} + N \mu^{-1-d/2} \|(\mathcal{M} f^2)^{1/2}\|_{L_p} \\
&= N\mu \|\mathcal{M} U^2\|_{L_{p/2}}^{1/2} + N \mu^{-1-d/2} \|\mathcal{M} f^2\|_{L_{p/2}}^{1/2} \\
&\le N\mu \|U^2\|_{L_{p/2}}^{1/2} + N \mu^{-1-d/2} \|f^2\|_{L_{p/2}}^{1/2} \\
&= N\mu \|U\|_{L_p} + N \mu^{-1-d/2} \|f\|_{L_p}.
\end{aligned}$$

By choosing a sufficiently small μ such that $N\mu \le 1/2$ and using

$$u_t = a_{ij} D_{ij} u - \lambda u + f,$$

we obtain the desired estimate in (i). The a priori estimate together with the method of continuity gives the unique solvability result in assertion (ii). For the case when $p \in (1, 2)$, we use a duality argument. Let $q = p/(p-1) \in (2, \infty)$. For any $g \in L_q$, let $v \in W_q^{1,2}$ be the solution of

$$-v_t - a_{ij} D_{ij} v + \lambda v = g \quad \text{in } \mathbb{R}^{d+1}.$$

The existence of v follows from the proof above and a change of variable $t \to -t$. As before, we may assume that $u \in C_0^\infty$. Then integration by parts yields

$$\begin{aligned}
\int_{\mathbb{R}^{d+1}} D^2 u g \, dx \, dt &= \int_{\mathbb{R}^{d+1}} D^2 u (-v_t - a_{ij} D_{ij} v + \lambda v) \, dx \, dt \\
&= \int_{\mathbb{R}^{d+1}} v (\partial_t - a_{ij} D_{ij} + \lambda) D^2 u \, dx \, dt \\
&= \int_{\mathbb{R}^{d+1}} D^2 v (\partial_t - a_{ij} D_{ij} + \lambda) u \, dx \, dt.
\end{aligned}$$

It then follows from the $W^{1,2}_q$ estimate proved above that

$$\left|\int_{\mathbb{R}^{d+1}} D^2 u g\,dx\,dt\right| \le \|D^2 v\|_{L_q}\|f\|_{L_p} \le N\|g\|_{L_q}\|f\|_{L_p}.$$

Since $g \in L_q$ is an arbitrary function, we deduce

$$\|D^2 u\|_{L_p} \le N\|f\|_{L_p}.$$

Similarly, we have

$$\sqrt{\lambda}\|Du\|_{L_p} + \lambda\|u\|_{L_p} \le N\|f\|_{L_p}.$$

The unique $W^{1,2}_p$ solvability is proved as before. Finally, we treat the case $T < \infty$. Certainly we may assume that $T = 0$ by shifting the t-coordinate. For $t > 0$, define $u(t,x) = u(-t,x)$. Then $u \in W^{1,2}_p$. For $\lambda > 0$, let $v \in W^{1,2}_p$ be the solution of

$$v_t - a_{ij}D_{ij}v + \lambda v = f 1_{t<0} \tag{2.7}$$

in $\mathbb{R}^{d+1}$. Then $w := u - v \in W^{1,2}_p$ satisfies

$$w_t - a_{ij}D_{ij}w + \lambda w = 0$$

in $\mathbb{R}^{d+1}_0$. By the property of the heat equation and the method of continuity, we conclude that $w = 0$ for $t < 0$. Thus

$$\begin{aligned}&\|u_t\|_{L_p(\mathbb{R}^{d+1}_0)} + \|D^2 u\|_{L_p(\mathbb{R}^{d+1}_0)} + \sqrt{\lambda}\|Du\|_{L_p(\mathbb{R}^{d+1}_0)} + \lambda\|u\|_{L_p(\mathbb{R}^{d+1}_0)}\\ &\le \|v_t\|_{L_p} + \|D^2 v\|_{L_p} + \sqrt{\lambda}\|Dv\|_{L_p} + \lambda\|v\|_{L_p} \le \|f\|_{L_p(\mathbb{R}^{d+1}_0)}.\end{aligned}$$

Sending $\lambda \to 0$, we obtain the estimate when $\lambda = 0$. Finally, the solvability is evident by the argument above in view of (2.7). The theorem is proved. □

To conclude this subsection, we describe how the argument above can be adapted to the divergence form equation

$$u_t - D_i(a_{ij}(t)D_j u) + \lambda u = f + \operatorname{div} g, \tag{2.8}$$

where $g = (g_1, \ldots, g_d)$. This equation is understood in the weak sense. For $p \in (1,\infty)$, we introduce the solution space

$$\mathcal{H}^1_p((S,T)\times\mathbb{R}^d) = (1-\Delta)^{1/2} W^{1,2}_p((S,T)\times\mathbb{R}^d)$$

and the Bessel potential space

$$\mathbb{H}^{-1}_p((S,T)\times\mathbb{R}^d) = (1-\Delta)^{1/2} L_p((S,T)\times\mathbb{R}^d).$$

It is well known that a function $u \in \mathcal{H}^1_p((S,T)\times\mathbb{R}^d)$ if and only if $u, Du \in L_p((S,T)\times\mathbb{R}^d)$ and $u_t \in \mathbb{H}^{-1}_p((S,T)\times\mathbb{R}^d)$. For a domain $\Omega \subset \mathbb{R}^d$, we define $\mathbb{H}^{-1}_p((S,T)\times\Omega)$ to be the set of functions f such that $f = g + \operatorname{div} h$ (in the weak sense) for some $g, h \in L_p((S,T)\times\Omega)$, and

$$\|f\|_{\mathbb{H}^{-1}_p((S,T)\times\Omega)} = \inf\left\{\|g\|_{L_p((S,T)\times\Omega)} + \|h\|_{L_p((S,T)\times\Omega)} : f = g + \operatorname{div} h\right\}.$$

It is easily seen that when $\Omega = \mathbb{R}^d$, these two definitions are equivalent.

Analogous to Lemma 1.1, we have

Lemma 2.6. *Let $T \in (-\infty, +\infty]$ and $\lambda \geq 0$. Suppose that $f, g \in L_2(\mathbb{R}_T^{d+1})$ and $u \in \mathcal{H}_2^1(\mathbb{R}_T^{d+1})$ satisfies* (2.8). *We assume that $f = 0$ when $\lambda = 0$. Then we have*

$$\|Du\|_{L_2(\mathbb{R}_T^{d+1})} + \sqrt{\lambda}\|u\|_{L_2(\mathbb{R}_T^{d+1})} \leq N\|g\|_{L_2(\mathbb{R}_T^{d+1})} + \frac{N}{\sqrt{\lambda}}\|f\|_{L_2(\mathbb{R}_T^{d+1})}.$$

By using this lemma and Corollary 2.4, we obtain the following L_p estimate for divergence form equations.

Theorem 2.7. (i) *Let $p \in (1, \infty)$, $\lambda \geq 0$, and $f, g \in L_p(\mathbb{R}_T^{d+1})$. Assume that $u \in \mathcal{H}_p^1(\mathbb{R}_T^{d+1})$ satisfies* (2.8). *We assume that $f = 0$ when $\lambda = 0$. Then we have*

$$\|Du\|_{L_p(\mathbb{R}_T^{d+1})} + \sqrt{\lambda}\|u\|_{L_p(\mathbb{R}_T^{d+1})} \leq N\|g\|_{L_p(\mathbb{R}_T^{d+1})} + \frac{N}{\sqrt{\lambda}}\|f\|_{L_p(\mathbb{R}_T^{d+1})},$$

where $N = N(d, \kappa, p) > 0$ is a constant.

(ii) *For any $f, g \in L_p(\mathbb{R}_T^{d+1})$ and $\lambda > 0$, there is a unique solution $u \in \mathcal{H}_p^1(\mathbb{R}_T^{d+1})$ to the equation* (2.8).

We leave the details of the proof as an exercise.

2.2 Equations with VMO$_x$ coefficients

In this subsection, we consider more general non-divergence form equation

$$u_t - (a_{ij}D_{ij}u + b_iD_iu + cu) + \lambda u = f. \tag{2.9}$$

We recall that a function has bounded mean oscillation (BMO) if $f^\# \in L_\infty$. We define the modulus of continuity of f by

$$\omega(R) = \sup_{r\in(0,R)} \sup_{(t,x)\in\mathbb{R}^{d+1}} \fint_{Q_r(t,x)} |f - (f)_{Q_r(t,x)}|\, dy\, ds.$$

We say that $f \in$ VMO if $\omega(R) \to 0$ as $R \to \infty$. In the sequel, we only require the coefficients to be VMO with respect to x and merely measurable with respect to t, which are also called VMO$_x$ coefficients. More precisely, let

$$\operatorname{osc}_x(a_{ij}, Q_r(t,x)) = \fint_{t-r^2}^{t} \fint_{B_r(x)} \left| a_{ij}(s,y) - \fint_{B_r(x)} a_{ij}(s,z)\, dz \right| dy\, ds.$$

Then we set

$$a_R^\# = \sup_{(t,x)\in\mathbb{R}^{d+1}} \sup_{r\leq R} \sup_{i,j} \operatorname{osc}_x(a_{ij}, Q_r(t,x)).$$

We impose on the leading coefficients the small mean oscillation condition with a parameter $\delta > 0$, which will be specified later.

Assumption 2.8 (δ). There is a constant $R_0 \in (0, 1]$ such that $a_{R_0}^\# \leq \delta$.

Clearly, if $a_{ij} \in \mathrm{VMO}$, then Assumption 2.8 is satisfied for any $\delta > 0$. The lower-order coefficients are assumed to be measurable and bounded

$$|b_i| \le 1/\kappa, \quad |c| \le 1/\kappa.$$

Theorem 2.9. *Let $p \in (1, \infty)$. Then there exists $\delta = \delta(d, \kappa, p) > 0$ such that under Assumption 2.8 (δ) the following are true.*

(i) *Suppose $f \in L_p(\mathbb{R}^{d+1}_T)$, and $u \in W^{1,2}_p(\mathbb{R}^{d+1}_T)$ satisfies (2.9) in $\mathbb{R}^{d+1}_T$. Then we have*

$$\begin{aligned} &\|u_t\|_{L_p(\mathbb{R}^{d+1}_T)} + \|D^2 u\|_{L_p(\mathbb{R}^{d+1}_T)} + \sqrt{\lambda}\|Du\|_{L_p(\mathbb{R}^{d+1}_T)} + \lambda\|u\|_{L_p(\mathbb{R}^{d+1}_T)} \\ &\le N(d, \kappa, p)\|f\|_{L_p(\mathbb{R}^{d+1}_T)} \end{aligned} \tag{2.10}$$

provided that $\lambda \ge \lambda_0$, where $\lambda_0 = \lambda_0(d, \kappa, R_0, p) \ge 0$.

(ii) *For any $f \in L_p(\mathbb{R}^{d+1}_T)$ and $\lambda > \lambda_0$, there is a unique solution $u \in W^{1,2}_p(\mathbb{R}^{d+1}_T)$ to the equation (2.9) in $\mathbb{R}^{d+1}_T$.*

We need to prove a few lemmas before the proof of Theorem 2.9. The lemma below shows that we can replace the L_2 norm on the right-hand side of (2.2) by any L_p norm.

Lemma 2.10. *Let $p \in [1, \infty]$. Suppose that $u \in C^\infty_b(Q_1)$ and satisfies*

$$u_t - a_{ij}(t) D_{ij} u + \lambda u = 0 \quad in \quad Q_1.$$

Then we have

$$[u]_{C^{1/2,1}(Q_{1/2})} \le N(d, \kappa)\|u\|_{L_p(Q_1)}. \tag{2.11}$$

Proof. By Hölder's inequality, we only need to show (2.11) for $p = 1$. From Lemma 2.2,

$$[u]_{C^{1/2,1}(Q_{1/2})} \le N\|u\|_{L_\infty(Q_{3/4})}.$$

Therefore, it remains to prove

$$\|u\|_{L_\infty(Q_{3/4})} \le \|u\|_{L_1(Q_1)}. \tag{2.12}$$

We use an iteration argument. First we apply Lemma 2.2 and use a scaling argument (Exercise) to derive for any $1/2 \le r < R \le 1$,

$$\|u\|_{L_\infty(Q_r)} \le \frac{N}{(R - r)^{d/2}}\|u\|_{L_2(Q_R)}. \tag{2.13}$$

Take $r_k = 1 - 2^{-k-1}$. Then clearly, $r_1 = 3/4$ and $r_k \to 1$ as $k \to \infty$. Denote $Q^k = Q_{r_k}$. Then by (2.13) and Hölder's inequality, for $k = 0, 1, 2, \ldots$,

$$\begin{aligned} \|u\|_{L_\infty(Q^k)} &\le N2^{kd/2}\|u\|_{L_2(Q^{k+1})} \le N2^{kd/2}\|u\|^{1/2}_{L_1(Q^{k+1})}\|u\|^{1/2}_{L_\infty(Q^{k+1})} \\ &\le \epsilon\|u\|_{L_\infty(Q^{k+1})} + \frac{N2^{kd}}{\epsilon}\|u\|_{L_1(Q^{k+1})}. \end{aligned}$$

Setting $\epsilon = 3^{-d}$ yields

$$\|u\|_{L_\infty(Q^k)} \le 3^{-d}\|u\|_{L_\infty(Q_{k+1})} + N3^d 2^{kd}\|u\|_{L_1(Q_{k+1})}.$$

Multiplying both sides by 3^{-kd} and taking the summation in k, we obtain

$$\sum_{k=1}^{\infty} 3^{-kd}\|u\|_{L_\infty(Q^k)} \le \sum_{k=1}^{\infty} 3^{-(k+1)d}\|u\|_{L_\infty(Q_{k+1})} + N\sum_{k=1}^{\infty}(2/3)^{kd}\|u\|_{L_1(Q_{k+1})}.$$

Since $u \in C_b^\infty(Q_1)$, all the summations above are finite. Absorbing the first summation to the left-hand side, we reach (2.12). The lemma is proved. □

Next we show a mean oscillation estimate similar to Lemma 2.5.

Lemma 2.11. *Let $r \in (0,\infty)$, $q \in (1,\infty)$, $\mu \in (0,1/4)$, $\lambda \ge 0$, $X_0 = (t_0, x_0) \in \mathbb{R}^{d+1}$, $\beta \in (1,\infty)$, and $\beta' = \beta/(\beta-1)$. Assume that $f \in L_{q,loc}(\mathbb{R}^{d+1})$ and $u \in C_{loc}^\infty$ satisfies*

$$\partial_t u - a_{ij}D_{ij}u + \lambda u = f$$

in $Q_r(X_0)$. Then under Assumption 2.8 (δ) we have

$$\begin{aligned}&\left(|U - (U)_{Q_{\mu r}(X_0)}|\right)_{Q_{\mu r}(X_0)}\\ &\le N\mu(|U|^q)^{\frac{1}{q}}_{Q_r(X_0)} + N\mu^{-\frac{d+2}{q}}(|f|^q)^{\frac{1}{q}}_{Q_r(X_0)} + N\mu^{-\frac{d+2}{q}}\delta^{\frac{1}{\beta' q}}(|U|^{\beta q})^{\frac{1}{\beta q}}_{Q_r(X_0)},\end{aligned}$$

where $N = N(d,\kappa) > 0$.

Proof. We take a cylinder $Q = Q_r(X_0)$ when $r < R_0$ and $Q = Q_{R_0}$ when $r \ge R_0$. Let $\bar a_{ij}(t)$ to be the average of a_{ij} in Q with respect to x. Clearly, u satisfies

$$\partial_t u - a_{ij}(t)D_{ij}u + \lambda u = (\bar a_{ij} - \bar a_{ij}(t))D_{ij}u + f := \tilde f$$

in $Q_r(X_0)$. By using the proof of Lemma 2.5 with L_q norms in place of the L_2 norms, it follows from Theorem 2.1 and Lemma 2.10 that

$$\left(|U - (U)_{Q_{\mu r}(X_0)}|\right)_{Q_{\mu r}(X_0)} \le N\mu(|U|^q)^{\frac{1}{q}}_{Q_r(X_0)} + N\mu^{-\frac{d+2}{q}}(|\tilde f|^q)^{\frac{1}{q}}_{Q_r(X_0)}. \tag{2.14}$$

By the triangle inequality and Hölder's inequality,

$$\begin{aligned}(|\tilde f|^q)^{\frac{1}{q}}_{Q_r(X_0)} &\le (|f|^q)^{\frac{1}{q}}_{Q_r(X_0)} + (|(a_{ij} - \bar a_{ij}(t))D_{ij}u|^q)^{\frac{1}{q}}_{Q_r(X_0)}\\ &\le (|f|^q)^{\frac{1}{q}}_{Q_r(X_0)} + (|(a_{ij} - \bar a_{ij}(t))1_Q|^{\beta' q})^{\frac{1}{\beta' q}}_{Q_r(X_0)}(|D_{ij}u|^{\beta q})^{\frac{1}{\beta q}}_{Q_r(X_0)}\\ &\le (|f|^q)^{\frac{1}{q}}_{Q_r(X_0)} + N\delta^{\frac{1}{\beta' q}}(|D_{ij}u|^{\beta q})^{\frac{1}{\beta q}}_{Q_r(X_0)},\end{aligned} \tag{2.15}$$

where in the second inequality we used the fact that u is supported on Q_{R_0} and in the third inequality we used Assumption 2.8 (δ). Combining (2.14) and (2.15) we obtain the desired estimate. The lemma is proved. □

Now we can finish

Proof of Theorem 2.9. Arguing as in the proof of Theorem 2.1, we only need prove the a priori estimate for $u \in C_0^\infty(\mathbb{R}^{d+1})$ when $T = \infty$. We divide the proof into several steps.

First, we assume $b = c = 0$ and $u \in C_0^\infty(Q_{R_0})$. We fix $q, \beta \in (1, \infty)$ such that $q\beta < p$. It follows from Lemma 2.11 that for any $X_0 \in \mathbb{R}^{d+1}$,

$$U^\#(X_0) \le N\mu(\mathcal{M}|U|^q)^{\frac{1}{q}}(X_0) + N\mu^{-\frac{d+2}{q}}(\mathcal{M}|f|^q)^{1/q}(X_0)$$
$$+ N\mu^{-\frac{d+2}{q}}\delta^{\frac{1}{\beta' q}}(\mathcal{M}|U|^{\beta q})^{\frac{1}{\beta q}}(X_0).$$

Recall that $1 < q < q\beta < p$. By the Fefferman–Stein theorem on sharp functions and the Hardy–Littlewood maximal function theorem,

$$\begin{aligned}
&\|U\|_{L_p} \le \|U^\#\|_{L_p}\\
&\le N\mu\|(\mathcal{M}|U|^q)^{1/q}\|_{L_p} + N\mu^{-\frac{d+2}{q}}\|(\mathcal{M}|f|^q)^{1/q}\|_{L_p}\\
&\quad + N\mu^{-\frac{d+2}{q}}\delta^{\frac{1}{\beta' q}}\|(\mathcal{M}|U|^{\beta q})^{\frac{1}{\beta q}}\|_{L_p}\\
&= N\mu\|\mathcal{M}|U|^q\|_{L_{p/q}}^{1/q} + N\mu^{-\frac{d+2}{q}}\|\mathcal{M}|f|^q\|_{L_{p/q}}^{1/q}\\
&\quad + N\mu^{-\frac{d+2}{q}}\delta^{\frac{1}{\beta' q}}\|\mathcal{M}|U|^{\beta q}\|_{L_{p/(\beta q)}}^{1/(\beta q)}\\
&\le N\mu\||U|^q\|_{L_{p/q}}^{1/q} + N\mu^{-\frac{d+2}{2}}\||f|^q\|_{L_{p/q}}^{1/q} + N\mu^{-\frac{d+2}{q}}\delta^{\frac{1}{\beta' q}}\||U|^{\beta q}\|_{L_{p/(\beta q)}}^{1/(\beta q)}\\
&= N\mu\|U\|_{L_p} + N\mu^{-\frac{d+2}{2}}\|f\|_{L_p} + N\mu^{-\frac{d+2}{q}}\delta^{\frac{1}{\beta' q}}\|U\|_{L_p}.
\end{aligned}$$

By choosing a sufficiently small μ and then a sufficiently small δ depending only on d, κ, and p, such that

$$N\mu + N\mu^{-\frac{d+2}{q}}\delta^{\frac{1}{\beta' q}} \le 1/2$$

and using

$$u_t = a_{ij}D_{ij}u - \lambda u + f,$$

we obtain the desired a priori estimate. Of course, by a translation of the coordinates, the same is true for $u \in C_0^\infty(Q_{R_0}(X_0))$ for any $X_0 \in \mathbb{R}^{d+1}$.

For general $u \in C_0^\infty$, we use the idea of partition of the unity. Take a nonnegative function $\zeta \in C_0^\infty(Q_{R_0})$ such that

$$\|\zeta\|_{L_p} = 1, \quad \|D\zeta\|_{L_p} \le N/R_0, \quad \|D^2\zeta\|_{L_p} + \|\zeta_t\|_{L_p} \le N/R_0^2,$$

where $N = N(d, p)$. For any $X_0 \in \mathbb{R}^{d+1}$, we denote $u^{X_0} = \zeta(\cdot - X_0)u$. Then $u^{X_0} \in C_0^\infty(Q_{R_0}(X_0))$ satisfies

$$\partial_t u^{X_0} - a_{ij}D_{ij}u^{X_0} + \lambda u^{X_0} = f^{X_0}$$

in $\mathbb{R}^{d+1}$, where

$$f^{X_0} = \zeta(\cdot - X_0)f + \zeta_t(\cdot - X_0)u - 2a_{ij}D_iuD_j\zeta(\cdot - X_0) - a_{ij}uD_{ij}\zeta(\cdot - X_0).$$

Thanks to the result in the first step, we have

$$\int_{\mathbb{R}^{d+1}} |\partial_t u^{X_0}|^p + |D^2 u^{X_0}|^p + \lambda^{p/2}|Du^{X_0}|^p + \lambda^p |u^{X_0}|^p \, dx\, dt$$
$$\le N \int_{\mathbb{R}^{d+1}} |f^{X_0}|^p \, dx\, dt,$$

which together with the triangle inequality implies that

$$\int_{\mathbb{R}^{d+1}} \Big(|\partial_t u|^p + |D^2 u|^p + \lambda^{p/2}|Du|^p + \lambda^p|u|^p \Big) \zeta^p(\cdot - X_0)\, dx\, dt$$
$$\le N \int_{\mathbb{R}^{d+1}} |f|^p \zeta^p(\cdot - X_0) + |Du|^p |D\zeta(\cdot - X_0)|^p$$
$$+ |u|^p \big(|\zeta_t(\cdot - X_0)|^p + |D^2\zeta(\cdot - X_0)|^p + \lambda^{p/2}|D\zeta(\cdot - X_0)|^p \big)\, dx\, dt.$$

We integrate both sides with respect to $X_0 \in \mathbb{R}^{d+1}$ and use Fubini's theorem to get

$$\int_{\mathbb{R}^{d+1}} \Big(|\partial_t u|^p + |D^2 u|^p + \lambda^{p/2}|Du|^p + \lambda^p|u|^p \Big)\, dx\, dt$$
$$\le N(d,\kappa,p) \int_{\mathbb{R}^{d+1}} \big(|f|^p + R_0^{-p}|Du|^p + (\lambda^{p/2}R_0^{-p} + R_0^{-2p})|u|^p \big)\, dx\, dt.$$

To get (2.10), it remains to take λ sufficiently large depending on d, κ, p, and R_0.

Finally, for general equations we move lower-order terms to the right-hand side to get

$$u_t - a_{ij}D_{ij}u + \lambda u = f + b_i D_i u + cu.$$

From the previous step, we have

$$\|u_t\|_{L_p} + \|D^2 u\|_{L_p} + \sqrt{\lambda}\|Du\|_{L_p} + \lambda\|u\|_{L_p}$$
$$\le N(d,\kappa,p)\big(\|f\|_{L_p} + \|Du\|_{L_p} + \|u\|_{L_p} \big).$$

Taking λ sufficiently large yields the desired estimate. The theorem is proved. □

We have a similar result for the divergence form equation

$$\partial_t u - \big(D_i(a_{ij}D_j u) + D_i(b_i u) + \tilde{b}_i D_i u + cu \big) + \lambda u = f + \operatorname{div} g. \tag{2.16}$$

We assume that the lower-order coefficients are assumed to be measurable and bounded

$$|b_i| \le 1/\kappa, \quad |\tilde{b}_i| \le 1/\kappa, \quad |c| \le 1/\kappa.$$

Theorem 2.12. *Let $p \in (1,\infty)$. Then there exists $\delta = \delta(d,\kappa,p) > 0$ such that under Assumption 2.8 (δ) the following are true.*

(i) *Suppose $f, g \in L_p(\mathbb{R}^{d+1}_T)$, and $u \in \mathcal{H}^1_p(\mathbb{R}^{d+1}_T)$ satisfies (2.16) in $\mathbb{R}^{d+1}_T$. Then we have*

$$\|Du\|_{L_p(\mathbb{R}^{d+1}_T)} + \sqrt{\lambda}\|u\|_{L_p(\mathbb{R}^{d+1}_T)} \le N(d,\kappa,p)\bigg(\|g\|_{L_p(\mathbb{R}^{d+1}_T)} + \frac{1}{\sqrt{\lambda}}\|f\|_{L_p(\mathbb{R}^{d+1}_T)} \bigg)$$

provided that $\lambda \geq \lambda_0$, *where* $\lambda_0 = \lambda_0(d, \kappa, R_0, p) \geq 0$.

(ii) *For any* $f \in L_p(\mathbb{R}^{d+1}_T)$ *and* $\lambda > \lambda_0$, *there is a unique solution* $u \in \mathcal{H}^1_p(\mathbb{R}^{d+1}_T)$ *to the equation* (2.16).

The proof of Theorem 2.12 is similar to that of Theorem 2.9. We leave it to the reader as an exercise.

We finish this section by a few remarks.

Remark 2.13. By a localization argument similar to the second step in the proof of Theorem 2.9, one can obtain the following local estimate. Let $p \in (1, \infty)$ and $0 < r < R < \infty$. Then there exists $\delta = \delta(d, \kappa, p) > 0$ such that under Assumption 2.8 (δ) the following is true. Suppose $f \in L_p(Q_R)$, and $u \in W^{1,2}_p(Q_R)$ satisfies (2.9) in Q_R. Then we have

$$\|u\|_{W^{1,2}_p(Q_r)} \leq N(d, \kappa, p, R_0, r, R)\big(\|f\|_{L_p(Q_R)} + \|u\|_{L_p(Q_R)} + \|Du\|_{L_p(Q_R)}\big).$$

To get an estimate without Du on the right-hand side, we can use multiplicative inequalities and an iteration argument similar to the proof of Lemma 2.10. Note that here we do not require λ to be sufficiently large (Exercise: why?).

Historical Remarks. The study of elliptic equations with VMO coefficients was initiated in [2, 3], and continued in [4, 6]. Their proof is based on certain estimates of the Calderón-Zygmund theorem and the Coifman-Rochberg-Weiss commutator theorem. The results in this sections were first proved in [18], where the class of VMO_x coefficients was introduced. Similar results also hold in the mixed norm space $L_q(L_p)$, where

$$\|f\|_{L_q(L_p)} := \left(\int \left(\int |f|^p\,dx\right)^{q/p} dt\right)^{1/q}.$$

In the non-divergence case, we require that $q \geq p$; see [19][1]. We also refer the reader to the recent developments in [15, 16, 10, 21, 8, 9, 11, 12] on equations and higher order systems with VMO_x or partially VMO coefficients.

3 Boundary L_p estimates

In this section, we discuss boundary L_p estimates for parabolic equations with time-irregular coefficients. We focus on the Dirichlet boundary condition. Nevertheless, most of the results here can be extended to the Neumann boundary value problem. For $T \in (-\infty, +\infty]$, denote

$$\Omega = \mathbb{R}^d_+ = \{(x_1, x') \in \mathbb{R}^d : x_1 > 0\}, \quad \Omega_T = (-\infty, T) \times \Omega.$$

Let $\lambda \geq 0$. We first consider the equation

$$\begin{cases} u_t - a_{ij}(t) D_{ij} u + \lambda u = f, & \text{in } (-\infty, T) \times \Omega, \\ u(t, 0, x') = 0, & \text{on } (-\infty, T) \times \partial\Omega. \end{cases} \tag{3.1}$$

[1]This condition was recently removed in [13].

Since u is not necessarily to be continuous, the boundary condition is in the sense of trace.

In the special case when $a_{ij} = \delta_{ij}$ so that $a_{ij}D_{ij}u = \Delta u$, we can use the method of odd/even extensions to apply the results proved in the first two sections. Indeed, suppose $f \in L_p(\Omega_T)$ and $u \in W_p^{1,2}(\Omega_T)$ satisfies (3.1). Then after the odd extensions

$$u(t, -x_1, x') = -u(t, x_1, x'), \quad f(t, -x_1, x') = -f(t, x_1, x') \quad \text{for } x_1 < 0,$$

we know that $f \in L_p(\mathbb{R}_T^{d+1})$ and $u \in W_p^{1,2}(\mathbb{R}_T^{d+1})$ satisfies

$$u_t - \Delta u + \lambda u = f \quad \text{in } \mathbb{R}_T^{d+1}. \tag{3.2}$$

By Theorem 2.1, we have

$$\begin{aligned}&\|u_t\|_{L_p(\mathbb{R}_T^{d+1})} + \|D^2u\|_{L_p(\mathbb{R}_T^{d+1})} + \sqrt{\lambda}\|Du\|_{L_p(\mathbb{R}_T^{d+1})} + \lambda\|u\|_{L_p(\mathbb{R}_T^{d+1})}\\ &\leq N(d, \kappa, p)\|f\|_{L_p(\mathbb{R}_T^{d+1})}.\end{aligned}$$

We then obtain the following a priori estimate:

$$\begin{aligned}&\|u_t\|_{L_p(\Omega_T)} + \|D^2u\|_{L_p(\Omega_T)} + \sqrt{\lambda}\|Du\|_{L_p(\Omega_T)} + \lambda\|u\|_{L_p(\Omega_T)}\\ &\leq N(d, \kappa, p)\|f\|_{L_p(\Omega_T)},\end{aligned}$$

which certainly implies the uniqueness of $W_p^{1,2}$-solution to (3.1). For the solvability of (3.1), we take the odd extension of f with respect to x_1 as above, and then solve (3.2) to get a unique solution $u \in W_p^{1,2}(\mathbb{R}_T^{d+1})$. It is easily seen that $v(t, x) := -u(t, -x_1, x')$ is another solution to the same equation (3.2). By the uniqueness of solutions in Theorem 2.1, we have

$$u(t, x) = v(t, x) = -u(t, -x_1, x').$$

Therefore, we conclude that the restriction of u in Ω_T satisfies (3.1) with the zero boundary condition.

Remark 3.1. The very same method also works in the case when $a_{i1} = a_{1i} = 0$ for $i = 2, \ldots, d$. In this case, we need to take the even extensions of a_{ij} with respect to x_1. In another special case when a_{ij} are constants, we can make a linear transformation to reduce to the case $a_{ij} = \delta_{ij}$.

However, in general if one of a_{1i} or a_{i1} is not identically zero, then after the extension u satisfies

$$u_t - \tilde{a}_{ij}D_{ij}u + \lambda u = f \quad \text{in } \mathbb{R}_T^{d+1},$$

where

$$a_{i1}(t, -x_1, x') = -a_{i1}(t), \quad a_{1i}(t, -x_1, x') = -a_{1i}(t) \quad \text{for } x_1 < 0,$$

which depend on both t and x, and discontinuous across $x_1 = 0$. Therefore, the L_p estimates proved in the previous section are not applicable.

Again we begin with the L_2 estimates.

Theorem 3.2. *Let $T \in (-\infty, +\infty]$ and $\lambda \geq 0$. Suppose that $f \in L_2(\Omega_T)$ and $u \in W_2^{1,2}(\mathbb{R}^2_T)$ satisfies (3.1). Then we have*

$$\|u_t\|_{L_2(\Omega_T)} + \|D^2 u\|_{L_2(\Omega_T)} + \sqrt{\lambda}\|Du\|_{L_2(\Omega_T)} + \lambda\|u\|_{L_2(\Omega_T)}$$
$$\leq N(d, \kappa)\|f\|_{L_2(\Omega_T)}.$$

Proof. We note that this theorem follows from Theorem 1.4 by using the extension technique described above. Here we give a different proof using integration by parts. By approximations, we may assume that $u \in C_0^\infty((-\infty, T] \times \bar{\Omega})$ and $u = 0$ on $\{x_1 = 0\}$. We multiply the equation by λu and integrate by parts. By using the zero boundary condition and the ellipticity condition,

$$\kappa\lambda \int_{-\infty}^T \int_\Omega |Du|^2\, dx\, dt + \lambda^2 \int_{-\infty}^T \int_\Omega u^2\, dx\, dt \leq \int_{-\infty}^T \int_\Omega \lambda u f\, dx\, dt,$$

which together with Young's inequality implies that

$$\sqrt{\lambda}\|Du\|_{L_2(\Omega_T)} + \lambda\|u\|_{L_2(\Omega_T)} \leq N\|f\|_{L_2(\Omega_T)}. \tag{3.3}$$

To estimate the second derivatives, we multiply the equation by $-D_{kk}u, k = 2, \ldots, d$ and integrate by parts. Because $D_k u = 0$ on $\{x_1 = 0\}$,

$$-\int_{-\infty}^T \int_\Omega u_t D_{kk}u\, dx\, dt = \int_{-\infty}^T \int_\Omega D_k u_t D_k u\, dx\, dt = \frac{1}{2}\int_\Omega |D_k u|^2\, dx\, dt \geq 0, \tag{3.4}$$

$$\int_{-\infty}^T \int_\Omega D_{ij}u D_{kk}u\, dx\, dt = \int_{-\infty}^T \int_\Omega D_{ijk}u D_k u\, dx\, dt = \int_\Omega D_{ik}u D_{jk}u\, dx\, dt, \tag{3.5}$$

and

$$-\int_{-\infty}^T \int_\Omega \lambda u D_{kk}u\, dx\, dt = \int_{-\infty}^T \int_\Omega \lambda D_k u D_k u\, dx\, dt \geq 0 \tag{3.6}$$

Using the ellipticity condition and Young's inequality, we get

$$\|DD_{x'}u\|_{L_2(\Omega_T)} \leq N\|f\|_{L_2(\Omega_T)}. \tag{3.7}$$

It remains to estimate $\|D_{11}u\|_{L_2(\Omega_T)}$. To this end, we multiply the equation by $D_{11}u$ and integrate by parts. Notice that because $u = u_t = 0$ on $\{x_1 = 0\}$, (3.4) and (3.6) still hold with $k = 1$. As to (3.5) with $k = 1$, when i or $j = 1$ we have

$$\int_{-\infty}^T \int_\Omega D_{ij}u D_{kk}u\, dx\, dt = \int_\Omega D_{ik}u D_{jk}u\, dx\, dt$$

without performing any integration by parts. On the other hand, if both i and $j > 1$, we have

$$\int_{-\infty}^T \int_\Omega D_{ij}u D_{kk}u\, dx\, dt = \int_{-\infty}^T \int_\Omega D_i u D_{jkk}u\, dx\, dt = \int_\Omega D_{ik}u D_{jk}u\, dx\, dt,$$

because $D_i u = 0$ on $\{x_1 = 0\}$. Consequently, by using the ellipticity condition and Young's inequality, we have

$$\|DD_1 u\|_{L_2(\Omega_T)} \le N\|f\|_{L_2(\Omega_T)}. \tag{3.8}$$

Combining (3.3), (3.7), and (3.8), the theorem is proved. □

Alternatively, to estimate $\|D_{11}u\|_{L_2(\Omega_T)}$ we can rewrite the equation into

$$u_t - a_{11}(t)D_{11}u - \sum_{i=2}^{d} D_{ii}u + \lambda u = \tilde{f},$$

where

$$\tilde{f} = f + \sum_{i \text{ or } j>1} a_{ij}D_{ij}u + \sum_{i=2}^{d} D_{ii}u.$$

By Remark 3.1, we have

$$\|D^2 u\|_{L_2(\Omega_T)} \le N\|\tilde{f}\|_{L_2(\Omega_T)} \le N\|f\|_{L_2(\Omega_T)} + N\|DD_{x'}u\|_{L_2(\Omega_T)}.$$

Using this and (3.7), we get the L_2 estimate of D^2u. Here the point is that this argument works equally well in the L_p setting for systems, and will be used in the proof of Theorem 3.7.

Below we present another proof of Theorem 3.2, in which we exploit the hidden divergence structure of the equation. This proof also works in the L_p setting provided that the corresponding L_p estimates for the divergence form equations are available.

Another proof of Theorem 3.2. As before, we may assume that

$$u \in C_0^\infty((-\infty, T] \times \bar{\Omega}) \quad \text{and} \quad u = 0 \quad \text{on } \{x_1 = 0\}.$$

We may also assume that $\lambda > 0$. We rewrite the equation into the divergence form equation

$$\begin{cases} u_t - D_i(a_{ij}(t)D_j u) + \lambda u = f, & \text{in } (-\infty, T) \times \Omega, \\ u(t, 0, x') = 0, & \text{on } (-\infty, T) \times \partial\Omega. \end{cases} \tag{3.9}$$

By the standard L_2 estimate for divergence form parabolic equations, we have

$$\|Du\|_{L_2(\Omega_T)} + \sqrt{\lambda}\|u\|_{L_2(\Omega_T)} \le \frac{N}{\sqrt{\lambda}}\|f\|_{L_2(\Omega_T)}.$$

Differentiate (3.9) with respect to $x_k, k = 2, \ldots, d$. Then $v_k := D_k u$ satisfies

$$\begin{cases} \partial_t v_k - D_i(a_{ij}(t)D_j v_k) + \lambda v_k = D_k f, & \text{in } (-\infty, T) \times \Omega, \\ v_k(t, 0, x') = 0, & \text{on } (-\infty, T) \times \partial\Omega. \end{cases}$$

Again by the standard L_2 estimate for divergence form parabolic equations, we get

$$\|Dv_k\|_{L_2(\Omega_T)} \le N\|f\|_{L_2(\Omega_T)},$$

which implies that

$$\|DD_{x'}u\|_{L_2(\Omega_T)} \le N\|f\|_{L_2(\Omega_T)}.$$

To estimate $D_{11}u$, we define

$$\tilde{a}_{11} = a_{11}, \quad \tilde{a}_{1j} = a_{1j} + a_{j1}, \quad \tilde{a}_{j1} = 0, \quad j = 2, \ldots, d,$$

and $\tilde{a}_{ij} = a_{ij}$ for $i, j \ge 2$. Then u satisfies

$$\begin{cases} u_t - D_i(\tilde{a}_{ij}(t)D_j u) + \lambda u = f, & \text{in } (-\infty, T) \times \Omega, \\ u(t, 0, x') = 0, & \text{on } (-\infty, T) \times \partial\Omega. \end{cases} \tag{3.10}$$

Since on $\{x_1 = 0\}$, $u = u_t = D_{ij}u = 0$ for $i, j \ge 2$, by the first equation in (3.10) and the continuity, we have $\tilde{a}_{1j}D_{1j}u = -f$. Therefore, $v_1 := D_1 u$ satisfies the conormal derivative boundary value problem

$$\begin{cases} \partial_t v_1 - D_i(\tilde{a}_{ij}(t)D_j v_1) + \lambda u = D_1 f, & \text{in } (-\infty, T) \times \Omega, \\ \tilde{a}_{1j}D_j v_1 = -f, & \text{on } (-\infty, T) \times \partial\Omega. \end{cases}$$

By the L_2 estimate for the conormal problem of divergence form parabolic equations, we have

$$\|Dv_1\|_{L_2(\Omega_T)} \le N\|f\|_{L_2(\Omega_T)},$$

which implies that

$$\|DD_1 u\|_{L_2(\Omega_T)} \le N\|f\|_{L_2(\Omega_T)}.$$

The theorem is proved. □

In the sequel, we denote $Q_r^+(X_0) = Q_r(X_0) \cap (\mathbb{R} \times \mathbb{R}^d_+)$. By localizing Theorem 3.2 as in Remark 2.13, we obtain

Corollary 3.3. *Let $\lambda \ge 0$. Suppose that $f \in L_2(Q_1^+)$ and $u \in W_2^{1,2}(Q_1^+)$ satisfies*

$$u_t - a_{ij}(t)D_{ij}u + \lambda u = f \tag{3.11}$$

in Q_1^+ and $u = 0$ on $\{x_1 = 0\} \cap Q_1$. Then we have

$$\|u\|_{W_2^{1,2}(Q_{1/2}^+)} \le N(d, \kappa)\big(\|f\|_{L_2(Q_1^+)} + \|u\|_{L_2(Q_1^+)}\big).$$

Lemma 3.4. *Let $\lambda \ge 0$. Assume that $u \in C_b^\infty(Q_1^+)$ satisfies $u = 0$ on $\{x_1 = 0\}$ and*

$$u_t - a_{ij}(t)D_{ij}u + \lambda u = 0 \tag{3.12}$$

in Q_1^+. Then we have

$$[u]_{C^{1/4,1/2}(Q_{1/2}^+)} \le N(d, \kappa)\|u\|_{L_2(Q_1^+)}.$$

Proof. Since $D^k_{x'}u, k \geq 1$, satisfies the same equation in Q_1^+ with the same boundary condition, by Corollary 3.3 and an iteration argument, we have

$$\|D^k_{x'}u\|_{W^{1,2}_2(Q^+_{\sqrt{2}/2})} \leq N(d,\kappa,k)\|u\|_{L_2(Q_1^+)}. \tag{3.13}$$

We use the following type of anisotropic Sobolev embedding theorem. Let

$$\Theta_r^+ = \{(t,x_1) \in (-r^2,0)\times(0,r)\}, \quad B'_r = \{x' \in \mathbb{R}^{d-1} : |x'| < r\}.$$

The triangle inequality gives

$$\sup_{\substack{(t,x),(s,y)\in Q^+_{1/2}\\(t,x)\neq(s,y)}} \frac{|u(t,x)-u(s,y)|}{|t-s|^{1/4}+|x-y|^{1/2}}$$

$$\leq \sup_{\substack{(t,x_1),(s,y_1)\in\Theta^+_{1/2},(t,x_1)\neq(s,y_1)\\x'\in B'_{1/2}}} \frac{|u(t,x_1,x')-u(s,y_1,x')|}{|t-s|^{1/4}+|x_1-y_1|^{1/2}}$$

$$+ \sup_{\substack{(s,y_1)\in\Theta^+_{1/2}\\x',y'\in B'_{1/2},x'\neq y'}} \frac{|u(s,y_1,x')-u(s,y_1,y')|}{|x'-y'|^{1/2}} := I_1 + I_2.$$

To estimate I_1, we view $u(t,x_1,x')$ as a function of (t,x_1) for a fixed $x' \in B'_{1/2}$. Then by the Sobolev embedding theorem

$$\sup_{\substack{(t,x_1),(s,y_1)\in\Theta^+_{1/2}\\(t,x_1)\neq(s,y_1)}} \frac{|u(t,x_1,x')-u(s,y_1,x')|}{|t-s|^{1/4}+|x_1-y_1|^{1/2}} \leq N\|u(\cdot,x')\|_{W^{1,2}_2(\Theta^+_{1/2})} \tag{3.14}$$

for each $x' \in B'_{1/2}$. On the other hand, there exists a positive integer k such that, for each $(t,x_1) \in \Theta^+_{1/2}$,

$$\sum_{j=0}^{2} \sup_{x'\in B'_{1/2}} |D_1^j u(t,x_1,x')| + \sup_{x'\in B'_{1/2}} |u_t(t,x_1,x')|$$

$$\leq N\sum_{j=0}^{2} \|D_1^j u(t,x_1,\cdot)\|_{W_2^k(B'_{1/2})} + N\|u_t(t,x_1,\cdot)\|_{W_2^k(B'_{1/2})}, \tag{3.15}$$

where $D_1^j u(t,x_1,x')$ and $u_t(t,x_1,x')$ are viewed as functions of $x' \in B'_1$. Combining (3.14) and (3.15) proves

$$I_1 \leq N \sum_{\substack{|\gamma|\leq k+2\\\gamma_1\leq 2}} \|D^\gamma u\|_{L_2(Q^+_{\sqrt{2}/2})} + N\sum_{\substack{|\vartheta|\leq k\\\vartheta_1=0}} \|D^\vartheta u_t\|_{L_2(Q^+_{\sqrt{2}/2})} \leq N\|u\|_{L_2(Q_1^+)}, \tag{3.16}$$

where the last inequality is due to (3.13).

For the estimate of I_2, we view $u(s, y_1, x')$ as a function of $x' \in B'_{1/2}$ for each $(s, y_1) \in \Theta^+_{1/2}$. Again by the Sobolev embedding theorem, there exists a sufficiently large integer k such that

$$\sup_{\substack{x',y' \in B'_{1/2} \\ x' \neq y'}} \frac{|u(s, y_1, x') - u(s, y_1, y')|}{|x' - y'|^{1/2}} \leq N \|u(s, y_1, \cdot)\|_{W_2^k(B'_{1/2})}.$$

Moreover, as a function of $(s, y_1) \in \Theta^+_{1/2}$, $D^j_{x'} u(s, y_1, x')$, $j = 0, \ldots, k$, satisfy

$$\sum_{j=0}^{k} \sup_{(s,y_1) \in \Theta^+_{1/2}} |D^j_{x'} u(s, y_1, x')| \leq N \sum_{j=0}^{k} \|D^j_{x'} u(\cdot, x')\|_{W_2^{1,2}(\Theta^+_{1/2})}$$

for each $x' \in B'_{1/2}$. From the above two inequalities, we obtain (3.16) with I_2 in place of I_1. The lemma is proved. □

Now we define a vector-valued function $\tilde{U} = (D^2_{x'} u, \sqrt{\lambda} D_{x'} u, \lambda u)$. Note that unlike the interior case, here the derivatives in x_1 cannot be included. As in Corollary 2.4, since $D_{x'} u$ and $D^2_{x'} u$ satisfy the same equation as u, under the assumptions of Lemma 3.4 we obtain

$$[\tilde{U}]_{C^{1/4,1/2}(Q^+_{1/2})} \leq N(d, \kappa) \|\tilde{U}\|_{L_2(Q^+_1)}. \tag{3.17}$$

By using (3.17), we further derive the following mean oscillation estimate.

Lemma 3.5. *Let $r \in (0, \infty)$, $\mu \in (0, 1/16)$, $\lambda \geq 0$, and $X_0 = (t_0, x_0) \in \overline{\mathbb{R}^{d+1}_+}$. Assume that $u \in C^\infty_{loc}(\overline{\mathbb{R}^{d+1}_+})$ satisfies (3.12) in $Q^+_r(X_0)$ and $u = 0$ on $\{x_1 = 0\}$. Then we have*

$$\left(|\tilde{U} - (\tilde{U})_{Q^+_{\mu r}(X_0)}|\right)_{Q^+_{\mu r}(X_0)} \leq N \mu^{\frac{1}{2}} (|\tilde{U}|^2)^{\frac{1}{2}}_{Q^+_r(X_0)} \tag{3.18}$$

where $N = N(d, \kappa) > 0$.

Proof. Using a scaling argument, it suffices to prove the inequality (3.18) only for $r = 8$. Now we consider two cases.

Case 1: *the first coordinate of $x_0 \geq 1$.* In this case, we see that $Q^+_{r/8}(X_0) = Q_{r/8}(X_0)$ and u satisfies the assumptions in Corollary 2.4. Hence with $Q_{r/8}(X_0)$ in place of $Q_r(X_0)$ (note that $8\kappa \in (0, 1/2)$), the left-hand side of (3.18) is controlled by

$$N\mu (|\tilde{U}|^2)^{\frac{1}{2}}_{Q_{r/8}(X_0)},$$

which is less than the right-hand side of (3.18).

Case 2: *the first coordinate of x_0 is in* $[0, 1]$. By denoting $Y_0 = (t_0, 0, x'_0)$, we have

$$Q^+_{\mu r}(X_0) \subset Q^+_2(Y_0) \subset Q^+_4(Y_0) \subset Q^+_r(X_0).$$

By (3.17) with 2 and 4 in place of 1/2 and 1, respectively (this case can be seen using a scaling argument as above), we have

$$
\begin{aligned}
&(|\tilde{U} - (\tilde{U})_{Q^+_{\mu r}(X_0)}|)_{Q^+_{\mu r}(X_0)} \le N\mu^{\frac{1}{2}}[\tilde{U}]_{C^{1/4,1/2}(Q^+_2(Y_0))} \\
&\le N\mu^{\frac{1}{2}}(|\tilde{U}|^2)^{\frac{1}{2}}_{Q^+_4(Y_0)} \le N\mu^{\frac{1}{2}}(|\tilde{U}|^2)^{\frac{1}{2}}_{Q^+_r(X_0)}.
\end{aligned}
$$

The lemma is proved. □

Following the proof of Lemma 2.5, using Theorem 3.2 and Lemma 3.5, we get

Lemma 3.6. *Let $r \in (0,\infty)$, $\mu \in (0,1/32)$, $\lambda \ge 0$, $X_0 = (t_0, x_0) \in \overline{\mathbb{R}^{d+1}_+}$, and $f \in L_{2,loc}(\overline{\mathbb{R}^{d+1}_+})$. Assume that $u \in C^\infty_{loc}(\overline{\mathbb{R}^{d+1}_+})$ satisfies (3.11) in $Q^+_r(X_0)$ and $u = 0$ on $\{x_1 = 0\}$. Then we have*

$$
\left(|\tilde{U} - (\tilde{U})_{Q^+_{\mu r}(X_0)}|\right)_{Q^+_{\mu r}(X_0)} \le N\mu^{\frac{1}{2}}(|\tilde{U}|^2)^{\frac{1}{2}}_{Q^+_r(X_0)} + N\mu^{-1-\frac{d}{2}}(|f|^2)^{\frac{1}{2}}_{Q^+_r(X_0)},
$$

where $N = N(d,\kappa) > 0$.

We now give the boundary L_p estimate for parabolic equations.

Theorem 3.7. *Let $T \in (-\infty, +\infty]$ and $\lambda \ge 0$. Suppose that $f \in L_p(\Omega_T)$ and $u \in W^{1,2}_p(\Omega_T)$ satisfies (3.1). Then we have*

$$
\begin{aligned}
&\|u_t\|_{L_p(\Omega_T)} + \|D^2u\|_{L_p(\Omega_T)} + \sqrt{\lambda}\|Du\|_{L_p(\Omega_T)} + \lambda\|u\|_{L_p(\Omega_T)} \\
&\le N(d,\kappa)\|f\|_{L_p(\Omega_T)}. \qquad (3.19)
\end{aligned}
$$

Moreover, for any $\lambda > 0$ and $f \in L_p(\Omega_T)$, there is a unique solution $u \in W^{1,2}_p(\Omega_T)$ to the equation (3.1).

Proof. As in the proof of Theorem 2.1, we only need to prove the a priori estimate assuming that $T = \infty$ and $u \in C^\infty_0(\overline{\mathbb{R}^{d+1}_+})$. We first consider the case when $p > 2$. By using the Hardy–Littlewood theorem on maximal functions and the Fefferman–Stein theorem on sharp functions in the half space, from Lemma 3.6 we obtain

$$
\|\tilde{U}\|_{L_p(\mathbb{R}^{d+1}_+)} \le N\|f\|_{L_p(\mathbb{R}^{d+1}_+)}.
$$

In particular,

$$
\|D^2_{x'}u\|_{L_p(\mathbb{R}^{d+1}_+)} \le N\|f\|_{L_p(\mathbb{R}^{d+1}_+)}. \qquad (3.20)
$$

To estimate the remaining terms, we use the idea illustrated after the proof of Theorem 3.2. We rewrite the equation into

$$
u_t - a_{11}(t)D_{11}u - \sum_{i=2}^d D_{ii}u + \lambda u = \tilde{f},
$$

where

$$\tilde{f} = f + \sum_{i \text{ or } j > 1} a_{ij} D_{ij} u - \sum_{i=2}^{d} D_{ii} u.$$

By Remark 3.1, we have

$$\begin{aligned}
&\|u_t\|_{L_p(\mathbb{R}^{d+1}_+)} + \|D^2 u\|_{L_p(\mathbb{R}^{d+1}_+)} + \sqrt{\lambda}\|Du\|_{L_p(\mathbb{R}^{d+1}_+)} + \lambda\|u\|_{L_p(\mathbb{R}^{d+1}_+)} \\
&\le N\|\tilde{f}\|_{L_p(\mathbb{R}^{d+1}_+)} \\
&\le N\|f\|_{L_p(\mathbb{R}^{d+1}_+)} + N\|D^2_{x'} u\|_{L_p(\mathbb{R}^{d+1}_+)} + N\|D_1 D_{x'} u\|_{L_p(\mathbb{R}^{d+1}_+)}. \qquad (3.21)
\end{aligned}$$

By using the L_p estimate for the Laplace equation in the whole space (which can be derived from Theorem 2.1 using, for instance, the proof of (1.12)) and the technique of odd/even extensions, we have

$$\|D_1 D_{x'} u\|_{L_p(\mathbb{R}^{d+1}_+)} \le N\|\Delta u\|_{L_p(\mathbb{R}^{d+1}_+)} \le N\|D_1^2 u\|_{L_p(\mathbb{R}^{d+1}_+)} + N\|D^2_{x'} u\|_{L_p(\mathbb{R}^{d+1}_+)}.$$

Then by a scaling argument as in (1.8), we get

$$\|D_1 D_{x'} u\|_{L_p(\mathbb{R}^{d+1}_+)} \le \epsilon\|D_1^2 u\|_{L_p(\mathbb{R}^{d+1}_+)} + \frac{N}{\epsilon}\|D^2_{x'} u\|_{L_p(\mathbb{R}^{d+1}_+)}. \qquad (3.22)$$

Inserting (3.22) with a small ϵ into (3.21), we obtain

$$\begin{aligned}
&\|u_t\|_{L_p(\mathbb{R}^{d+1}_+)} + \|D^2 u\|_{L_p(\mathbb{R}^{d+1}_+)} \le N\|\tilde{f}\|_{L_p(\mathbb{R}^{d+1}_+)} \\
&\le N\|f\|_{L_p(\mathbb{R}^{d+1}_+)} + N\|D^2_{x'} u\|_{L_p(\mathbb{R}^{d+1}_+)},
\end{aligned}$$

which together with (3.20) yields the a priori estimate (3.19).

For the remaining case when $p \in (1, 2)$, we use a duality argument. Let $q = p/(p-1) \in (2, \infty)$. For any $g \in L_q(\mathbb{R}^{d+1}_T)$, let $v \in W^{1,2}_q(\mathbb{R}^{d+1}_T)$ be the solution of

$$\begin{cases} -v_t - a_{ij} D_{ij} v + \lambda v = g, & \text{in } \mathbb{R}^{d+1}_+, \\ u(t, 0, x') = 0, & \text{on } \partial\mathbb{R}^{d+1}_+. \end{cases}$$

The existence of v follows from the proof above and a change of variable $t \to -t$. As before, we may assume that $u \in C_0^\infty(\overline{\mathbb{R}^{d+1}_+})$ and $u = 0$ on $\{x_1 = 0\}$. Then integration by parts yields

$$\begin{aligned}
\int_{\mathbb{R}^{d+1}_+} D^2_{x'} u g \, dx\, dt &= \int_{\mathbb{R}^{d+1}_+} D^2_{x'} u(-v_t - a_{ij} D_{ij} v + \lambda v) \, dx\, dt \\
&= \int_{\mathbb{R}^{d+1}_+} v(\partial_t - a_{ij} D_{ij} + \lambda) D^2_{x'} u \, dx\, dt \\
&= \int_{\mathbb{R}^{d+1}_+} D^2_{x'} v(\partial_t - a_{ij} D_{ij} + \lambda) u \, dx\, dt.
\end{aligned}$$

It then follows from the $W_q^{1,2}$ estimate proved above that

$$\left| \int_{\mathbb{R}^{d+1}_+} D^2_{x'} u g \, dx \, dt \right| \le \|D^2_{x'} v\|_{L_q(\mathbb{R}^{d+1}_+)} \|f\|_{L_p(\mathbb{R}^{d+1}_+)}$$
$$\le N \|g\|_{L_q(\mathbb{R}^{d+1}_+)} \|f\|_{L_p(\mathbb{R}^{d+1}_+)}.$$

Since $g \in L_q(\mathbb{R}^{d+1}_+)$ is an arbitrary function, we deduce

$$\|D^2_{x'} u\|_{L_p} \le N \|f\|_{L_p}.$$

Following the argument in the previous step, we can estimate the remaining terms and conclude the proof. □

Remark 3.8. (1) By following the proof of Theorem 2.9 using the perturbation argument, we can extend the result Theorem 3.7 to equations with VMO_x coefficients. Similar results also hold for divergence form equations. We refer the reader to [12] for the proofs, where more general higher-order divergence and non-divergence parabolic systems are considered.

(2) By using the techniques of partition of the unity and locally flattening of the boundary, we can extend the results to equations in $C^{1,1}$ domain for the non-divergence case, and in Lipschitz domains with small Lipschitz constant for the divergence case; see, for instance, [9, 11].

(3) Under the conditions of Lemma 3.4, we can obtain a better estimate by using Theorem 3.7. In fact, by using the $W_p^{1,2}$ estimate instead of the $W_2^{1,2}$ estimate and the proof of Lemma 2.10, we obtain that

$$[u]_{C^{(1+\alpha)/2, 1+\alpha}(Q^+_{1/2})} \le N(d, \kappa, \alpha) \|u\|_{L_1(Q^+_1)}$$

for any $\alpha \in (0, 1)$. Moreover, by differentiate the equation in the tangential direction, we can estimate the $C^{\alpha/2,\alpha}$ norm of $D_{x'} Du$ for any $\alpha \in (0, 1)$. On the other hand, in general there is no a priori Hölder estimate for $D_{11} u$ near the boundary when the coefficients are merely measurable in t. See, for instance, [23, §3] and [22, §16].

Historical Remarks. For $p \ge 2$, Theorem 3.7 was first proved in [14]. The restriction on p was removed in [8]. Here we mainly followed the argument in [12].

References

[1] BERNSTEIN S.: Sur la généralisation du problème de Dirichlet, *Math. Ann.* **69** (1910), no. 1, 82–136.

[2] CHIARENZA F., FRASCA M., LONGO P.: Interior $W^{2,p}$ estimates for nondivergence elliptic equations with discontinuous coefficients, *Ricerche Mat.* **40** (1991), 149–168.

[3] CHIARENZA F., FRASCA M., LONGO P.: $W^{2,p}$-solvability of the Dirichlet problem for nondivergence elliptic equations with VMO coefficients, *Trans. Amer. Math. Soc.* **336** (1993), no. 2, 841–853.

[4] CHIARENZA F., FRANCIOSI M., FRASCA M.: L^p-estimates for linear elliptic systems with discontinuous coefficients, *Atti Accad. Naz. Lincei Cl. Sci. Fis. Mat. Natur. Rend. Lincei (9) Mat. Appl.* **5** (1994), no. 1, 27–32.

[5] CAMPANATO S.: Un risultato relativo ad equazioni ellittiche del secondo ordine di tipo non variazionale (Italian), *Ann. Scuola Norm. Sup. Pisa (3)* **21** (1967), 701–707.

[6] DI FAZIO G.: L^p estimates for divergence form elliptic equations with discontinuous coefficients (Italian summary), *Boll. Un. Mat. Ital. A (7)* **10** (1996), no. 2, 409–420.

[7] DONG H., KRYLOV N. V.: Second-order elliptic and parabolic equations with $B(\mathbb{R}^2, \mathrm{VMO})$ coefficients, *Trans. Amer. Math. Soc.* **362** (2010), no. 12, 6477–6494.

[8] DONG H.: Solvability of second-order equations with hierarchically partially BMO coefficients, *Trans. Amer. Math. Soc.* **364** (2012), no. 1, 493–517.

[9] DONG H., KIM D.: Elliptic equations in divergence form with partially BMO coefficients, *Arch. Ration. Mech. Anal.* **196** (2010), no. 1, 25–70.

[10] DONG H., KIM D.: Parabolic and elliptic systems with VMO coefficients, *Methods Appl. Anal.* **16** (2009), no. 3, 365–388.

[11] DONG H., KIM D.: L_p solvability of divergence type parabolic and elliptic systems with partially BMO coefficients, *Calc. Var. Partial Differential Equations* **40** (2011), no. 3–4, 357–389.

[12] DONG H., KIM D.: On the L_p-solvability of higher order parabolic and elliptic systems with BMO coefficients, *Arch. Rational Mech. Anal.* **199** (2011), no. 3, 889–941.

[13] DONG H., KIM D.: On L_p-estimates for elliptic and parabolic equations with A_p weights, preprint (2015).

[14] KIM D.: Parabolic equations with partially BMO coefficients and boundary value problems in Sobolev spaces with mixed norms, *Potential Anal.* **33** (2010), no. 1, 17–46.

[15] KIM D., KRYLOV N. V.: Elliptic differential equations with coefficients measurable with respect to one variable and VMO with respect to the others, *SIAM J. Math. Anal.* **39** (2007), no. 2, 489–506.

[16] KIM D., KRYLOV N. V.: Parabolic equations with measurable coefficients, *Potential Anal.* **26** (2007), no. 4, 345–361.

[17] KRYLOV. N. V.: On equations of minimax type in the theory of elliptic and parabolic equations in the plane, *Matematicheski Sbornik* **81** (1970), no. 1, 3–22 in Russian; English translation in *Math. USSR Sbornik* **10** (1970), 1–20.

[18] KRYLOV N. V.: Parabolic and elliptic equations with VMO coefficients, *Comm. Partial Differential Equations* **32** (2007), no. 1–3, 453–475.

[19] KRYLOV N. V.: Parabolic equations with VMO coefficients in spaces with mixed norms, *J. Funct. Anal.* **250** (2007), no. 2, 521–558.

[20] KRYLOV N. V.: *Lectures on Elliptic and Parabolic Equations in Sobolev Spaces*, American Mathematical Society, 2008.

[21] KRYLOV N. V.: Second-order elliptic equations with variably partially VMO coefficients, *J. Funct. Anal.* **257** (2009), 1695–1712.

[22] LIEBERMAN G.: Intermediate Schauder Theory For second order parabolic

equations. IV. Time irregularity and regularity, *Differential and Integral Equations* **5** (1992), no. 6, 1219–1236.

[23] SINESTRARI E., VON WAHL W.: On the solutions of the first boundary value problem for the linear parabolic equations, *Proc. Roy. Soc. Edinburgh Sect. A* **108** (1988), no. 3-4, 339–355.

[24] TALENTI G.: Sopra una classe di equazioni ellittiche a coefficienti misurabili, *Ann. Mat. Pura Appl. (4)* **69** (1965), 285–304.

Lectures on the Analysis of Nonlinear Partial Differential Equations Vol. 4
MLM 4, pp. 93–113

A Simple Introduction to Hardy-Littlewood Circle Method

Xiaochun Li*

Abstract

This note is written for giving a short course at Institute of Mathematics, Chinese Academy of Science, in May 2014. The purpose of this course is to provide a quick introduction on Hardy-Littlewood circle method, designed for students who have only limited knowledge on number theory.

1 Poisson summation and Dirichlet principle

We start with an important tool—Poisson summation formula. For any integrable function f, $\hat{f}$ stands for the Fourier transform, that is,

$$\hat{f}(\xi) = \int_{\mathbb{R}} f(x)e^{-2\pi i x\xi}\, dx. \tag{1.1}$$

Theorem 1.1 (Poisson summation, [1]). *Let $f \in L^1(\mathbb{R}^d)$ and $|f(x)| + |\hat{f}(x)| \le \frac{C_{d,\delta}}{(1+|x|)^{d+\delta}}$ for some $\delta > 0$. Then $\sum\limits_{n\in\mathbb{Z}^d} \hat{f}(n) = \sum\limits_{n\in\mathbb{Z}^d} f(n)$.*

Proof. Define $F(x) = \sum\limits_{n\in\mathbb{Z}^d} f(x+n)$. Then F is periodic and $F \in L^1(\mathbb{T}^d)$, since $f \in L^1(\mathbb{R}^d)$ and $|f(x)| \le \frac{C_{d,\delta}}{(1+|x|)^{d+\delta}}$. Let $C_n = \int_{\mathbb{T}^d} F(x)e^{-2\pi i n\cdot x}dx$.

$$\begin{aligned} C_n &= \int_{\mathbb{T}^d} \sum_{m\in\mathbb{Z}^d} f(x+m)e^{-2\pi i n x}dx \\ &= \sum_{m\in\mathbb{Z}^d} \int_{\mathbb{T}^d} f(x+m)e^{-2\pi i n x}dx \\ &= \sum_{m\in\mathbb{Z}^d} \int_{\mathbb{T}^d+m} f(x)e^{-2\pi i n x}dx \end{aligned}$$

*Department of Mathematics, University of Illinois at Urbana-Champaign

$$= \int_{\bigcup_m \mathbb{T}^d+m} f(x)e^{-2\pi inx}dx$$
$$= \hat{f}(n).$$

Thus $F(x) = \sum_n \hat{f}(n)e^{2\pi inx}$, since $F \in L^1(\mathbb{T}^d)$ and F is a continuous function. Let $x = 0, \Rightarrow F(0) = \sum_n \hat{f}(n) = \sum_n f(n)$. □

A simple application of this formula is the so-called Gauss identity.

Theorem 1.2 (Gauss). *Let η be any complex number with* $\mathrm{Re}(\eta) > 0$. *Then*

$$\sum_{n\in\mathbb{Z}} e^{-2\pi n^2\eta} = \frac{1}{\sqrt{2\eta}} \sum_{n\in\mathbb{Z}} e^{-\frac{\pi n^2}{2\eta}}. \tag{1.2}$$

Proof. Let $f(x) = e^{-2\pi x^2\eta}$. Then we have

$$\begin{aligned}\hat{f}(x) &= \int_{\mathbb{R}} e^{-2\pi x^2\eta}e^{-2\pi i\xi\cdot x}dx\\ &= \frac{1}{\sqrt{\eta}}e^{-\frac{\pi\xi^2}{2\eta}} \int_{\mathbb{R}} e^{-\pi\left(y+\frac{i\xi^2}{\sqrt{2\eta}}\right)^2}dy\\ &= \frac{1}{\sqrt{\eta}}e^{-\frac{\pi\xi^2}{2\eta}} \int_{\mathbb{R}} e^{-\pi y^2}dy\\ &= \frac{1}{\sqrt{\eta}}e^{-\frac{\pi\xi^2}{2\eta}}.\end{aligned}$$

If $\mathrm{Re}(\eta) > 0$, both sides of (1.1) converges absolutely. Thus $\sum_{n\in\mathbb{Z}} e^{-2\pi n^2\eta} = \frac{1}{\sqrt{2\eta}} \sum_{n\in\mathbb{Z}} e^{-\frac{\pi n^2}{2\eta}}$. □

In the end of this section, we introduce the Principle of Dirichlet in number theory. There are several ways to prove it. One classical method is to use the Farey number disections of order N. Here we use the pigeonhole principle to prove it.

Theorem 1.3 (Dirichlet principle). *For any $0 < \theta \le 1$, and any positive integer N , there exists non-negative integers p, q with $(p, q) = 1$ and $1 \le q \le N$, s.t.* $\left|\theta - \frac{p}{q}\right| \le \frac{1}{qN}$.

Proof. Consider the set

$$\left\{n\theta : n = [\theta^{-1}], 2[\theta^{-1}], \cdots, N[\theta^{-1}]\right\}.$$

We use $[x]$ to represent the biggest integer not exceeding x. Usually $[x]$ is called the integer part of x. So

$$\left\{n\theta - [n\theta] : n = 0, [\theta^{-1}], 2[\theta^{-1}], \cdots, N[\theta^{-1}]\right\}$$

is a collection of $N+1$ numbers between $[0,1)$. By the Pigeonhole Principle, there exists $n_1, n_2 \in \{0,1,\cdots,N\}$ such that

$$\text{a)}\quad n_1 \neq n_2, \qquad \text{b)}\quad |(n_1\theta - [n_1\theta]) - (n_2\theta - [n_2\theta])| \leq \frac{1}{N}.$$

By b), we get $\left|\theta - \frac{[n_1\theta]-[n_2\theta]}{n_1-n_2}\right| \leq \frac{1}{N|n_1-n_2|}$. Let $\frac{[n_1\theta]-[n_2\theta]}{n_1-n_2} = \frac{p}{q}$ in lowest terms. Because $|n_1-n_2| \geq q$, we then obtain $\left|\theta - \frac{p}{q}\right| \leq \frac{1}{qN}$. Moreover, q can be chosen to be positive, which yields p is nonnegative. □

The Dirichlet principle supplies a nice rational approximation of any real number, which is commonly employed in analytic number theory, for instance, in the study of the exponential sums and Waring's problem.

2 Major arcs and minor arcs

In this section, we shall introduce important concepts–major arcs and minor arcs, that serves as one of main ingredients in the circle method. For given $N \in \mathbb{N}$, and any $1 \leq a < q \leq N$ and $(a,q) = 1$, we define

$$J_{\frac{a}{q}} := \left\{\theta \in (0,1) : \left|\theta - \frac{a}{q}\right| \leq \frac{q}{N}\right\}.$$

If $q \leq \frac{10}{N}$, $J_{\frac{a}{q}}$ is called a major arc, otherwise $J_{\frac{a}{q}}$ is called a minor arc.

Theorem 2.1. *Major arcs are pairwise disjoint.*

Proof. We proof $J_{\frac{a_1}{q_1}} \bigcap J_{\frac{a_2}{q_2}} = \emptyset$ if $q_1, q_2 \leq \frac{N}{10}$ and $\frac{a_1}{q_1} \neq \frac{a_2}{q_2}$. Without loss of generality, we can assume $q_1 \geq q_2$. Assume $J_{\frac{a_1}{q_1}} \bigcap J_{\frac{a_2}{q_2}} \neq \emptyset$. Then there exists a $\theta \in J_{\frac{a_1}{q_1}} \bigcap J_{\frac{a_2}{q_2}}$, so

$$\begin{cases} \left|\theta - \dfrac{a_1}{q_1}\right| & \leq \dfrac{1}{q_1 N}, \\ \left|\theta - \dfrac{a_2}{q_2}\right| & \leq \dfrac{2}{q_2 N} \end{cases}$$

$$\Rightarrow \left|\frac{a_1}{q_1} - \frac{a_2}{q_2}\right| \leq \left|\theta - \frac{a_1}{q_1}\right| + \left|\theta - \frac{a_2}{q_2}\right| \leq \frac{2}{q_2 N}$$

$$\Rightarrow \left|a_1q_2 - a_2q_1\right| \leq \frac{2q_1q_2}{q_2N} = \frac{2q_1}{N} \leq \frac{1}{5}$$

$$\Rightarrow a_1q_2 = a_2q_1, \text{ since } a_1q_2 - a_2q_1 \in \mathbb{Z}.$$

Thus, $\frac{a_1}{q_2} = \frac{a_2}{q_1}$, a contradiction. □

The minor arcs are NOT disjoint, however, they enjoy the finite overlapping property, that is, for any given point, there are at most finitely many (say, no more than 100) minor arcs covering it.

The circle method is designed for handling the size of integrals of some periodic function on the torus $\mathbb{T}$. Very often, the integral is the Fourier coefficient of the periodic function. In order to get a desired approximation of the integral, the circle $\mathbb{T}$ is partitioned by using major arcs plus the minor part. In the lattice problem we will discuss in this note, the reader will notice that the contribution is mainly from the major arcs, because minor part gives no significant size. However the situation in Waring's problem is different, in there the minor arcs are the main enemy remained.

Gauss sum

In the end of this section, we introduce the Gauss sum. Let a, k, q be integers, $(a, q) = 1$ and $q > 1$. The Gauss sum is given by

$$S(a,k,q) = \sum_{l=0}^{q-1} e^{2\pi i \frac{a}{q} l^2} e^{2\pi i \frac{k}{q} l} . \tag{2.1}$$

We use $S(a, q)$ to denote $S(a, 0, q)$, that is

$$S(a,q) = \sum_{l=1}^{q-1} e^{2\pi i \frac{a}{q} l^2} . \tag{2.2}$$

Lemma 2.1. *Let $a, q \in \mathbb{Z}, q > 1, (a, q) = 1, k \in \mathbb{Z}$. Then*

$$|S(a,k,q)| \leq \sqrt{2q}.$$

Proof. Let $S = S(a, k, q)$. Then we have

$$\begin{aligned}
|S|^2 = S\bar{S} &= \sum_{l_1=0}^{q-1}\sum_{l_2=0}^{q-1} e^{2\pi i \frac{a}{q}(l_1^2-l_2^2)} e^{2\pi i \frac{k}{q}(l_1-l_2)} \\
&= \sum_{l_2=0}^{q-1}\left(\sum_{l_1=l_2}^{q-1+l^2} e^{2\pi i \frac{a}{q}(l_1^2-l_2^2)} e^{2\pi i \frac{k}{q}(l_1-l_2)}\right) \\
&= \sum_{l_2=0}^{q-1}\left(\sum_{l_1=l_2}^{q-1+l_2} e^{2\pi i \frac{a}{q}(l_1+l_2)(l_1-l_2)} e^{2\pi i \frac{k}{q}(l_1-l_2)}\right) \\
&= \sum_{l_2=0}^{q-1}\sum_{m=0}^{q-1} e^{2\pi i \frac{a}{q}(2l_2+m)m} e^{2\pi i \frac{k}{q}m} \\
&= \sum_{m=0}^{q-1} e^{2\pi i \frac{a}{q}m^2} e^{2\pi i \frac{k}{q}m}\left(\sum_{l_2=0}^{q-1} e^{2\pi i \frac{a}{q}(2l_2 m)}\right) .
\end{aligned}$$

Notice

$$\sum_{l_2=0}^{q-1} e^{2\pi i \frac{a}{q}(2l_2 m)} = \begin{cases} q, & q|2ma, \\ 0, & q\nmid 2ma. \end{cases}$$

Since

$$q|2ma \Rightarrow q|2m \Rightarrow m = 0 \text{ or } m = \frac{q}{2},$$

there are at most two non-zero terms in the sum $\sum_{m=0}^{q-1}(\dots)$. Thus

$$|S|^2 \le 2q \Rightarrow |S| \le \sqrt{2q}. \qquad \square$$

3 Lattice points on sphere

We state the main theorem on the lattice points. Let $d \in \mathbb{N}$ and $r_d(R)$ denotes the number of solutions in integer of the equation $n_1^2 + n_2^2 + \cdots + n_d^2 = R^2$.

Theorem 3.1. *If $d \ge 5$, there are positive constants $c_1(d), c_2(d)$, s.t.*

$$c_1(d)R^{d-2} \le r_d(R) \le c_2(d)R^{d-2}. \tag{3.1}$$

Remark. For the lower dimensional cases, say $d = 2, 3, 4$, $r_d(R) \le CR^{d-2+\varepsilon}$.

Singular series

We define the singular series

$$\mathfrak{S}_d(R) = \sum_{q=1}^{\infty} \sum_{a \in \mathcal{P}_q} e^{-2\pi i R^2 \frac{a}{q}} \frac{S(a,q)^d}{q^d}$$

where $\mathcal{P}_q = \{a : 1 \le a \le q, (a,q) = 1\}$, and $S(a,q)$ is the Gauss sum defined as in (2.2).

The singular series can be used to approximate $r_d(R)$. More precisely, we have

Theorem 3.2. *If $d \ge 2$, then we have*

$$r_d(R) = C_d R^{d-2} \mathfrak{S}_d(R) + O(R^{\frac{d}{2}}). \tag{3.2}$$

Here C_d is non-zero constant.

Sections 3–5 are devoted to prove (3.2). In the rest of this section, we aim to build up the upper and lower bounds for the singular series. First we have the upper bound.

Lemma 3.1 (Upper bound of $\mathfrak{S}_d(R)$). *If $d \ge 5$, $|\mathfrak{S}_d(R)| \lesssim C_d$.*

Proof.

$$
\begin{aligned}
|\mathfrak{S}_d(R)| &\le \sum_{q=1}^{\infty}\sum_{a\in\mathcal{P}_q}\frac{|S(a,q)|^d}{q^d}\\
&\le \sum_{q=1}^{\infty}\sum_{a\in\mathcal{P}_q}\frac{\sqrt{2q}^d}{q^d}\\
&\le \sum_{q=1}^{\infty}\left(\frac{2}{q}\right)^{\frac{d}{2}}\cdot\phi(q)\\
&\le 2^{\frac{d}{2}}\sum_{q=1}^{\infty}\frac{1}{q^{\frac{d}{2}-1}}\\
&= C_d
\end{aligned}
$$

where ϕ is the Euler function. $\phi(n)$ is the number of integers which are relatively prime to n between 1 and n. □

We now turn to the lower bound.

Lemma 3.2 (Lower bound of $|\mathfrak{S}_d(R)|$). *If* $d \ge 5$, *then* $\left|\mathfrak{S}_d(R)\right| \ge C_d(>0)$.

Proof. Let us define singular Kloosterman Sum

$$K_{d,q}(R) = \frac{1}{q^d}\sum_{a\in\mathcal{P}_q} e^{-2\pi i R^2\frac{a}{q}} S(a,q)^d.$$

Then,

$$\mathfrak{S}_d(R) = \sum_{q=1}^{\infty} K_{d,q}(R), \qquad K_{d,1}(R) = S(1,1)^d = 1,$$

$$K_{d,2}(R) = e^{-\pi i R^2} 2^{-d} S(1,2)^d = e^{-\pi i R^2} 2^{-d}\left(\sum_{l=1}^{2} e^{2\pi i l^2\frac{1}{2}}\right) = 0,$$

$$
\begin{aligned}
\sum_{q=3}^{\infty} K_{d,q}(R) &\le \sum_{q=3}^{\infty}\frac{2^{\frac{d}{2}}\phi(q)}{q^{\frac{d}{2}}}\\
&\le \sum_{q=3}^{\infty}\frac{2^{\frac{d}{2}}}{q^{\frac{d}{2}-1}}\\
&\le \sum_{q=3}^{\infty} 2^{\frac{d}{2}}\int_2^{\infty} q^{1-\frac{d}{2}}\,dq\\
&= \frac{8}{d-4}.
\end{aligned}
$$

Hence, if $d > 12, |\mathfrak{S}_d(R)| \ge K_{d,1} - \frac{8}{d-4} > 0$. We omit the case $5 \le d \le 12$ (Kloosterman's refinements), since it can be done similarly. □

Corollary 3.1. *Theorem* 3.2 *implies Theorem* 3.1

Proof. From Theorem 3.2,

$$r_d(R) = C_d R^{d-2}\mathfrak{S}_d(R) + O(R^{\frac{d}{2}}).$$

Since $\mathfrak{S}_d(R) \sim C(d)$ for $d \geq 5, r_d(R) \sim R^{d-2}$ (because $\frac{d}{2} < d-2 \Leftrightarrow 5 \leq d$). □

4 Theta function $\Theta(z)$

We start to prove the approximation $r_d(R) = C_d R^{d-1}\mathfrak{S}_d(R) + O(R^{\frac{d}{2}})$. In fact, write

$$\begin{aligned} r_d(R) &= \int_0^1 \sum_{n_1,\ldots,n_d} e^{2\pi i(n_1^2+\cdots+n_d^2)\theta} e^{-2\pi i R^2\theta} d\theta \\ &= \int_0^1 e^{-2\pi i R^2\theta} \sum_{n_1,\ldots,n_d} e^{2\pi i(n_1^2+\cdots+n_d^2)\theta} d\theta. \end{aligned}$$

For any $\epsilon > 0$ (ϵ will be chosen as $\epsilon = R^2$),

$$\begin{aligned} r_d(R) &= e^{2\pi\epsilon R^2} \int_0^1 e^{-2\pi i R^2\theta} \sum_{n_1,\ldots,n_d} e^{2\pi i(n_1^2+\cdots+n_d^2)(\theta+i\epsilon)} d\theta \\ &= e^{2\pi\epsilon R^2} \int_0^1 e^{-2\pi i R^2\theta} \Big(\sum_n e^{2\pi n^2(\theta+i\epsilon)}\Big)^d d\theta \\ &= e^{2\pi\epsilon R^2} \int_0^1 e^{-2\pi i R^2\theta} \Big(\sum_n e^{-2\pi n^2(\epsilon-i\theta)}\Big)^d d\theta \\ &= e^{2\pi\epsilon R^2} \int_0^1 e^{-2\pi i R^2\theta} \big(\Theta(\epsilon-i\theta)\big)^d d\theta, \end{aligned}$$

where $\Theta(z) = \sum\limits_{n\in\mathbb{Z}} e^{-2\pi n^2 z}$ with $\mathrm{Re}(z) > 0$. Θ is called Theta function.

Lemma 4.1. *Let* $\theta \in (0,1)$ *and* $\theta = \frac{a}{q} + \beta, (a,q) = 1$. *Then*

$$\Theta(\epsilon - i\theta) = \frac{1}{q\sqrt{2\epsilon - i\beta}} \sum_{k\in\mathbb{Z}} \left(\sum_{l=0}^{q-1} e^{2\pi i \frac{a}{q} l^2} e^{2\pi i \frac{k}{q} l} \right) e^{-\frac{\pi k^2}{2q^2(\epsilon - i\beta)}}.$$

Proof.

$$\begin{aligned} \Theta(\epsilon - i\theta) &= \sum_n e^{-2\pi n^2(\epsilon - i\theta)} \\ &= \sum_n e^{-2\pi n^2(\epsilon - i\beta - i\frac{a}{q})} \\ &= \sum_{l=0}^{q-1} \sum_{k\in\mathbb{Z}} e^{-2\pi(kq+l)^2(\epsilon - i\beta - \frac{a}{q})} \\ &= \sum_{l=0}^{q-1} e^{2\pi i \frac{a}{q} l^2} \Big(\sum_k e^{-2\pi(kq+l)^2(\epsilon - i\beta)}\Big) \quad (*) \end{aligned}$$

Let $g(x) = e^{-2\pi(xq+l)^2(\epsilon-i\beta)}$. Then

$$
\begin{aligned}
\hat{g}(\xi) &= \int g(x)e^{-2\pi i x\xi}dx \\
&= \int e^{-2\pi(xq+l)^2(\epsilon-i\beta)}e^{-2\pi i x\xi}dx \\
&= \frac{e^{2\pi i\frac{a}{q}l}}{q\sqrt{2(\epsilon-i\beta)}}\int e^{-\pi\nu^2}e^{-2\pi i\xi\nu\frac{1}{q\sqrt{2(\epsilon-i\beta)}}}d\nu \quad (\text{because} \quad \nu=\sqrt{2(\epsilon-i\beta)}(xq+l)) \\
&= \frac{e^{2\pi i\frac{a}{q}l}}{q\sqrt{2(\epsilon-i\beta)}}e^{-\frac{\pi\xi^2}{2q^2(\epsilon-i\beta)}}.
\end{aligned}
$$

Applying Poisson summation formula,

$$
\sum_k e^{-2\pi(kq+l)^2(\epsilon-i\beta)} = \sum_k \hat{g}(k) = \sum_k e^{2\pi i\frac{k}{q}l}\frac{1}{q\sqrt{2(\epsilon-i\beta)}}e^{-\frac{\pi k^2}{2q^2(\epsilon-i\beta)}}.
$$

Returning to $(*)$, we get

$$
\Theta(\epsilon-i\theta) = \frac{1}{q\sqrt{2(\epsilon-i\beta)}}\sum_{k\in\mathbb{Z}}\left(\sum_{l=0}^{q-1}e^{2\pi i\frac{a}{q}l^2}e^{2\pi i\frac{k}{q}l}\right)e^{-\frac{\pi k^2}{2q^2(\epsilon-i\beta)}}. \tag{4.1}
$$

□

The principal contribution of R.H.S. in (4.1) is the term with $k=0$. To see this, we set the error term $\mathcal{E}(\Theta)$ to be

$$
\mathcal{E}(\Theta) = \frac{1}{q\sqrt{2(\epsilon-i\beta)}}\sum_{k\neq 0}\left(\sum_{l=0}^{q-1}e^{2\pi\frac{a}{q}l^2}e^{2\pi i\frac{k}{q}l}\right)e^{-\frac{\pi k^2}{2q^2(\epsilon-i\beta)}}.
$$

Then we write

$$
\Theta(\epsilon-i\theta) = \frac{S(a,q)}{q\sqrt{2(\epsilon-i\beta)}} + \mathcal{E}(\Theta). \tag{4.2}
$$

Here the first term is the principal contribution and the error term $\mathcal{E}(\Theta)$ is small. To control the error term, we have the following lemma.

Lemma 4.2. *Suppose* $|\beta| \leq \frac{\sqrt{\epsilon}}{q}$ *and* $\epsilon < \frac{1}{q^2}$. *Then* $\exists\delta>0$ *s.t.*

$$
\left|\mathcal{E}(\Theta)\right| \leq \frac{C_\delta}{\sqrt{q}(\epsilon^2+\delta^2)^{\frac{1}{4}}}e^{-\frac{8\epsilon}{q^2(\epsilon^2+\beta^2)}} \leq C\cdot\epsilon^{-\frac{1}{4}} = O(\epsilon^{-\frac{1}{4}}).
$$

Proof.

$$
\begin{aligned}
|\mathcal{E}(\Theta)| &\le \frac{1}{q|\sqrt{2(\epsilon-i\beta)}|}\sum_{k\neq 0}\left|\sum_{l=0}^{q-1} e^{2\pi i\frac{a}{q}l^2}e^{2\pi i\frac{k}{q}l}\right| \left|e^{-\frac{\pi k^2}{2q^2(\epsilon-i\beta)}}\right| \\
&\le \frac{\sqrt{2q}}{q\sqrt{2}(\epsilon^2+\beta^2)^{\frac{1}{4}}}\sum_{k\neq 0} e^{-\frac{\pi k^2}{2q(\epsilon-i\beta)}} \quad \text{(by Gauss sum estimate)} \\
&\le \frac{1}{\sqrt{q}(\epsilon^2+\beta^2)^{\frac{1}{4}}}\sum_{k\neq 0} e^{-\frac{\pi k^2}{2q(\epsilon-i\beta)}} \\
&\le \frac{C_\delta}{\sqrt{q}(\epsilon^2+\beta^2)^{\frac{1}{4}}} e^{-\frac{\delta\epsilon}{2q^2(q^2+\beta^2)}} \quad \text{(because } k\neq 0, \text{ for instance } \delta = \frac{1}{2}\text{)}. \qquad (**)
\end{aligned}
$$

We set $X = \frac{\epsilon}{q^2(\epsilon^2+\beta^2)}$,

$$\frac{C_\delta}{\sqrt{q}(\epsilon^2+\beta^2)^{\frac{1}{4}}} e^{-\frac{\delta\epsilon}{2q^2(\epsilon^2+\beta^2)}} = C_\delta \epsilon^{-\frac{1}{4}} X^{\frac{1}{4}} e^{-\frac{\delta}{2}X}.$$

From $|\beta| \le \frac{\sqrt{\epsilon}}{q}$, $\epsilon^2+\beta^2 \le \epsilon^2 + \frac{\epsilon}{q^2} \le \frac{2\epsilon}{q^2}$ $(\epsilon < \frac{1}{q^2})$

$$\Rightarrow X \ge \frac{1}{q^2\cdot\frac{2q}{q^2}} = \frac{1}{2} \Rightarrow X^{\frac{1}{4}} e^{-\frac{\delta}{2}X} \le C'_\delta.$$

Thus $\mathcal{E}(\Theta) = O(\epsilon^{\frac{1}{4}})$. □

Finally, we are ready to get the estimate of the d-th power of the Theta function.

Lemma 4.3. *Suppose* $|\beta| \le \frac{\sqrt{\epsilon}}{q}$ *and* $\epsilon < \frac{1}{q^2}$. *Then*

$$\left(\Theta\left(\epsilon - i\left(\beta+\frac{a}{q}\right)\right)\right)^d = \frac{S(a,q)^d}{q^d(\sqrt{2(\epsilon-i\beta)})^d} + O(\epsilon^{-\frac{d}{4}}). \tag{4.3}$$

Proof. From (4.2) and the binomial theorem, we have

$$
\begin{aligned}
\left(\Theta\left(\epsilon - i\left(\beta+\frac{a}{q}\right)\right)\right)^d &= \left(\frac{1}{q\sqrt{2(\epsilon-i\beta)}}S(a,q)+\mathcal{E}(\Theta)\right)^d \\
&= \sum_{j=0}^{d}\binom{d}{j}\left(\frac{S(a,q)}{q\sqrt{2(\epsilon-i\beta)}}\right)^{d-j}(\mathcal{E}(\Theta))^j \\
&:= \frac{S(a,q)}{q^d(\sqrt{2(\epsilon-i\beta)})^d} + I(\Theta).
\end{aligned}
$$

We claim that: $I(\Theta) = O(\epsilon^{-\frac{d}{4}})$, then we get the conclusion.

To see the Claim on the upper bound of $I(\Theta)$, we estimate

$$\begin{aligned}\left|I(\Theta)\right| &\leq \sum_{j=1}^{d}\binom{d}{j}\Big(\frac{\left|S(a,q)\right|}{q\sqrt{2}(\epsilon^2+\beta^2)^{\frac{1}{4}}}\Big)^{d-j}(\mathcal{E}(\Theta))^j \\ &\leq \sum_{j=1}^{d}\binom{d}{j}\frac{1}{q^{\frac{d}{2}}(\epsilon^2+\beta^2)^{\frac{d}{4}}}q^{\frac{j}{2}}(\epsilon^2+\beta^2)^{\frac{j}{4}}(\mathcal{E}(\Theta))^j \\ &\leq \sum_{j=1}^{d}\binom{d}{j}\frac{1}{q^{\frac{d}{2}}(\epsilon^2+\beta^2)^{\frac{d}{4}}}q^{\frac{j}{2}}(\epsilon^2+\beta^2)^{\frac{j}{4}}+\frac{C_\delta}{q^{\frac{j}{2}}(\epsilon^2+\beta^2)^{\frac{j}{4}}}e^{-\frac{\delta\epsilon j}{2q^2(\epsilon^2+\beta^2)}}\ \text{(by (**))} \\ &\leq \sum_{j=1}^{d}\binom{d}{j}\frac{C_\delta}{q^{\frac{d}{2}}(\epsilon^2+\beta^2)^{\frac{d}{4}}}e^{-\frac{\delta\epsilon j}{2q^2(\epsilon^2+\beta^2)}}.\end{aligned}$$

Set $X=\frac{\epsilon}{q^2(\epsilon^2+\beta^2)}$

$$\begin{aligned}\left|I(\Theta)\right| &\leq C_\delta\cdot\epsilon^{-\frac{d}{4}}X^{\frac{d}{4}}\sum_{j=1}^{d}\binom{d}{j}e^{-\frac{\delta}{2}Xj} \\ &\leq C'_{\delta,d}\epsilon^{-\frac{d}{4}}X^{\frac{d}{4}}e^{-\frac{\delta}{2}X}=O(\epsilon^{-\frac{d}{4}}).\end{aligned}$$

Therefore we finish the proof. □

5 Major arc and minor arc estimates

Recall

$$\begin{aligned}r_d(R) &= e^{2\pi\epsilon R^2}\int_0^1 d^{-2\pi iR^2\theta}\Big(\Theta(\epsilon-i\beta)\Big)^d d\theta, \\ J_{\frac{a}{q}} &= \Big\{\theta\in(0,1):\Big|\theta-\frac{a}{q}\Big|\leq\frac{1}{q[N]}\Big\}\quad \text{(We choose } N=[R]).\end{aligned}$$

Let

$$\begin{aligned}\mathcal{M}_R &= \bigcup_{q\geq\frac{[R]}{10}}\bigcup_{a\in\mathcal{P}}J_{\frac{a}{q}}, \\ \mathcal{M}_R^{\complement} &= (0,1)\setminus\mathcal{M}_R.\end{aligned}$$

So

$$\begin{aligned}r_d(R) &= e^{2\pi\epsilon R^2}\int_{\mathcal{M}_R}d^{-2\pi iR^2\theta}\Big(\Theta(\epsilon-i\beta)\Big)^d d\theta \\ &\quad + e^{2\pi\epsilon R^2}\int_{\mathcal{M}_R^{\complement}}d^{-2\pi iR^2\theta}\Big(\Theta(\epsilon-i\beta)\Big)^d d\theta \\ &:= I_{\mathcal{M}_R}+I_{\mathcal{M}_R^{\complement}}.\end{aligned}$$

Here $\mathcal{M}_R$ is the union of all major arcs.

Lemma 5.1 (Minor arc estimate). *Let* $\epsilon=R^{-2}$. *Then* $I_{\mathcal{M}_R^{\complement}}=O(R^{\frac{d}{2}})$.

Proof. For any $\theta \in I_{\mathcal{M}_R^{\complement}}$, by Dirichlet principle, $\exists\ a, q$ *s.t.*

$$\begin{cases} (a,q)=1, & \dfrac{[R]}{10} \le q \le [R], \\ \left|\theta - \dfrac{a}{q}\right| \le \dfrac{1}{[R]q}. \end{cases}$$

Inserting the absolute vales in (4.3) and applying the triangle inequality, we have

$$\begin{aligned} \left|\Theta(\epsilon - i\theta)\right|^d &\le C\frac{\left|S(a,q)\right|^d}{q^d(\epsilon^2+\beta^2)^{\frac{d}{4}}} + C\epsilon^{-\frac{d}{4}} \\ &\le \frac{C}{q^{\frac{d}{2}}\epsilon^{\frac{d}{2}}} + CR^{\frac{d}{2}} \\ &= O(R^{\frac{d}{2}}). \end{aligned}$$

□

We now turn to the major arc estimate. For simplicity, we assume $R \in \mathbb{N}$. Then

$$\begin{aligned} I_{\mathcal{M}_R} &= e^{2\pi}\int_{\bigcup\limits_{q \ge \frac{R}{10}} J_{\frac{a}{q}}} e^{-2\pi i R^2\theta}\Big(\Theta(\epsilon - i\theta)\Big)^d d\theta \\ &= e^{2\pi}\sum_{q=1}^{\frac{R}{10}}\sum_{a\in\mathcal{P}_q}\int_{J_{\frac{a}{q}}} e^{-2\pi i R^2\theta}\Big(\Theta(\epsilon - i\theta)\Big)^d d\theta. \end{aligned}$$

Let $\theta = \frac{a}{q} + \beta$ and apply (4.3),

$$\begin{aligned} I_{\mathcal{M}_R} &= e^{2\pi}\sum_{q=1}^{\frac{R}{10}}\sum_{a\in\mathcal{P}_q} e^{-2\pi i R^2\frac{a}{q}}\int_{J_{\frac{a}{q}}} e^{-2\pi i R^2\beta}\big(\Theta(\epsilon - i\theta)\big)^d d\theta \\ &= e^{2\pi}\sum_{q=1}^{\frac{R}{10}}\sum_{a\in\mathcal{P}_q} e^{-2\pi i R^2\frac{a}{q}}\int_{J_{\frac{a}{q}}} e^{-2\pi i R^2\beta}\Big(\frac{S(a,q)^d}{q^d\big(\sqrt{2(\epsilon - i\beta)}\big)^d}\Big) + O(R^{\frac{d}{2}}) \\ &= e^{2\pi}\sum_{q=1}^{\frac{R}{10}}\sum_{a\in\mathcal{P}_q} e^{-2\pi i R^2\frac{a}{q}}\int_{J_{\frac{a}{q}}} e^{-2\pi i R^2\beta}\frac{S(a,q)^d}{q^d\big(\sqrt{2(\epsilon - i\beta)}\big)^d} d\theta \\ &\quad + e^{2\pi}\sum_{q=1}^{\frac{R}{10}}\sum_{a\in\mathcal{P}_q} e^{-2\pi i R^2\frac{a}{q}}\int_{J_{\frac{a}{q}}} e^{-2\pi i R^2\beta} O(R^{\frac{d}{2}}) d\theta \\ &:= \mathrm{II}_{\mathcal{M}_R} + O_{\mathcal{M}_R}. \end{aligned}$$

The second term $O_{\mathcal{M}_R}$ is easy to handle. Indeed, we have the following lemma.

Lemma 5.2. $O_{\mathcal{M}_R} = O(R^{\frac{d}{2}})$.

Proof. Since major arcs are disjoint, we have

$$\begin{aligned}
\left|O_{\mathcal{M}_R}\right| &\leq e^{2\pi}\Big(\sum_{q=1}^{\frac{R}{10}}\sum_{a\in\mathcal{P}_q} e^{-2\pi i R^2\frac{a}{q}}\int_{J_{\frac{a}{q}}} e^{-2\pi i R^2\beta}\Big)\,O(R^{\frac{d}{2}})\\
&\leq e^{2\pi}\left|\bigcup_{\frac{a}{q}} J_{\frac{a}{q}}\right| O(R^{\frac{d}{2}})\\
&= O(R^{\frac{d}{2}}).
\end{aligned}$$

□

For the major arc part, we have

Lemma 5.3.

$$\mathrm{II}_{\mathcal{M}_R} = C_d R^{d-2}\sum_{q=1}^{\frac{R}{10}}\sum_{a\in\mathcal{P}_q} e^{-2\pi i R^2\frac{a}{q}}\frac{S(a,q)^d}{q^d} + O(R^{\frac{d}{2}}) \tag{5.1}$$

where C_d is a constant depending on d ($C_d \neq 0$).

Proof. First write

$$\mathrm{II}_{\mathcal{M}_R} = e^{2\pi}\sum_{q=1}^{\frac{R}{10}}\sum_{a\in\mathcal{P}_q} e^{-2\pi i R^2\frac{a}{q}}\frac{S(a,q)^d}{q^d}\int_{|\beta|\leq\frac{1}{qR}}\frac{e^{-2\pi i R^2\beta}}{(\sqrt{2(\epsilon-i\beta)})^d}d\beta\,.$$

Set $I_{R,q} = \int_{|\beta|\leq\frac{1}{qR}}\frac{e^{-2\pi i R^2\beta}}{(\sqrt{2(\epsilon-i\beta)})^d}d\beta$ and change variables $\beta = R^{-2}\alpha$. We then have

$$\begin{aligned}
I_{R,q} &= R^{d-2}\int_{-\frac{R}{q}}^{\frac{R}{q}}\frac{e^{-2\pi i\alpha}}{(\sqrt{2(1-i\alpha)})^d}d\alpha\\
&= R^{d-2}\int_{-\infty}^{\infty}\frac{e^{-2\pi i\alpha}}{(\sqrt{2(1-i\alpha)})^d}d\alpha\\
&\quad + R^{d-2}\int_{-\frac{R}{q}}^{\frac{R}{q}}\frac{e^{-2\pi i\alpha}}{(\sqrt{2(1-i\alpha)})^d}d\alpha\\
&\quad - R^{d-2}\int_{-\infty}^{\infty}\frac{e^{-2\pi i\alpha}}{(\sqrt{2(1-i\alpha)})^d}d\alpha.
\end{aligned}$$

Notice that the integrals in second term and last term are dominated by

$$\begin{aligned}
\left|\int_{-\frac{R}{q}}^{\frac{R}{q}}\ldots d\alpha - \int_{-\infty}^{\infty}\ldots d\alpha\right| &\leq \left|\int_{-\infty}^{-\frac{R}{q}}\ldots d\alpha + \int_{-\frac{R}{q}}^{\infty}\ldots d\alpha\right|\\
&\leq 2\int_{\frac{R}{q}}^{\infty}\frac{C}{\alpha^{\frac{d}{2}}}d\alpha\\
&= O\left(\frac{q^{\frac{d}{2}-1}}{R^{\frac{d}{2}-1}}\right).
\end{aligned}$$

Let $C_0 = \int_{-\infty}^{\infty} \frac{e^{-2\pi i\alpha}}{(\sqrt{2(1-i\alpha)})^d} d\alpha$. Thus we write

$$I_{R,q} = C_0 R^{d-2} + O\big((qR)^{\frac{d}{2}-1}\big). \tag{5.2}$$

Remark. C_0 is a non-zero constant! In the case of $d = 1$, $C_0 = i\sqrt{2}\int_a^{\infty} \frac{e^{-2\pi y}}{\sqrt{y-1}} dy$. Thus (5.2) allows us to represent

$$\begin{aligned}
\amalg_{\mathcal{M}_R} &= e^{2\pi} \sum_{q=1}^{\frac{R}{10}} \sum_{a \in \mathcal{P}_q} e^{-2\pi i R^2 \frac{a}{q}} \frac{S(a,q)^d}{q^d} I_{R,q} \\
&= e^{2\pi} C_0 R^{d-2} \sum_{q=1}^{\frac{R}{10}} \sum_{a \in \mathcal{P}_q} e^{-2\pi i R^2 \frac{a}{q}} \frac{S(a,q)^d}{q^d} \\
&+ e^{2\pi} \sum_{q=1}^{\frac{R}{10}} \sum_{a \in \mathcal{P}_q} e^{-2\pi i R^2 \frac{a}{q}} \frac{S(a,q)^d}{q^d} O\big((qR)^{\frac{d}{2}-1}\big) \\
&:= I + II.
\end{aligned}$$

The second term satisfies

$$\begin{aligned}
II &\le C R^{\frac{d}{2}-1} \sum_{q=1}^{R} \sum_{a \in \mathcal{P}_q} \frac{|S(a,q)|^d}{q^{\frac{d}{2}+1}} \\
&\le C R^{\frac{d}{2}-1} \sum_{q=1}^{R} \sum_{a \in \mathcal{P}_q} \frac{1}{q} \\
&= O(R^{\frac{d}{2}}).
\end{aligned}$$

□

Recall the singular series

$$\mathfrak{S}_d(R) = \sum_{q=1}^{\infty} \sum_{a \in \mathcal{P}_q} e^{-2\pi i R^2 \frac{a}{q}} \frac{S(a,q)^d}{q^d}.$$

Hence we aim to replace the first term in (5.1) by the singular series. In fact, we have

Lemma 5.4.

$$\amalg_{\mathcal{M}_R} = C_d R^{d-2} \mathfrak{S}_d(R) + O(R^{\frac{d}{2}}), \quad \textit{if } d \ge 5. \tag{5.3}$$

Proof.

$$\begin{aligned}
&\left|\sum_{q\geq\frac{R}{10}}\sum_{a\in\mathcal{P}_q} e^{-2\pi i R^2\frac{a}{q}}\frac{S(a,q)^d}{q^d}\right| \\
&\leq C\sum_{q\geq\frac{R}{10}}\sum_{a\in\mathcal{P}_q}\frac{q^{\frac{d}{2}}}{q^d} \\
&\leq C\sum_{q\geq\frac{R}{10}}\frac{1}{q^{\frac{d}{2}-1}} \\
&= O(R^{2-\frac{d}{2}})\leq CR^{\frac{d}{2}}.
\end{aligned}$$

Thus (5.3) follows from (5.1).

□

From the lemma above we obtain (3.1) immediately. Henceforth we have $r_d(R)\sim R^{d-2}$ for $d\geq 5$. This finishes the proof of the main theorem: **Theorem 3.1**.

6 Weyl sum

Let $P(x)$ be a polynomial of degree d with real coefficients, $S_N=\sum_{n=1}^{N}e^{2\pi iP(n)}$ is called a Weyl sum. The pointwise estimate of this type of exponential sum is an important but only partially understood topic in additive number theory. The first non-trivial estimate was obtained by Weyl. Later Vinogradov improved the higher-order cases by using a refined circle method (now is called the Vinogradov method). We only introduce Weyl's differencing method in this course.

The simplest case is the linear polynomial case. In this case the Weyl sum can be treated easily. For $\alpha\in\mathbb{R}$, let $\|\alpha\|$ denote the distance from α to its nearest integer.

Lemma 6.1. *For any* $\alpha\in\mathbb{R}$, $2\|\alpha\|\leq|\sin(\pi\alpha)|$.

Proof. Let n_α be the nearest integer to α, then

$$\begin{aligned}
&\alpha=n_\alpha+\|\alpha\| \text{ (or } \alpha=n_\alpha-\|\alpha\|) \\
\Rightarrow&|\sin(\pi\alpha)|=|\sin(\pi n_\alpha\pm\pi\|\alpha\|)|=|\sin(\pi\|\alpha\|)|\geq 2\|\alpha\|,
\end{aligned}$$

because $|2x|\leq|\sin(\pi x)|$ if $0\leq x\leq\frac{1}{2}$.

□

Theorem 6.1. $\left|\sum_{n=1}^{N}e^{2\pi i\alpha n}\right|\leq\min\left\{N,\ \frac{1}{2\|\alpha\|}\right\}$.

Proof.

$$\begin{aligned}\sum_{n=1}^{N} e^{2\pi i\alpha n} &= \frac{e^{2\pi i(N+1)\alpha} - e^{2\pi i\alpha}}{e^{2\pi i\alpha} - 1}\\ &= \frac{e^{2\pi i\alpha}\left(e^{2\pi iN\alpha} - 1\right)}{e^{2\pi i\alpha} - 1}\\ &= \frac{e^{2\pi i\alpha}e^{2\pi iN\alpha}\left(e^{\pi iN\alpha} - e^{-\pi iN\alpha}\right)}{e^{\pi i\alpha}\left(e^{\pi i\alpha} - e^{-\pi i\alpha}\right)}\\ &= e^{\pi i(N+1)\alpha}\frac{\sin(\pi N\alpha)}{\sin(\pi\alpha)}.\end{aligned}$$

Applying the previous lemma, we get

$$\left|\sum_{n=1}^{N} e^{2\pi i\alpha n}\right| \le \frac{1}{|\sin(\pi\alpha)|} \le \frac{1}{2\|\alpha\|}.$$

On the other hand, clearly the sum is bounded by N. Putting them together, we have $\left|\sum\limits_{n=1}^{N} e^{2\pi i\alpha n}\right| \le \min\left\{N, \frac{1}{2\|\alpha\|}\right\}$. □

We now consider quadratic Weyl sum

$$S = \sum_{n=1}^{N} e^{2\pi i(\alpha n^2 + \beta n)}.$$

To get a non-trivial estimate, we need a technical lemma first.

Lemma 6.2. *Suppose* $\alpha \in \mathbb{R}$ *satisfies* $\left|\alpha - \frac{a}{q}\right| \le \frac{1}{q^2}$ *for some* $a \in \mathbb{Z}$, $q \in \mathbb{N}$ *with* $(a,q) = 1$. *Then*

$$\sum_{h=1}^{H} \min\left\{N, \frac{1}{\|\alpha h\|}\right\} \le C\Big(\frac{HN}{q} + H\ln q + N + q\ln q\Big). \tag{6.1}$$

Proof. **Case 1**: $H \le q$,

$$\begin{aligned}\sum_{h=1}^{H} \min\left\{N, \frac{1}{\|\alpha h\|}\right\} &\le \sum_{h=1}^{q} \min\left\{N, \frac{1}{\|\alpha h\|}\right\}\\ &= N + \sum_{h=1}^{q-1} \min\left\{N, \frac{1}{\|\alpha h\|}\right\}.\end{aligned}$$

Claim. $\sum\limits_{h=1}^{q-1} \min\left\{N, \frac{1}{\|\alpha h\|}\right\} \le 2(N + q\ln q)$.

Proof. From $\left|\alpha - \frac{a}{q}\right| \le \frac{1}{q^2}$, $\alpha = \frac{a}{q} + \frac{\theta}{q^2}$ for some $|\theta| \le 1$. Thus $\alpha h = \frac{ah}{q} + \frac{\theta h}{q^2}$.

Let $\epsilon = \frac{\theta h}{q^2}$. We have $|\epsilon| < \frac{1}{q}$ since $h < q$, $|\theta| \le 1$. Observe that if h runs over $1, 2, \ldots, q-1$ (all non-zero residues modulo q), then ah runs over all non-zero residues modulo q. Thus

$$\begin{aligned}\sum_{h=1}^{q-1} \min\left\{N, \frac{1}{\|\alpha h\|}\right\} &= \sum_{h=1}^{q-1} \min\left\{N, \frac{1}{\|\frac{ah}{q} + \epsilon\|}\right\} \\ &= \sum_{m=1}^{q-1} \min\left\{N, \frac{1}{\|\frac{m}{q} + \epsilon\|}\right\} \\ &\le 2N + \sum_{m=2}^{q-2} \frac{1}{\|\frac{m}{q} + \epsilon\|}.\end{aligned}$$

Because $|\epsilon| < \frac{1}{q}$,

$$\begin{aligned}|\frac{m}{q} + \epsilon| &\ge \frac{m}{q} - |\epsilon| \ge \frac{m-1}{q} > 0 \text{ since } m \ge 2, \\ |\frac{m}{q} + \epsilon| &\le \frac{m}{q} + |\epsilon| \le \frac{m+1}{q} < 1 \text{ since } m \le q-2,\end{aligned}$$

$$\begin{aligned}\sum_{h=1}^{q-1} \min\left\{N, \frac{1}{\|\alpha h\|}\right\} &\le 2N + \sum_{m=2}^{[\frac{q}{2}]} \frac{q}{m-1} + \sum_{m=[\frac{q}{2}]+1}^{q-2} \frac{1}{1 - \frac{m+1}{q}} \\ &\le 2N + \sum_{m=2}^{[\frac{q}{2}]} \frac{q}{m-1} + \sum_{m=[\frac{q}{2}]+1}^{q-2} \frac{q}{q = m+1} \\ &\le 2N + q\sum_{m=1}^{[\frac{q}{2}]} \frac{1}{m} + q\sum_{m=1}^{\frac{q}{2}} \frac{1}{m} \\ &\le 2(N + q\log q).\end{aligned}$$

Hence we finish the proof of the claim. □

The Case 1 now follows immediately from this claim.

Case 2: $H \ge q$. Without loss of generality, we can assume $q \mid H$.

$$\sum_{h=1}^{H} \min\left\{N, \frac{1}{\|\alpha h\|}\right\} = \sum_{k=0}^{\frac{H}{q}-1} \sum_{h=kq+1}^{kh+q} \min\left\{N, \frac{1}{\|\alpha h\|}\right\},$$

$$\sum_{h=kq+1}^{kq+q} \min\left\{N, \frac{1}{\|\alpha h\|}\right\} \le C(N + \log q) \qquad \text{(can be done as before)}$$

$$\Rightarrow \sum_{h=1}^{H} \min\left\{N, \frac{1}{\|\alpha h\|}\right\} \le \frac{CH}{q}(N + q\log q).$$ □

Theorem 6.2 (Weyl). *Suppose* $\alpha \in \mathbb{R}$ *satisfies* $\left|\alpha - \frac{a}{q}\right| \le \frac{1}{q^2}$ *for some* $a \in \mathbb{Z},\ q \in \mathbb{N}$ *with* $(a, q) = 1$. *Then*

$$\left|\sum_{n=1}^{N} e^{2\pi i(\alpha n^2+\beta n)}\right| \le C\left(\frac{N}{\sqrt{q}} + \sqrt{N\log q} + \sqrt{q\log q}\right),$$

here C *is an absolute constant independent of* α, β, N, a, q.

Proof. Let $S = \sum_{n=1}^{N} e^{2\pi i(\alpha n^2+\beta n)}$ and $P(n) = \alpha n^2 + \beta n$. Then

$$\begin{aligned}
|S|^2 = S\bar{S} &= \sum_{n=1}^{N}\sum_{m=1}^{N} e^{2\pi i(P(m)-P(n))} \\
&= \sum_{n=1}^{N}\sum_{h=1-n}^{N-n} e^{2\pi i(P(n+h)-P(n))} \quad \text{(because } m = n+h\text{)} \\
&= N + \sum_{h=1}^{N-1}\sum_{n=1}^{N-h} e^{2\pi i(P(n)-P(n+h))} + \sum_{h=1}^{N-1}\sum_{n=1}^{N-h} e^{2\pi i(P(n+h)-P(n))}.
\end{aligned}$$

We use $P(n+h) - p(n) = (2\alpha h)n + \alpha h^2 + \beta h$. Thus

$$\begin{aligned}
|S|^2 &= N + \sum_{h=1}^{N-1} e^{-2\pi i(\alpha h^2+\beta h)}\left(\sum_{n=1}^{N-h} e^{2\pi i(-2\alpha h)n}\right) \\
&\quad + \sum_{h=1}^{N-1} e^{2\pi i(\alpha h^2+\beta h)}\left(\sum_{n=1}^{N-h} e^{2\pi i(-2\alpha h)n}\right) \\
&\le N + \sum_{h=1}^{N-1}\left|\left(\sum_{n=1}^{N-h} e^{2\pi i(-2\alpha h)n}\right)\right| + \sum_{h=1}^{N-1}\left|\sum_{h=1}^{N-1} e^{2\pi i(\alpha h^2+\beta h)}\right| \\
&\le N + 2\sum_{h=1}^{N-1} \min\left\{N, \frac{1}{2\|2\alpha h\|}\right\} \\
&\le N + 2 \sum_{h:\text{even between } 2 \text{ and } 2N} \min\left\{N, \frac{1}{2\|\alpha h\|}\right\} \\
&\le N + 2\sum_{h=1}^{2N} \min\left\{N, \frac{1}{\|\alpha h\|}\right\}.
\end{aligned}$$

Now we apply (6.1) to get

$$|S|^2 \le N + C\left(\frac{N^2}{q} + N\log q + q\log q\right).$$

Taking the square root for both sides, we finish the proof. □

7 Weyl sum—High order cases

In this section, let $P(x) = \sum_{k=0}^{d} \alpha_k x^k = \alpha_d x + \cdots + \alpha_1 x + \alpha_0, \alpha_j \in \mathbb{R}$ and $\left|\alpha_d - \frac{a}{q}\right| \leq \frac{1}{q^2}$ for some a, q with $(a, q) = 1$. We study the Weyl sum

$$S = \sum_{n=1}^{N} e^{2\pi i P(n)} .$$

In order to estimate the higher order Weyl sum, we need some technical lemmata.

Lemma 7.1.

$$(a+b)^m \leq 2^m (a^m + b^m), \;\; m \in \mathbb{N}, \;\; a, b \geq 0 . \tag{7.1}$$

Proof. The proof is easy. We leave it as an exercise. □

Lemma 7.2.

$$\left(\sum_{k=1}^{M} a_k\right)^{2^D} \leq M^{2^{D-1}} \left(\sum_{k=1}^{M} a_k^{2^D}\right) \tag{7.2}$$

where $M, \; D \in \mathbb{N}, \; a_k \geq 0$.

Proof. We use mathematical induction on D. First for $D = 1$, $\left(\sum_{k=1}^{M} a_k\right)^2 \leq M \sum_{k=1}^{M} a_k^2$ by Cauchy-Schwarz inequality. Assume the lemma holds for D, then

$$\begin{aligned}
\left(\sum_{k=1}^{M} a_k\right)^{2^{D+1}} &= \left(\left(\sum_{k=1}^{M} a_k\right)^2\right)^{2^D} \\
&\leq \left(M \sum_{k=1}^{M} a_k^2\right)^{2^D} \\
&= M^{2^D} \left(\sum_{k=1}^{M} a_k^2\right)^{2^D} \\
&\leq M^{2^D} M^{2^{D-1}} \left(\sum_{k=1}^{M} a_k^2\right)^{2^{D+1}} \\
&\leq M^{2^{D+1}} \left(\sum_{k=1}^{M} a_k^2\right)^{2^{D+1}} .
\end{aligned}$$

Thus the lemma follows. □

Lemma 7.3. *Let P be a real polynomial of degree $d \geq 2$ and $S = \sum_{n=1}^{N} e^{2\pi i P(n)}$. Then*

$$\left|S\right|^{2^{d-1}} \leq C_d \left(N^{2^{d-1}-1} + N^{2^{d-1}-d} \sum_{h_1,\ldots,h_{d-1}} \left| \sum_{n} e^{2\pi i P_{h_1,\ldots,h_{d-1}}(n)} \right| \right) \tag{7.3}$$

where $P_{h_1,\ldots,h_{d-1}}(n)$ is a linear polynomial of n with leading coefficient $d!h_1,\ldots,h_{d-1}\alpha_d$.

Proof. Again we use the induction on d. The $d = 2$ case was done in §6 for the quadratic case. Assume that (7.3) holds for any real polynomial of degree d. Let

$$P(n) = \sum_{k=0}^{d+1} \alpha_k n^k, \quad S = \sum_{n=1}^{N} e^{2\pi i P(n)}.$$

We show that

$$|S|^{2^d} \leq C_{d+1} \left(N^{2^d-1} + N^{2^d-(d+1)} \sum_{h_1,\ldots,h_d} \left| \sum_{n} e^{2\pi i P_{h_1,\ldots,h_d}(n)} \right| \right)$$

where $P_{h_1,\ldots,h_d}(n) = (d+1)!h_1 \cdots h_d \alpha_{d+1} n + \beta_0$ for some $\beta_0 \in \mathbb{R}$.

First, as we did in the previous section, one sees

$$|S|^2 \leq C \left(N + \sum_{h=1}^{N-1} \left| \sum_{n=1}^{N-h} e^{2\pi i \left(P(n+h) - P(n)\right)} \right| \right). \tag{7.4}$$

Notice that

$$P(n+h) - P(n) = \sum_{k=0}^{d+1} \alpha_k \left((n+h)^k - n^k \right),$$

which is a polynomial (of n) of degree d with leading coefficient $(d+1)\alpha_{d+1}h$. Then we estimate the $|S|^{2^d}$ by

$$\begin{aligned}
|S|^{2^d} &= \left(|S|^2\right)^{2^{d-1}} \\
&\leq C_d \left(N + \sum_{h=1}^{N-1} \left| \sum_{n=1}^{N-h} e^{2\pi i \left(P(n+h)-P(n)\right)} \right| \right)^{2^{d-1}} \\
&\leq C_d \left(N^{2^{d-1}} + \tilde{C}_d \left(\sum_{h=1}^{N-1} \left| \sum_{n=1}^{N-h} e^{2\pi i \left(P(n+h)-P(n)\right)} \right| \right)^{2^{d-1}} \right) \quad \text{(by (7.1))} \\
&\leq C_d \left(N^{2^{d-1}} + N^{2^{d-2}} \sum_{h=1}^{N-1} \left| \sum_{n=1}^{N-h} e^{2\pi i \left(P(n+h)-P(n)\right)} \right|^{2^{d-1}} \right) \quad \text{(by (7.2))}.
\end{aligned}$$

The induction hypothesis yields that

$$\left|\sum_{n=1}^{N-h} e^{2\pi i\left(P(n+h)-P(n)\right)}\right|^{2^{d-1}} \leq C_d\left(N^{2^{d-1}-1}+N^{2^{d-1}-d}\sum_{h_1,\ldots,h_{d-1}}\left|\sum_{n} e^{2\pi i P_{h_1,\ldots,h_d}(n)}\right|\right)$$

where $P_{h_1,\ldots,h_{d-1}}(n)=(d+1)!h_1\cdots h_{d-1}\alpha_{d+1}n+\beta_1$. Let $h=h_d$, henceforth

$$|S|^{2^d} \leq C_d\left(N^{2^{d-1}}+N^{2^{d-2}}N^{2^{d-1}}+N^{2^{d-2}}N^{2^{d-1}-d}\sum_{h_1,\ldots,h_d}\left|\sum_{n} e^{2\pi i P_{h_1,\ldots,h_d}(n)}\right|\right)$$

$$\leq C_{d+1}\left(N^{2^d-1}+N^{2^d-(d+1)}\sum_{h_1,\ldots,h_d}\left|\sum_{n} e^{2\pi i P_{h_1,\ldots,h_d}(n)}\right|\right)$$

because $2^{d-1}\leq 2^d-1$ and $2^{d-2}+2^{d-1}\leq 2^d-1$ whenever $d\geq 2$. Therefore we are done by induction. □

Theorem 7.1. *Let* $P(x)=\sum_{k=0}^{d}\alpha_k x^k$ *be real polynomial. Suppose* $\left|\alpha_d-\frac{a}{q}\right|\leq\frac{1}{q^2}$ *for some* $a\in\mathbb{Z}$, $q\in\mathbb{N}$ *with* $(a,q)=1$. *Then*

$$\left|\sum_{n=1}^{N} e^{2\pi i P(n)}\right| \leq C_{d,\delta}N^{1+\epsilon}\left(\frac{1}{q}+\frac{1}{N}+\frac{q}{N^d}\right)^{\delta} \tag{7.5}$$

for any ϵ, *here* $\delta=2^{1-d}$.

Remarks. If $q\geq N^d$, the R.H.S $\ll N$. So it is a trivial bound. Only for $q\geq N^d$, the R.H.S gives a non-trivial bound.

Proof. Assume $q\geq N^d$. Let $\delta=2^{1-d}$. By (7.3) and the estimate $\left|\sum_{n=1}^{N}e^{2\pi i n\alpha}\right|\leq C\min\left\{N,\frac{1}{\|\alpha\|}\right\}$.

$$|S|^{\frac{1}{\delta}}=|S|^{2^{d-1}}=C_d\left(N^{2^{d-1}-1}+N^{2^{d-1}-d}\sum_{h_1,\ldots,h_{d-1}}\min\left\{N,\frac{1}{\|d!h_1\cdots h_{d-1}\alpha_d\|}\right\}\right)$$

Notice $1\leq d!h_1\cdots h_{d-1}\leq d!N^{d-1}$ since $1\leq h_j\leq N$. So

$$\sum_{h_1,\ldots,h_{d-1}}\min\left\{N,\frac{1}{\|d!h_1\cdots h_{d-1}\alpha_d\|}\right\}=\sum_{h=1}^{d!N^{d-1}}\sum_{\substack{h_1,\ldots,h_{d-1}\\ d!h_1\cdots h_{d-1}\alpha_d=h}}\min\left\{N,\frac{1}{\|d!h\alpha_d\|}\right\}$$

$$=\sum_{h=1}^{d!N^{d-1}}\min\left\{N,\frac{1}{\|d!h\alpha_d\|}\right\}\#\left\{(h_1,\ldots,h_{d-1}):d!h_1\cdots h_{d-1}\alpha_d=h\right\}.$$

Here $\#(S)$ denotes the number of elements in the set S. From the equation $d!h_1 \cdots h_{d-1}\alpha_d = h$, we see that each h_j $(j = 1, \ldots, d-1)$ is a divisor of h. Thus we have

$$\#\{(h_1, \ldots, h_{d-1}) : d!h_1 \cdots h_{d-1}\alpha_d = h\} \leq (\tau(h))^{d-1},$$

where $\tau(h)$ is the number of divisors of h. It is well known that $\tau(h) \leq C_\epsilon h^\epsilon$ for any $\epsilon > 0$. Henceforth,

$$\#\{(h_1, \ldots, h_{d-1}) : d!h_1 \cdots h_{d-1}\alpha_d = h\} \leq C_\epsilon h^\epsilon \leq C_\epsilon N^\epsilon . \tag{7.6}$$

On the other hand, (6.1) yields

$$\sum_{h=1}^{d!N^{d-1}} \min\left\{N, \frac{1}{\|d!h\alpha_d\|}\right\} \leq C\left(\frac{d!N^d}{q} + d!N^{d-1}\log q + N + q\log q\right). \tag{7.7}$$

By utilizing (7.6) and (7.7), we then obtain

$$\sum_{h_1,\ldots,h_{d-1}} \min\left\{N, \frac{1}{\|d!h_1 \cdots h_{d-1}\alpha_d\|}\right\} \leq CN^\epsilon\left(\frac{d!N^d}{q} + d!N^{d-1}\log q + N + q\log q\right),$$

which implies

$$\begin{aligned} |S|^{\frac{1}{\delta}} &\leq C_d\left(N^{2^{d-1}-1+2\epsilon} + \frac{N^{2^{d-1}+2\epsilon}}{q} + N^{2^{d-1}-d+2\epsilon}q\right) \\ &= C_{d,\epsilon}N^{2^{d-1}+2\epsilon}\left(\frac{1}{N} + \frac{1}{q} + \frac{q}{N^d}\right). \end{aligned}$$

Take δ-power for both sides and we have

$$\begin{aligned} |S| &\leq C_{d,\epsilon}N^{1+2\delta\epsilon}\left(\frac{1}{N} + \frac{1}{q} + \frac{q}{N^d}\right)^\delta \\ &\leq C_{d,\epsilon}N^{1+\epsilon}\left(\frac{1}{N} + \frac{1}{q} + \frac{q}{N^d}\right)^\delta . \end{aligned}$$

□

Acknowledgement. Sincere gratitude goes to Professors Xi, Professor Yang, Professor Zhang and Professor Wang for their kind invitation and the warm hospitality I received while staying at CAS in this May. Finally I would like to thank Yang Zhang for his typing the note.

References

[1] Hugh L. Montgomery, Ten Lectures on the Interface Between Analytic Number Theory and Harmonic Analysis, CBMS 84, American Mathematical Society, 1994.

Lectures on the Analysis of Nonlinear Partial Differential Equations Vol. 4
MLM 4, pp. 115–193

Lectures on Elliptic Free Boundary Problems

Fanghua Lin

Abstract

This series of lectures is devoted to the study of some classical elliptic free boundary problems. The understanding of free boundary problems has been a fascinating and beautiful topic in the analysis and partial differential equations. Here we discuss some general existence and regularity theorems for solutions, as well as the results concerning the regularity of free boundaries. In particular, the classical obstacle problem are carefully examined. The study of singular sets of free boundaries was a very difficult and challenging task for many years. Here the important break through in late 1990s by Caffarelli is also presented in detail. One believes these ideas and techniques may be applied to a much wide class of problems.

1 Introduction

This write-up covers a set of ten lectures (of total 20 hours) selected from a course given by the author at the Courant Institute of Mathematical Sciences over the years under the title "Elliptic Free Boundary Problems". The author had given similar lectures in the summer schools at Tsinghua University in 2009, and at the East China Normal University in 2010. The literature on this subject is large, and many beautiful ideas and techniques have been developed over the last half century. The author has no intention here that this set of lectures would be fully comprehensive or relatively complete description of the whole theory. Rather, one would select a few important and elegant papers that illustrate well some of the key ideas involved in handling a couple of basic questions such as the existence and regularity of classical obstacle problems, and the regularity of free boundaries arising in the solutions. The basic references for these lectures are the book by D. Kinderlehrer and G. Stampacchia [13], the book by A. Friedman [5] and the book by L. Caffarelli and S. Salsa [3]. Otherwise, in each chapter I shall provide more references that the lectures are based on.

Examples of Free Boundary Problems

Let us first describe a few classical free boundary value problems.

Example 1.1 (A potential problem). Consider the following PDE in a bounded domain $\Omega \subseteq \mathbb{R}^n$

$$\begin{cases} -\Delta u = 1 \quad \text{in } \Omega, \\ u|_{\partial\Omega} = 0, \\ \left.\dfrac{\partial u}{\partial \nu}\right|_{\partial\Omega} = \text{constant}. \end{cases} \tag{1.1}$$

In general, this equation is overdetermined if Ω is given, but we shall ask what is (u, Ω) if they are both unknowns. The answer is that Ω is a ball and u is a radial quadratic function. The proof for this result was given by J. Serrin and H. Weinberger separately in 1972. Serrin's proof adopted the idea of moving plane. Weinberger's proof simply used integration by parts in a tricky way, which does not rely on the maximum principle. Hence, it could be generalized to some cases of elliptic systems. □

Example 1.2 (Schiffer's conjecture). Consider the following eigenvalue problem

$$\begin{cases} -\Delta u = \lambda u \quad \text{in } \Omega, \\ \left.\dfrac{\partial u}{\partial \nu}\right|_{\partial\Omega} = 0, \\ u|_{\partial\Omega} = \text{constant} \neq 0, \end{cases} \tag{1.2}$$

where λ is positive and is not necessarily the first eigenvalue for $-\Delta$ in Ω. Treating (u, Ω) as unknowns, Schiffer's conjecture asks: is Ω a ball?

Schiffer's conjecture is closely related to the Pompeiu problem, in the sense that the failure of the Pompeiu property is equivalent to the existence of a non-trivial solution of (1.2). A domain Ω is said to satisfy *Pompeiu property*, if and only if $f = 0$ is the only continuous function in $\mathbb{R}^n$ such that, for any isometry R of $\mathbb{R}^n$, $\int_{R(\Omega)} f(x)\,dx = 0$. It is known that if a bounded Lipschitz domain $\Omega \subseteq \mathbb{R}^n$ has a connected boundary $\partial\Omega$, and if Ω fails to have the Pompeiu property, then $\partial\Omega$ is real analytic. □

Example 1.3 (Minimizing the k-th Dirichlet eigenvalue of $-\Delta$ on a domain; Polya-Szegö problem). For $k \geq 1$, find out

$$\inf\{\lambda_k(\Omega) : \Omega \subseteq \mathbb{R}^n, |\Omega| = 1, \text{ and } \Omega \text{ is bounded, connected and Lipschitz}\},$$

where λ_k is the k-th Dirichlet eigenvalue of $-\Delta$ on Ω.

For $k = 1$, this problem is solved by Faber-Krahn inequality. It is well-known that the first Dirichlet eigenvalue of $-\Delta$ on Ω is given by

$$\lambda_1(\Omega) = \inf\left\{\int_\Omega |\nabla u|^2 : u \in H_0^1(\Omega) \text{ with } \int_\Omega u^2 = 1\right\}.$$

Faber-Krahn inequality states that for $n \geq 2$,

$$\lambda_1(B) \leq \lambda_1(\Omega), \qquad \text{for } \forall\, \Omega \subseteq \mathbb{R}^n \text{ with } |\Omega| = |B|,$$

where B is a ball in $\mathbb{R}^n$. The equality holds if and only if Ω is a ball.

For $k > 1$, the problem is still open. We remark that choosing the right conditions on domains is critical. The infimum may not be obtained over the class of bounded connected Lipschitz domains. In fact, we consider $\lambda_2(\Omega)$, where Ω is a bounded connected Lipschitz domain. Since the second eigenfunction u_2 must change sign in Ω, we define Ω_+ (respectively Ω_-) to be the subset of Ω on which $u_2 > 0$ (respectively $u_2 < 0$). Then $u|_{\Omega_+}$ and $u|_{\Omega_-}$ are the first eigenfunctions on Ω_+ and Ω_- respectively. By Faber-Krahn inequality, the corresponding eigenvalue would decrease if $\Omega_\pm$ are replaced by two identical balls $B_\pm$ whose total volume is equal to $|\Omega|$. We denote $\lambda_* = \lambda_1(B_+) = \lambda_1(B_-)$. Clearly, λ_* can not be achieved as $\lambda_2(\Omega)$ for some bounded connected Lipschitz domain Ω. We can further show that there exists a sequence of bounded connected Lipschitz domains Ω_k such that $\lambda_2(\Omega_k) \to \lambda_*$ as $k \to +\infty$. Therefore, the infimum is not obtained among bounded connected Lipschitz domains. □

An important class of free boundary problems is the obstacle problem, about which we shall discuss a lot.

Example 1.4 (Obstacle problem). Consider a domain Ω whose boundary is decomposed into two parts $\partial\Omega = \Gamma_s \cup \Gamma_f$ with $\Gamma_s \cap \Gamma_f$ intersects only on their boundaries (of Hausdorff dimension less than that of $\partial\Omega$). We fix Γ_s and leave Γ_f as the free boundary. We shall solve for (u, Γ_f) in the following equation:

$$\begin{cases} -\Delta u = f & \text{in } \Omega, \\ u|_{\Gamma_s} = g \geq 0, \\ u|_{\Gamma_f} = 0, \\ \left.\dfrac{\partial u}{\partial \nu}\right|_{\Gamma_f} = 0. \end{cases} \tag{1.3}$$

The equation (1.3) comes from the following model problem in $\mathbb{R}$. Given $\psi \in C^0[p,q]$, $\psi(p) < 0$, $\psi(q) < 0$ and ψ is positive somewhere on (p,q), let $\Omega = \{(x,y) : x \in [p,q], y \geq \psi(x)\}$. Solve

$$\min\left\{l(\gamma) : \gamma(0) = p, \gamma(1) = q, \gamma \subseteq \Omega\right\}, \tag{1.4}$$

where γ is a curve connecting p and q and $l(\gamma)$ is its length. ψ is called the obstacle function.

Suppose u is the solution. We would find $u(x) \geq \psi(x)$, $u(p) = u(q) = 0$, and $u_{xx} = 0$ if $u > \psi$. Denote $q' = \sup\{x : u(x) = \psi(x)\}$. It is easy to show $p < q' < q$, and

$$\begin{cases} u_{xx} = 0 & \text{in } (q', q), \\ u(q') = \psi(q'), \\ u_x(q') = \psi_x(q'), \\ u(q) = 0. \end{cases} \tag{1.5}$$

Note that q' is not a priori known here. Let $\omega = u - \psi$. Then ω solves

$$\begin{cases} \omega_{xx} = -\psi_{xx} \quad \text{in } (q', q), \\ \omega > 0 \quad \text{in } (q', q), \\ \omega(q') = \omega_x(q') = 0, \\ \omega(q) = -\psi(q), \end{cases} \tag{1.6}$$

where q' is to be determined. This recovers the form of (1.3) with $\Gamma_f = \{q'\}$ and $\Gamma_s = \{q\}$. To fully solve (1.4), we need to solve (1.5) or (1.6) in a similar manner on many intervals with free boundaries. The landscape could be rather complicated. □

Example 1.5 (Harmonic modulo). Consider a homeomorphism from $B_1 \backslash B_R$ to $B_1 \backslash B_r$, where R, $r < 1$. In 2D, in order to make the map conformal, it is required that $R = r$ and the map is in the form of $az + bz^{-1}$ with a and b to be determined.

For higher dimensional case, we consider $(u_1, u_2, \cdots, u_n) \triangleq u : B_1^n \backslash B_R^n \to B_1^n \backslash B_r^n$, where $\mathbf{u}$ is a homeomorphism and satisfies $\Delta \mathbf{u} = 0$. For example, a specific form of u is $ax + \frac{bx}{|x|^n}$. Then the question is whether such condition provides any constraints between R and r. The answer is that, given R, $\exists\, r(n, R) < 1$, s.t. $0 < r \le r(n, R)$, i.e. r can be arbitrarily small but can not get too close to 1. The upper bound can be proved easily by Harnack's inequality. Consider $1 - u_1$ on $\Gamma_R \triangleq \partial B^n_{(1+R)/2} \subset\subset B_1^n \backslash B_R^n$. By assumption it is a nonnegative function on $B_1^n \backslash B_R^n$. We can show through topological argument that $1 - u_1 \ge 1$ at some point on Γ_R. By Harnack's inequality,

$$1 \le \sup_{\Gamma_R}(1 - u_1) \le C(n, R) \inf_{\Gamma_R}(1 - u_1),$$

which implies $r < \sup_{\Gamma_R} u_1 \le 1 - C(n, R)^{-1} \triangleq r(n, R) < 1$.

This problem is related to the obstacle problem in the following sense. Consider

$$\min_u \left\{ \int_{B_1^n(0)} |\nabla u|^2 \, dx : u(x) = x \text{ on } \partial B_1^n(0), |u(x)| \ge r \right\}. \tag{1.7}$$

Suppose u is the solution. It is easy to see that $\Delta u = 0$ at x if $|u(x)| > r$. If $\Delta u = 0$ everywhere in B_1^n, u is smooth. Hence, $P \circ u$ is a continuous function from B_n^1 onto S_1^{n-1}, where $P(x) = \frac{x}{|x|}$ which is well-defined and continuous in $B_1^n(0) \backslash B_r^n(0)$. This leads to a contradiction. Therefore, there is a domain $\Omega \subset B_1^n$ to be determined, in which $|u(x)| = r$; $\Delta u = 0$ outside Ω and $|u(x)| > r$.

A natural guess of Ω would be a ball with radius R. It is better to choose R as small as possible since u is harmonic outside Ω and harmonic function obtains minimum energy of the functional in (1.7). A lower bound for possible R is given by the estimate we have just obtained in harmonic modulo. □

Example 1.6 (Mullins-Sekerka problem). The following equation is used to formulate solidification and liquidification of materials near zeros specific heat:

$$\begin{cases} \partial_t u - \Delta u = 0 & \text{in } \Omega\backslash\Gamma_t,\ t \in [0,T], \\ \partial_\nu u = 0 & \text{on } \partial\Omega \times [0,T], \\ u = -\alpha H & \text{on } \Gamma_t,\ t \in [0,T], \\ [\partial_n u] = v & \text{on } \Gamma_t,\ t \in [0,T], \\ \Gamma_t|_{t=0} = \Gamma_0. \end{cases} \tag{1.8}$$

In the above equation, Γ_t is the free boundary with v being its velocity in normal direction. $\partial_\nu u$ is the outer unit normal vector along $\partial\Omega$ and $[\partial_n u] = \partial_n^+ u - \partial_n^- u$ is the jump of normal derivatives between two sides of Γ_t. We note that n here is the outward unit normal vector with respect to the domain enclosed by Γ_t. H is the mean curvature of Γ_t s.t. it is positive for a sphere.

Let Ω_t^- be the region enclosed by Γ_t and $\Omega_t^+ = \Omega\backslash(\Gamma_t \cup \Omega_t^-)$. Define $Q^\pm = \bigcup_{0\le t\le T} \Omega_t^\pm \times \{t\}$ and $u^\pm = u|_{\overline{Q^\pm}}$. If we replace u^+ (or u^-) by a constant then (1.8) is often referred as *Hele-Shaw problem*, which studies pressure of immiscible fluid in air. That u^+ (respectively u^-) is constant means that the pressure in the air is a constant everywhere, and $u^- = -\alpha H$ (respectively $u^+ = -\alpha H$) is the surface tension of the fluid in air (respectively of an air bubble in the fluid). Equations for Hele-Shaw problem write as follows:

$$\begin{cases} \partial_t u - \Delta u = 0 & \text{in } \Omega, \\ u = -\alpha H & \text{on } \partial\Omega, \\ \dfrac{\partial u}{\partial n} = v_n & \text{on } \partial\Omega. \end{cases}$$

Another version for the static case is that

$$\begin{cases} -\Delta u = \delta_0 & \text{in } \Omega, \\ u = -\alpha H & \text{on } \partial\Omega, \\ \dfrac{\partial u}{\partial n} = v_n & \text{on } \partial\Omega. \end{cases}$$

□

Example 1.7 (Stefan problem).

$$\begin{cases} u_t - u_{xx} = 0 & \text{for } 0 < x < s(t),\ t > 0, \\ u(0,t) = f(t) \ge 0 \quad \text{or} \quad u_x(0,t) = g(t) & \text{for } t > 0, \\ u(x,0) = \varphi(x) \ge 0 & \text{for } 0 < x < b,\ \varphi(b) = 0, \\ u(s(t),t) = 0 & \text{for } t > 0, \\ \dfrac{\mathrm{d}s(t)}{\mathrm{d}t} = -u_x(s(t),t) \quad \text{and} \quad s(0) = b. \end{cases}$$

A typical situation Stefan problem models is a 1D system in which water and ice coexist. $u(x)$ characterizes the temperature distribution of the system. $s(t)$ is the interface between water and ice and $[0, s(t)]$ is the water domain. The propagation of the interface is governed by the heat flux at the interface. On the left boundary at $x = 0$, we assign the temperature by setting the Dirichlet boundary condition or assign the heat flux by setting the Neumann boundary condition.

In higher dimensional case, the Stefan problem is as follows:

$$\begin{cases} u_t - \Delta u = 0 \quad \text{in } \bigcup\limits_{0\le t\le T} \Omega_t \times \{t\},\ \Omega_t \text{ is enclosed by} \Gamma_t, \\ u(x,0) = \varphi(x) \ge 0 \quad \text{for } x \in \Omega_0, \\ u = -H \quad \text{on } \Gamma_t \times [0,T], \\ \dfrac{\partial u}{\partial n} = v_n, \end{cases}$$

where H is again the mean curvature along Γ_t and v_n is the normal propagation velocity of Γ_t. □

Example 1.8 (Two fluids with interfaces). The following equation characterizes the dynamics of two incompressible Newtonian fluids with interface Γ_t:

$$\begin{cases} \rho(v_t + v\cdot\nabla v) = -\nabla p \quad \text{in } \mathbb{R}^n\backslash\Gamma_t, \\ \operatorname{div} v = 0 \quad \text{in } \mathbb{R}^n\backslash\Gamma_t, \\ (\partial_t + v\cdot\nabla v) \text{ is tangential to } \Gamma_t,\ t > 0, \\ p_+(t,x) - p_-(t,x) = \varepsilon H_+(t,x), \end{cases}$$

Again H_+ is the mean curvature along the interface, v is the velocity field of the fluid and p is the pressure. That $(\partial_t + v\cdot\nabla v)$ is tangential to Γ_t means that the fluids do not penetrate the interface.

The weak formulation of two fluids with interfaces is that

$$\begin{cases} (\rho v)_t + v\cdot\nabla(\rho v) = -\nabla p + \varepsilon H(t,x)\delta_{\Gamma_t}, \\ \operatorname{div} v = 0. \end{cases}$$

□

Example 1.9 (3D gravity water waves). The water wave equation characterizes the dynamics of the water flow and the water-air interface in the presence of gravity. Let $h(t,x)$ denote the height of water. $\Omega_t = \{(x,z) : x\in\mathbb{R}^2,\ z\le h(t,x)\}$ is the water domain and $\Gamma_t = \{(x, h(t,x)) : x\in\mathbb{R}^2\}$ is the water-air interface

$$\begin{cases} v_t + v\cdot\nabla v = -\nabla p - g\mathbf{e}_3 \quad \text{in } \Omega_t, \\ \operatorname{div} v = 0 \quad \text{in } \Omega_t, \\ \partial_t h = v\cdot\nabla_{(x,z)}(h - z) = 0 \quad \text{for } x\in\mathbb{R}^2, \\ p|_{\Gamma_t} = 0. \end{cases}$$

□

Example 1.10 (Immersed boundary). The immersed boundary method basically formulates the interaction between the fluid and the elastic object immersed in it. Let Γ_t be the immersed elastic object. The governing equation is

$$\begin{cases} v_t + v \cdot \nabla v + \nabla p = \mu \Delta v + \varepsilon H_{\Gamma_t} \delta_{\Gamma_t}, \\ \operatorname{div} v = 0, \\ \dfrac{\mathrm{d}x}{\mathrm{d}t} = v(t,x) \quad \text{on } x \in \Gamma_t, \end{cases}$$

where H_{Γ_t} is the mean curvature of Γ_t. Although the immersed boundary method is popular and successful as a numerical method in biological and physical simulation, there are not so many theoretical results concerning this equation. □

Example 1.11 (Water-dam problem). In the water-dam problem, the dam is modeled as a porous media that water can go through. Because of the high pressure of upstream water, the water will permeate through the dam slowly and form wet region at the lower part inside the dam. The flow in the dam follows the Darcy's law. Let $\Omega \subseteq \mathbb{R}^2$ be the wet region inside the dam and let Γ be the interface between dry region and wet region, i.e. the upper boundary of the wet region. Denote the bottom of the dam by Γ_b; denote the underwater parts on the upstream and downstream surface of the dam by Γ_1 and Γ_2 respectively. In general, the height of water inside the dam at the downstream surface is higher than the downstream water level. Denote the wet wall above the water level on the downstream side by Γ_s, where water keeps seeping out. Then the equation with unknowns (u, Ω) for water-dam problem is as follows

$$\begin{cases} -\Delta u = 0 \quad \text{in } \Omega, \\ u = g_i \quad \text{on } \Gamma_i, \\ \dfrac{\partial u}{\partial \nu} \geq 0 \quad \text{on } \Gamma_s, \\ \dfrac{\partial u}{\partial \nu} = 0 \quad \text{on } \Gamma_b, \\ u(x,y) = y \quad \text{on } \Gamma, \\ \dfrac{\partial u}{\partial \nu} = 0 \quad \text{on } \Gamma, \end{cases}$$

where ν is the outer normal vector with respect to Ω. □

More examples and applications of free boundary problems can be found in [5]. Many of these examples described here are famous, and one can find huge volumes of literature addressing these problems by simple searchings on internet some key words.

2 General theory of existence and uniqueness

2.1 Riesz lemma and theory of existence in Hilbert space

Lemma 2.1 (Riesz). *Let K be a closed convex set in a real Hilbert space H and let $y_0 \in H$. Then there is a unique $x_0 \in K$ (denoted by $x_0 = \mathrm{Proj}_K y_0$) which is the nearest to y_0, i.e.*

$$\|x_0 - y_0\| = \inf_{x \in K} \|x - y_0\|. \tag{2.1}$$

Moreover, it is equivalent to the following variational inequality:

$$\langle x_0 - y_0, x - x_0 \rangle \geq 0, \quad \forall x \in K. \tag{2.2}$$

Proof. We first prove the equivalence between (2.1) and (2.2). Suppose that $x_0 \in K$ satisfies (2.1). For $\forall x \in K$, $\varepsilon \in [0,1]$, let $x_\varepsilon = x_0 + \varepsilon(x - x_0) = \varepsilon x + (1-\varepsilon)x_0 \in K$. By assumption, $\|x_\varepsilon - y_0\|^2 \geq \|x_0 - y_0\|^2$. This gives

$$0 \leq \left.\frac{\mathrm{d}}{\mathrm{d}\varepsilon}\right|_{\varepsilon=0} \|x_\varepsilon - y_0\|^2 = \langle x_0 - y_0, x - x_0 \rangle,$$

which proves (2.2). It could be shown in a similar manner that (2.1) holds if $x_0 \in K$ satisfies (2.2).

If there are x_1 and x_2 satisfying (2.2), we find that for $\forall x \in K$,

$$\begin{aligned} &\langle x - x_1, x_1 - y_0 \rangle \geq 0, \\ &\langle x - x_2, x_2 - y_0 \rangle \geq 0. \end{aligned}$$

Hence,

$$\begin{aligned} &\langle x_2 - x_1, x_1 - y_0 \rangle \geq 0, \\ &\langle x_1 - x_2, x_2 - y_0 \rangle \geq 0. \end{aligned}$$

This implies $\langle x_2 - x_1, x_1 - x_2 \rangle \geq 0$, and thus $x_1 = x_2$.

To prove the existence of such x_0, we take a minimizing sequence $\{x_n\} \subseteq K$, s.t. $\|x_n - y_0\| \to \inf_{x \in K} \|x - y_0\|$. Note that due to the parallelogram identity

$$\begin{aligned} 2\|x_n - y_0\|^2 + 2\|x_m - y_0\|^2 &= \|x_n - x_m\|^2 + 4\left\|\frac{x_n + x_m}{2} - y_0\right\|^2 \\ &\geq \|x_n - x_m\|^2 + 4 \inf_{x \in K} \|x - y_0\|^2. \end{aligned}$$

This gives $\lim_{n,m\to\infty} \|x_n - x_m\| = 0$, i.e. $\{x_n\}$ is a Cauchy sequence. Hence we can define $x_0 \in K$ as its limit (since K is closed), which is the point satisfying (2.1). □

Let $\Omega \subseteq \mathbb{R}^n$ be a smooth, bounded domain. The Sobolev space $H^1(\Omega)$ is defined by $H^1(\Omega) = \{u \in L^2(\Omega) : \nabla u \in L^2(\Omega)\}$, associated with the inner product

$$\langle u, v \rangle = \int_\Omega \nabla u \cdot \nabla v + \int_\Omega uv. \tag{2.3}$$

$H_0^1(\Omega)$ is the closure of $C_0^\infty(\Omega)$ in $H^1(\Omega)$ with respect to (2.3).

Definition 2.1. Let $v \in H^1(\Omega)$. We say $v(x_0) > 0$ if there $\exists \rho > 0$ and $\phi \in C^1(B_\rho(x_0))$, with $\phi > 0$ in $B_\rho(x_0)$, s.t. $v(x) \geq \phi(x)$ a.e. in $B_\rho(x_0)$.

For properties of H^1 (or in general $W^{1,p}$) functions and their boundary values, one can read many standard texts on Sobolev Functions.

We consider functions $a_{ij} \in L^\infty(\Omega)$ for $i, j = 1, \cdots, n$, s.t.

$$\begin{aligned} &a_{ij} = a_{ji} \in L^\infty(\Omega), \\ &\lambda|\xi|^2 \leq a_{ij}(x)\xi_i\xi_j \leq \Lambda|\xi|^2, \quad \forall \xi \in \mathbb{R}^n, \forall x \in \Omega \end{aligned} \tag{2.4}$$

with some $0 < \lambda < \Lambda$. We define

$$L = -\partial_{x_i}(a_{ij}(x)\partial_{x_j}), \tag{2.5}$$

$$a(u, v) = \int_\Omega a_{ij}(x)u_i(x)v_j(x), \quad u, v \in H^1(\Omega). \tag{2.6}$$

Note that $a(u, v)$ is an inner product on $H_0^1(\Omega)$ which is equivalent to (2.3) due to uniform ellipticity and Poincaré inequality.

Consider $\psi \in H^1(\Omega)$ with $\psi \leq 0$ on $\partial\Omega$. Define

$$K \equiv K_\psi = \{v \in H_0^1(\Omega) : v \geq \psi \text{ on } \Omega\}. \tag{2.7}$$

K is a closed, strictly convex set in $H = H_0^1(\Omega)$.

Theorem 2.1. *For $v \in H_0^1(\Omega)$, let*

$$I(v) = a(v, v) - 2\langle f, v\rangle,$$

where $f \in H^{-1}(\Omega) \triangleq (H_0^1(\Omega))^*$. Then there $\exists!\, u \in K$, s.t. $I(u) = \min_{v \in K} I(v)$. Moreover, this is equivalent to the following variational inequality:

$$a(u, v - u) \geq \langle f, v - u\rangle, \quad \forall v \in K. \tag{2.8}$$

A typical example of the above variational problem is

$$\min\left\{\int_\Omega |\nabla u|^2 \, dx : u \in H_0^1(\Omega),\ u \geq \psi \text{ on } \Omega\right\}.$$

Proof. By Riesz representation theorem, $\exists g \in H_0^1(\Omega)$, s.t. $\langle f, v\rangle = a(g, v)$, $\forall v \in H_0^1(\Omega)$. Hence,

$$I(v) = a(v, v) - 2\langle f, v\rangle = a(v - g, v - g) - a(g, g).$$

Then

$$u \text{ minimizes } I(\cdot) \text{ over } K \Leftrightarrow u \text{ minimizes } a(v - g, v - g) \text{ among } v \in K.$$

Recall that $a(u, v)$ defines an equivalent inner product with (2.3) in $H_0^1(\Omega)$. We denote $H_0^1(\Omega)$ equipped with the inner product $a(u, v)$ by $\tilde{H}$. Hence, u minimizes $\|v - g\|_{\tilde{H}}^2$ in K. By Lemma 2.1, there $\exists!\, u \in K$, s.t. $I(u) = \min_{v \in K} I(v)$; moreover, it is equivalent to solve

$$a(u - g, v - u) \geq 0, \quad \forall v \in K$$

or

$$a(u, v - u) \geq a(g, v - u) = \langle f, v - u\rangle, \quad \forall v \in K. \qquad \square$$

2.2 Some properties of the solution

Definition 2.2. $g \in H_0^1(\Omega)$ is a supersolution of $L - f$ if

$$\langle Lg - f, \xi \rangle = a(g, \xi) - \langle f, \xi \rangle \geq 0, \quad \forall\, 0 \leq \xi \in H_0^1(\Omega),$$

where L is defined in (2.5).

Fact 2.1. *u is a supersolution of $L - f$.*

Proof. Let $\xi \in H_0^1(\Omega)$, $\xi \geq 0$. Consider $u_\varepsilon = u + \varepsilon\xi \in K$, $\forall\, \varepsilon \in (0,1)$.

$$I(u) \leq I(u_\varepsilon) \Rightarrow \left.\frac{\mathrm{d}}{\mathrm{d}\varepsilon}\right|_{\varepsilon=0} I(u_\varepsilon) \geq 0 \Rightarrow a(u, \xi) \geq \langle f, \xi \rangle,$$

which completes the proof. □

Fact 2.2. *u is a minimum supersolution in the sense that, for $\forall$ supersolution y of $L - f$ s.t. $y \geq \psi$ and $y \geq 0$ on $\partial\Omega$, we have $u \leq y$.*

Proof. Let $\xi = \min(u, y) \in K$. By (2.8),

$$a(u, \xi - u) \geq \langle f, \xi - u \rangle.$$

Since $\xi - u \leq 0$ in Ω and $\xi - u = 0$ on $\partial\Omega$, one has (as y is a supersolution)

$$a(y, \xi - u) \leq \langle f, \xi - u \rangle.$$

Thus,

$$\begin{aligned} 0 \geq a(y - u, \xi - u) &= \int_\Omega a_{ij}(y - u)_i(\xi - u)_j \,\mathrm{d}x \\ &= \int_{\Omega \cap \{x:\, y(x) > u(x)\}} a_{ij}(y - u)_i(\xi - u)_j \,\mathrm{d}x \\ &\quad + \int_{\Omega \cap \{x:\, y(x) \leq u(x)\}} a_{ij}(y - u)_i(\xi - u)_j \,\mathrm{d}x. \end{aligned}$$

The first term on the right hand side vanishes since $\xi = u$ on $\Omega \cap \{x : y(x) > u(x)\}$; the second term has to be nonnegative as $\xi = y$ on $\Omega \cap \{x : y(x) \leq u(x)\}$ and a_{ij} is uniformly elliptic. Therefore, $u \leq y$ on Ω. □

Corollary 2.1. *Let u be a solution of (2.8) with $f = 0$. Suppose $\psi \leq M$. Then $\psi \leq u \leq M$.*

Definition 2.3. For given $\psi \in H^1(\Omega)$, $\psi \leq 0$ on $\partial\Omega$, let $u \in K$ be the solution of (2.8). We define

$$\mathcal{N} = \{x \in \Omega : u(x) > \psi(x)\}$$

as the non-coincidence set, which is open by Definition 2.1. We also define the coincidence set

$$\Lambda = \{x \in \Omega : u(x) = \psi(x)\} = \Omega \backslash \mathcal{N},$$

and the free boundary $\Gamma = \partial\Lambda$.

It is not difficult to see that $Lu = f$ in $\mathcal{N}$ while $u = \psi$ in Λ. By Fact 2.1, we also know that

$$\langle Lu - f, \xi \rangle = a(u, \xi) - \langle f, \xi \rangle \geq 0, \quad \forall 0 \leq \xi \in H_0^1(\Omega).$$

This gives a nonnegative Radon measure $\mathrm{d}\mu \geq 0$ s.t.

$$a(u, \xi) - \langle f, \xi \rangle = \int_\Omega \xi \, \mathrm{d}\mu, \quad \forall \xi \in H_0^1(\Omega).$$

Hence we obtain the following

Theorem 2.2. *Let u be the solution of* (2.8). *Then there is a Radon measure $\mathrm{d}\mu \geq 0$ s.t. $Lu = f + \mathrm{d}\mu$ in Ω with* $\mathrm{spt}(\mathrm{d}\mu) \subseteq \Lambda$.

In the sequel, we shall assume $a_{ij} = \delta_{ij}$ and $f = 0$.

Lemma 2.2. *Let u be the solution of* (2.8) *with $a_{ij} = \delta_{ij}$ and $f = 0$. Assume $B_R(0) \triangleq B_R \subseteq \Omega$. Let $l(x)$ be a linear function and u_0 be the solution of*

$$\begin{cases} \Delta u_0 = 0 & \text{in } B_R, \\ u_0 = u - \psi \geq 0 & \text{on } \partial B_R. \end{cases} \tag{2.9}$$

Then

$$u_0 + l - \omega(R) \leq u \leq u_0 + l + \omega(R),$$

where

$$\omega(r) = \sup_{B_r(0)} |\psi(x) - l(x)|, \quad 0 \leq r \leq R_0.$$

In particular, for $l(x) = \psi(0)$, we have

$$|u_0(x) + \psi(0) - u(x)| \leq \sup_{B_R(0)} |\psi(x) - \psi(0)|.$$

Proof. It is known that

$$0 = -\Delta(u_0 + l - \omega(R)) \leq -\Delta u \quad \text{in } \mathscr{D}'(B_R)$$

and

$$u_0 + l - \omega(R) \leq u \quad \text{on } \partial B_R.$$

By maximum principle, $u_0 + l - \omega(R) \leq u$ on B_R.

On the other hand, since $u(x) \leq u_0 + l + \omega(R)$ on ∂B_R and $\mathrm{spt}(\Delta(u - u_0 - l - \omega(R))) \subseteq \Lambda \cap B_R$, we have

$$\sup_{B_R}(u - u_0 - l - \omega(R)) \leq \max\left\{0, \sup_{\Lambda \cap B_R}(u - u_0 - l - w(R))\right\} \leq 0.$$

This proves the lemma. $\square$

Lemma 2.3. *Assume $0 \in \Lambda$ and let u, l and ω be defined as in Lemma 2.2. Then we have*

$$\sup_{B_{R/2}} |u - l| \leq C(n)\omega(R),$$

where $C(n)$ is a positive constant only depending on n.

Proof. Since $u \geq \psi$ on B_R, we find $u - l \geq -\omega(R)$ on B_R. This proves one side of the estimate.

By Lemma 2.2,

$$\sup_{B_{R/2}} (u(x) - l(x)) \leq \sup_{B_{R/2}} u_0(x) + \omega(R). \tag{2.10}$$

Note that u_0 defined in (2.9) is a nonnegative harmonic function on B_R. By Harnack's inequality,

$$\sup_{B_{R/2}} u_0(x) \leq C(n)u_0(0) \leq C(n)(u(0) - l(0) + \omega(R)).$$

The second inequality again follows from Lemma 2.2. Since $u(0) = \psi(0)$ (as $0 \in \Lambda$),

$$\sup_{B_{R/2}} u_0(x) \leq C(n)(\psi(0) - l(0) + \omega(R)) \leq C(n)\omega(R).$$

Note that the definition of $C(n)$ changes in the last inequality. This, together with (2.10), proves the other side of the lemma. □

Corollary 2.2. *If $\psi \in C^0(\overline{\Omega})$ in the definition of K (see (2.7)), and u is the solution of (2.8) with $a_{ij} = \delta_{ij}$ and $f = 0$, then $u \in C^0(\Omega)$.*

Proof. For $x_0 \in \mathcal{N}$, u is locally a harmonic function and is thus continuous. For $x_0 \in \Lambda$, we apply Lemma 2.3 with $l(x) \equiv \psi(x_0)$ and $B_R(x_0) \subseteq \Omega$ and find

$$\operatorname*{osc}_{B_{R/2}(x_0)} u(x) \leq C(n) \operatorname*{osc}_{B_R(x_0)} \psi.$$

The right hand side goes to 0 as $R \to 0$ since $\psi \in C^0(\Omega)$, which implies that u is continuous at x_0 as well. □

Definition 2.4. Given $E \subset\subset \Omega$, define

$$K_E = \{v \in H_0^1(\Omega) : v \geq 1 \text{ on } E\}.$$

Then the capacity of E with respect to Ω is defined by

$$\mathrm{Cap}_\Omega(E) = \inf_{v \in K_E} \int_\Omega |\nabla v|^2 \, dx.$$

There is a simple fact about sets with zero capacity.

Fact 2.3.

$$\mathrm{Cap}_\Omega(F) = 0 \Leftrightarrow \mathrm{Cap}_\Omega(E) = 0, \quad \forall E \subset\subset F.$$

It can be shown that there $\exists!\, v_E$, with $-\Delta v_E \geq 0$, $0 \leq v_E \leq 1$ in Ω and $v_E = 1$ on E, s.t.

$$\begin{cases} -\Delta v_E = 0 & \text{in } \Omega \backslash E, \\ -\Delta v_E = \mathrm{d}\mu & \text{in } \Omega, \end{cases}$$

where $\mathrm{d}\mu$ is a nonnegative Radon measure on Ω. v_E has the following integral representation:

$$v_E(x) = \int_\Omega G(x,y)\,\mathrm{d}\mu(y),$$

where $G(x,y)$ is Green's function of $-\Delta$ on Ω. Moreover,

$$\begin{aligned} \mathrm{Cap}_\Omega(E) &= \int_\Omega |\nabla v_E|^2\,\mathrm{d}x = \int_\Omega (-\Delta v_E) v_E\,\mathrm{d}x = \int_\Omega v_E\,\mathrm{d}\mu \\ &= \int_\Omega \int_\Omega G(x,y)\,\mathrm{d}\mu(x)\mathrm{d}\mu(y) \end{aligned}$$

and

$$\mu(E) \leq C(\mathrm{Cap}_\Omega(\mathrm{E}))^{\frac{1}{2}}, \quad \forall E \subset\subset \Omega.$$

2.3 Monotone operators and theory in reflexive Banach space

Next we shall extend the theory on Hilbert space to the case of reflexive Banach space. A good reference for this is the book by H. Brezis [2].

Definition 2.5. Let X be a real reflexive Banach space with dual X^*. $\langle\cdot,\cdot\rangle$ denotes the pairing between X^* and X. A mapping A from $\mathcal{D}(A) \subseteq X$ to X^* is called monotone if

$$\langle Au - Av, u - v\rangle \geq 0, \quad \forall\, u,v \in \mathcal{D}(A).$$

Example 2.1. Let $X = H_0^1(\Omega)$ and $X^* = H^{-1}(\Omega)$. Define $Au = -\partial_{x_i}(a_{ij}(x)\partial_{x_j}u) - f$ with a_{ij}'s satisfying (2.4). Then A is monotone. □

Definition 2.6. When $\mathcal{D}(A)$ is convex, A is called hemicontinuous if $\forall\, u,v \in \mathcal{D}(A)$, the map

$$t \mapsto \langle A(tu + (1-t)v), u - v\rangle$$

is continuous for $t \in [0,1]$.

Definition 2.7. Let $M \subset X$ be a finite dimensional subspace. Let $j : M \hookrightarrow X$ be the embedding and $j^* : X^* \to M^*$ be the dual map, i.e. j^*f is the restriction of $f \in X^*$ to M. If j^*Aj is continuous on $M \cap \mathcal{D}(A)$ for any such M, then we say A is continuous on finite dimensional subspace of $\mathcal{D}(A)$.

Theorem 2.3. *Suppose that $A : \mathcal{D}(A) = X \to X^*$ is monotone and hemicontinuous. Then for any bounded closed convex subset K of X, there exists a solution $u_0 \in K$ to the variational inequality*

$$\langle Au_0, v - u_0\rangle \geq 0, \quad \forall\, v \in K. \tag{2.11}$$

Theorem 2.4. *Let K be a bounded closed convex subset of X and suppose that S : $\mathcal{D}(A) \to X^*$ is monotone, $K \subseteq \mathcal{D}(A)$, and A is continuous on finite dimensional subspace of $D(A)$. Then there exists a $u_0 \in K$, s.t.* (2.11) *holds.*

Lemma 2.4 (Minty). *Assume K to be any closed convex subset of X. Let A : $K \to X^*$ be monotone and hemicontinuous. Then u_0 satisfying* (2.11) *iff*

$$\langle Av, v - u_0\rangle \geq 0, \quad \forall\, v \in K. \tag{2.12}$$

Proof. By monotonicity of A,

$$0 \leq \langle Av - Au_0, v - u_0\rangle = \langle Av, v - u_0\rangle - \langle Au_0, v - u_0\rangle.$$

Hence (2.11) implies (2.12). To show the converse, note that for $\forall\, w \in K, t \in (0,1)$,

$$v = tw + (1-t)u_0 = u_0 + t(w - u_0) \in K.$$

By (2.12) one has

$$\langle A(u_0 + t(w - u_0)), w - u_0\rangle \geq 0.$$

Let $t \to 0^+$ and (2.11) holds for $\forall\, v = w \in K$ by hemicontinuity. □

Proof of Theorem 2.4. We first consider the case where X is a finite dimensional space, i.e. $X = \mathbb{R}^m$. Let $K \subseteq \mathbb{R}^m$ be bounded, closed and convex, and $A : K \to \mathbb{R}^m$. We assume that A is continuous (but we do not require that it is monotone). The variational inequality (2.11) can be written in the form

$$\langle u_0, v - u_0\rangle \geq \langle u_0 - Au_0, v - u_0\rangle, \quad \forall\, v \in K. \tag{2.13}$$

For $w \in K$, let $Tw = w - Aw$. It is clear that in the finite dimensional space, there is a unique $u \in K$, s.t.

$$\begin{aligned} u &= \mathrm{Proj}_K(w - Aw) = \mathrm{Proj}_K Tw, \\ \langle u, v - u\rangle &\geq \langle Tw, v - u\rangle, \quad \forall\, v \in K. \end{aligned}$$

It is also easy to show that Proj_K is continuous on $\mathbb{R}^m$. Since $\mathrm{Proj}_K \circ (I - A) : K \to K$ is continuous and K is bounded, closed and convex, by Brouwer's fixed-point theorem, there is $u_0 \in K$, s.t. $\mathrm{Proj}_K \circ (I - A)u_0 = u_0$. This proves (2.13).

Now for any finite-dimensional subspace $M \subseteq X$, by assumption, j^*Aj : $K \cap M \to M^*$ is continuous. Hence, there is a $u_M \in K \cap M$ solving

$$\langle j^*Aju_M, v - u_M\rangle \geq 0, \quad \forall\, v \in K \cap M.$$

Since $\langle j^*Aju_M, \xi\rangle = \langle Au_M, \xi\rangle$, for $\forall\, \xi \in K \cap M$, we have

$$\langle Au_M, v - u_M\rangle \geq 0, \quad \forall\, v \in K \cap M.$$

By Minty's lemma,

$$\langle Av, v - u_M\rangle \geq 0, \quad \forall\, v \in K \cap M.$$

For $\forall\, v \in K$, define $S(v) = \{u \in K : \langle Av, v-u\rangle \geq 0\}$. Since $S(v)$ is weakly closed subset of K and since K is bounded and X is reflexive, $S(v)$ is weakly compact by Banach-Alaoglu theorem. We claim that for $\forall\, m$ and $\forall\, v_1, v_2, \cdots, v_m \in K$, $\bigcap_{i=1}^m S(v_i) \neq \varnothing$. In fact, given $M = \text{span}\{v_1, \cdots, v_m\}$, there exists $u_M \in K \cap M$ satisfying

$$\langle Av, v - u_M\rangle \geq 0, \quad \forall\, v \in K \cap M,$$

which implies $u_M \in \bigcap_{i=1}^m S(v_i)$. Hence, $\bigcap_{v\in K} S(v) \neq \varnothing$ by the definition of compactness. Indeed, $S(v_0)\backslash S(v)$ is relatively open set for fixed $v_0 \in K$ and $\forall\, v \in K$, where the topology on $S(v_0)$ is induced by weak topology on X. If $\bigcap_{v\in K} S(v) = \varnothing$, $\{S(v_0)\backslash S(v)\}_{v\in K}$ forms an open covering of $S(v_0)$. Since $S(v_0)$ is compact, there exists a finite subcovering, denoted by $\{S(v_0)\backslash S(v_i)\}_{i=1}^n$. Hence, $\bigcap_{i=0}^n S(v_i) = \varnothing$, which is a contradiction. Then $u_0 \in \bigcap_{v\in K} S(v)$ solves (2.12) and thus (2.11). □

Proof of Theorem 2.3. We shall prove that if A is monotone and hemicontinuous, A is continuous on finite dimensional space. Let $M \subseteq X$ be a finite dimensional subspace and let j and j^* be defined as before.

First we will show that j^*Aj maps bounded sets of M into bounded sets of M^*. Suppose that $\exists\, \{v_n\} \subseteq M$, s.t. $\|v_n\| \leq C_0$ and

$$\|j^*Ajv_n\| = \|j^*Av_n\| \to +\infty.$$

By monotonicity, one has

$$\langle j^*Av_n - j^*Au, v_n - u\rangle \geq 0, \quad \forall\, u \in M.$$

Hence

$$\left\langle y_n - \frac{j^*Au}{\|j^*Av_n\|}, v_n - u\right\rangle \geq 0,$$

where $y_n = \frac{j^*Av_n}{\|j^*Av_n\|}$. Since M and M^* are of finite dimension, there exist subsequences, also denoted by $\{v_n\} \subseteq M$ and $\{y_n\} \subseteq M^*$, such that

$$v_n \to v \in M, \quad y_n \to y \in M^*.$$

Note that $\|y\| = 1$. By taking $n \to +\infty$, one find

$$\langle y, v - u\rangle \geq 0, \quad \forall\, u \in M.$$

Here we use $\|j^*Av_n\| \to +\infty$. This implies $y = 0$, which contradicts with $\|y\| = 1$.

Next we shall prove j^*Aj is continuous on M. Suppose not, there is a sequence $\{v_n\} \subseteq M$, s.t.

$$v_n \to v, \quad j^*Av_n \to w, \quad \text{and} \quad w \neq j^*Av.$$

Note that $\{v_n\}$, and thus $\{j^*Av_n\}$ are bounded. This gives the existence of a limiting point w and the convergence of $\{j^*Av_n\}$ up to a subsequence. Since $\langle j^*Av_n - j^*Au, v_n - u\rangle \geq 0$,

$$\langle w - j^*Au, v - u\rangle \geq 0, \quad \forall\, u \in M.$$

Following the proof of Minty's lemma, we can show that this implies

$$\langle w - j^* Av, v - u\rangle \geq 0, \quad \forall\, u \in M.$$

Hence, we get $w - j^* Av = 0$, which is a contradiction.

Then Theorem 2.3 is a direct corollary of Theorem 2.4 and the result we just proved. □

Remark 2.1. Theorems 2.3 and 2.4 can be generalized to the case where K is unbounded. We shall assume that A is coercive: there is $\phi_0 \in K$, s.t.

$$\frac{1}{\|v\|}\langle Av - A\phi_0, v - \phi_0\rangle \to \infty, \qquad \text{if } v \in K,\ \|v\| \to +\infty. \tag{2.14}$$

Theorem 2.5. *Let K be unbounded closed convex set and let A be as in Theorems 2.4 or 2.3. If A is also coercive, then there is a solution of the variational inequality (2.11).*

Remark 2.2. An observation is that $\|u\|$ should be bounded. Suppose $u \in K$ is a solution of (2.11), i.e.

$$\langle Au, v - u\rangle \geq 0, \forall v \in K.$$

This gives

$$\frac{1}{\|u\|}\langle Au - Av, u - v\rangle \leq \left\langle Av, \frac{v-u}{\|u\|}\right\rangle, \quad \forall\, v \in K.$$

In particular, for $v = \phi_0$, one has

$$\frac{1}{\|u\|}\langle Au - A\phi_0, u - \phi_0\rangle \leq \left\langle A\phi_0, \frac{\phi_0 - u}{\|u\|}\right\rangle \leq C_0.$$

This means $\|u\|$ has an a priori upper bound.

Proof. For $\forall\, R > 0$, let $K_R = K \cap \{\|u\| \leq R\}$ and let $u_R \in K_R$ be the solution of (2.11) corresponding to the convex set K_R.

If $\|u_R\| < R$ for some $R > 0$ then we are done. Indeed, $\forall\, v \in K$, there $\exists \varepsilon > 0$ depending on v such that $w = u_R + \varepsilon(v - u_R) \in K_R$. Hence,

$$0 \leq \langle Au_R, w - u_R\rangle = \varepsilon\langle Au_R, v - u_R\rangle,$$

which implies u_R is a solution of (2.11) in K.

If $\|u_R\| = R$ for $\forall\, R > 0$, let $R > \|\phi_0\|$ and we find $\langle Au_R, u_R - \phi_0\rangle \leq 0$. On the other hand, by the definition of coerciveness

$$\langle Au_R - A\phi_0, u_R - \phi_0\rangle \geq C_R\|u_R\|,$$

where $C_R \to +\infty$ as $R \to +\infty$. Hence

$$\langle Au_R, u_R - \phi_0\rangle \geq (C_R - \|A\phi_0\|)\|u_R - \phi_0\|.$$

The right hand side is positive if R is sufficiently large, which is a contradiction. □

To obtain uniqueness of the solution, we need the following definition.

Definition 2.8. A is called strictly monotone if A is monotone and if

$$\langle Au - Av, u - v\rangle = 0 \quad \Leftrightarrow \quad u = v.$$

Theorem 2.6. *If A is strictly monotone, then there is at most one solution of* (2.11).

Proof. Suppose there are two solutions u_1 and u_2. Then

$$\langle Au_i, v - u_1\rangle \geq 0, \quad \forall v \in K, \text{ for } i = 1, 2.$$

Then

$$\begin{aligned}&\langle Au_1, u_2 - u_1\rangle \geq 0 \leq \langle Au_2, u_1 - u_2\rangle\\ \Rightarrow &\langle Au_2 - Au_1, u_2 - u_1\rangle \leq 0\\ \Rightarrow &u_1 = u_2.\end{aligned}$$

□

If we assume

$$\langle Au - Av, u - v\rangle \geq \alpha\|u - v\|^{1+\delta} \tag{2.15}$$

for some α, $\delta > 0$, we can obtain stability estimate.

Theorem 2.7. *Let A be as in Theorems* 2.4 *or* 2.3, *and let* (2.15) *hold. Let $u_i \in K$ $(i = 1, 2)$ be such that*

$$\langle Au_i, v - u_i\rangle \geq \langle f_i, v - u_i\rangle, \quad \forall v \in K,$$

where $f_i \in X^*$. Then

$$\|u_2 - u_1\| \leq \left(\frac{1}{\alpha}\|f_1 - f_2\|\right)^{\frac{1}{\delta}}.$$

Proof.

$$\alpha\|u_2 - u_1\|^{1+\delta} \leq \langle Au_2 - Au_1, u_2 - u_1\rangle \leq |\langle f_2 - f_1, u_2 - u_1\rangle| \leq \|u_2 - u_1\|\|f_2 - f_1\|.$$

□

2.4 A special class of problems

Assume H to be a Hilbert space and let $K \subseteq H$ be a closed, convex set. Let $f \in H^* \simeq H$. Define $a(u, v)$ to be a bilinear form on $H \times H$, s.t.

$$\|a(u, v)\| \leq C\|u\|\|v\|, \quad \forall u, v \in H; \quad \text{(boundedness)} \tag{2.16}$$

$$\|a(u, u)\| \geq \alpha\|u\|^2, \quad \forall u \in H. \quad \text{(coerciveness)} \tag{2.17}$$

A typical example is that $H = H_0^1(\Omega)$ and

$$a(u, v) = \int_\Omega a_{ij}(x) u_i(x) v_j(x)\, dx,$$

where

$$a_{ij} \in L^\infty(\Omega), \text{ and } a_{ij}(x)\xi_i\xi_j \geq \alpha|\xi|^2, \quad \forall \xi \in \mathbb{R}^n.$$

Note that there is no symmetry assumption as in (2.4) so that $a(u,v)$ is not necessarily an inner product on H.

Theorem 2.8. *There exists a unique solution u of the variational inequality*

$$a(u, v-u) \geq \langle f, v-u\rangle, \quad \forall v \in K,$$

where $\langle \cdot,\cdot\rangle$ is the inner product on H. Furthermore, the map $f \mapsto u$ is continuous, and

$$\|u_2 - u_1\| \leq \frac{1}{\alpha}\|f_2 - f_1\|, \tag{2.18}$$

where u_i $(i = 1, 2)$ is the solution corresponding to f_i.

Proof. If $a(u,v)$ is symmetric, define $(u,v) = a(u,v)$ to be a new inner product on H. Let $\tilde{H}$ be the Hilbert space H equipped with $(\cdot,\cdot)$. By (2.16) and (2.17), $H \simeq \tilde{H}$. For $\forall f \in H$, by Riesz representation theorem, there $\exists \tilde{f} \in \tilde{H}$, s.t.

$$(\tilde{f}, v) = \langle f, v\rangle, \quad \forall v \in H.$$

Then $u = \mathrm{Proj}_K \tilde{f}$ in the metric given by $(\cdot,\cdot)$.

For general case, we shall use the continuity method. Define

$$S(u,v) = \frac{1}{2}(a(u,v) + a(v,u)), \quad \sigma(u,v) = \frac{1}{2}(a(u,v) - a(v,u)),$$
$$a_t(u,v) = S(u,v) + t\sigma(u,v), \quad 0 \leq t \leq 1.$$

Let

$$E = \{t \in [0,1] : \text{The variational inequality } a_t(u, v-u) \geq \langle f, v-u\rangle,$$
$$\forall v \in K \text{ has a unique solution and (2.18) holds}\}.$$

It is known that $0 \in E$. We assume $t_0 \in E$ and take $t \in [0,1]$ close to t_0, with $\delta = t_0 - t$. Since $\sigma(u,v)$ is a bounded linear functional in v, by Riesz representation theorem, one has $Tu \in H$, s.t.

$$\langle Tu, v\rangle = \sigma(u,v), \quad \forall v \in H.$$

It is easy to show the map $T : u \mapsto Tu$ is linear and bounded. Denote $f_u^\delta = f + \delta Tu$. We notice that

$$a_t(u, v-u) \geq \langle f, v-u\rangle \Leftrightarrow a_{t_0}(u, v-u) \geq \langle f, v-u\rangle + (t_0 - t)\sigma(u, v-u)$$
$$\Leftrightarrow a_{t_0}(u, v-u) \geq \langle f_u^\delta, v-u\rangle.$$

For $\forall w \in K$, let $\mathcal{L}w$ be the unique solution $z \in K$ of the variational inequality

$$a_{t_0}(z, v-z) \geq \langle f_w^\delta, v-z\rangle, \quad \forall v \in K.$$

We have

$$\|\mathcal{L}w_1 - \mathcal{L}w_2\| \leq \frac{1}{\alpha}\|f^\delta_{w_1} - f^\delta_{w_2}\| \leq C|\delta|\|w_1 - w_2\| < \frac{1}{2}\|w_1 - w_2\|.$$

The last inequality is true if we take $|\delta| < \delta_* = \frac{1}{2C}$. Note that δ_* is independent of t_0. Hence, by contraction mapping theorem, there $\exists!\, u \in K$, s.t. $\mathcal{L}u = u$. This u solves the variational inequality associated with $a_t(\cdot,\cdot)$. If we can further show that for $\forall\, t$, $|t - t_0| < \delta_*$ one has (2.18), then we can start from $0 \in E$ and conclude that $E = [0, 1]$, which proves the theorem.

To show (2.18), we let t s.t. $|t_0 - t| < \delta_*$. Let $u_i \in K$ $(i = 1, 2)$ be the solution of

$$a_t(u_i, v - u_i) \geq \langle f_i, v - u_i\rangle, \quad \forall v \in K.$$

One immediately has

$$a_t(u_1, u_2 - u_1) \geq \langle f_1, u_2 - u_1\rangle,$$
$$a_t(-u_2, u_2 - u_1) \geq \langle -f_2, u_2 - u_1\rangle,$$

which gives

$$\alpha\|u_1 - u_2\|^2 \leq a_t(u_1 - u_2, u_1 - u_2) \leq \langle f_1 - f_2, u_1 - u_2\rangle \leq \|u_1 - u_2\|\|f_1 - f_2\|.$$

This proves (2.18). □

Much of discussions in this section can be found in the references A. Friedman [5], D. Kinderlehrer and G. Stampacchia [13] and H. Brezis [2].

3 Regularity of solutions

3.1 Schauder estimates

In Section 2.2 we already proved as a corollary to Lemma 2.3 that if the obstacle ψ is continuous, then the solution u is continuous in the interior. Next we state and prove the interior Schauder estimate.

Theorem 3.1 (Interior estimate). *Let u be the solution of*

$$\min\left\{\int_\Omega |\nabla u|^2 \, dx : u \in H^1_0(\Omega), \; u \geq \psi \text{ on } \Omega\right\},$$

where $\psi \leq 0$ on $\partial\Omega$.

(i) *If $\psi \in C^{0,\alpha}(\Omega)$, then $u \in C^{0,\alpha}(\Omega)$.*
(ii) *If $\psi \in C^{1,\alpha}(\Omega)$, then $u \in C^{1,\alpha}(\Omega)$.*

Here α can be any number in $[0, 1]$. If $\alpha = 0$, then it simply means continuous; and $\alpha = 1$ it means either u or ∇u is Lipschitz continuous.

Proof.

Part (i). Assume $0 \in \Lambda$, the set of coincidence. Then we already have

$$|u(x) - u(0)| \leq C\omega\left(|x|\right) \leq C|x|^{\alpha}, \quad \forall x \in B_R(0),$$

by choosing $l(x) = \psi(0) = u(0)$ in Lemma 2.3, and noting that $\omega(r) \leq Cr^{\alpha}$, if $\psi \in C^{0,\alpha}(\Omega)$. We need to show

$$|u(x) - u(y)| \leq C|x-y|^{\alpha}, \quad \forall x, y \in B_{R/4}(0). \tag{3.1}$$

Case (a). Either x or $y \in \Lambda$. Then (3.1) follows directly from Lemma 2.3.

Case (b). $x, y \notin \Lambda$, and $|x-y| \geq \frac{1}{4}\min\{\text{dist}(x,\Lambda), \text{dist}(y,\Lambda)\}$. Assume for $x^* \in \Lambda$, $|x - x^*| = \min\{\text{dist}(x,\Lambda), \text{dist}(y,\Lambda)\}$. Then

$$\begin{aligned}|u(x) - u(y)| &\leq |u(x) - u(x^*)| + |u(y) - u(x^*)| \\ &\leq C\left(|x - x^*|^{\alpha} + |y - x^*|^{\alpha}\right) \leq C|x-y|^{\alpha}.\end{aligned}$$

Case (c). $x, y \notin \Lambda$, and $|x-y| < \frac{1}{4}\min\{\text{dist}(x,\Lambda), \text{dist}(y,\Lambda)\}$.

Assume x^* as above and let $r = |x - x^*|$. Note u is harmonic in $B_r(x)$. We define

$$h(\xi) = \frac{u(r\xi + x) - u(x)}{r^{\alpha}}, \quad \xi \in B_1(0).$$

Since

$$\begin{cases} \Delta h(\xi) = 0, \\ |h(\xi)| \leq C, \quad \forall \xi \in B_1(0), \end{cases}$$

the gradient estimate for harmonic functions gives

$$\|\nabla h\|_{L^{\infty}(B_{1/2})} \leq C.$$

Let $y = r\xi_0 + x$ for some $|\xi_0| \leq 1/4$. Then

$$|h(0) - h(\xi_0)| \leq C|\xi_0|,$$

i.e.

$$|u(x) - u(y)| \leq Cr^{\alpha}\,|\xi_0| \leq C|x-y|^{\alpha}.$$

Part (ii). Choose $l(x) = \psi(0) + \nabla\psi(0)x$ in Lemma 2.3. By the assumption $0 \in \Lambda$, $u(0) = \psi(0)$. Then

$$|u(x) - \psi(0) - \nabla\psi(0)x| \leq C\omega(|x|) \leq C|x|^{1+\alpha}, \quad 0 \leq |x| \leq R. \tag{3.2}$$

Hence $\nabla u(0)$ is well-defined. Similarly one can show that $\nabla u(x)$ can be defined pointwise on $B_R(0)$. We need to check that

$$|\nabla u(x) - \nabla u(y)| \leq C|x-y|^{\alpha}, \quad \forall x, y \in B_{R/4}(0).$$

Case (a). Either x or $y \in \Lambda$. Assume $y \in \Lambda$. By (3.2), $\nabla u(y) = \nabla\psi(y)$. Let $x^* \in \Lambda$, s.t. $r^* \triangleq |x-x^*| = \inf_{z\in\Lambda}|x-z|$. Set $v(x) = u(x)-\psi(x^*)-\nabla\psi(x^*)(x-x^*)$. Note that v is harmonic in $B_{r^*}(x)$, and for $\xi \in B_1(0)$ we have

$$\left|\frac{v(r^*\xi + x)}{(r^*)^{\alpha+1}}\right| \le C.$$

Once again the gradient estimate for harmonic functions yields

$$|\nabla v(x)| = |\nabla u(x) - \nabla\psi(x^*)| \le C(r^*)^\alpha.$$

Thus, we obtain

$$|\nabla u(y) - \nabla u(x)| \le |\nabla\psi(x^*) - \nabla\psi(y)| + C(r^*)^\alpha \le Cr^\alpha,$$

as $|x^* - y| \le r^* + |x| \le 2|x|$.

Cases (b) and (c) can be shown in a similar way, combining this argument with those in the first part. □

Remark 3.1. There is a natural generalization of the above Schauder estimates to the variational inequality

$$\langle Au, v-u\rangle \ge \langle f, v-u\rangle,$$

where $f \in C^{0,\alpha}$, and A as in Theorem 2.3.

Remark 3.2. One can consider adapting the above estimates to the following non-linear problems:

1. Let u be the minimizer of $\left\{\int_\Omega |\nabla u|^p \,\mathrm{d}x : u \ge \psi \text{ in } \Omega\right\}$, where $\psi \le 0$ on $\partial\Omega$, and $p > 1$. Then

$$-\mathrm{div}\left(|\nabla u|^{p-2}\nabla u\right) = \mathrm{d}\mu \ge 0 \quad \text{in} \quad \mathscr{D}'(\Omega), \tag{3.3}$$

where $\mathrm{spt}(\mu) \subseteq \Lambda$, the set of coincidence. We can introduce the following regularization of the energy:

$$I_\varepsilon(v) = \int_\Omega \left(|\nabla v|^2 + \varepsilon\right)^{p/2} \mathrm{d}x,$$

which corresponds to the variational inequality

$$-\mathrm{div}\left(\left(|\nabla u_\varepsilon|^2 + \varepsilon\right)^{\frac{p-2}{2}} \nabla u_\varepsilon\right) \ge 0,$$

or equivalently $-a_{ij}(x)u_{ij} \ge 0$, where, having dropped the ε-dependence in the notation:

$$a_{ij}(x) = \delta_{ij} + (p-2)\frac{u_i u_j}{\varepsilon + |\nabla u|^2}.$$

It is not difficult to see that the matrix (a_{ij}) satisfies (2.4). Hence we can treat the above problem as a linear one that is uniformly elliptic. Note that this gain comes at the expense of the variational structure, since the equation is no longer in the divergence form. Nevertheless, we can apply Krylov-Safanov's Harnack inequality (see [6] or [10]) to obtain interior $C^{1,\alpha}$ estimates, where $\alpha \le \alpha_0$, for some maximal $\alpha_0 < 1$. For the corresponding boundary estimates we need to resort to methods due to Krylov and Caffarelli, see [6] or [10].

2. The same techniques do not work in the context of the minimal surface equation for a graph, that is

$$\operatorname{div}\left(\frac{\nabla u}{\sqrt{1+|\nabla u|^2}}\right) = 0. \tag{3.4}$$

Equivalently we have

$$\left(\delta_{ij} - \frac{u_i u_j}{1+|\nabla u|^2}\right) u_{ij} = 0.$$

This time the corresponding matrix (a_{ij}) has its smallest eigenvalue tending to 0 when $|\nabla u|^2$ tends infinity; hence, the problem is degenerate elliptic only unless one has an uniform bound on $|\nabla u|$. Therefore, the corresponding obstacle problem

$$\min_{u\in H_0^1(\Omega)} \left\{ \int_\Omega \sqrt{1+|\nabla u|^2}\, \mathrm{d}x : u \ge \psi,\ u|_{\partial\Omega} = \psi \right\} \tag{3.5}$$

for $\psi \le 0$ on $\partial\Omega$ requires some additional geometric considerations. Nonetheless the principle idea presented in the proof of Theorem 3.1 can be made to work, see [15] and references therein.

3.2 $W^{2,p}$-estimates

In this section, we shall discuss $W^{2,p}$-estimates of solutions to the obstacle problems due to G. Stampacchia and J. L. Lions.

Let $\Omega \subseteq \mathbb{R}^n$ be a bounded and smooth domain. Define

$$Au = -a_{ij}(x)u_{ij} + b_i(x)u_i + c(x)u,$$

where there exist $\lambda,\ \Lambda > 0$, s.t.

$$\begin{aligned} &\lambda I \le (a_{ij}) \le \lambda^{-1} I, \\ &\sum_{i,j} \|a_{ij}\|_\alpha + \sum_i \|b_i\|_\alpha + \|c\|_\alpha \leqslant \Lambda, \\ &c \ge 0. \end{aligned} \tag{3.6}$$

Consider the solution u to the variational inequality

$$\langle Au, v-u\rangle \geqslant \langle f, v-u\rangle, \quad \forall v \in K, \tag{3.7}$$

where

$$K = \{v : v \geq \psi \text{ in } \Omega,\ v|_{\partial\Omega} = \phi\} \quad \text{is a subset of } W^{2,p}.$$

We also assume that for some $\alpha \in (0,1)$,

$$\|f\|_\alpha < \infty, \quad \|\psi\|_{2+\alpha} < \infty, \quad \|\phi\|_{2+\alpha} < \infty, \quad \psi|_{\partial\Omega} \leq \phi. \tag{3.8}$$

The technique to obtain $W^{2,p}$-estimates of u is called the penalty method. Consider the penalized problem

$$\begin{cases} Au + \beta_\varepsilon(u - \psi) = f & \text{in } \Omega, \\ u = \phi \quad \text{on } \partial\Omega. \end{cases} \tag{3.9}$$

Here $\{\beta_\varepsilon(t)\}_{\varepsilon\in(0,1)}$ are non-decreasing smooth functions in t, s.t.

$$\begin{aligned} &\lim_{\varepsilon\to 0^+} \beta_\varepsilon(t) \to -\infty \quad \text{if } t < 0, \\ &\lim_{\varepsilon\to 0^+} \beta_\varepsilon(t) \to 0 \quad \text{if } t > 0, \\ &\beta_\varepsilon(0) = 0, \\ &\beta_\varepsilon(t) \leq B, \quad \forall t \in \mathbb{R},\ \forall \varepsilon \in (0,1) \end{aligned} \tag{3.10}$$

for some positive constant B independent of ε.

Remark 3.3. In this section, the differentiability of β_ε is irrelevant; we may choose β_ε to be merely continuous, e.g.

$$\beta_\varepsilon(t) = \begin{cases} \dfrac{t}{\varepsilon}, & t < 0, \\ 0, & t \geq 0. \end{cases}$$

We will see later that the differentiability of β_ε is indispensable in establishing $C^{1,1}$-estimates (Section 3.3) and BV-estimates (Section 3.4).

Remark 3.4. The basic idea of the penalty method is to replace "rigid" constraint $u \geq \psi$ by some "soft" penalty functions characterized by $\{\beta_\varepsilon(t)\}_{\varepsilon\in(0,1)}$, so that the original problem (3.7) is turned into more regular penalized problem (3.9). We are then able to obtain various estimates for the penalized problem using classic elliptic regularity theory. If some bounds could be preserved as we make the penalty functions more and more rigid by taking $\varepsilon \to 0$, e.g. they are independent of ε, we may find out the solution to the original problem and a similar estimate with the help of certain compactness. The penalty functions are chosen so that one would pay a lot for violating the original constraint. Hence the behavior of penalized problem would be close to the origin problem.

Lemma 3.1. *There is a solution of* (3.9), *denoted by* u_ε.

Proof. The existence is a direct result of the method of monotone iteration. That is, if we have

$$\begin{cases} -a_{ij}u_{ij} + b_i u_i = F(x,u) & \text{in } \Omega, \\ u|_{\partial\Omega} = \phi, \end{cases}$$

with $\frac{\partial F}{\partial z}(x,z) \leq 0$, then there exists a solution u, see [10, 15]. □

We shall use however a different approach with desired estimates. For $\forall N > 0$, we define

$$\beta_{\varepsilon,N}(t) = (\beta_\varepsilon(t) \wedge N) \vee (-N).$$

Consider

$$\begin{cases} Au + \beta_{\varepsilon,N}(u - \psi) = f & \text{in } \Omega, \\ u = \phi & \text{on } \partial\Omega. \end{cases} \tag{3.11}$$

It is known that for each $v \in L^p(\Omega)$, $1 < p < \infty$, there is a unique $w = Tv \in W^{2,p}(\Omega)$ solving

$$\begin{cases} Aw = f - \beta_{\varepsilon,N}(v - \psi) & \text{in } \Omega, \\ w = \phi & \text{on } \partial\Omega, \end{cases}$$

with

$$\|w\|_{2,p} \leq C(\varepsilon, N, f, \phi, B) \equiv R_0.$$

Note that the constant on the right hand side does not depend on ψ since the L^p-bound of $\beta_{\varepsilon,N}$ only depends on N and Ω. Hence, if we define

$$T : B_R(0) \subseteq L^p(\Omega) \to B_R(0)$$

for $R > R_0$, then T is a compact map from a bounded closed convex set $B_R(0) \subseteq L^1(\Omega)$ into itself. By Schauder's fixed-point theorem, there $\exists\, u_{\varepsilon,N}$, s.t. $Tu_{\varepsilon,N} = u_{\varepsilon,N} \in W^{2,p}(\Omega)$, $1 < p < \infty$, which solves (3.11). Recall that

Theorem 3.2 (Schauder's fixed-point theorem). *A continuous mapping T from a closed convex set S of a Banach space into a compact subset of S has a fixed point.*

Remark 3.5. In fact, $u_{\varepsilon,N} \in C^{2,\alpha}(\overline{\Omega})$ if we further require β_ε to be sufficiently smooth, say Lipschitz. This is a consequence of classical Schauder's theory, see [10, 15] or [6].

Now we want to find a lower bound for $\beta_{\varepsilon,N}(u_{\varepsilon,N} - \psi)$ that is independent of N and ε. Let

$$\mu_{\varepsilon,N} = \min_{\Omega} \beta_{\varepsilon,N}(u_{\varepsilon,N} - \psi).$$

We only consider the case when $\mu_{\varepsilon,N} < 0 = \beta_{\varepsilon,N}(0)$; otherwise $\mu_{\varepsilon,N}$ is already bounded from below. We may assume $\mu_{\varepsilon,N}$ is obtained at $x_0 \in \Omega$, i.e. $x_0 \notin \partial\Omega$. Indeed, for $x \in \partial\Omega$, $\beta_{\varepsilon,N}(u_{\varepsilon,N}(x) - \psi(x)) = \beta_{\varepsilon,N}(\phi(x) - \psi(x)) \geq 0$. Then $A(u_{\varepsilon,N} - \psi)(x_0) \leq 0$. Hence,

$$\begin{aligned} \mu_{\varepsilon,N} &= \beta_{\varepsilon,N}(u_{\varepsilon,N}(x_0) - \psi(x_0)) = -Au_{\varepsilon,N}(x_0) + f(x_0) \\ &\geq -A\psi(x_0) + f(x_0) \geq -C(\lambda, \Lambda, \psi, f). \end{aligned}$$

In particular, the constant C does not depend on ε and N. By this together with the assumptions on β_ε, we conclude that

$$\|Au_{\varepsilon,N}\|_0 \leq \|f\|_0 + \|\beta_{\varepsilon,N}(u_{\varepsilon,N} - \psi)\|_0 \leq C(\lambda, \Lambda, \psi, f, B).$$

This implies

$$\|u_{\varepsilon,N}\|_{2,p} \le C(n,p,\lambda,\Lambda,\psi,\phi,f,B), \quad \forall p \in (1,\infty).$$

Note that C is independent of ε and N. Now taking $N \to \infty$, there $\exists\, u_\varepsilon \in W^{2,p}(\Omega)$, s.t. $u_{\varepsilon,N} \to u_\varepsilon$ weakly in $W^{2,p}(\Omega)$ as $N \to \infty$, and

$$\|u_\varepsilon\|_{2,p} \le C(n,p,\lambda,\Lambda,\psi,\phi,f,B), \tag{3.12}$$

where C does not depend on ε. Then taking $\varepsilon \to 0^+$, there $\exists\, u \in W^{2,p}(\Omega)$, s.t. $u_\varepsilon \to u$ weakly in $W^{2,p}(\Omega)$ as $\varepsilon \to 0^+$, and $\|u\|_{2,p} \le C(n,p,\lambda,\Lambda,\psi,\phi,f,B)$. Note that this holds for $\forall p \in (1,\infty)$. Hence this implies $u_\varepsilon \to u$ strongly in $C^\alpha(\Omega)$ for some $\alpha \in (0,1)$.

Now we shall prove $u \in W^{2,p}(\Omega)$ is a desired solution of (3.7). It is easy to show $|\beta_\varepsilon(u_\varepsilon - \psi)| \le C(\lambda,\Lambda,\psi,f,B)$. This immediately implies $u \ge \psi$ by (3.10). Also $\beta_\varepsilon(u_\varepsilon - \psi) \to 0$ on the set $\{x : u(x) - \psi(x) > 0\}$ since u_ε is uniformly bounded. Therefore,

$$\limsup_{\varepsilon\to 0^+} \beta_\varepsilon(u_\varepsilon - \psi) \le 0.$$

It implies

$$\begin{cases} Au = f & \text{on } \{u > \psi\}, \\ Au \ge f & \text{a.e. in } \Omega. \end{cases}$$

Thus we have proved

Theorem 3.3. *There is a solution of the variational inequality* (3.7), *i.e.*

$$\begin{cases} Au - f \ge 0, \\ u \ge \psi, \\ (Au - f)(u - \psi) = 0, \end{cases}$$

holds for a.e. in Ω and $u = \phi$ on $\partial\Omega$. Moreover, $u \in W^{2,p}(\Omega)$, $\forall p \in (1,\infty)$, and

$$\|u\|_{2,p} \le C(n,p,\lambda,\Lambda,\psi,\phi,f).$$

Note that we omit the dependence on B since it can be chosen arbitrarily in (3.10).

Theorem 3.4. *Let u_1 and u_2 be solutions in $C(\overline{\Omega}) \cap W^{2,2}(\Omega)$ of* (3.7) *corresponding to f_1 and f_2 respectively. If $f_1 \ge f_2$, then $u_1 \ge u_2$.*

Corollary 3.1. *Under the assumptions* (3.6) *and* (3.8), *the solution of* (3.7) *in $C(\overline{\Omega}) \cap W^{2,2}(\Omega)$ is unique.*

Proof of Theorem 3.4. Suppose $O = \{x \in \Omega : u_2(x) > u_1(x)\}$ is nonempty; O is open. We know that $u_2 > u_1 \ge \psi$ on O and thus $Au_1 \ge f_1$, and $Au_2 = f_2$. Hence,

$$A(u_1 - u_2) \ge f_1 - f_2 \ge 0.$$

This together with $u_1 = u_2$ on $\partial\Omega$ implies that $u_1 \ge u_2$ in O. Contradiction! □

3.3 Interior $C^{1,1}$-estimates

In this section, we shall prove interior $C^{1,1}$-estimates for the solutions to the obstacle problem, again by the penalty method, due to several authors including Caffarelli, Frehse and Kinderlehrer. We should note that $C^{1,1}$-regularity is optimal for general obstacle problems with a_{ij}, b_i, c, ψ, ϕ and Ω sufficiently smooth, i.e. the solution is not necessarily in $C^2(\Omega)$ even if a_{ij}, b_i, c, ψ, ϕ and Ω are all C^∞.

Example 3.1. Let $\Omega = (-1,1)$ and $\psi(x) = \frac{3}{4} - x^2$. We consider the following obstacle problem:

$$\langle -\Delta u, v - u\rangle \geq 0, \quad u \in K = \{w \in H_0^1(\Omega) : w \geq \psi\}, \ \forall v \in K.$$

It is easy to see that

$$u(x) = \begin{cases} x+1, & x \in \left(-1, -\dfrac{1}{2}\right), \\ \psi(x), & x \in \left[-\dfrac{1}{2}, \dfrac{1}{2}\right], \\ 1-x, & x \in \left(\dfrac{1}{2}, 1\right) \end{cases}$$

is the unique solution. However, it is not C^2 at $\pm\frac{1}{2}$. □

For simplicity, we shall consider in the sequel the following variational inequality:

$$\begin{aligned} &\langle -\Delta u - f, v - u\rangle \geq 0, \quad \forall v \in K = \{w : w \geq \psi, \ w|_{\partial\Omega} = \phi\}, \\ &u \in K, \ u|_{\partial\Omega} = \phi, \end{aligned} \tag{3.13}$$

where f, ψ and ϕ satisfy (3.8) and we also assume $f \in W^{2,p}(\Omega)$ for some $p \in (n, \infty)$. Note the regularity assumption on f can be further reduced. The penalized problem is

$$\begin{cases} -\Delta u_\varepsilon + \beta_\varepsilon(u_\varepsilon - \psi) = f & \text{in } \Omega, \\ u_\varepsilon|_{\partial\Omega} = \phi, \end{cases} \tag{3.14}$$

where $\{\beta_\varepsilon\}_{\varepsilon\in(0,1)}$ are non-decreasing smooth functions of t satisfying (3.10) and

$$\beta_\varepsilon' > 0, \quad \beta_\varepsilon'' \leq 0.$$

It is known that the solution u_ε for (3.14) is in $W^{2,q}(\Omega)$, $\forall q \in (1,\infty)$. By $-\Delta u_\varepsilon = f - \beta_\varepsilon(u_\varepsilon - \psi) \in W^{2,p}(\Omega)$, we know that $u_\varepsilon \in W^{4,p}(K')$, $\forall K' \subset\subset \Omega$. Note that p comes from the regularity assumption on f. Then the right hand side becomes of $W^{2,p}(K')$, which implies $u_\varepsilon \in W^{4,p}(K)$, $\forall K \subset\subset K'$. For given $K \subset\subset \Omega$, we define a cut-off function $\eta \in C_0^\infty(\Omega)$, s.t.

$$\eta \in [0,1] \text{ in } \Omega, \text{ and } \eta = 1 \text{ in } K. \tag{3.15}$$

Choose an arbitrary vector $\tau \in \mathbb{S}^{n-1}$ as the direction for taking derivative. Our task now is to obtain L^∞-bound of $u_{\varepsilon,\tau\tau}$ in K which is uniform in ε and $\tau \in \mathbb{S}^{n-1}$. Then we take $\varepsilon \to 0$ so that $u_\varepsilon \to u$. By virtue of the weak* - compactness of $C^{1,1}(K)$, we know that $u \in C^{1,1}(K)$ and its norm share the same uniform bound.

Take pure second partial derivative of (3.14) in the direction of τ and we find

$$\begin{aligned}&-\Delta u_{\varepsilon,\tau\tau} + \beta'_\varepsilon(u_\varepsilon - \psi)(u_{\varepsilon,\tau\tau} - \psi_{\tau\tau}) + \beta''_\varepsilon(u_\varepsilon - \psi)(u_{\varepsilon,\tau} - \psi_\tau)^2 = f_{\tau\tau}\\ &\Rightarrow \Delta u_{\varepsilon,\tau\tau} \le \beta'_\varepsilon(u_\varepsilon - \psi)(u_{\varepsilon,\tau\tau} - \psi_{\tau\tau}) - f_{\tau\tau}.\end{aligned}$$

We turn to look at

$$\begin{aligned}\Delta(\eta u_{\varepsilon,\tau\tau}) &= \eta\Delta u_{\varepsilon,\tau\tau} + \Delta\eta u_{\varepsilon,\tau\tau} + 2\nabla\eta\nabla u_{\varepsilon,\tau\tau}\\ &\le \eta(-f_{\tau\tau} + \beta'_\varepsilon(u_\varepsilon - \psi)(u_{\varepsilon,\tau\tau} - \psi_{\tau\tau})) + u_{\varepsilon,\tau\tau}\Delta\eta + 2\nabla\eta\nabla u_{\varepsilon,\tau\tau}\\ &= \eta(-f_{\tau\tau} + \beta'_\varepsilon(u_\varepsilon - \psi)(u_{\varepsilon,\tau\tau} - \psi_{\tau\tau})) - u_{\varepsilon,\tau\tau}\Delta\eta + 2\mathrm{div}(u_{\varepsilon,\tau\tau}\nabla\eta).\end{aligned}$$

Let w_ε solve

$$\begin{cases}\Delta w_\varepsilon = -\eta f_{\tau\tau} - u_{\varepsilon,\tau\tau}\Delta\eta + 2\mathrm{div}(u_{\varepsilon,\tau\tau}\nabla\eta) & \text{in } \Omega,\\ w_\varepsilon|_{\partial\Omega} = 0.\end{cases}$$

Since $\eta f_{\tau\tau} \in L^p(\Omega)$, $u_{\varepsilon,\tau\tau}\nabla\eta \in L^p(\Omega)$ and $u_{\varepsilon,\tau\tau}\Delta\eta \in L^p(\Omega)$, we know that

$$\|w_\varepsilon\|_{C^\alpha(\Omega)} \le C\left(\|f\|_{W^{2,p}(\Omega)} + \|u_\varepsilon\|_{W^{2,p}(\Omega)}\right) \le C(n, p, \psi, \phi, f, B, \eta) \tag{3.16}$$

for some $\alpha \in (0,1)$ (cf. Theorem A.7 for equations of non-divergence form). In particular, the positive constant C does not depend on ε due to (3.12). Also we find

$$\Delta(\eta u_{\varepsilon,\tau\tau} - w_\varepsilon) \le \eta\beta'_\varepsilon(u_\varepsilon - \psi)(u_{\varepsilon,\tau\tau} - \psi_{\tau\tau}). \tag{3.17}$$

Consider the minimum of $\eta u_{\varepsilon,\tau\tau} - w_\varepsilon$ on $\overline{\Omega}$. We assume that it is obtained at $x_0 \in \overline{\Omega}$.

If $\eta(x_0) = 0$, $\eta u_{\varepsilon,\tau\tau}(x) - w_\varepsilon(x) \ge \eta(x_0)u_{\varepsilon,\tau\tau}(x_0) - w_\varepsilon(x_0) = -w_\varepsilon(x_0)$ in Ω. Hence, combined with (3.16), $u_{\varepsilon,\tau\tau}(x) = \eta(x)u_{\varepsilon,\tau\tau}(x) \ge w_\varepsilon(x) - w_\varepsilon(x_0) \ge -C$ in K.

If $\eta(x_0) > 0$, $x_0 \in \Omega$. By virtue of (3.17),

$$\begin{aligned}\Delta(\eta u_{\varepsilon,\tau\tau} - w_\varepsilon)(x_0) \ge 0 &\Rightarrow \eta\beta'_\varepsilon(u_\varepsilon - \psi)(x_0)(u_{\varepsilon,\tau\tau} - \psi_{\tau\tau})(x_0) \ge 0\\ &\Rightarrow (u_{\varepsilon,\tau\tau} - \psi_{\tau\tau})(x_0) \ge 0.\end{aligned}$$

Hence,

$$\eta u_{\varepsilon,\tau\tau} - w_\varepsilon \ge \eta(x_0)u_{\varepsilon,\tau\tau}(x_0) - w_\varepsilon(x_0) \ge \eta(x_0)\psi_{\tau\tau}(x_0) - w_\varepsilon(x_0)$$

in Ω, which implies $u_{\varepsilon,\tau\tau} \ge -C$ in K. Note that $\eta \equiv 1$ in K.

Therefore, in either case, we find for $\forall\tau \in \mathbb{S}^{n-1}$ and $\varepsilon \in (0,1)$, $u_{\varepsilon,\tau\tau} \ge -C(n, p, \psi, \phi, f, B, \eta)$ in K.

On the other hand,

$$\Delta u_\varepsilon = \beta_\varepsilon(u_\varepsilon - \psi) - f \le B + \|f\|_0.$$

This gives

$$u_{\varepsilon,\tau\tau} \leq (n-1)C(n,p,\psi,\phi,f,B,\eta) + B + \|f\|_0 = C(n,p,\psi,\phi,f,B,\eta) \quad \text{in } K.$$

Hence we have proved

$$\|u_{\varepsilon,\tau\tau}\|_{L^\infty(K)} \leq C(n,p,\psi,\phi,f,B,\eta), \quad \forall \varepsilon \in (0,1),\ \tau \in \mathbb{S}^{n-1}.$$

Next, we take the limit $\varepsilon \to 0$, since $u_\varepsilon \to u$ uniformly in Ω, we can easily obtain the desired $C^{1,1}$ estimate for u.

Remark 3.6. The method above can be extended to general second-order linear elliptic operators of non-divergence type with suitably smooth coefficients.

3.4 BV-estimates

In this section, we shall deal with obstacle problems with equation of divergence type. We wish to show later that the coincidence set Λ is of "good properties". *Formally* it can be stated as $\|\nabla\chi_\Lambda\|_{L^1(\Omega)} \leq C$. To rigorously formulate it, we introduce the notion of BV-functions.

Definition 3.1. For $u \in L^1(\Omega)$, define

$$\|u\|_{BV(\Omega)} = \|u\|_{L^1(\Omega)} + \sup\left\{\int_\Omega u \operatorname{div}\xi \,\mathrm{d}x : \xi \in (C_0^1(\Omega))^n,\ \|\xi\|_0 \leq 1\right\}.$$

We shall say $u \in BV(\Omega)$ iff $\|u\|_{BV(\Omega)} < \infty$.

Example 3.2. It is easy to show that for $v \in W^{1,1}(\Omega)$, $\|v\|_{BV(\Omega)} \leq \|v\|_{W^{1,1}(\Omega)}$. □

Example 3.3. Let $E \subset\subset \Omega$ be a bounded smooth open set and $v = \chi_E$. Then

$$\left|\int_\Omega v \operatorname{div}\xi \,\mathrm{d}x\right| = \left|\int_E \operatorname{div}\xi \,\mathrm{d}x\right| = \left|\int_{\partial E} \xi\cdot n \,\mathrm{d}\mathcal{H}^{n-1}\right| \leq \|\xi\|_0 \mathcal{H}^{n-1}(\partial E),$$

where the equality holds for $\xi|_{\partial E} = n$. Hence we have

$$\|v\|_{BV(\Omega)} = |E| + \mathcal{H}^{n-1}(\partial E).$$ □

Remark 3.7. If one can show $\chi_\Lambda \in BV(\Omega)$, then by a theorem of De Giorgi (see [16], [7] and [4]) the reduced boundary $\partial * \Lambda$ is rectifiable, i.e. it is contained in a countable union of C^1-hypersurfaces. In this way, we are able to discuss the regularity of coincidence set later in this notes.

Let $\Omega \subset \mathbb{R}^n$ be a bounded Lipschitz domain. Define

$$Au = -\partial_i\left(a_{ij}(x)\partial_j u\right), \tag{3.18}$$

where $a_{ij} \in W^{2,\infty}(\Omega)$ and

$$\begin{gathered}\sum_{i,j=1}^n \|a_{ij}\|_{W^{2,\infty}(\Omega)} \leq \Lambda,\\ \lambda|\xi|^2 \leq a_{ij}\xi_i\xi_j \leq \lambda^{-1}|\xi|^2 \quad \forall \xi \in \mathbb{R}^n\end{gathered} \tag{3.19}$$

for some positive Λ and λ. Consider the variational inequality

$$\langle Au - f, v - u\rangle \geq 0, \quad u \in K_\psi \triangleq \{w \in H_0^1(\Omega : w \geq \psi\}, \ \forall v \in K\Psi, \tag{3.20}$$

where $f \in C^{0,1}(\overline{\Omega})$ and $\psi \in W^{3,1}(\Omega) \cap C^{2,\alpha}(\overline{\Omega})$ with $\psi|_{\partial\Omega} \leq 0$.

Define penalty function $\{\beta_\varepsilon\}_{\varepsilon\in(0,1)}$ as before. Namely they are non-decreasing smooth functions in t satisfying (3.10). Let u_ε solve the penalized problem

$$\begin{cases} Au_\varepsilon + \beta_\varepsilon(u_\varepsilon - \psi) = f & \text{in } \Omega, \\ u_\varepsilon|_{\partial\Omega} = 0. \end{cases} \tag{3.21}$$

We shall prove that $Au_\varepsilon \in BV_{loc}(\Omega)$. For $\forall K \subset\subset \Omega$, as before, we define a cut-off function $\eta \in C_0^\infty(\Omega)$ associated with K satisfying (3.15). We also define $\{\gamma_\delta(t)\}_{\delta\in(0,1)}$ as smooth approximations of sgn(t), s.t.

$$|\gamma_\delta(t)| \leq 1, \ \gamma_\delta'(t) \geq 0, \ \gamma_\delta(0) = 0, \ \text{and} \ \lim_{\delta\to 0^+} \gamma_\delta(t) = \text{sgn}(t).$$

The last convergence is in the pointwise sense.

Let $\tau \in \mathbb{S}^{n-1}$ be an arbitrary unit vector. We take derivative of (3.21) in the direction of τ

$$\frac{\partial}{\partial\tau} Au_\varepsilon + \beta_\varepsilon'(u_\varepsilon - \psi)\frac{\partial}{\partial\tau}(u_\varepsilon - \psi) = f_\tau. \tag{3.22}$$

Define $\gamma = \gamma_\delta(u_{\varepsilon,\tau} - \psi_\tau)$. We calculate

$$\begin{aligned}
\int_\Omega \frac{\partial}{\partial\tau}(Au_\varepsilon)\eta\gamma &= \int_\Omega \frac{\partial}{\partial\tau}(a_{ij}(x)u_{\varepsilon,j})(\gamma_i\eta + \eta_i\gamma) \\
&= \int_\Omega a_{ij}(x)(u_{\varepsilon,\tau} - \psi_\tau)_j\gamma_i\eta + \int_\Omega a_{ij}(x)\psi_{j\tau}\gamma_i\eta \\
&\quad + \int_\Omega \frac{\partial}{\partial\tau}(a_{ij}(x)u_{\varepsilon,j})\eta_i\gamma + \int_\Omega \left(\frac{\partial}{\partial\tau}a_{ij}(x)\right) u_{\varepsilon,j}\gamma_i\eta \\
&\triangleq \mathrm{I} + \mathrm{II} + \mathrm{III} + \mathrm{IV}.
\end{aligned}$$

Notice that $\gamma_\delta' \geq 0$,

$$\mathrm{I} = \int_\Omega [a_{ij}(u_{\varepsilon,\tau} - \psi_\tau)_i(u_{\varepsilon,\tau} - \psi_\tau)_j] \cdot \gamma_\delta'(u_{\varepsilon,\tau} - \psi_\tau)\eta \geq 0.$$

We also have

$$\begin{aligned}
|\mathrm{II}| &= \left| -\int_\Omega [a_{ij}(x)\psi_{j\tau}\eta]_i\gamma \right| \leq C(\Lambda,\eta)\|\psi\|_{W^{3,1}(\Omega)}, \\
|\mathrm{IV}| &= \left| -\int_\Omega [(a_{ij})_\tau u_{\varepsilon,j}\eta]_i\gamma \right| \leq C(\Lambda,\eta)\|u_\varepsilon\|_{W^{2,p}(\Omega)}, \\
|\mathrm{III}| &\leq C(\Lambda,\eta)\|u_\varepsilon\|_{W^{2,p}(\Omega)}.
\end{aligned}$$

In sum,

$$-\int_\Omega \frac{\partial}{\partial\tau}(Au_\varepsilon)\eta\gamma \leq C(\Lambda,\eta)\left(\|\psi\|_{W^{3,1}(\Omega)} + \|u_\varepsilon\|_{W^{2,p}(\Omega)}\right) \leq C(n,p,\lambda,\Lambda,\eta,\psi,f).$$

The last inequality comes from the $W^{2,p}$-estimate. By (3.22), we obtain

$$\int_\Omega \beta'_\varepsilon(u_\varepsilon - \psi)(u_{\varepsilon,\tau} - \psi_\tau)\gamma_\delta(u_{\varepsilon,\tau} - \psi_\tau)\eta = \int_\Omega f_\tau \gamma\eta - \int_\Omega \frac{\partial}{\partial\tau}(Au_\varepsilon)\eta\gamma$$
$$\leq C(n, p, \lambda, \Lambda, \eta, \psi, f).$$

Note that

$$\lim_{\delta\to 0^+} \beta'_\varepsilon(u_\varepsilon - \psi)(u_{\varepsilon,\tau} - \psi_\tau)\gamma_\delta(u_{\varepsilon,\tau} - \psi_\tau) = \left|\frac{\partial}{\partial\tau}\beta_\varepsilon(u_\varepsilon - \psi)\right| \text{ pointwise.}$$

For fixed ε, by bounded convergence theorem,

$$\int_\Omega \eta\left|\frac{\partial}{\partial\tau}\beta_\varepsilon(u_\varepsilon - \psi)\right| \leq C(n, p, \lambda, \Lambda, \eta, \psi, f).$$

Again by (3.22),

$$\int_\Omega \eta\left|\frac{\partial}{\partial\tau}(Au_\varepsilon)\right| \leq \int_\Omega \eta\left|\frac{\partial}{\partial\tau}\beta_\varepsilon(u_\varepsilon - \psi)\right| + \int_\Omega \eta|f_\tau| \leq C(n, p, \lambda, \Lambda, \eta, \psi, f)$$
$$\Rightarrow \int_\Omega \eta\,|\nabla(Au_\varepsilon)| \leq C(n, p, \lambda, \Lambda, \eta, \psi, f)$$
$$\Rightarrow \int_K |\nabla(Au_\varepsilon)| \leq C(n, p, \lambda, \Lambda, K, \psi, f).$$

Note that the last inequality, together with

$$\int_K |Au_\varepsilon| \leq \int_K |f| + \int_K |\beta_\varepsilon(u_\varepsilon - \psi)| \leq C(\lambda, \Lambda, \psi, f, B),$$

implies that $Au_\varepsilon \in BV_{loc}(\Omega)$, and $\|Au_\varepsilon\|_{BV(K)}$ is independent of ε. By taking $\varepsilon \to 0^+$ and using the compactness of $BV_{loc}(\Omega)$, we know that there $\exists\, v \in BV_{loc}(\Omega)$ and a subsequence of $\{Au_\varepsilon\}$, denoted by $\{Au_{\varepsilon_k}\}$, s.t.

$$Au_{\varepsilon_k} \to v \text{ in } L^1_{loc}(\Omega).$$

Moreover,

$$\|v\|_{BV(K)} \leq \|Au_{\varepsilon_k}\|_{BV(K)}, \quad \forall K \subset\subset \Omega.$$

Recall that u is the weak limit of u_ε in $W^{2,p}(\Omega)$, $\forall p \in (1, \infty)$. In particular,

$$Au_\varepsilon \to Au \text{ in } L^1_{loc}(\Omega).$$

Therefore, $Au = v \in BV_{loc}(\Omega)$, and

$$\|Au\|_{BV(K)} \leq \|Au_{\varepsilon_k}\|_{BV(K)} \leq C(K), \quad \forall K \subset\subset \Omega.$$

Thus we have proved

Theorem 3.5. *Suppose that* $\psi \in W^{3,1}(\Omega) \cap C^{2,\alpha}(\overline{\Omega})$, $f \in C^{0,1}(\overline{\Omega})$. *Let* $u \in K_\psi$ *be the solution of* (3.20), *where* A *is given by* (3.18) *and* (3.19). *Then* $Au \in BV_{loc}(\Omega)$.

Corollary 3.2. *Suppose, in addition, that $A\psi - f \neq 0$. Then the coincidence set $\Lambda = \{x \in \Omega : u(x) = \psi(x)\}$ is of locally finite perimeter, i.e. $\chi_\Lambda \in BV_{loc}(\Omega)$.*

Remark 3.8. The condition $A\psi - f \neq 0$ rules out the case that the landscape of coincidence set could be rather complicated.

Proof. By assumption, $A\psi - f \in W^{1,1}(\Omega)$ and $A\psi - f \neq 0$. We claim that

$$\chi_\Lambda = \frac{Au - f}{A\psi - f} \quad \text{a.e. in } \Omega.$$

Hence, $\chi_\Lambda \in BV_{loc}(\Omega)$. □

The Schauder estimates presented in this chapter were due to Caffarelli and Kinderlehrer: Potential methods in variational inequalities, J. Analyse Math. 37 (1980), 285–295. The BV-estimate in the last section was due to Brezis and Kinderlehrer: The smoothness of solutions to nonlinear variational inequalities, Indiana Univ. Math. J. 23 (1973/74), 831–844. The $W^{2,p}$-estimates are classical and are taken from the book [5]. The additional $C^{1,1}$-estimate via the penalty method was observed by the author.

4 Regularity of free boundary

4.1 Examples of free boundary with singularities

Example 4.1. Consider an analytic curve $\Gamma \subset \mathbb{R}^2$ with a cusp at one point, say 0, having the figure ∞. Let Λ be the closed set enclosed by Γ, V an open neighborhood of Γ, and $N = V \backslash \overline{\Lambda}$. We consider the boundary value problem:

$$\begin{cases} \Delta \tilde{u} = 1 & \text{in } N, \\ \tilde{u} = 0, \quad \dfrac{\partial \tilde{u}}{\partial \nu} = 0 & \text{on } \Gamma. \end{cases} \tag{4.1}$$

Set $\tilde{u} \equiv 0$ in Λ. By the Cauchy-Kowalevski Theorem, (4.1) has a unique analytic solution in $N \cup \Gamma$, provided that the neighborhood V is small enough. Since $\frac{\partial^2 \tilde{u}}{\partial \nu^2} = 1$ on Γ, shrinking V further if necessary, we have $\tilde{u} > 0$ in N. Let $u = \tilde{u} - \frac{|x|^2}{4}$, $\psi = -\frac{|x|^2}{4}$. Then the function u is a solution in $\Omega = \overline{\Lambda} \cup N$ of the obstacle problem:

$$\Delta u \leq 0, \quad u \geq \psi, \quad (u - \psi)\Delta u = 0 \text{ in } \Omega,$$

with strictly concave obstacle ψ. The free boundary Γ has a cusp at 0.

Example 4.2. Let Ω be a convex C^2 domain which is symmetric with respect to the hypersurface $P = \{x |\, x_n = 0\}$. Let $E \subset F \subset \Omega \cap P$ be such that:

(i) E is open, and F is relatively closed in $\Omega \cap P$,
(ii) $E = \bigsqcup_i E_i$, $F = \bigsqcup_i F_i$, i.e. disjoint unions of their components E_i and F_i respectively,

with $E_i \subset F_i$ for every i. Let $\psi \in C^2(\overline{\Omega})$, $\psi < 0$ on $\partial\Omega$, and u be the solution of

$$\min \int_\Omega |\nabla v|^2 \, dx, \quad v \in K_\psi.$$

We set $\Lambda = \{u = \psi\}$, and write

$$\Lambda = \bigsqcup_j \Lambda_j,$$

$$\mathring{\Lambda} = \bigsqcup_j \mathring{\Lambda}_j,$$

the disjoint unions of closed and open components respectively.

Theorem 4.1. *There is a C^∞ superharmonic function ψ in $\overline{\Omega}$ with $\psi \le 0$ such that $E_j = \mathring{\Lambda}_j \cap P$, and $F_j = \Lambda_j \cap P$.*

Remark 4.1. If $(x_0', 0)$ is an accumulation point of components F_j, then in any neighborhood of $(x_0', 0)$, there are infinitely many components of the free boundary. If $n = 2$ and ψ is concave, then N is topologically an annulus. If, in addition ψ is analytic, then Γ is a Jordan curve with analytic parametrization. This latter statements were due to H. Lewy, see [13] and references therein.

Proof of Theorem 4.1. Firstly note that for any open set $O \in \mathbb{R}^n$, there is a non-negative, smooth function $\alpha(x)$ such that $O = \{x \in \mathbb{R}^n : \alpha(x) > 0\}$. Indeed, let $\{B_{r_i}(x_i)\}_i$ be a covering of O by balls contained in O, with $r_j < 1/2$. Let $\xi \in C^\infty$ be such that $\xi \equiv 0$, if $|x| \ge 1$; and $\xi(x) > 0$, if $|x| < 1$. Then let

$$\alpha(x) = \sum_j \alpha_j \xi\left(\frac{x - x_j}{r_j}\right), \qquad \text{where } \alpha_j = \frac{(r_j)^j}{j!}.$$

We let $O = E \subset P$ and construct $\alpha(x')$ as above. Let

$$\begin{aligned}
&\Gamma_+ = \left\{(x', \alpha(x')) : x' \in \Omega \cap \mathbb{R}^{n-1}\right\}, \quad \Gamma_- = \left\{(x', -\alpha(x')) : x' \in \Omega \cap \mathbb{R}^{n-1}\right\};\\
&\Omega_+ = \{x : x_n > \alpha(x')\}, \quad \Omega_- = \{x : x_n < -\alpha(x')\};\\
&\Lambda = \{x : -\alpha(x') \le x_n \le \alpha(x')\};\\
&\Omega = \Omega_+ \cup \Omega_- \cup \Lambda.
\end{aligned}$$

Notice that $E \subset \mathring{\Lambda} \cap P$. Consider the boundary value problem:

$$\begin{cases} \Delta v = 1 + h & \text{in } \Omega_+,\\ v = 0, \quad \dfrac{\partial v}{\partial x_n} = 0 & \text{on } \Gamma_+. \end{cases} \tag{4.2}$$

Lemma 4.1. *There exist functions v, h in $C^\infty(\overline{\Omega}_+)$, for which (4.2) hold, and h vanishes on Γ_+ as well as its derivatives; furthermore, $v > 0$, $\Delta v > 0$ in Ω_+.*

Proof. Exercise, see for example discussions in [5]. □

We extend the definition of v into Ω as:

$$v(x) = \begin{cases} v\left(x', -x_n\right) & \text{in } \Omega_-, \\ 0 \quad \text{in } \Lambda. \end{cases}$$

Letting

$$f = \begin{cases} \Delta v = 1 + h & \text{in } \Omega_+ \cup \Omega_-, \\ 1 \quad \text{in } \Lambda, \end{cases}$$

we have $f \in C^\infty\left(\overline{\Omega}\right)$ and $f > 0$. Consider

$$\begin{cases} \Delta g = -f & \text{in } \Omega, \\ g = -v & \text{on } \partial\Omega. \end{cases}$$

Then $u = v + g$ satisfies

$$\text{(i)} \begin{cases} \Delta u = 0 & \text{in } \Omega\backslash\Lambda, \\ u = 0 & \text{on } \partial\Omega; \end{cases}$$

$$\text{(ii)} \begin{cases} u = g, \quad \nabla u = \nabla g & \text{on } \partial\Lambda, \\ u > \psi & \text{in } \Omega\backslash\Lambda, \end{cases}$$

i.e. u is a solution to the obstacle problem with obstacle ψ. Moreover, for every j we have

$$\begin{cases} F_j = \Lambda_j \cap P, \\ E_j = \overset{\circ}{\Lambda}_j \cap P. \end{cases}$$

□

4.2 Free boundary regularity I: basic estimates

Let u_0 be the unique solution of

$$\min_K \int_\Omega |\nabla v|^2 \, dx, \quad K = \{v \in H^1(\Omega) : v \geq \psi \text{ in } \Omega, \ v = \phi \text{ on } \partial\Omega\}.$$

Then

$$\Delta u_0 \leq 0, \quad u_0 \geq \psi, \quad \text{and} \quad (u_0 - \psi)\,\Delta u_0 = 0.$$

We consider $u = u_0 - \psi$ and $f = -\Delta\psi$. Hence we have

$$-\Delta u + f \geq 0, \quad u \geq 0, \quad \text{and} \quad u(-\Delta u + f) = 0.$$

We assume $\Omega \subseteq \mathbb{R}^n$ to be a bounded domain and $f \in C^{0,\alpha}(\Omega)$, $0 \leq \alpha \leq 1$. We have $u \in C^{1,1}(\Omega)$. We define

$$\begin{aligned} N &= \{x \in \Omega : u(x) > 0\}, \\ \Lambda &= \{x \in \Omega : u(x) = 0\}, \\ \Gamma &= \partial N \cap \Omega. \end{aligned}$$

We first want to show:

(i) Γ has Lebesgue measure zero;
(ii) If $y \in \Gamma$, then

$$\liminf_{\substack{x \to y \\ x \in N}} D_{ii}(u(x)) \geq 0$$

for any i (direction).

Lemma 4.2 (Nondegeneracy of u). *Suppose that $f \geq \lambda > 0$ (i.e. $\Delta\psi \leq -\lambda$) in Ω, and let $x_0 \in \overline{N}$. Then for any ball $B_r(x_0) \in \Omega$,*

$$\sup_{B_r(x_0)} (u(x) - u(x_0)) \geq \frac{\lambda r^2}{2n}. \tag{4.3}$$

Proof. First we assume $x_0 \in N$, and consider the auxiliary function

$$w(x) = u(x) - u(x_0) - \frac{\lambda}{2n} |x - x_0|^2 .$$

Then $\Delta w \geq 0$ in N and $w(x_0) = 0$. Hence by the maximum principle,

$$\sup_{B_r(x_0) \cap N} w \geq 0,$$

and it is attained on the boundary $\partial(N \cap B_r(x_0))$. Since $w \leq 0$ on $\partial N \cap B_r(x_0)$, there exists an $x_1 \in \partial B_r(x_0) \cap \overline{N}$ such that $w(x_1) \geq 0$. Hence (4.3) follows.

If $x_0 \in \overline{N} \backslash N$, then we can take a sequence of $\{x_m\} \subset N$ converging to x_0 and apply the above argument. □

Lemma 4.3. *If $x \in N$, $\operatorname{dist}(x, \Gamma) < \varepsilon$, $\operatorname{dist}(x, \partial\Omega) \geq 1$, then*

$$u(x) \leq M\varepsilon^2, \qquad \text{where } M = \left\|D^2 u\right\|_{L^\infty(\Omega)}, \tag{4.4}$$

$$|\nabla u(x)| \leq C_M \varepsilon. \tag{4.5}$$

Proof. Since $u = 0$, $|\nabla u| = 0$ on Γ and $u \in C^{1,1}$, (4.4) follows. For (4.5) we will have a Taylor expansion argument. For $x \in N$, we have the expansion

$$0 \leq u(x + se_i) = u(x) + u_i(x)s + \int_0^s \int_0^\tau D_{ij} u(x + te_i)\, dt\, d\tau. \tag{4.6}$$

Choose e_i so that $-u_i(x) = |\nabla u|(x)$ and let $s = \varepsilon$ to get

$$0 \leq M\varepsilon^2 - |\nabla u|(x)\varepsilon + \frac{\varepsilon^2}{2} M,$$

and rearranging yields the claim. □

Theorem 4.2. *If $f \geq \lambda > 0$ in Ω and $x_0 \in \Gamma$, $\operatorname{dist}(x_0, \partial\Omega) \geq \delta_0 > 0$, then there exists an ε_0 and r such that*

$$\frac{|B_\varepsilon(x_0) \cap N|}{|B_\varepsilon|} \geq r > 0, \quad \forall \varepsilon \in (0, \varepsilon_0), \tag{4.7}$$

where r, ε_0 depend only on λ, δ_0 and $M = \left\|D^2 u\right\|_{L^\infty(\Omega_{\delta_0})}$.

Corollary 4.1. *If $f \geq \lambda > 0$, then Γ has Lebesgue measure zero.*

Proof of Corollary 4.1. If E is a measurable subset of $\mathbb{R}^n$ with positive measure (note that Γ is measurable), then

$$DE(x_0) = \lim_{\varepsilon \to 0^+} \frac{|B_\varepsilon(x_0) \cap E|}{|B_\varepsilon|} = 1 \quad \text{for a.e. } x_0 \in E.$$

Then (4.7) implies for almost every $x_0 \in \Gamma$, $D\Gamma(x_0) < 1$. Thus, we conclude $|\Gamma| = 0$. □

Proof. By Lemma 4.2 there is a point $x \in \partial B_\varepsilon(x_0)$ such that

$$u(x) \geq \frac{\lambda \varepsilon^2}{2n}.$$

Consider $y \in B_{\delta\varepsilon}(x)$ with δ small. Since $\|\nabla u\|_{L^\infty(B_{\delta\varepsilon}(x))} \leq C\varepsilon$, we have

$$|u(x) - u(y)| \leq \delta\varepsilon \cdot C\varepsilon = \delta C \varepsilon^2.$$

Thus,

$$u(y) \geq \frac{\lambda\varepsilon^2}{2n} - C\delta\varepsilon^2 > 0, \qquad \text{for } \delta < \frac{\lambda}{2nC},$$

i.e. $B_{\delta\varepsilon}(x) \subset N$. □

Theorem 4.3. *Let f be a constant function, $y \in \Gamma$, $\operatorname{dist}(y, \partial\Omega) \geq \delta_0 > 0$. Then if $x \in N$*

$$D_{ii}u(x) \geq \frac{-C}{|\log|x-y||^\varepsilon}, \qquad \varepsilon = \frac{1}{2(n-1)}, \tag{4.8}$$

where $C = C(\delta_0, n, M)$ and $M = \|D^2 u\|_{C^{1,1}(\Omega_{\delta_0})}$.

Remark 4.2. (a) Note that

$$D_{ii}u(x) \geq \frac{-C}{|\log(\operatorname{dist}(x, \Gamma))|^\varepsilon}$$

follows from (4.8), and hence

$$\liminf_{x \in N,\, x \to y \in \Gamma} D_{ii}u(x) \geq 0.$$

(b) In general cases $Au - f \geq 0$, $u \geq 0$ and $u(Au - f) = 0 \in \Omega$, if $f \in C^\alpha(\Omega)$, $a_{ij} \in C^\alpha(\Omega)$, then

$$\liminf_{x \in N,\, x \to y \in \Gamma} D_{ii}u(x) \geq 0 \tag{4.9}$$

as well.

First we shall show (4.9) in the case Δ as a corollary of Theorem 4.3.

Proof of (4.9) *with* $A = \Delta$. (Blow-up argument) We will prove this lower bound by contradiction. Suppose (4.8) is not true. Then there exist $\{x_m\} \in N$, $x_m \to 0 = y \in \Gamma$ such that

$$D_{ii}u(x_m) \leq -\varepsilon_0 < 0,$$

and $\operatorname{dist}(x_m, \Gamma) = |x_m - x_m^*|$, where $x_m^* \in \Gamma$. Now consider

$$v_m(x) = \frac{u(x_m^* + \lambda_m x)}{\lambda_m^2},$$

where $x \in B_2(0)$ and $\lambda_m = \frac{1}{\delta}|x_m - x_m^*| \to 0$ with $\delta < 1$ chosen sufficiently small so that

$$\frac{C}{|\log \delta|^{\varepsilon}} < \frac{\varepsilon_0}{2}.$$

Then $v_m \geq 0$, and $\|D^2 v_m\|_{L^\infty(B_2(0))} \leq M$. Letting $f_m(x) = f(x_m^* + \lambda_m x)$, we have $\|f_m\|_{C^\alpha} \leq A$. There exist $\xi_m = \lambda_m^{-1}(x_m - x_m^*) \in N(v_m)$ with $|\xi_m| = \delta$, such that $B_\delta(\xi_m) \subset N(v_m) \cap B_2(0)$. We may assume, up to a subsequence, $\xi_m \to \xi_*$. Also $0 \in \Gamma_m$. Furthermore, we have the interior estimate $\|v_m\|_{C^{2,\alpha}(B_{\delta/2}(\xi_*))} \leq B$.

Once we could establish $v_m \to v_*$ in $C^{1,1}$, $f_m \to f_* = f(0) = \text{constant}$, $0 \in N$ and $B_{\delta/2}(\xi_*) \subset N(v^*)$, we would have

$$D_{ii}v_*(\xi_*) \geq \frac{-C}{|\log \delta|^{\varepsilon}} > -\varepsilon_0/2$$

due to Theorem 4.3, which will give the desired contradiction with

$$D_{ii}v_m(\xi_*) \leq -\varepsilon_0 < 0.$$

To this end, we need the following notion and the compactness result. We define the set of functions $P_1(M, A, \lambda_0)$ such that $u \in P_1(M, A, \lambda_0)$ if and only if

$$\begin{cases} \Delta u = f & \text{in } N(u) = \{u > 0\} \cap B_1(0) \\ & \text{for some } f \geq \lambda_0 > 0 \text{ with } \|f\|_{C^\alpha(B_1(0))} \leq A, \\ u \geq 0 & \text{in } B_1(0), \quad 0 \in \partial N(u) = \Gamma_u, \quad \|u\|_{C^{1,1}(B_1(0))} \leq M. \end{cases}$$

Lemma 4.4 (Compactness of $P_1(M, A, \lambda_0)$). *For any* $\{u_m\} \subset P_1(M, A, \lambda_0)$, *there is a subsequence re-labeled as* $\{u_{m'}\}$ *converging to* $u \in P_1(M, A, \lambda_0)$ *in* $C^{1,1}$ *in the weak-$*$ sense, and we have*

$$\Gamma_m \to \Gamma, \quad \textit{i.e. } \overline{\lim_m} N(u_m) \subseteq \overline{N(u)}, \textit{ and } \overline{\lim_m} \Lambda(u_m) \subseteq \Lambda(u).$$

Proof of Lemma 4.4. Note that $u_{m'} \to u$ in C^0 and $D^2 u_{m'} \to D^2 u$ in the weak-$*$ topology of L^∞, as we can pass to the limit in the equations. Hence, the weak-$*$ convergence in $C^{1,1}$ follows. The nonnegativity condition in the limit follows by using the equations satisfied by u_m and the uniform interior estimate. Hence, it remains to prove convergence of the free boundary only.

$\overline{\lim}_m \Lambda(u_m) \subset \Lambda(u)$ holds by definition and the uniform convergence of $\{u_m\}$. For the remaining part, let $x_0 \notin \overline{N(u)}$. Then $u(x) \equiv 0$ for $x \in B_\rho(x_0)$. If there is a sequence of points $\{x_m\}$ converging to x_0 such that $u_m(x_m) > 0$, then using the nondegeneracy lemma 4.2 we obtain

$$\sup_{\partial B_{\rho/2}(x_m)} u_m(x) \geq \frac{\lambda_0}{2n}(\rho/2)^2 .$$

By the uniform convergence $u_m \to u_0$, we have

$$\sup_{B_{2\rho/3}(x_0)} u_0(x) \geq \frac{\lambda_0}{2n}(\rho/2)^2$$

which is a contradiction with $u(x) \equiv 0$ in $B_\rho(x_0)$. □

In order to complete the proof of the theorem, there remains to prove the lower bound for $D_{ii}u$. Define

$$-M_k = \inf_{B_{2^{-k}}(0)} D_{ii}u.$$

We shall estimate M_k inductively for $M_k > 0$. Otherwise, if $M_k \leq 0$ for some k, we have nothing to prove. Note that $|M_k| \leq M$ uniformly, and $D_{ii}u \geq -M$.

Let $x \in N$ with $|x| \leq 2^{-(k+1)}$, and let $B_s(x)$ be the largest ball in N. Then $s \leq 2^{-(k+1)}$, and $B_s(x) \subset B_{2^{-k}}(0)$. Let $y_0 \in \Gamma \cap \partial B_s(x)$. Let y_1 be a point lying in the segment between x and y_0 such that $|y_1 - y_0| = \delta s$, for small enough δ to be chosen later. Then $u(y_1) \leq \frac{M}{2}(\delta s)^2$.

In each axis direction of the coordinate, we can choose unit vector e_i, s.t. $\langle e_i, x - y_0\rangle \geq 0$. Then $D_{ii}u = D_{-i,-i}u = D_{e_i,e_i}u$, and $y_1 + te_i \in B_s(x) \subset N$ for $0 \leq t \leq \sqrt{\delta}s$. Let $h = \sqrt{\delta}s/2$. Since

$$0 \leq u(y_1 + he_i) = u(y_1) + u_i(y_1)h + \int_0^h \int_0^\tau D_{ii}u(y_1 + te_i)\,\mathrm{d}t\,\mathrm{d}\tau,$$

we have

$$\int_0^{\sqrt{\delta}s/2} \int_0^\tau D_{ii}u(y_1 + te_i)\,\mathrm{d}t\,\mathrm{d}\tau \geq -\frac{M}{2}(\delta s)^2 - CM\delta s \cdot \sqrt{\delta}s \geq -\delta^{3/2}s^2CM.$$

Hence

$$\sup_{0<t<\frac{\sqrt{\delta}s}{2}} D_{ii}u(y_1 + te_i) \geq -\delta^{1/2}M.$$

Consider $w = D_{ii}u + M_k$. It satisfies

$$\Delta w = 0 \quad \text{in} \quad B_s(x), \quad w \geq 0.$$

Moreover

$$w(\tilde{y}) \geq M_k\left(1 - \delta^{1/2}\right)$$

for some $\tilde{y} \in B_{s(1-\delta/2)}(x)$. Now we apply Harnack's inequality to w to get

$$w(x) \geq w(\tilde{y}) \cdot c\delta^{n-1},$$

equivalently

$$D_{ii}u(x) \geq -M_k + c\delta^{n-1}\left(M_k - c\delta^{1/2}\right).$$

Choosing δ such that $c\delta^{1/2} = \varepsilon_0 M_k$ (and $\delta < 1/2$ if ε_0 is small) we have

$$D_{ii}u(x) \geq -M_k + CM_k^{2n-1}.$$

Since x is arbitrary, this gives the set of inequalities

$$-M_{k+1} \geq -M_k + CM_k^{2n-1},$$

or equivalently

$$M_{k+1} \leq M_k - CM_k^{2n-1}.$$

Hence, M_k is positive and decreasing in k. It is easy to show

$$0 < M_k \leq M_0 - C\sum_{i=0}^{k-1} M_i^{2n-1} \leq M_0 - CkM_k^{2n-1}.$$

This gives $CkM_k^{2n-1} \leq M_0$ and therefore

$$M_k \leq Ck^{-\varepsilon},$$

where $\varepsilon = \frac{1}{2(n-1)}$. For $|x| = 2^{-k}$, we conclude that

$$D_{ii}u(x) \geq -\frac{C}{(\log|x|)^{\varepsilon}}. \qquad \square$$

4.3 Free boundary regularity II: main theorems

To this end, we are able to study the regularity of the free boundary Γ. For convenience, we introduce the following notation as a special case of previously defined $P_1(M, A, \lambda_0)$.

Definition 4.1. We say $u \in P_R(M)$ if and only if $u \in C^{1,1}(B_R)$ with $\|D^2u\|_{L^\infty(B_R)} \leq M$, s.t.

$$\begin{cases} \Delta u = 1 & \text{in } \{x : u > 0\} \cap B_R, \\ u \geq 0, & 0 \in \Gamma = \partial\{x : u > 0\}. \end{cases}$$

Moreover, if $u \in P_R(M)$ is also convex, we say $u \in P_R^*(M)$.

Remark 4.3. (a) (Compactness) If $\{u_m\} \subset P_R(M)$, there is a subsequence $\{u_{m'}\}$, s.t. $u_{m'} \to u_\infty \in P_R(M)$. The convergence is in the sense of weak-$*$ in $C^{1,1}(B_R)$. We shall omit this in the sequel if no confusion would be made.

(b) Given $\{u_m\} \subset P_R(M)$ and $\lambda_m \to 0^+$, define

$$v_m(x) = \frac{u_m(\lambda_m x)}{\lambda_m^2} \in P_R(M).$$

Then there exists a subsequence $v_{m'} \to v_0 \in P_R(M)$. Moreover, $v_0 \in P_R^*(M)$.

We also recall the following lemma on the convergence of the free boundary as a special case of Lemma 4.4.

Lemma 4.5. *Suppose $\{u_m\} \subset P_R(M)$ and $u_m \to u_0 \in P_R(M)$. Then*

$$\overline{\lim_m} N(u_m) \subset \overline{N(u_0)}, \quad \overline{\lim_m} \Lambda(u_m) \subset \Lambda(u_0).$$

Hence,

$$\overline{\lim_m} N(u_m) = \overline{N(u_0)}, \quad \overline{\lim_m} \Lambda(u_m) = \Lambda(u_0).$$

Lemma 4.6. *Given $u \in P_R(M)$ and $\lambda_m \to 0^+$, let*

$$u_{\lambda_m}(x) = \frac{u(\lambda_m x)}{\lambda_m^2} \in P_R(M).$$

Then there exists $u_0 \in P_R^*(M)$, s.t. $u_{\lambda_m} \to u_0$.

Lemma 4.7. *Given $u \in P_1^*(M)$ and $\lambda_m \to 0^+$, then $u_{\lambda_m} \to u_0(x)$ where*

$$u_0(x) = \sum_{i=1}^n \frac{a_i}{2} x_i^2, \qquad \textit{with } a_i \geq 0, \sum_i a_i = 1,$$

or

$$u_0(x) = \frac{1}{2}(x_n^+)^2.$$

Proof. It is easy to see that $u_0 \in P_\infty^*(M)$. If $\mathring{\Lambda}(u_0) = \varnothing$, then $\Lambda(u_0)$ is contained in a hyperplane.

We know that

$$\begin{cases} \Delta u_0 = 1 & \text{a.e. in } \mathbb{R}^n, \\ u_0 \geq 0, & \|D^2 u_0\|_{L^\infty} \leq M. \end{cases}$$

Hence $\Delta u_0 \equiv 0$ in $\mathbb{R}^n$ by an easy proof of removable singularities in this case. The latter fact implies that

$$u_0(x) = \sum_{i=1}^n \frac{a_i}{2} x_i^2, \qquad \text{with } a_i \geq 0, \sum_i a_i = 1.$$

Suppose $\mathring{\Lambda}(u_0) \neq \varnothing$. Since $u_0 = \lim_{m\to\infty} \frac{u(\lambda_m x)}{\lambda_m^2}$,

$$\Lambda(u_0) = \lim_m \Lambda(u_m) \triangleq C_0(\Lambda(u))$$

is a subset in $\mathbb{R}^n$ and a convex set. Hence, either $\Lambda(u_0) = \{x \in \mathbb{R}^n : x_n \leq 0\} \triangleq C_*$ or $\Lambda(u_0) \subsetneqq C_*$. Consider $v = \frac{\partial u_0}{\partial x_n}$.

In the first case where $\Lambda(u_0) = C_*$,

$$\begin{cases} \Delta v = 0 & \text{in } \{x_n > 0\}, \\ v = 0 & \text{on } \{x_n = 0\}, \\ v > 0 & \text{in } \{x_n > 0\} \text{ grows linearly at } \infty. \end{cases}$$

The last property is due to the scaling law $u(\lambda_m x)/\lambda_m^2$ and Lemma 4.2. Therefore, this implies $v = Cx_n$ in $\{x_n > 0\}$ and $u_0 = \frac{1}{2}(x_n^+)^2$.

In the second case where $\Lambda(u_0) \neq C_*$, again we have

$$\begin{cases} \Delta v = 0 & \text{in } N(u_0), \\ v = 0 & \text{on } \Gamma(u_0), \\ v > 0 & \text{in } N(u_0). \end{cases}$$

Recall that u takes off from $\Gamma(u)$ quadratically due to Lemma 4.2. Consider the Dirichlet eigenvalue problem of $-\Delta_{\mathbb{S}^{n-1}}$ on $N(u_0) \cap \mathbb{S}^{n-1}$, i.e.

$$\begin{cases} \Delta_{\mathbb{S}^{n-1}} \phi + \mu\phi = 0 & \text{on } N(u_0) \cap \mathbb{S}^{n-1}, \\ \phi = 0 & \text{on } \partial N(u_0) \cap \mathbb{S}^{n-1}. \end{cases}$$

From $\mathbb{S}_+^{n-1} \subsetneqq N(u_0) \cap \mathbb{S}^{n-1}$, we know that the first eigenvalue $\mu < n-1$, where $(n-1)$ is the first Dirichlet eigenvalue of $\mathbb{S}_+^{n-1}$. Now we look for a positive solution of

$$\Delta h = 0 \quad \text{in } N(u_0), \quad h = 0 \quad \text{on } \partial N(u_0)$$

in the form of $h(x) = r^\alpha \phi(\omega)$, where ϕ is the eigenfunction corresponding to μ and where ω is the coordinate on $\mathbb{S}^{n-1}$. This would give $\alpha \in (0,1)$. It can be proved by maximum principle that $v \geq \varepsilon h$ for some small $\varepsilon > 0$ on $N(u_0) \cap B_\delta(0)$ with sufficiently small δ. Hence v is not Lipschitz, which is a contradiction with $u_0 \in C^{1,1}$. □

Remark 4.4. In the proof above, we essentially show that if $u \in P_1^*(M)$ and $\mathring{\Lambda}(u) \neq \varnothing$, then

$$C_0(\Lambda(u)) = \{x : x_n \leq 0\}.$$

Definition 4.2. The minimum diameter of a set $\Omega \subset \mathbb{R}^n$, denoted by $MD(\Omega)$, is defined by

$$MD(\Omega) = \inf\{D : \Omega \text{ can be put between two parallel hyperplanes } \Pi_1 \text{ and } \Pi_2 \text{ with distance } D\}.$$

Lemma 4.8. *Let*

$$\delta_r(\Lambda(u)) = \frac{MD(\Lambda(u) \cap B_r)}{2r}.$$

For $\forall\, \varepsilon > 0$ and $\delta > 0$, there is $\lambda = \lambda(\varepsilon, \delta) > 0$, s.t. if $u \in P_1^*(M)$ and $\delta_1(\Lambda(u)) > \varepsilon$, then in a suitable coordinate system, one has

(a) $\Lambda(u) \supset B_\lambda \cap \{x : \alpha(x, -e_n) < \frac{\pi}{2} - \delta\}$,
(b) $N(u) \supset B_\lambda \cap \{x : \alpha(x, e_n) < \frac{\pi}{2} - \delta\}$,

where $\alpha(x, y)$ is the angle between vectors x and y.

Proof. First we show that (a) implies (b). In fact, if (a) is true and if $\exists p \in \Lambda(u) \cap B_\lambda \cap \{x : \alpha(x, e_n) < \frac{\pi}{2} - \delta\}$, then by the convexity of u (and thus $\Lambda(u)$), we know that there is a neighborhood of 0 contained in $\Lambda(u)$, which contradicts with $0 \in \partial\{u > 0\}$.

We shall prove (a) by contradiction. Suppose that there exists positive ε_0 and δ_0, a sequence of $\lambda_m \to 0^+$, and a sequence of $\{u_m\}$ satisfying $\delta_1(\Lambda(u_m)) > \varepsilon_0$, s.t. (a) is not true for u_m on the scale of λ_m for any choice of e_n.

On the other hand, we know that $u_m \to u_0 \in P_1^*(M)$ up to a subsequence, with $\delta_1(\Lambda(u_0)) \geq \varepsilon_0 > 0$. Hence, $\mathring{\Lambda}(u_0) \neq \varnothing$. This implies $C_0(\Lambda(u)) = \{x : x \leq 0\}$ for a suitable choice of coordinate (cf. Remark 4.4). This leads to the contradiction by Lemma 4.5. □

Remark 4.5. Let $u \in P_1^*(M)$. If $a \in \Gamma = \partial\{u > 0\}$, s.t.

$$\liminf_{r \to 0^+} \frac{L^n(\Lambda(u) \cap B_r(a))}{r^n} \geq \delta_0 > 0, \tag{4.10}$$

then

$$\frac{MD(\Lambda(u) \cap B_r)}{2r} \geq C(\delta_0) > 0$$

for r sufficiently small. Therefore, we can replace the condition $\delta_1(\Lambda(u)) \geq \varepsilon_0$ by a density hypothesis in the form of (4.10).

To this end, one might ask whether a similar result of Lemma 4.8 holds at points in a neighborhood of 0. We start from the following observation. Assume $\delta_1(\Lambda(u)) > \varepsilon$ and $y \in \Gamma_u = \partial\Lambda(u)$ with $|y| < \frac{1}{4}$. We notice that

$$\begin{aligned} &\delta_{\frac{1}{2}}\left(\Lambda(u) \cap B_{\frac{1}{2}}(y)\right) = MD\left(\Lambda(u) \cap B_{\frac{1}{2}}(y)\right) \geq MD\left(\Lambda(u) \cap B_{\frac{1}{4}}(0)\right) \\ \Rightarrow &\delta_{\frac{1}{2}}\left(\Lambda(u) \cap B_{\frac{1}{2}}(y)\right) \geq \frac{1}{4}\delta_{\frac{1}{2}}(\Lambda(u)). \end{aligned}$$

Corollary 4.2. *For $\forall \varepsilon > 0$ and $\delta > 0$, there is $\lambda = \lambda(\varepsilon, \delta) > 0$, s.t. if $u \in P_1^*(M)$ and $\delta_{\frac{1}{4}}(\Lambda(u)) > \varepsilon$, then for $\forall y \in \Gamma \cap B_{\frac{1}{4}}(0)$, in a suitable coordinate system (depending on y), one has*

(a) $\Lambda(u) \supset B_\lambda(y) \cap \{x : \alpha(x - y, -e_n(y)) < \frac{\pi}{2} - \delta\}$,
(b) $N(u) \supset B_\lambda(y) \cap \{x : \alpha(x - y, e_n(y)) < \frac{\pi}{2} - \delta\}$,

where $\alpha(x, y)$ is the angle between vectors x and y.

In the notation of Corollary 4.2, one can show by simple geometric argument that

$$\alpha(e_n(0), e_n(y)) \leq C\delta.$$

This gives

Theorem 4.4 (Lipschitz regularity). *Assume the same hypothesis as in Corollary 4.2. Then $\Gamma \cap B_{\frac{1}{4}}(0)$ is the graph of a Lipschitz function f with $\|\nabla f\|_{L^\infty} \leq C\delta$, i.e. in a suitable coordinate, $\Gamma \cap B_{\frac{1}{4}}(0)$ can be represented by $x_n = f(x')$.*

In the general case, we assume $u \in P_1(M)$ satisfies $\delta_\rho(\Lambda(u)) \geq \varepsilon_0 > 0$. Note that there is no convexity assumption on u. By a blow-up argument, we can show in a suitable coordinate,

$$\Lambda(u) \supset B_{\rho\frac{\lambda}{2}} \cap \{x : -x_n \geq 2\rho\lambda\delta\}$$

for sufficiently small ρ. One can also show a similar result for $N(u)$ using non-degeneracy.

A critical issue in this case is that, the coordinate on each scale ρ is "scale-dependent". This means when it comes to a finer scale, i.e. when ρ becomes smaller, the coordinate has to (and it has the freedom) rotate to fit the free boundary. We need to control this rotation from scale to scale, so that the set $B_{\rho\frac{\lambda}{2}} \cap \{x : -x_n \geq 2\rho\lambda\delta\}$ can "converge" in a particular sense. Such bounds on the rotation come from the convexity estimates on the solution. Recall that we only have logarithmic decay of pure second derivative when it approaches the free boundary (cf. Theorem 4.3 and Remark 4.2). This would give a logarithmically decaying sequence of the rotation angles when we come to finer and finer scales, which is not convergent. Therefore, we need to improve the convexity to obtain Hölder decay, i.e. the rotation angle behaves like r^α, where r is the size of scale. Now the sequence is summable!

General case: Now we consider the general case $u \in P_1(M)$, hence removing the convexity assumption.

Lemma 4.9. *Given $\epsilon > 0$ and $\delta > 0$, there is a $\rho_0 = \rho_0(\epsilon, \delta) > 0$ such that if $u \in P_1(M)$ and $\delta_\rho(\Lambda(u)) \geq \epsilon$ for some $0 < \rho < \rho_0$, then in a suitable system of coordinates*

$$\begin{cases} \Lambda(u) \supset B_{\rho\lambda/2} \cap \{x : -x_n \geq 2\rho\lambda\delta\}, \\ N(u) \supset B_{\rho\lambda/2} \cap \{x : x_n \geq 2\rho\lambda\delta\}, \end{cases} \tag{4.11}$$

where $\lambda = \lambda(\epsilon, \delta)$ is as defined in Lemma 4.8.

Proof. If one of the assertions is not true, then there are sequences $\{u^{(m)}\} \in P_1(M)$ and $\rho_m > 0$, where $\rho_m \to 0^+$ such that (4.11) is not valid for any system of coordinates. By taking a subsequence, if necessary, we have

$$u^{(m)}_{\rho_m} \to u \in P_1^*(M), \quad \Lambda(u) \supset \overline{\lim}\Lambda\left(u^{(m)}_{\rho_m}\right), \quad \overline{N(u)} \supset \overline{\lim}N\left(u^{(m)}_{\rho_m}\right);$$

hence we obtain a contradiction. □

Note that from (4.11) one has

$$\delta_{\rho\lambda/2}(\Lambda(u)) \geq 1 - 4\delta > \frac{1}{2},$$

if $\delta < \frac{1}{8}$. Hence one may apply (4.11) again to replace $\rho\lambda/2$ with ρ, if $\epsilon \leq \frac{1}{8}$ in Lemma 4.9, which may be assumed.

Lemma 4.10. *Let* $\epsilon < \frac{1}{8}$, $\delta < \frac{1}{8}$, $\lambda = \lambda(\epsilon, \delta)$, $\rho_0 = \rho_0(\epsilon, \delta)$ *as before. If* $u \in P_1(M)$, $\delta_\rho(\Lambda(u)) \geq \epsilon$, $\rho < \rho_0$, *then there exists a sequence of coordinates* $\left\{e_i^{(k)} : 1 \leq i \leq n\right\}$ *such that*

$$\begin{cases} \Lambda(u) \supset B_{\rho\left(\frac{\lambda}{2}\right)^k} \cap \left\{x : -x_n^{(k)} \geq \rho\,(\lambda/2)^k\, 4\delta\right\}, \\ N(u) \supset B_{\rho\left(\frac{\lambda}{2}\right)^k} \cap \left\{x : x_n^{(k)} \geq \rho\,(\lambda/2)^k\, 4\delta\right\}, \end{cases} \tag{4.12}$$

$\forall\, k \in \mathbb{Z}_+$.

Proof. Applying Lemma 4.9 iteratively gives the claim. □

Lemma 4.11. *There is* $\mu \in (0,1)$ *such that*

$$D_{ii}u \geq -M\mu^{k-1} \qquad in\ N(u) \cap B_{\rho(\lambda/2)^k}.$$

Proof. We prove by induction. The base case $k = 1$ follows from the uniform second derivative estimate. Consider $\Delta h = 0$ in B_1 such that

$$h = \begin{cases} 0 & \text{on } \partial B_1 \cap \{x_n < -1/2\}, \\ -1 & \text{on } \partial B_1 \cap \{x_n \geq -1/2\}. \end{cases}$$

Let $\mu = -\inf_{B_{1/2}} h$, and note $0 < \mu < 1$. We have the inductive hypothesis

$$D_{ii}u(x) \geq -M\mu^{k-1} \quad \text{in } N(u) \cap B_{\rho(\lambda/2)^k}.$$

Consider the auxiliary function

$$w(x) = M\mu^{k-1} h\left(\frac{x}{\rho\,(\lambda/2)^k}\right) \qquad \text{in } N(u) \cap B_{\rho(\lambda/2)^k}.$$

Then both w and $D_{ii}u$ are harmonic in $N(u) \cap B_{\rho(\lambda/2)^k}$, and $w \leq D_{ii}u$ on $N(u) \cap \partial B_{\rho(\lambda/2)^k}$ and on $\partial N(u) \cap B_{\rho(\lambda/2)^k}$. Hence $D_{ii}u \geq -M\mu^k$ on $B_{\rho(\lambda/2)^{k+1}} \cap N(u)$ by the maximum principle. □

Lemma 4.12. *If* $u \in C^{1,1}(B_R)$, $u \geq 0$, $u(0) = 0$, $D_{ii}u \geq -\tau$ *on* $\overline{0x_0}$, *where* $|x_0| < R$, $\tau \geq 0$, *then for* $0 < t < 1$,

$$u\,(tx_0) \leq \frac{1}{2}\,|x_0|^2\,\tau + u\,(x_0)\,.$$

Proof. Let $u\,(t_0x_0) = \max_{0 \leq t \leq 1} u\,(tx_0)$. We may assume that $0 < t_0 < 1$, as otherwise the claim becomes trivial. So $(D_iu)\,(t_0x_0) = 0$, where $i = \frac{x_0}{|x_0|}$. Hence by Taylor expansion

$$u\,(x_0) = u\,(t_0x_0) + \int D_{ii}u \geq u\,(t_0x_0) - \tau\,|x_0|^2/2,$$

and rearranging gives the claim. □

Lemma 4.13. *Let* $\theta_k = (3Mn)^{1/2}\mu^{(k-2)/2}\rho(\lambda/2)^{k-1}$, *and* $x \in N(u) \cap \left(\overline{B}_{\rho(\lambda/2)^k}\backslash\overline{B}_{\rho(\lambda/2)^{k+1}}\right)$. *Then there is a point* y *such that* $|x-y| \le \theta_k$ *and the points* sy *are in* $N(u)$ *for all* s *with* $s \ge 1$ *and* $|sy| \le \rho(\lambda/2)^{k-1}$.

Proof. There is y such that $|x-y| \le \theta_k$ and $u(y) \ge \frac{\theta_k^2}{2n}$. Let s be as above. Take $x_0 = sy$, with $t = 1/s$ in Lemma 4.12. Note that by Lemma 4.11, $D_{ii}u \ge -M\mu^{k-2}$ on $\overline{0x_0}$. We have

$$\frac{\theta_k^2}{2n} \le u(y) \le \frac{1}{2}|sy|^2 M\mu^{k-2} + u(sy)$$

by the definition of θ_k. Hence, $u(sy) > 0$, i.e. $sy \in N(u)$. □

Note that in the previous lemma we have

$$x \in N(u) \cap \left(\overline{B}_{\rho(\lambda/2)^k}\backslash B_{\rho(\lambda/2)^{k+1}}\right), \quad |x-y| \le \theta_k = \sqrt{C(n)M}\mu^{\frac{k-2}{2}}\rho(\lambda/2)^{k-1}.$$

Then

$$\alpha(x,y) \le C(n,M)\frac{\mu^{\frac{k-2}{2}}}{\lambda^2} \to 0, \quad \text{as} \quad k \to \infty.$$

Lemma 4.14. *Under the assumptions of Lemma* 4.10 *there exists a* $k_0 = k_0(\epsilon,\delta)$ *and* $C = C(\epsilon,\delta) > 0$ *such that in a suitable system of coordinates*

$$\begin{cases} \Lambda(u) \supset B_{\rho\left(\frac{\lambda}{2}\right)^{k_0}} \cap \{x : \alpha(x,-e_n) < \pi/2 - C\delta\}, \\ N(u) \supset B_{\rho\left(\frac{\lambda}{2}\right)^{k_0}} \cap \{x : \alpha(x,+e_n) < \pi/2 - C\delta\}. \end{cases} \tag{4.13}$$

Proof. Take e_n as $e_n^{(k)}$ with $k = k_0 - 1$ as in Lemma 4.10. Suppose that the first statement in (4.13) is not true, then there is a point x_j such that $x_j \in N(u) \cap \left(B_{\rho(\lambda/2)^j}\backslash B_{\rho(\lambda/2)^{j+1}}\right)$ and for $j \ge k_0$,

$$\alpha\left(x_j, -e_n^{(k_0)}\right) < \pi/2 - C\delta.$$

By Lemma 4.13 there exists a y_j and a segment sy_j in $N(u)$, where $s \ge 1$, $|sy_j| \le \rho(\lambda/2)^{j-1}$. Taking $x_{j-1} = s_j y_j$ so that

$$|s_j y_j| = \rho\left(\frac{\lambda}{2}\right)^{j-1}.$$

Then $x_{j-1} \in N(u) \cap \left(\overline{B}_{\rho(\lambda/2)^{j-1}}\backslash B_{\rho(\lambda/2)^j}\right)$, and

$$\alpha(x_j, x_{j-1}) \le \frac{C}{\lambda^2}\mu^{\frac{j-2}{2}}.$$

Repeating this procedure, we find a sequence of points $\{x_i\}$, $k_0 \le i < j$ such that

$$x_i \in N(u) \cap \partial B_{\rho(\lambda/2)^i}, \text{ and } \alpha(x_0, x_{i-1}) \le \frac{C}{\lambda^2}\mu^{(i-2)/2}.$$

Thus, $x = x_{k_0}$ satisfies

$$x \in N(u) \cap \partial B_{\rho(\lambda/2)^{k_0}}, \text{ and } \alpha(x, x_j) \le \tilde{C}(\lambda, \mu)\mu^{\frac{k_0-2}{2}}.$$

We let $C \ge 5$ in the first statement in (4.13) and k_0 sufficiently large, then

$$\alpha(x, -e_n) < \frac{\pi}{2} - 4\delta,$$

which is a contradiction to Lemma 4.10. Finally, we note that the first statement in (4.13) implies the latter by a blow-up argument. □

Remark 4.6. Note that Lemma 4.14 implies the free boundary regularity for the general case as in the convex case, cf. Theorem 4.4. One can see further that we have actually proved that the free boundary is C^1 under the same assumptions.

Remark 4.7. Finally, we consider a different approach to the regularity of the free boundary. In particular, we will show that a flat free boundary is Lipschitz. Given

$$\left| u - \frac{1}{2}\left(x_n^+\right)^2 \right|_{C^{1,\alpha}(B_1)} \ll 1,$$

consider

$$h = u_{x_n} - \frac{1}{2}\left(u - \frac{|x - x_0|^2}{2n}\right), \quad x_0 \in N \cap B_{1/4}.$$

Then

$$\begin{cases} \Delta h = 0 & \text{in } N \cap B_1, \\ h \ge 0 & \text{on } \Gamma \cap B_1 \text{ and } \partial B_1 \cap N. \end{cases}$$

Therefore, $h \ge 0$ in $N \cap B_1$. Hence, $\left(u_{x_n} - \frac{1}{2}u\right)(x_0) \ge 0$ at every point in $B_{1/4} \cap N$. Similarly, $u_e - \frac{1}{2}u \ge 0$ in $B_{1/4} \cap N$, whenever e is close to e_n. We conclude that Γ is a Lipschitz graph in $B_{1/4}$. A more careful argument would lead to that Γ is C^1. Next we shall present yet another approach to show that the Lipschitz regularity of Γ implies that $\Gamma \in C^{1,\alpha}$.

The following two theorems are also well-known results for the regularity of free boundary when there is no convexity assumption on u. Rather one assumes the free boundary is a graph in some direction.

Theorem 4.5 (Alt). *Let $u \in P_1(M)$, $\frac{\partial u}{\partial x_1} \ge 0$ in B_1 and $\frac{\partial u}{\partial x_1} > 0$ in $N(u) \cap B_1$. Let*

$$\partial N \triangleq \Gamma = \{x : x_1 = f(x_2, \cdots, x_n)\}.$$

Then f is Lipschitz continuous, i.e. Γ is a Lipschitz graph.

Theorem 4.6 (Athanasopoulos-Caffarelli). *Under the assumptions of Theorem 4.5, if f is Lipschitz continuous, then f is locally $C^{1,\alpha}$.*

Proof of Theorem 4.5. First we want to show there is a cone $C \subseteq \mathbb{R}^n_+ = \{x \in \mathbb{R}^n : x_1 > 0\}$ such that for $\forall\, y_0 \in B_{\frac{1}{2}}(0) \cap \Gamma$, whenever $\xi \in C \cap \mathbb{S}^{n-1}$,

$$u_\xi(x) = \nabla u(x) \cdot \xi \geq 0 \quad \text{for } x \in B_{\frac{1}{4}}(y_0),$$
$$u_\xi(x) > 0 \quad \text{for } x \in B_{\frac{1}{4}}(y_0) \cap N(u).$$

Let

$$R(x) = \begin{cases} 0, & \text{if } |x - y_0| < \dfrac{1}{4}, \\ \left[|x - y_0|^2 - \dfrac{1}{4^2} \right]^2, & \text{if } |x - y_0| \geq \dfrac{1}{4}, \end{cases}$$

and assume $0 \neq \xi \in \mathbb{R}^n$, and $\varepsilon_0 > 0$ to be chosen later. We consider

$$v(x) = u_\xi(x) - u(x) + \varepsilon_0 R(x) \quad \text{in } B_{\frac{1}{2}}(y_0).$$

We choose ε_0 sufficiently small so that

$$\Delta v(x) = 0 - 1 + \varepsilon_0 \Delta R(x) \leq -\frac{1}{2} < 0 \quad \text{in } N(u) \cap B_{\frac{1}{2}}(y_0).$$

We want to choose ξ properly so that $v \geq 0$ on $\partial\left(N(u) \cap B_{\frac{1}{2}}(y_0)\right)$.

First, one observes that on $\partial N(u) \cap B_{\frac{1}{2}}(y_0)$, $u(x) = u_\xi(x) = 0$ and $R(x) \geq 0$. Next we let $\sum_{j=2}^n |\xi_j| \leq 1$. Hence

$$\sum_{j=2}^n |\xi_j u_{x_j}| \leq C_0 \text{ and } |u| \leq C_0 \quad \text{in } B_1.$$

It is possible to choose ξ_1 large so that

$$u_\xi(x) - u(x) + \varepsilon_0 R(x) \geq 0 \quad \text{on } \partial B_{\frac{1}{2}}(y_0) \cap N(u).$$

Indeed, since $|\nabla u| = u = 0$ on Γ and $u \in C^{1,1}(B_1)$,

$$\begin{aligned} u_\xi(x) - u(x) + \varepsilon_0 R(x) &\geq -\sum_{j=2}^n |\xi_j u_{x_j}| - u(x) + \varepsilon_0 R(x) \\ &\geq \varepsilon_0 \left(\frac{3}{16}\right)^2 - \sum_{j=2}^n |\xi_j u_{x_j}| - u(x) \geq \left(\frac{1}{16}\right)^2 \varepsilon_0 \end{aligned}$$

if $\operatorname{dist}(x, \Gamma) < \delta_0$ for some small $\delta_0 > 0$. On the other hand, if $\operatorname{dist}(x, \Gamma) \geq \delta_0$, then $u_{x_1}(x) \geq C(\delta_0) u_{x_1}(\frac{1}{2}, 0)$ by Harnack's inequality. Hence, we choose

$$\xi_1 \geq \frac{3C_0}{C(\delta_0) u_{x_1}(\frac{1}{2}, 0)} \triangleq M_0,$$

which gives

$$u_\xi(x) - u(x) + \varepsilon_0 R(x) \geq -2C_0 + \xi_1 u_{x_1} \geq C_0 > 0.$$

Therefore, for $\forall \xi \in C \cap \mathbb{S}^{n-1} = \left\{(\xi_1, \cdots, \xi_n) \in \mathbb{S}^{n-1} : \xi_1 \geq M_0 \sum_{j=2}^n |\xi_j|\right\}$,

$$u_\xi(x) - u(x) + \varepsilon_0 R(x) \geq 0 \quad \text{for } x \in N(u) \cap B_{\frac{1}{2}}(y_0).$$

In particular,

$$u_\xi(x) \geq u(x) > 0 \quad \text{in } N(u) \cap B_{\frac{1}{4}}(y_0), \qquad \text{and } u_\xi(x) \geq u(x) \geq 0 \quad \text{in } B_{\frac{1}{4}}(y_0).$$

The geometric consequence of the above fact is that for $\forall\, y_0 \in \Gamma$ with $B_{\frac{1}{2}}(y_0) \subseteq B_1$,

$$\Lambda_+(y_0) = \left\{y \in B_{\frac{1}{4}}(y_0) : \frac{y - y_0}{|y - y_0|} \in C \cap \mathbb{S}^{n-1}\right\} \subseteq N(u) \cap B_{\frac{1}{4}}(y_0),$$
$$\Lambda_-(y_0) = \left\{y \in B_{\frac{1}{4}}(y_0) : -\frac{y - y_0}{|y - y_0|} \in C \cap \mathbb{S}^{n-1}\right\} \subseteq \Lambda(u) \cap B_{\frac{1}{4}}(y_0).$$

Hence $\Gamma = \{x_1 = f(x_2, \cdots, x_n)\}$ is a Lipschitz graph. □

To prove Theorem 4.6, we need the following lemma.

Lemma 4.15. *Let $L = \nabla \cdot (A(y)\nabla)$ be a uniformly elliptic operator in B_1 with bounded measurable coefficients. If*

$$Lv = Lw = 0 \quad \text{in } B_1^+,$$
$$w > 0 \quad \text{in } B_1^+,$$
$$v = w = 0 \quad \text{on } \{y_1 = 0\},$$

then $\frac{v}{w} \in C^\alpha\left(\overline{B_{1/2}^+}\right)$.

Proof of Theorem 4.6. Since $u(f(x_2, \cdots, x_n), x_2, \cdots, x_n) \equiv 0$,

$$u_{x_j} + u_{x_1} f_{x_j} = 0, \quad j = 2, \cdots, n.$$

Thus $f_{x_j} = -\left.\frac{u_{x_j}}{u_{x_1}}\right|_\Gamma$. Note that $\frac{u_{x_j}}{u_{x_1}}$ is well-defined in $N(u)$.

Next we observe that $\Delta u_{x_j} = 0$ in $N(u)$ for $j = 1, \cdots, n$. We define $T : x \mapsto y$ by

$$y_1 = x_1 - f(x_2, \cdots, x_n), \text{ and } y_i = x_i \text{ for } i = 2, \cdots, n.$$

By change of variable, $w_j(y) \triangleq u_{x_j}(T^{-1}y)$, we find

$$Lw_j = 0 \quad \text{in } \{y_1 > 0\},\ j = 1, \cdots, n,$$

for some uniformly elliptic operator $L = \nabla \cdot (A(y)\nabla)$ with bounded measurable coefficients. Since $w_1 > 0$ in $\{y_1 > 0\}$ and $w_j \equiv 0$ on $\{y_1 = 0\}$ for $j = 1, \cdots, n$. This is true because $u \in C^{1,1}(B_1)$ (or $C^{1,\alpha}$ would be OK).

Now applying Lemma 4.15, we obtain $\left.\frac{w_j}{w_1}\right|_{y_1=0}$ is Hölder continuous, i.e. $f_{x_j} = -\left.\frac{u_{x_j}}{u_{x_1}}\right|_\Gamma$ is Hölder continuous. □

There remains to prove Lemma 4.15.

Lemma 4.16 (Harnack's principle at the boundary). *Let L be a uniformly elliptic operator with bounded measurable coefficients, i.e.*

$$L = \frac{\partial}{\partial x^i}\left(a^{ij}(x)\frac{\partial}{\partial x^j}\right),$$
$$a^{ij} \in L^\infty, \quad \lambda|\xi|^2 \le a^{ij}\xi_i\xi_j \le \lambda^{-1}|\xi|^2.$$

Suppose $Lv = Lw = 0$ in B_1^+, with $v = w = 0$ on $B_1 \cap \{x_1 = 0\}$ and $v, w > 0$ in B_1^+. Then there is a constant $C(n, \lambda)$, s.t.

$$\sup_{B_{1/2}^+} w \le Cw\left(\frac{1}{2}, 0\right),$$
$$\sup_{B_{1/2}^+} \frac{v}{w} \le C\frac{\sup_{B_{1/2}^+} v}{\sup_{B_{1/2}^+} w}.$$

For convenience, we recall some classic estimates in elliptic equations.

Lemma 4.17 (Moser's Harnack inequality). *Let L be defined as in Lemma 4.16. For any positive solution of $Lu = 0$ in B_1, and for $\forall\, R \in (0, 1)$,*

$$\sup_{B_R} u \le C(R, n, \lambda) \inf_{B_R} u.$$

Lemma 4.18 (De Giorgi-Nash). *Let L be defined as in Lemma 4.16. Let $Lu = 0$ in B_1. Then for $\forall\, R \in (0, 1)$, $\forall\, x, y \in B_R$,*

$$|u(x) - u(y)| \le C(R, n, \lambda, \alpha)\|u\|_{L^2(B_1)}|x - y|^\alpha.$$

Define

$$A = \{x = (x_1, \cdots, x_n) : 0 < x_n < 8,\ |x_j| < 4,\ \forall j = 1, \cdots, n-1\},$$
$$A_0 = \{x = (x_1, \cdots, x_n) : 0 < x_n < 2,\ |x_j| < 1,\ \forall j = 1, \cdots, n-1\},$$

and let z_0 be the center of A_0, i.e. $z_0 = (0, \cdots, 0, 1)$.

Lemma 4.19. *Let L be defined as in Lemma 4.16. There is a constant $C = C(\lambda, n)$, s.t. for $u \in H^1(A)$, $u > 0$ in A solving $Lu = 0$ in A and $u(x', 0) = 0$, we have*

$$\sup_{A_0} u \le Cu(z_0).$$

Proof. We may assume $u(z_0) = 1$. Also by reflecting u across $\{x_n = 0\}$ as an odd function in x_n, we may assume $\tilde{L}u = 0$ in $\tilde{A} \triangleq \{x : |x_n| < 8,\ |x_j| < 4,\ j = 1, \cdots, n-1\}$. We note that $\tilde{L}$ share the same ellipticity constant λ with L. Define $\omega_Q(u) = \mathrm{osc}_Q u$. Then Moser's Harnack inequality gives

$$\omega_Q(u) \ge \rho\omega_{\frac{1}{2}Q}(u)$$

with constant $\rho = \rho(n, \lambda) > 1$. Also, by Moser's Harnack inequality,

$$u(x', x_n) \le Ku(x', 2x_n) \quad \text{for } |x'| \le 2, 0 < x_n \le \frac{3}{2},$$
$$u(x', x_n) \le Ku(z_0) \quad \text{for } |x'| \le 2, 1 \le x_n \le 3,$$

where $K = K(n, \lambda) > 1$.

Suppose there is a point $Y_0 \in A_0$ such that $u(Y_0) \ge K^{k_0+2}$ with k_0 to be determined. Then $\operatorname{dist}(Y_0, \mathbb{R}^{n-1}) \le 2^{-k_0}$. (Otherwise, $u(Y_0) \le K^{k_0+1}u(z_0)$.) Also,

$$\omega_{Q(Y_0, 2^{-k_0+N})}(u) \ge \rho^N \omega_{Q(Y_0, 2^{-k_0})}(u) \ge \rho^N K^{k_0+2},$$

where $Q(x, r)$ is the cube in $\mathbb{R}^n$ centered at x with edge length $2r$. Now we choose N, which will be fixed for all, so that $\rho^N \ge 2K^2$. Hence, $\omega_{Q(Y_0,2)}(u) \ge 2K^{k_0+4}$. Since u is an odd function, there is a $Y_1 \in Q(Y_0, 2^{-k_0+N}) \cap \mathbb{R}^N_+$, s.t. $u(Y_1) \ge K^{k_0+4}$. Thus, $\operatorname{dist}(Y_1, \mathbb{R}^{n-1}) \le 2^{-k_0-2}$ and

$$\omega_{Q(Y_1, 2^{-k_0-2+N})}(u) \ge \rho^N K^{k_0+4} \ge 2K^{k_0+6}.$$

Therefore, $\exists Y_2 \in Q(Y_1, 2^{-k_0-2+N}) \cap \mathbb{R}^n_+$, s.t. $u(Y_2) \ge K^{k_0+6}$. Repeating this process, we will obtain a sequence $\{Y_k\}$ such that for $\forall k = 1, 2, \cdots$,

(i) $\operatorname{dist}(Y_k, \mathbb{R}^{n-1}) \le 2^{-k_0-2k}$;
(ii) $Y_k \in Q(Y_{k-1}, 2^{-k_0-2(k-1)+N}) \cap \mathbb{R}^n_+$;
(iii) $u(Y_k) \ge K^{k_0+2(k+1)}$.

We want to show all Y_k's are strictly contained in A. Indeed, if k_0 is sufficiently large, e.g. taking $k_0 = N + 3$, then for $\forall i = 1, \cdots, n-1$,

$$|Y_{k,i}| \le \sum_{j=1}^{k} |Y_{j,i} - Y_{j-1,i}| + |Y_{0,i}| \le \left(\sum_{j=1}^{k} 2^{-2(j-1)} \right) 2^{-k_0+N} + 1 < \frac{3}{2}.$$

This leads to the contradiction since $u(Y_k) \to +\infty$. □

Lemma 4.20. *Let u and w be positive solutions of $Lu = Lw = 0$ in A with $u(x', 0) = w(x', 0) = 0$. Then*

$$\sup_{A_0} \frac{u}{w} \le C(n, \lambda) \frac{u(z_0)}{w(z_0)}.$$

Proof. Let $A_1 = \{x = (x_1, \cdots, x_n) : 0 < x_n < 4,\ |x_j| < 2,\ \forall j = 1, \cdots, n-1\}$. We introduce the notation

$$\alpha = \partial A_1 \cap \{x_n > 1\}, \quad \beta = \partial A_1 \cap \{x_n > 0\}.$$

Consider

$$\begin{cases} Lu_1 = 0 & \text{in } A_1, \\ u_1 = \chi_\alpha & \text{on } \partial A_1, \end{cases}$$

and

$$\begin{cases} Lu_2 = 0 & \text{in } A_1, \\ u_2 = \chi_\beta & \text{on } \partial A_1. \end{cases}$$

Note that $u(x) \le Cu(z_0)$ for $x \in A_1$ by the preceding lemma. Hence for $x \in A_1$,

$$u(x) \le Cu(z_0)\, u_2(x).$$

Also $w(x) \ge Cw(z_0)$ for $x \in \alpha$, and hence for $x \in A_1$,

$$w(x) \ge cw(z_0)\, u_1(x).$$

<u>Claim 1:</u> $u_2(x) \le Cu_1(x)$ for every $x \in A_0$. Hence,

$$\sup_{A_0} \frac{u}{w} \le c(n, \lambda) \frac{u(z_0)}{w(z_0)}.$$

Proof of Claim 1. Let $0 \le \phi \le 1$, $\phi = 1$ on β and $\phi \in C_0^\infty\left(\mathbb{R}^n \backslash \overline{A}_{1/2}\right)$. For every $x \in A_0$, one has for G the Green function of L on A_1,

$$\begin{aligned} u_2(x) \le & \int_{\partial A_1} P(x,y)\phi(y)\,\mathrm{d}y = \int_{\partial A_1} \langle A(y)\nabla_y G(x,y), \nu(y)\rangle\, \phi(y)\,\mathrm{d}y \\ \le & C\left(\int_{A_1 \backslash A_{1/2}} \left|\nabla_y G(x,y)\right|^2 \mathrm{d}y\right)^{1/2} \\ \le & C\left(\int_{A_1 \backslash A_{1/4}} G^2(x,y)\,\mathrm{d}y\right)^{1/2} \le CG(x, y_0), \end{aligned}$$

where the last step again follows from Lemma 4.19. To conclude the proof of the claim, we need the following estimate:

<u>Claim 2:</u> $|G(x, y_0)| \le C\,|x - y_0|^{-(n-2)} \le C$ on $A_1 \backslash B_{1/2}(y_0)$.

Note that this second claim implies $G(x, y_0) \le Cu_1(x)$ for every $x \in A_1 \backslash B_{1/2}(y_0)$ by Harnack applied to u_1 and the comparison principle, after noting that away from its singularity G is a solution associated with the operator L. Consequently, we can conclude that $u_2(x) \le Cu_1(x)$.

This second claim follows from a simple duality argument using the fact that G is the Green's function associated with L, and bounding $\|G(x,\cdot)\|_{L^{\frac{2n}{n-2}}}$ by duality implies the claim. Consider the problem

$$\begin{cases} Lu = f \in L^{\frac{2n}{n-2}}(A_1), \\ u = 0 \quad \text{on } \partial A_1. \end{cases}$$

Then by standard elliptic estimates, $\nabla u \in L^2(A_1)$, and by the Poincaré inequality, $u \in L^2(A_1)$. Assume that the support of f is in $B_{1/4}(y_0)$. We have the representation formula

$$u(\xi) = \int_{A_1} G(\xi, y) f(y)\,\mathrm{d}y = \int_{B_{1/4}(y_0)} G(\xi, y) f(y)\,\mathrm{d}y.$$

Note that $\|u\|_{L^2(A_1)} \le C\|f\|_{L^{\frac{2n}{n-2}}}$ implies

$$u(x) \le \|u\|_{L^\infty\left(A_1 \setminus B_{1/2}(y_0)\right)} \le C\|f\|_{L^{\frac{2n}{n+2}}}.$$

Hence we have

$$u(x) = \int_{B_{1/4}(y_0)} G(\xi, y) f(y)\, \mathrm{d}y \le C\|f\|_{L^{\frac{2n}{n-2}}},$$

which by duality gives the desired bound for the $L^{\frac{2n}{n-2}}$-norm of $G(x,\cdot)$. Thus, the proof of Lemma 4.20 is complete. □

Note that Lemmas 4.19 and 4.20 precisely amount to Lemma 4.16. Now we are ready complete the proof of the main lemma 4.15.

Proof of Lemma 4.15. Let $\alpha_k = \sup_{B^+_{2^{-k}}} \frac{v}{w}$, and $\beta_k = \inf_{B^+_{2^{-k}}} \frac{v}{w}$ for $k = 1, 2, \cdots$. We want to show that for some $\delta < 1$, we have $\alpha_k - \beta_k \le M_0 \delta^{k-1}$ for $k = 1, 2, \cdots$. Here

$$M_0 = \sup_{B^+_{1/2}} \frac{v}{w} - \inf_{B^+_{1/2}} \frac{v}{w} \le C \left[\frac{\sup_{B^+_{1/2}} v}{\sup_{B^+_{1/2}} w} - \frac{\sup_{B^+_{1/2}} (-v)}{\sup_{B^+_{1/2}} w} \right] = C \frac{\operatorname{osc}_{B^+_{1/2}} v}{\sup_{B^+_{1/2}} w}.$$

We argue by induction on k. The base case $k = 1$ holds by the preceding lemma. Suppose that the claim holds for k. Then we prove the case for $k+1$. We assume

$$\frac{v\left(0, 2^{-k-1}\right)}{w\left(0, 2^{-k-1}\right)} \ge \beta_k + \frac{1}{2}\left(\alpha_k - \beta_k\right),$$

and consider $v^*(x) = \left(v - \beta_k w\right)\left(2^{-k}x\right) \ge 0$ on B^+_1, where $w^*(x) = w\left(2^{-k}x\right)$. Then applying Lemma 4.16 to v^* and w^* yields

$$\sup_{B^+_{1/2}} \frac{w^*}{v^*} \le C \frac{\sup_{B^+_{1/2}} w^*}{\sup_{B^+_{1/2}} v^*} \le \tilde{C} \frac{w^*\left(0, \frac{1}{2}\right)}{v^*\left(0, \frac{1}{2}\right)} \le \frac{\tilde{C}}{\alpha_k - \beta_k}.$$

Now observe that

$$\sup_{B_{2^{-k-1}}} \frac{v}{w} = \alpha_{k+1} \le \alpha_k,$$

and

$$\inf_{B_{2^{-k-1}}} \frac{v}{w} = \beta_{k+1} = \inf_{B^+_{1/2}} \frac{v^*}{w^*} + \beta_k \ge \beta_k + \frac{1}{\tilde{C}}\left(\alpha_k - \beta_k\right).$$

Therefore,

$$\alpha_{k+1} - \beta_{k+1} \le \left(1 - \frac{1}{\tilde{C}}\right)\left(\alpha_k - \beta_k\right) = \delta\left(\alpha_k - \beta_k\right).$$

Hence, we have proved the claim. □

4.4 Application to minimal surface and prescribed mean curvature problems

In this section we will follow [15] to show how the boundary behavior of solutions of the nonparametric least area and prescribed mean curvature problem can be tackled by a reduction to obstacle problems, for which the regularity has been established in the preceding section.

We first consider the Dirichlet problem for the minimal surface equation in a bounded C^2-domain Ω of $\mathbb{R}^2$:

$$\begin{cases} \operatorname{div}\left(\dfrac{u}{\sqrt{1+|\nabla u|^2}}\right) = 0 \quad \text{in} \quad \Omega, \\ u|_{\partial\Omega} = \varphi. \end{cases} \tag{4.14}$$

When Ω is convex, (4.14) has a unique solution $u \in C^2(\Omega) \cap C^0\left(\overline{\Omega}\right)$ for every $\varphi \in C^0(\partial\Omega)$. In contrast, in the case of nonconvex φ one can construct a smooth φ for which (4.14) fails to have a solution. See for instance [6] for the issues concerning the solvability of (4.14).

However, one can instead study the solution to the following variational problem:

$$\begin{cases} \min I(v) \quad \text{for} \quad v \in BV(\Omega), \\ I(v) = \displaystyle\int_\Omega \sqrt{1+|\nabla v|^2} + \int_{\partial\Omega} |v-\varphi|, \end{cases} \tag{4.15}$$

which is called the nonparametric least area problem. Here we characterize $BV(\Omega)$ as the space of integrable functions that can be approximated in $L^1(\Omega)$ norm by a sequence of Lipschitz functions v_i that satisfy

$$\limsup_i \int_\Omega |Dv_i| < \infty.$$

Note that in case the limit v is also Lipschitz, $I(v)$ corresponds to the area of the Lipschitz surface obtained by taking the union of the graph of v over Ω and the part of the boundary cylinder $\partial\Omega \times \mathbb{R}$ enclosed by the curves $\{(x,\varphi(x)) : x \in \partial\Omega\}$ and $\{(x,v(x)) : x \in \partial\Omega\}$.

Giusti showed in [7] the existence of a unique solution $u \in C^2(\Omega) \cap BV(\Omega)$ to the problem (4.15). Furthermore, we can decompose $\partial\Omega$ as

$$\partial\Omega = \partial_+(\Omega) \cup \partial_-(\Omega) \cup \Gamma,$$

where

$$\begin{cases} \partial_+(\Omega) = \operatorname{int}\{x \in \partial\Omega : H_{\partial\Omega}(x) \geq 0\}, \\ \partial_-(\Omega) = \{x \in \partial\Omega : H_{\partial\Omega}(x) < 0\}, \\ \Gamma = \partial\Omega \backslash \left(\partial_+(\Omega) \cup \partial_-(\Omega)\right). \end{cases}$$

Here $H_{\partial\Omega}(x)$ denotes the mean curvature of $\partial\Omega$ with respect to inward normal at $x \in \partial\Omega$. Then $u = \varphi$ on $\partial_+(\Omega)$ and u is Hölder continuous at each point of $\partial_+(\Omega)$ provided φ is Lipschitz continuous.

In fact, in [17] L. Simon proved that if $\partial\Omega$ is C^4, then u is Hölder continuous at each point of $\partial_-(\Omega)$ too, and the restriction of u to $\partial_-(\Omega)$ is locally Lipschitz continuous, as long as φ is Lipschitz continuous there.

Here we aim to show the higher regularity of the trace of u over the part of $\partial_-(\Omega)$ where $u \neq \varphi$ and obtain the $C^{1/2}$-Hölder estimate of u near such points. Our strategy is to reduce the problem to a free boundary one and make use of the regularity theory for free boundaries we developed in the preceding chapter, as well as the higher regularity results that we will prove in the last chapter of these notes.

Since the proofs in the higher dimension cases are identical to that in the two dimension, for simplicity, we shall discuss below on the two dimensional case. For Ω bounded, C^2 domain in $\mathbb{R}^2$ and $\varphi \in C^0(\partial\Omega)$, we study the trace of the unique solution u to the variational problem (4.15). The trace of u over $\partial\Omega$ is defined by the requirement

$$\lim_{\rho\to 0+} \rho^{-2} \int_{\Omega\cap\{\xi\,:\,|\xi-x|<\rho\}} |u(\xi)-u(x)|\,\mathrm{d}\xi = 0$$

for all $x \in \partial\Omega$, where such a value $u(x)$ exists. Let $\gamma = \gamma_- \cup \gamma_+$, where

$$\begin{cases} \gamma_+ = \{x \in \partial_-(\Omega) \ : \ \varphi(x) > u(x)\}, \\ \gamma_- = \{x \in \partial_-(\Omega) \ : \ \varphi(x) < u(x)\}. \end{cases}$$

Note that the sets γ_+, γ_- and γ are well-defined up to linear measure zero sets.

Theorem 4.7. *Let Ω be a $C^{2,\alpha}$ domain in $\mathbb{R}^2$ for $0 < \alpha < 1$, and let φ, u, γ_+ be as above. Suppose $x_0 \in \gamma_+$ so that $(\partial\Omega \cap B_{r_0}(x_0))\setminus\gamma_+$ is a set of one-dimensional Hausdorff measure zero for some $r_0 > 0$. Then u restricted to $\partial\Omega \cap B_{r_0/2}$ is a $C^{1,\alpha}$ function. Furthermore, its $C^{1,\alpha}$-norm depends only on the $C^{2,\alpha}$-norm of $\partial\Omega$, $L^\infty(\Omega)$-norm of φ, r_0, and*

$$\inf\left\{|H_{\partial\Omega}(x)| \ : \ x \in \partial\Omega \cap B_{r_0}(x_0)\right\}.$$

The analogous statement holds for γ_- in place of γ_+ as well.

Corollary 4.3. *Under the hypothesis of Theorem 4.7, u is Hölder continuous in $B_{r_0/2}(x_0) \cap \overline{\Omega}$ with Hölder exponent exactly equal to $\frac{1}{2}$. If, in addition, $\partial\Omega$ is of class $C^{k,\alpha}$ or analytic, then u restricted to $B_{r_0/2}(x_0) \cap \partial\Omega$ is $C^{k-1,\alpha}$ for $k = 3, 4, \cdots$ and $0 < \alpha < 1$, or analytic respectively.*

For the proof, we refer to [15]. In order to restrict our attention to the free boundary aspect of the problem, we also state the following measure-theoretic results without proof.

Lemma 4.21. *Let Q be the multiplicity-one integral 2-current in $\mathbb{R}^3$ with $\operatorname{spt}(Q) \subseteq \partial\Omega \times \mathbb{R}$ and $\partial Q = T - Tu$, where*

$$\begin{cases} T = \{(x, \varphi(x)) \ : \ x \in \partial\Omega\}, \\ Tu = \{(x, u(x)) \ : \ x \in \partial\Omega\}. \end{cases}$$

Then $\Delta = \operatorname{graph}(u) + Q$ *is the unique area minimizing integral whose boundary is the given* T *and whose support lies in* $\overline{\Omega} \times \mathbb{R}$.

Corollary 4.4. $\Delta \backslash T$ *is a* $C^{1,1}$ *is a* $C^{1,1}$ *surface with its* $C^{1,1}$*-character depending only on the* $L^\infty(\partial\Omega)$ *norm of* φ, $C^{1,1}$*-character of* $\partial\Omega$, *and the distance to* T.

Lemma 4.22. *Let* φ, u, Ω *be as above. We define*

$$\varphi^* = \begin{cases} \varphi & \text{on} \quad \partial\Omega \backslash B_{r_0}(x_0), \\ \varphi + 2r_0 & \text{on} \quad \partial\Omega \cap B_{r_0}(x_0), \end{cases}$$

and u^* *as the solution of* (4.15) *with boundary data* φ^*. *Then* $u \equiv u^*$.

Proof. Use Corollary 4.4 and the generalized maximum principle of R. Finn, see for example Theorem 13.10 in [6]. □

Proof of Theorem 4.7. Now we will restrict ourselves to a neighborhood of an arbitrary point in

$$\gamma_* = \left\{ (x, u(x)) \ : \ x \in \gamma_+ \cap B_{r_0/2}(x_0) \right\}$$

and study the local behavior of Δ at such an arbitrary point. By Lemma 4.22 $\operatorname{dist}(\gamma_*, \partial\Delta) \geq r_0/2$. Fix a point $p \in \gamma_*$ and introduce a coordinate system $(y, y_3) = (y_1, y_2, y_3)$ such that:

(i) p is the origin of $\mathbb{R}^3$,
(ii) y_1-axis is the same as x_3-axis,
(iii) $y_2 y_3$-plane is the tangent plane of $\partial\Omega \times \mathbb{R}$ at p,
(iv) e_3 is the unit normal of $\partial\Omega \times \mathbb{R}$ at p, hence normal to Δ at p by Corollary 4.4.

By Lemma 4.22 and Corollary 4.4, Δ can be locally represented as the graph of a $C^{1,1}$-function $u = u(y)$, say over $D_\delta = \left\{ y \in \mathbb{R}^2 \ : \ |y| < \delta \right\}$, where $\delta = \delta(r_0, \partial\Omega) > 0$. Likewise, over D_δ. $\partial\Omega \times \mathbb{R}$ can be represented as the graph of a $C^{2,\alpha}$-function $w(y) = w(y_2)$. By scaling we can assume $\delta = 1$. Let

$$K = \left\{ v \in C^{0,1}(D_1) \ : \ v \geq w \quad \text{in} \quad B_1 \quad \text{and} \quad v = u \quad \text{on} \quad \partial B_1 \right\}.$$

We note that $u \in K$ satisfies the variational inequality

$$\int_{D_1} a_j(\nabla u)(v - u)_j \, dy \geq 0 \tag{4.16}$$

for all $v \in K$, where $a_j(\nabla u) = u_j / \sqrt{1 + |\nabla u|^2}$. $u_j = u_{y_j}$.

Let $h = u - w$ so that $h \in C^{1,1}(B_1)$. Define

$$\begin{aligned} \Lambda(h) &= \{ y \in B_1 \ : \ h(y) = 0 \}, \\ \Omega(h) &= \{ y \in B_1 \ : \ h(y) > 0 \}, \\ F(h) &= \partial(\Omega(h)) \cup \partial(\Lambda(h)). \end{aligned}$$

Clearly. $u(F(h)) = \gamma_*$ in $B_1 \times \mathbb{R}$. There remains to show that γ_* is the graph of a $C^{1,\alpha}$-function over $\gamma_+ \cap B_{r_0/2}(x_0)$. In other words, we will first show that γ_* is a $C^{1,\alpha}$-curve by showing that $F(h)$ is $C^{1,\alpha}$ (and hence, so is its image). Next, we will show that γ_* is C^1-graph over $\gamma_+ \cap B_{r_0/2}(x_0)$.

Consider $p = 0 \in F(h)$. Either 0 is an isolated point of $F(h)$, or it has positive density with respect to $\Lambda(h)$, or it has zero density with respect to $\Lambda(h)$. Note that by the regularity analysis in the preceding section, if 0 has positive density with respect to $\Lambda(h)$, then it is a regular point of the free boundary and $F(h)$ is a locally C^1 curve. Hence, $C^{1,\alpha}$ regularity would follow by the higher regularity theory we will establish in the ultimate chapter of these notese.

On the other hand. $\gamma_* = u(F(h))$ is the trace of a $BV(\Omega)$ function, hence 0 cannot be an isolated point of $F(h)$. Thus, it suffices to show that 0 cannot be a point of zero density with respect to $\Lambda(h)$. For this purpose, consider the Jacobi-field equation on the minimal surfaces:

$$\Delta_M V + |A_M|^2 V = 0, \tag{4.17}$$

where $|A_M|^2$ is the squared length of the second fundamental form and Δ_M the Laplace-Beltrami operator, both on the surface M. In our coordinate system: $V = -u_{y_1}/\sqrt{1+|du|^2}$, Lipschitz on $\overline{\Omega(h)}$, and satisfying

$$\begin{cases} \Delta_M V + |A_M|^2 V = 0 \quad \text{in} \quad \Omega(h), \quad V > 0 \quad \text{in} \quad \Omega(h), \\ V|_{F(h)} = 0. \end{cases} \tag{4.18}$$

Recall that in the case 0 is a point of zero density with respect to $\Lambda(h)$, by the blow-up analysis of the preceding section:

$$\Lambda(h) \cap B_\eta(0)\mathbb{C} = \{y, \theta(y, (\pm, 0)) \leq \Phi^{-1}(h)\},$$

and $h(y) = ay_2^2 + o(y_1^2 + y_2^2)$ for $(y_1, y_2) \in (\Omega(h) \cap B_\eta(0)) \backslash \mathbb{C}$. In particular, $\Lambda(h) \cap B_\eta(0) \subset \mathbb{C}_1$, where $C_1 = \{(y_1, y_2) \in B_\eta(0) \ : \ y_1 \geq |y_2|\}$.

On $\mathbb{C}_1$ consider $U_0(y) = r^\alpha \cos(\alpha\theta)$, where $\alpha = 2/3$, $\theta(y_1, y_2)$ is the angle between vectors (y_1, y_2) and $(-1, 0)$. Note that U_0 solves

$$\begin{cases} \Delta_M U_0 = 0, \quad U_0 > 0 \quad \text{in} \quad \mathbb{C}_1, \\ U_0|_{\partial \mathbb{C}_1} = 0. \end{cases} \tag{4.19}$$

Hence, by the maximum principle we have: $V \geq \epsilon U_0$ in $\mathbb{C}_1 \cap B_\eta(0)$ for some $\epsilon > 0$. Consequently, V cannot be Lipschitz at 0, a contradiction. So 0 cannot be a point of zero density with respect to $\Lambda(h)$. So $F(h)$ is C^1 near 0. By higher regularity theory for free boundaries to be addressed in the ultimate chapter, we have: $\gamma_* - u(F(h))$ is a $C^{1,\alpha}$-curve.

There remains to show that γ_* is the graph of a $C^{1,\alpha}$-function. We argue by contradiction. Note that $u \in C^2$ uniformly in $\overline{\Omega(h)} \cap B_{1/2}$. If γ_* is not a uniformly C^1-graph over γ_+, then we would obtain a point, say 0 by another change of coordinates, in $F(h) \cap D_{1/2}$ such that the second order blow-up of h at 0 is of form

$$h_0(y) = ay_2^2, \quad 2a = -H_{\partial\Omega}(0) > 0, \tag{4.20}$$

i.e. $h(y) = ay_2^2 + o(y_1^2 + y_2^2)$. However, (4.18) and the Hopf boundary-point Lemma give: $\partial_{y_2} V \neq 0$ at 0. Hence: $\partial_{y_1}\partial_{y_2} h(0) \neq 0$, a contradiction with (4.20). Hence, the proof is complete. □

In the remaining part of this section, we consider the equation for surfaces with prescribed mean curvature in a bounded C^2 domain $\Omega \subset \mathbb{R}^2$, that is

$$\operatorname{div}\left(\frac{u}{\sqrt{1+|\nabla u|^2}}\right) = H(x). \tag{4.21}$$

We begin with stating a few general facts regarding the solvability of (4.21), for which we refer to [9]. A necessary condition for (4.20) to have a solution $u \in C^2(\Omega)$ is

$$\left|\int_A H \, \mathrm{d}x\right| < \mathcal{H}_1(\partial A), \tag{4.22}$$

for every Caccioppoli set $A \subset \Omega$ such that $\emptyset \neq A \neq \Omega$. Here $\mathcal{H}_1$ denotes the one-dimensional Hausdorff measure.

In [9] the extremal case of equality in (4.22) is studied. In [15] his result is partially recovered, alongside a proof of higher regularity for u restricted to the part of $\partial\Omega$ where $H_{\partial\Omega}(x) < H(x)$ whenever H and $\partial\Omega$ are smooth.

Let Ω be a bounded, C^2 domain in $\mathbb{R}^2$. and let H be a Lipschitz function on $\overline{\Omega}$. Consider the unique solution u (up to an additive constant) of (4.21) with H satisfying (4.22) with equality. Let $\Omega_1 \subsetneq \Omega$ with $\operatorname{per}(\Omega_1) < \infty$. Then the set $U = \{(x_1, x_2, x_3) \in \Omega_1 \times \mathbb{R} \, : \, x_3 < u(x)\}$ is a minimizer of the functional

$$\int |\nabla X_U| \, \mathrm{d}x + \int H X_U \, \mathrm{d}x \mathrm{d}x_3 \quad \text{in} \quad \Omega_1 \times \mathbb{R} \tag{4.23}$$

in the sense that for every set $V \subset \Omega_1 \times \mathbb{R}$, coinciding with U outside some compact set $K \subset \Omega_1 \times \mathbb{R}$, we have

$$\int_K |\nabla X_U| \, \mathrm{d}x + \int_K H X_U \, \mathrm{d}x \mathrm{d}x_3 \leq \int_K |\nabla X_V| \, \mathrm{d}x + \int_K H X_V \, \mathrm{d}x \mathrm{d}x_3. \tag{4.24}$$

Moreover, if in a neighborhood of $x_0 \in \partial\Omega$, the inequality $H_{\partial\Omega}(x) < H(x)$ is satisfied, then $u(x)$ is bounded from above on this neighborhood. u is always bounded from below in this case. Hence, $u \geq 0$ can be assumed. See [9] for further discussion.

Suppose there exist positive constants C_0, r_0 such that

$$H_{\partial\Omega}(x) \leq H(x) - C_0, \quad \forall x \in B_{r_0}(x_0) \cap \partial\Omega \tag{4.25}$$

for some $x_0 \in \partial\Omega$. We will show that u restricted to $B_{r_0/2}(x_0) \cap \partial\Omega$ is smooth in the following sense:

Theorem 4.8. *Let Ω be a bounded, $C^{2,\alpha}$ domain in $\mathbb{R}^2$, and let H be a Lipschitz function on $\overline{\Omega}$ which satisfies* (4.22) *with equality. Suppose* (4.25) *holds. Then the restriction of u to $B_{r_0/2}(x_0) \cap \partial\Omega$ is uniformly $C^{1,\alpha}$. Moreover, u is $C^{0,1/2}$ in $\overline{\Omega} \cap B_{r_0/2}(x_0)$.*

Once again, using the higher regularity theory to be addressed in the ultimate chapter of these notes, one would deduce from Theorem 4.8:

Corollary 4.5. *In Theorem 4.8:*

(i) *if Ω is $C^{k,\alpha}$, and if H is $C^{k-2,\alpha}$, then the restriction of u to $B_{r_0/2}(x_0) \cap \partial\Omega$ is $C^{k-1,\alpha}$ for $k = 3, 4, \cdots$, $0 < \alpha < 1$;*

(ii) *if Ω and H are analytic, then u is also analytic on $B_{r_0/2}(x_0) \cap \partial\Omega$.*

Let Q be the multiplicity-one integral current in $\mathbb{R}^3$ with $\text{spt}(Q) \subseteq (\partial\Omega \cap B_{r_0}(x_0)) \times \mathbb{R}$ and $\partial Q = T - Tu$, where:

$$T = \{(x, u^*(x)) \ : \ x \in \partial\Omega\}, \quad Tu = \{(x, u(x)) \ : \ x \in \partial\Omega\},$$

and

$$u^*(x) = \begin{cases} u(x). & \text{if} \quad x \in \partial\Omega \backslash B_{r_0}(x_0), \\ u(x) + \eta(x). & \text{if} \quad x \in \partial\Omega \cap B_{r_0}(x_0). \end{cases}$$

Above $\eta(x)$ is a smooth function such that $0 \leq \eta \leq M_0$, and $\eta(x) \equiv M_0 = \sup\{u(x) \ : \ x \in \partial\Omega \cap B_{r_0}(x_0)\}$ for $x \in \partial\Omega \cap B_{r_0/2}(x_0)$. Then we have the following lemma analogous to Lemma 4.21:

Lemma 4.23. $\Delta = [\text{graph}(u) + Q] \cap (B_{r_0}(x_0) \times \mathbb{R})$ *is the unique integral current which minimizes*

$$M(\mathbb{S}) + \int_{B_{r_0}(x_0) \cap \Omega} H(x) \eta_{\mathbb{S}}(x) \, dx,$$

where $\eta_{\mathbb{S}}(x)$ is the slice $< \mathbb{S}, P, x >$, where $P(x_1, x_2) = x$, and $\partial\mathbb{S} = \partial\Delta$.

For measure-theoretic definitions and facts we refer to [4].

The argument for proving Theorem 4.8 is completely analogous to that for proving Theorem 4.7. Yet this time the Jacobi-field equation is

$$\Delta_M V + \left(|A_M|^2 - H^2\right) V = \delta_3 H, \tag{4.26}$$

where $V = 1/\sqrt{1 + |\nabla u|^2}$ and $|\delta_3 H| \leq |\nabla H| V$. Nevertheless, the comparison argument employed in the case of nonparemetric least area and the resulting estimate are still valid.

In [9], E. Giusti also showed that if $H(x) \equiv H_{\partial\Omega}(x)$ in an open arc $l \subset \partial\Omega$, then:

$$\lim_{x \to x_0} u(x) = +\infty \quad \text{uniformly in } x_0 \in K \subset\subset l.$$

This can also be deduced from Lemma 4.23 and the regularity theory for obstacle problems. We briefly sketch the idea:

If there is a sequence $x_i \in K$ with

$$\lim_{x_i \to x_0} u(x_i) = u(x_0) < \infty,$$

then $(x_0, u(x_0))$ will be a point in the support of the minimizer. Choose a ball $\partial\Omega \cap B_{r_0}(x_0, t)$ and move it upward from $t < 0$ towards $(x_0, u(x_0))$. Here we choose r_0 small enough that $B_{r_0}(x, t) \subset K \times \mathbb{R}$, Then it must touch the free boundary at some point p, possibly different from $(x_0, u(x_0))$. Then in a neighborhood of p, the vertical cylinder $\partial\Omega \times \mathbb{R}$ and the graph(u) are graphs over the tangent plane to $\partial\Omega \times \mathbb{R}$ at p. Now we may apply the Hopf boundary-point lemma at p to obtain a contradiction.

For further discussion on the solvability of the Dirichlet problem for (4.21) and the extension of the analysis that leads to Theorems 4.7 and 4.8 to the case of parametric elliptic integrals, we refer to [15].

The regularity theory of free boundaries in higher dimensions were first established by Luis Caffarelli: The regularity of free boundaries in higher dimensions, Acta Math. 139 (1977), no. 3–4, 155–184. The presentation here follows the approach used by Caffarelli in: Compactness methods in free boundary problems, Comm. Partial Differential Equations 5 (1980), no. 4, 427–448. Further developments using the boundary Harnack inequality approach may be found in a much detailed exposition [3]. See also [5].

5 Regularity of singular sets of free boundary

5.1 Singular set of the free boundary

Definition 5.1. Let $u \in P_1(M)$ and let $N(u)$, $\Lambda(u)$ and Γ_u (or Γ) be defined as before. $x_0 \in \Gamma_u$ is said to be in the singular set of Γ_u, denoted by $x_0 \in \operatorname{sing} u$ if for $\forall r > 0$, $\delta_r(\Lambda(u)) \le \sigma(r)$, where $\delta_r(\Lambda(u))$ is defined in Lemma 4.8 and $\sigma(r)$ is some monotone increasing function defined on $\mathbb{R}^+$ satisfying $\sigma(0^+) = 0$. That is to say, for $\forall r > 0$, $\Lambda(u) \cap B_r(x_0)$ is contained in a strip of width $r\sigma(r)$.

Remark 5.1. The existence of such function $\sigma(r)$ can be abstracted out from the Caffarelli's regularity theorem for free boundaries so that if for some $r_0 > 0, \delta_{r_0}(\Lambda(u)) \ge \sigma(r_0)$, then Γ_u is a C^1 hypersurface inside $B_{r_0/2}$. Thus $\operatorname{sing} u$ is a relative closed subset of Γ_u.

The main result of this chapter is the following:

Theorem 5.1. *Given* $0 = x_0 \in \Gamma$ *and* $x_0 \in \operatorname{sing} u$, *there is an* $M_{x_0} \ge 0$, *s.t.*

(i) $Q_{x_0}(x) = \frac{x^T M_{x_0} x}{2} \ge 0$ *with* $\Delta Q_{x_0} = \operatorname{tr}(M_{x_0}) = 1$;
(ii) $|u(x) - Q_{x_0}(x)| \le |x|^2 \tilde{\sigma}(|x|)$ *for some universal* $\tilde{\sigma}(r)$.

Moreover, M_{x_0} *is continuous in* x_0 *for* $x_0 \in \operatorname{sing} u \cap \Gamma$. *If* $\dim \ker M_{x_0} = k$, *then* $\operatorname{sing} u$ *is contained in a* k*-dimensional* C^1*-manifold in a neighborhood of* x_0. *The size of the neighborhood depends on the smallest nonzero eigenvalue of* M_{x_0}.

Lemma 5.1. *For* $\forall \varepsilon > 0$, $\exists N_\varepsilon = N$, *s.t. if* $w \in P_N$ *is a solution in* B_N, $0 \in \Gamma \cap \operatorname{sing} w$, *then there is* $Q(x) = \frac{x^T A x}{2}$, *s.t.* $A \ge 0$, $\Delta Q = 1$ *and* $\|Q - w\|_{L^\infty(B_1)} \le \varepsilon$.

Proof. Suppose not. Then $\exists \{w_k\}$ solutions on B_k with $k \to +\infty$, s.t. $0 \in \Gamma_k \cap \operatorname{sing} w_k$, and there is no such Q_k that $\|Q_k - w_k\|_{L^\infty(B_1)} \le \varepsilon_0$ for some $\varepsilon_0 > 0$. By compactness, there is an entire solution $w_\infty \in P_\infty(M)$, s.t. $0 \in \Gamma_\infty$ and $w_k \rightharpoonup w_\infty$ in $C^{1,1}_{loc}$. We claim that w_∞ must be a quadratic polynomial.

We first note that Λ_∞ has no interior. If not, since Λ_∞ is convex for entire solutions, $\Lambda_\infty \cap B_1$ contains a ball $B_{r_0}(a)$ where $w_\infty \equiv 0$. Again by the convexity,

Λ_∞ would contain a cone of a fixed size with the vertex at the origin. From nondegeneracy (see Lemma 4.2), since $w_k \to w_\infty$ uniformly on $B_{r_0}(a)$, $w_k \equiv 0$ on $B_{r_0/2}(a)$ for sufficiently large k. Consequently, via nondegeneracy again, $w_k \equiv 0$ on a truncated cone (with $|x| > r_0/2$) that contains the truncated central one third of the cone (again in the region $|x| > r_0/2$) described above. Here r_0 is a fixed, but small positive constant. Since $0 \in \operatorname{sing} w_k \cap \Gamma_k$, we find $\Lambda(w_k) \cap B_{r_0} \triangleq \Lambda_k \cap B_{r_0}$ must be contained in a strip with width $r_0 \sigma r_0$. In fact, if we define $v_k(x) = k^{-2} w_k(kx) \in P_1(M)$ for $x \in B_1$, $0 \in \operatorname{sing} v_k$. This implies $\Lambda(v_k) \cap B_{r_0} \subset S_{r_0\sigma(r_0)}$, the latter is impossible to contain a truncated cone of fixed size as above when r_0 is suitably small. Hence, Λ_∞ has empty interior, which implies w_∞ is a quadratic polynomial $\frac{1}{2}x^T Bx$ with $B \geq 0$; moreover, $\Delta w_\infty = 1$. Then $\|Q_k - w_k\|_{L^\infty(B_1)} \leq \varepsilon$ must be true if we take $Q_k \equiv w_\infty$ when k is large, which contradicts with our assumption. So we proved the lemma. □

Corollary 5.1. *Let $w \in P_1(M)$ be a solution, with $0 \in \Gamma_w \cap \operatorname{sing} w$. Then $\forall \varepsilon > 0$, $\exists r_0(\varepsilon)$, s.t. $\forall 0 < r < r_0$, $\exists Q^r = \frac{x^T A_r x}{2}$, satisfying $A_r \geq 0$, $\Delta Q^r \equiv 1$ and $\|w - Q^r\|_{L^\infty(B_r)} \leq \varepsilon r^2$.*

Remark 5.2. One may write in another way $\varepsilon = \tilde{\sigma}(r)$, where $\tilde{\sigma}(r)$ has the same properties as $\sigma(r)$ described above.

Now the problem of proving Theorem 5.1 is the uniqueness of the limit of such Q^r's as $r \to 0^+$. If, for example, $\|w - Q^r\|_{L^\infty(B_r)} \leq Cr^\alpha \cdot r^2$ for some $\alpha > 0$, then the limit of $\{Q^r\}$ is unique and we are done! In order to show the uniqueness, we introduce the following monotonicity approach due to Alt, Caffarelli and Friedman.

Theorem 5.2 (Monotonicity). *Let u_1 and u_2 be two continuous functions in $B_1(0) \subset \mathbb{R}^n$, s.t.*

(a) $u_1 \cdot u_2 = 0$,
(b) $u_1(0) = u_2(0) = 0$,
(c) $u_i \Delta u_i \geq 0$, $i = 1, 2$.

Then

$$I(R) = \left(\frac{1}{R^2}\int_{B_R} \frac{|\nabla u_1|^2(x)}{r^{n-2}}\,\mathrm{d}x\right)\left(\frac{1}{R^2}\int_{B_R} \frac{|\nabla u_2|^2(x)}{r^{n-2}}\,\mathrm{d}x\right) \triangleq \frac{1}{R^2}I_1(R) \cdot \frac{1}{R^2}I_2(R)$$

is increasing in R.

Remark 5.3. It is easy to check the following properties of $I(R)$:

(i) If $u_1(x) = \alpha x_1^+$, $R^{-2}I_1(R) = C(n)\alpha^2$.
(ii) If $u_{i,\lambda}(x) \triangleq u_i(\lambda x)/\lambda$, $I\left(\frac{R}{\lambda}; u_{1,\lambda}, u_{2,\lambda}\right) = I(R; u_1, u_2)$.
(iii) Suppose the supports of u_1 and u_2 are separated by an $(n-1)$-dimensional hypersurface denoted by Γ. If Γ is smooth and $D_\nu u_i(x)$ exists, then $R^{-2}I_i(R) \to C(n)|D_\nu u_i|^2$ as $R \to 0^+$. Thus, by Theorem 5.2,

$$C(n)^2 |D_\nu u_1|^2 |D_\nu u_2|^2 \leq I\left(\frac{1}{2}\right).$$

The following lemma, which is easy to show, would be used in the proof of Theorem 5.2.

Lemma 5.2. *Given a set* $\Sigma \subset \mathbb{S}^{n-1}$, *let* C_Σ *be the cone over* Σ. *Let* $h(x) = r^\alpha f(\theta)$, *s.t.*

$$\begin{cases} \Delta h(x) = 0 \quad \text{in } C_\Sigma, \\ h(x)|_{\partial C_\Sigma} = 0, \\ \Delta_\theta f(\theta) + \lambda_1 f(\theta) = 0 \quad \text{in } \Sigma, \\ f(\theta)|_{\partial\Sigma} = 0. \end{cases}$$

Then $\lambda_1 = \alpha(\alpha + n - 2)$.

Proof of Theorem 5.2. By rescaling, we only need to prove $I'(1) \geq 0$, i.e.

$$I'(1) = I_1'(1)I_2(1) + I_1(1)I_2'(1) - 4I_1(1)I_2(1) \geq 0.$$

So we need to verify

$$\frac{I_1'(1)}{I_1(1)} + \frac{I_2'(1)}{I_2(1)} \geq 4. \tag{5.1}$$

We notice that

$$\begin{aligned} I_i(1) &= \int_{B_1} \frac{|\nabla u_i|^2}{r^{n-2}}\,\mathrm{d}x \leq \int_{B_1} \frac{\Delta(u_i^2/2)}{r^{n-2}}\,\mathrm{d}x = \int_{B_1} \frac{\Delta(u_i^2/2)}{r^{n-2}} - \frac{u_i^2}{2}\Delta\left(\frac{1}{r^{n-2}}\right)\mathrm{d}x \\ &= \int_{\Sigma_i} \left[u_i D_r u_i + \frac{n-2}{2}u_i^2\right]\mathrm{d}A(x), \end{aligned}$$

where $\Sigma_i = \{x \in \partial B_1 : u_i(x) \neq 0\}$. We used Green's identity in the last equation. On the other hand,

$$I_i'(1) = \int_{\Sigma_i} |\nabla u_i|^2\,\mathrm{d}A(x) = \int_{\Sigma_i} u_{i,r}^2 + u_{i,\sigma}^2\,\mathrm{d}A(x),$$

where $u_{i,\sigma}$ is the gradient along ∂B_1. Hence, by Cauchy-Schwarz inequality,

$$\begin{aligned} \frac{I_i'(1)}{I_i(1)} &\geq \frac{\int_{\Sigma_i} u_{i,r}^2 + u_{i,\sigma}^2\,\mathrm{d}A}{\int_{\Sigma_i} u_i u_{i,r} + \frac{n-2}{2}u_i^2\,\mathrm{d}A} \geq \frac{\int_{\Sigma_i} u_{i,r}^2 + u_{i,\sigma}^2\,\mathrm{d}A}{\int_{\Sigma_i}\left(\frac{A_i}{2}u_i^2 + \frac{1}{2A_i}u_{i,r}^2\right) + \frac{n-2}{2}u_i^2\,\mathrm{d}A} \\ &= 2\frac{\int_{\Sigma_i} u_{i,r}^2 + u_{i,\sigma}^2\,\mathrm{d}A}{\int_{\Sigma_i} \frac{1}{A_i}u_{i,r}^2 + (A_i + n - 2)u_i^2\,\mathrm{d}A}. \end{aligned}$$

Since

$$\frac{\int_{\Sigma_i} u_{i,\sigma}^2\,\mathrm{d}A}{\int_{\Sigma_i} u_i^2\,\mathrm{d}A} \geq \lambda_{i,1},$$

where $\lambda_{i,1}$ is the first Dirichlet eigenvalue of $\Delta_{\mathbb{S}^{n-1}}$ on Σ_i, we find

$$\frac{1}{A_i} = \frac{A_i + n - 2}{\lambda_{i,1}}$$

would be optimal choice of A_i's, i.e. $A_i(A_i+n-2) = \lambda_{i,1}$. In the notation of Lemma 5.2, $A_i = \alpha_i$. This gives $I_i'(1)/I_i(1) \geq 2A_i$. Note that all above become equality when we choose $u_i|_{C_{\Sigma_i}}$ to be the homogeneous extension of degree A_i of $u_i|_{\Sigma_i}$. Now proving (5.1) reduces to showing $A_1 + A_2 \geq 2$. To be more precise, given two disjoint cones $C_i \subset \mathbb{R}^n$, $i = 1, 2$, let h_i be the corresponding homogeneous function of degree A_i, which is harmonic and nonnegative on C_i and vanishes on ∂C_i. Then we want to show $A_1 + A_2 \geq 2$.

First we note that, by symmetrization argument, among all domains $\Sigma \subset \mathbb{S}^{n-1}$ for given $|\Sigma|$, the one minimizing $\lambda_1(\Sigma)$ is a spherical cap with volume $|\Sigma|$. Hence, one reduces to the case where

$$\Sigma_1 = \mathbb{S}^{n-1} \cap \{x_1 > a\}, \quad \Sigma_2 = \mathbb{S}^{n-1} \cap \{x_1 < a\}.$$

Since Σ_i's are radially symmetric with respect to x_1-axis, the corresponding eigenfunctions are radial, i.e. $u_i = u_i(x_1)$. We calculate $\lambda_{1,1}$ by Rayleigh quotient on Σ_1

$$\lambda_{1,1} = \inf_u \frac{\int_{\Sigma_1} |\nabla_{\mathbb{S}^{n-1}} u|^2 \, \mathrm{d}A}{\int_{\Sigma_1} u^2 \, \mathrm{d}A} = \inf_u \frac{\int_a^1 u_x^2 (1-x^2)^{\frac{n-1}{2}} \, \mathrm{d}x}{\int_a^1 u^2 (1-x^2)^{\frac{n-3}{2}} \, \mathrm{d}x}.$$

One can then further show that the above quantities are monotone decreasing in n. When the dimension becomes infinity, a proper normalization would lead to a limiting eigenvalue problem with Gaussian weight on the entire space, and where certain convexity of eigenvalues can be established as in the earlier work of S. Friedland and W. K. Haymen: Comment. Math. Helv. 51 (1976), 133–161. One can also check a slightly general form in the paper by Caffarelli-Kenig-Jerison: Some new monotonicity theorems with applications to free boundary problems, Ann. of Math. (2) 155 (2002), no. 2, 369–404. □

Proof of Theorem 5.1. Let $0 \in \Gamma \cap \operatorname{sing} u$. Then by Corollary 5.1, for $\forall r > 0$,

(a) $\Lambda \cap B_r \subset S_{r\sigma(r)}$;

(b) $\exists Q^r = \frac{1}{2} x^T M_r x$ with $\Delta Q^r = 1$, $M_r \geq 0$ and

$$\|u - Q^r\|_{L^\infty(B_r)} \leq r^2 \tilde{\sigma}(r), \quad \|\nabla(u - Q^r)\|_{L^\infty(B_1)} \leq \tilde{\sigma}^{\frac{1}{2}}(r).$$

Let $u^r(x) = r^{-2} u(rx)$, and $\Lambda^r = \Lambda(u^r)$. Then $\Lambda^r \cap B_1 \subset S_{\sigma(r)}$, and

$$\|u^r - Q_r\|_{L^\infty(B_1)} \leq \tilde{\sigma}(r), \quad \|\nabla(u^r - Q_r)\|_{L^\infty(B_r)} \leq r\tilde{\sigma}^{\frac{1}{2}}(r).$$

Hence, outside $S_{\sigma(r)^{\frac{1}{3}}}$, one has

$$|D_{ij}(u^r - Q^r)| \leq \frac{1}{\tilde{\sigma}(r)^{\frac{2}{3}}} \operatorname{osc}(u^r - Q^r) \leq \tilde{\sigma}(r)^{\frac{1}{3}}.$$

Let $u_e = D_e u$, where e is a unit vector to be determined later. $\Delta u_e = 0$ in $\{u > 0\}$, and $u_e \equiv 0$ on Λ. Define $u_e^+ = \max\{u_e, 0\}$ and $u_e^- = \min\{u_e, 0\}$. We apply monotonicity formula to $u_e^\pm$, i.e.

$$I(R) = \frac{1}{R^2} \int_{B_R} \frac{|\nabla u_e^+|^2}{r^{n-2}} \, \mathrm{d}x \cdot \frac{1}{R^2} \int_{B_R} \frac{|\nabla u_e^-|^2}{r^{n-2}} \, \mathrm{d}x$$

is increasing in R. We calculate

$$I^+(R,e) = \frac{1}{R^2}\int_{\{D_eu>0\}\cap B_R} \frac{|DD_eu|^2}{r^{n-2}}\,\mathrm{d}x = \int_{\{D_ew^R>0\}\cap B_1} \frac{|DD_eu^R|^2}{r^{n-2}}\,\mathrm{d}x.$$

We are going to replace DD_eu^R by DD_eQ^R and split $I^+(R,e)$ into the integral in $S_{\sigma(R)^{\frac{1}{3}}}$ and that in its compliment. Noticing that $D_jD_eQ^R = M_j^Re = e_j^TM^Re$, we have

$$I^+(R,e) = \int_{\{D_eu^R>0\}\cap B_1} \frac{\|M^Re\|^2}{r^{n-2}}\,\mathrm{d}x + O\left(\tilde{\sigma}(r)^{\frac{1}{3}}\right).$$

Let $D^\pm = \{D_eQ^R = x^TM^Re > \pm\tilde{\sigma}^{\frac{1}{2}}(R)\}$ and $D = \{D_eQ^R = x^TM^Re > 0\}$. Then

$$D^+ \subset \{D_eu^R > 0\} \subset D^-,$$
$$\int_{D^+} \frac{\mathrm{d}x}{r^{n-2}}\|M^Re\|^2 - C\tilde{\sigma}(R)^{\frac{1}{3}} \le I^+(R,e) \le \int_{D^-} \frac{\mathrm{d}x}{r^{n-2}}\|M^Re\|^2 + C\tilde{\sigma}(R)^{\frac{1}{3}},$$
$$\left|\int_{D^\pm} \frac{\mathrm{d}x}{r^{n-2}} - \int_D \frac{\mathrm{d}x}{r^{n-2}}\right| \le \frac{C\tilde{\sigma}(R)^{\frac{1}{2}}}{\|M^Re\|}. \tag{5.2}$$

The last estimate is based on the fact that $D^-\backslash D^+$ is contained in a strip with $O(\tilde{\sigma}(R)^{\frac{1}{2}}/\|M^Re\|)$. Let $C^* = \int_{B_1^+} \frac{\mathrm{d}x}{r^{n-2}}$ and we have

$$I^+(R,e) = \|M^Re\|^2\left[C^* + O\left(\frac{\tilde{\sigma}(R)^{\frac{1}{2}}}{\|M^Re\|}\right)\right] + O\left(\tilde{\sigma}(R)^{\frac{1}{3}}\right) = C^*\|M^Re\|^2 + O\left(\tilde{\sigma}(R)^{\frac{1}{3}}\right)$$

Similarly,

$$I^-(R,e) = C^*\|M^Re\|^2 + O\left(\tilde{\sigma}(R)^{\frac{1}{3}}\right). \qquad \square$$

To this end, we need the following results to complete the proof. As a direct result of the monotonicity (Theorem 5.2) and the calculation in the above proof, we have

Corollary 5.2.

$$\|M^{R_1}e\|^2 \le \|M^{R_2}e\|^2 + O\left(\tilde{\sigma}(R_2)^{\frac{1}{3}}\right) \qquad \textit{for } \forall\, 0 < R_1 < R_2.$$

Lemma 5.3.

$$\|M^{R_1} - M^{R_2}\| \le O\left(\tilde{\sigma}(R_2)^{\frac{1}{6}}\right) \qquad \textit{for } \forall\, 0 < R_1 < R_2.$$

Proof. Let $N = M^{R_2} - M^{R_1}$; N is symmetric and $\mathrm{tr}(N) = 0$. By Corollary 5.2,

$$\|(M^{R_2} - N)e\|^2 \le \|M^{R_2}e\|^2 + O\left(\tilde{\sigma}(R_2)^{\frac{1}{3}}\right)$$
$$\Rightarrow -2\langle M^{R_2}e, Ne\rangle + \|Ne\|^2 \le O\left(\tilde{\sigma}(R_2)^{\frac{1}{3}}\right).$$

Let e be the eigenvector corresponding to the smallest eigenvalue of N, denoted by λ. Note that $\lambda \le 0$. Then

$$-2\lambda e^T M^{R_2} e + \lambda^2 \le O\left(\tilde{\sigma}(R)^{\frac{1}{3}}\right).$$

It immediately gives $|\lambda| \le O\left(\tilde{\sigma}(R)^{\frac{1}{6}}\right)$, as $e^T M^{R_2} e \ge 0$. Since $\operatorname{tr}(N) = 0$, we find $\|N\| \le O\left(\tilde{\sigma}(R)^{\frac{1}{6}}\right)$. □

Corollary 5.3. *As a result of Lemma 5.3, we have*

(i) *As $R \to 0$, $M^R \to M^0$;*

(ii) $\|M^R - M^0\| \le O\left(\tilde{\sigma}(R)^{\frac{1}{6}}\right)$*;*

(iii) $$\left\|u - \tfrac{1}{2}x^T M^0 x\right\|_{L^\infty(B_R)} \le \|u - Q^R\|_{L^\infty(B_R)} + \|Q^R - Q^0\|_{L^\infty(B_R)} \le R^2 O\left(\tilde{\sigma}(R)^{\frac{1}{6}}\right).$$

This readily proves (i) and (ii) in Theorem 5.1.

Proof of Theorem 5.1, *continued.* To show the continuity of M_x on $x \in \Gamma \cap \operatorname{sing} u$, we have

$$\|Q^R_{x_0} - Q^R_{x_1}\|_{B_R(x_0)\cap B_R(x_1)} \le 2R^2 \tilde{\sigma}(R)^{\frac{1}{6}}, \qquad \text{whenever } |x_1 - x_0| \le \frac{R}{2}.$$

In particular, if x_0, $x_1 \in \Gamma \cap \operatorname{sing} u$, and $|x_0 - x_1| \le \delta R$ with some small δ, we will have

$$\|M_{x_1} - M_{x_0}\| \le \tilde{\sigma}(R)^{\frac{1}{6}}.$$

The rest of conclusions of the theorem follows from elementary geometric considerations, and we leave these to readers. This completes the proof of Theorem 5.1. □

5.2 Uniqueness of second order blow-up

In this section following [14], we will prove that the quadratic polynomial Q_{x_0} in the blow ups is unique. This would give an alternative proof of the uniqueness of second order blow ups established in Theorem 5.1. The other conclusions of the Theorem 5.1 thus follow also. Our proof is based on a monotonicity formula due to G. S. Weiss [20].

Remark 5.4. Almgren [1] proved the following monotonicity formula for harmonic functions. Let u be harmonic in $B_1 \subset \mathbb{R}^n$. Define for $0 < r < 1$:

$$D(r) = \int_{B_r} |\nabla u|^2 \, dx, \quad H(r) = \int_{\partial B_r} u^2, \quad N(r) = \frac{rD(r)}{H(r)}.$$

Then $N(r)$ is monotone increasing in r. Indeed, we calculate the logarithmic derivative of $N(r)$:

$$\frac{N'(r)}{N(r)} = \frac{1}{r} + \frac{D'(r)}{D(r)} - \frac{H'(r)}{H(r)} = \frac{n-1}{r} + \frac{2\int (u_r)^2}{D(r)} - \frac{H'(r)}{H(r)}.$$

from

$$D'(r) = \int_{\partial B_r} \left((u_r)^2 + \frac{(u_\theta)^2}{r^2} \right),$$

which is obtained by expressing $|\nabla u|^2$ in polar coordinates, and the Rellich identity

$$\int_{\partial B_r} \frac{(u_\theta)^2}{r^2} = \frac{n-2}{r} D(r) + \int_{\partial B_r} (u_r)^2,$$

Similarly, we have

$$H'(r) = \frac{n-1}{r} H(r) + 2 \int_{\partial B_r} u u_r.$$

Using these expressions for $D'(r)$ and $H'(r)$, we obtain

$$\frac{N'(r)}{N(r)} = \frac{2 \int_{\partial B_r} (u_r)^2}{D(r)} - \frac{2 \int_{\partial B_r} u u_r}{\int_{\partial B_r} u^2}.$$

Finally, note that

$$D(r) = \int_{B_r} \Delta \left(\frac{u^2}{2} \right) = \int_{\partial B_r} u u_r,$$

where the first equality is due to the harmonicity of u and the latter due to the divergence theorem. Hence, applying the Cauchy-Schwarz inequality to the denominator of the positive term and the numerator of the second term, the claim follows.

In particular $k = N(0^+) = \lim_{r \downarrow 0} N(r)$ exists, and k is the vanishing order of u at 0. Assuming u vanishes at 0 with order $k \in \{1, 2, \cdots\}$, Almgren's monotonicity formula implies:

$$\frac{D(r)}{r^{n-2+2k}} - k \frac{H(r)}{r^{n-1+2k}}, \quad 0 < r < 1$$

is also a monotone increasing and non-negative function of $r \in (0,1)$.

With a similar proof, Weiss [20] showed that the following monotonicity formula holds in the context of the obstacle problem.

Lemma 5.4 (Weiss Monotonicity). *Let $u \in P_1(M)$, $x_0 \in \Gamma_u$ such that $B_R(x_0) \subseteq B_1$, and*

$$\Phi(x_0, r, u) = \frac{\int_{B_r(x_0)} \left[|\nabla u|^2(x) + 2u(x) \right] \mathrm{d}x}{r^{n+2}} - 2 \frac{\int_{\partial B_r(x_0)} u^2}{r^{n+3}}.$$

Then we have

$$\frac{\mathrm{d}}{\mathrm{d}r} \Phi(x_0, r, u) = \frac{2}{r^{n+2}} \int_{\partial B_r(x_0)} \left(\frac{\partial u^\lambda}{\partial \rho} - \frac{2u}{\rho} \right)^2 \geq 0,$$

for $0 < r \leq R$. Hence it is monotone increasing.

An easy consequence of this monotonicity formula is the following:

Corollary 5.4 (Existence of Homogeneous Blow-Ups). *Let $u \in P_1(M)$. Then for any sequence λ_m, $\lambda_m \downarrow 0$, there is a subsequence, relabeled as λ_m, such that*

$$u^{\lambda_m}(x) = \frac{u(\lambda_m x)}{(\lambda_m)^2} \to u_0(x)$$

uniformly, where

$$u_0(x) = |x|^2 u_0\left(\frac{x}{|x|}\right).$$

Proof of Corollary. Since $u^\lambda \in P_1(M)$, we apply Lemma 5.4 to obtain

$$\begin{aligned}\Phi\left(\underline{0}, 1, u^\lambda\right) - \Phi\left(\underline{0}, 0^+, u^\lambda\right) &= \int_0^1 \frac{2}{r^{n+2}} \int_{\partial B_r(\underline{0})} \left(\frac{\partial u^\lambda}{\partial \rho} - \frac{2u^\lambda}{\rho}\right)^2 \mathrm{d}r \\ &= \int_0^\lambda \frac{2}{r^{n+2}} \int_{\partial B_r(\underline{0})} \left(\frac{\partial u}{\partial \rho} - \frac{2u}{\rho}\right)^2 \mathrm{d}r \to 0^+,\end{aligned}$$

as $\lambda \to 0^+$. Thus, for a subsequence, relabeled as u^{λ_m}, such that $u^{\lambda_m} \to u_0(x)$ in $C^{1,\beta}$ for any $0 < \beta < 1$ (due to $u^{\lambda_m} \in P_1(M)$) with $u_0 \in P_1(M)$, we have

$$\begin{aligned}&\int_0^1 \frac{2}{r^{n+2}} \int_{\partial B_r(\underline{0})} \left(\frac{\partial u_0}{\partial \rho} - \frac{2u_0}{\rho}\right)^2 \mathrm{d}r \\ &= \lim_{\lambda_m \to 0^+} \int_0^1 \frac{2}{r^{n+2}} \int_{\partial B_r(\underline{0})} \left(\frac{\partial u^\lambda}{\partial \rho} - \frac{2u^\lambda}{\rho}\right)^2 \mathrm{d}r \\ &= \lim_{\lambda_m \to 0^+} \int_0^{\lambda_m} \left(\frac{\partial u}{\partial \rho} - \frac{2u}{\rho}\right)^2 \mathrm{d}r = 0.\end{aligned}$$

Hence, $\left(\frac{\partial u_0}{\partial \rho} - \frac{2u_0}{\rho}\right)$ vanishes for almost every $0 < r < 1$. Thus, we conclude $u_0(x) = |x|^2 u_0\left(\frac{x}{|x|}\right)$. □

Remark 5.5. Let $\mathcal{F} = \{\Gamma_u : u \in P_1(M)\}$. Then:

(1) $\forall E \in \mathcal{F}$, $a \in E$, $E_{a,\lambda} \in \mathcal{F}$, where $E_{a,\lambda} = \left(\frac{E-a}{\lambda}\right) \cap B_1$, $0 < \lambda \leq 1 - |a|$.
(2) $\forall E \in \mathcal{F}$, $a \in E$, $\{\lambda_m\} \downarrow 0$, there is a subsequence, relabeled as $\{\lambda_m\}$ such that

$$E_{a,\lambda_m} \rightharpoonup T \text{ with } T_{\underline{0},\lambda} \equiv T \text{ for} 0 < \lambda < 1,$$

i.e. there is a tangent cone of E at each point $a \in E$.

The first property is due to the invariance of $P_1(M)$ under scaling and translation, whereas the latter follows directly from Corollary 5.4. The notation $E_i \rightharpoonup F$ means: for any $\epsilon > 0$ and for all sufficiently large $i, i \geq i(\epsilon)$, E_i is contained in the ϵ-neighborhood of F.

For $E \in \mathcal{F}$, let

$S_j = \{a \in E$: the invariant dimension of $T \leq j$, for all tangent cones T of$E$ at $a\}$

for $j = 0, 1, 2, \cdots, n$.

Definition 5.2. For a tangent cone T of E at a, that is: $T_{0,\lambda} \equiv T = \lim_{\lambda_m} E_{a,\lambda_m}$ for a sequence $\lambda_m \downarrow 0$, $0 < \lambda \leq 1$, a linear subspace V of $\mathbb{R}^n$ is called an invariant space of T, if $(T + v) \cap B_1 \subset T$ for all $v \in V$. The maximum dimension of all such invariant spaces V is called the invariant dimension of T.

The following result is a version of the dimension reduction principle of Federer [5] and its improvement due to Almgren [1], [19]. Its proof can be simplified after establishing the uniqueness of the homogeneous degree 2 blow-ups at singular points of Γ_u.

Theorem 5.3 (Reduction and Stratification Principle). (i) *For every $E \in \mathcal{F}$, the Hausdorff dimension of E, $\dim_H E \leq n - 1$. Moreover, there is an $(n-1)$-dimensional hyperplane $T \in \mathcal{F}$.*

(ii) *$\dim_H S_j \leq j$, for $j = 0, 1, 2, \cdots, n-1$, $S_0 \subset S_1 \subset S_2 \subset \cdots \subset S_{n-1} = E$. Moreover, S_0 consists of isolated points.*

Instead of pursuing this direction we resume working towards the uniqueness of the second order blow-ups. Let v be one of the second order blow-ups of $u \in P_1(M)$, and let $w = u - v$. We define

$$D(w, r) = \int_{B_r} |\nabla w|^2 \, dx, \quad H(w, r) = \int_{\partial B_r} w^2.$$

Lemma 5.5 (Generalized Almgren-Weiss Monotonicity).

$$\frac{d}{dr}\left[\frac{D(w,r)}{r^{n+2}} - 2\frac{H(w,r)}{r^{n+3}}\right] = 2\int_{\partial B_r} \frac{\left|\rho\frac{\partial w}{\partial \rho} - 2w\right|}{r^{n+4}} \geq 0.$$

Proof. A direct calculation yields the following formula:

$$\frac{d}{dr}\left[\frac{D(w,r)}{r^{n+2}} - 2\frac{H(w,r)}{r^{n+3}}\right] = 2\int_{\partial B_r} \frac{\left|\rho\frac{\partial w}{\partial \rho} - 2w\right|}{r^{n+4}} + \epsilon(r),$$

where the error term $\epsilon(r)$ is

$$\epsilon(r) = \frac{2}{r^{n+3}} \int_{B_r} \left(2w - \rho\frac{\partial w}{\partial \rho}\right) \Delta w \, dx.$$

First, we observe that

$$2w - \rho\frac{\partial w}{\partial \rho} = 2u - \rho\frac{\partial u}{\partial \rho},$$

because v is a homogeneous function of order 2. So we consider $\left(2u - \rho\frac{\partial u}{\partial \rho}\right)(\Delta u - \Delta v)$. If $x \in \Lambda(u)$, then $u \equiv 0$, $\nabla u(x) = 0$. Hence the first factor vanishes. If $x \in N(u)$ instead, then $\Delta u(x) = \Delta v(x) = 1$. Hence the second factor vanishes. Therefore $\epsilon(r) = 0$ in either case. □

Lemma 5.6 (Convexity).

$$\frac{\mathrm{d}}{\mathrm{d}r}\left[\frac{\int_{\partial B_r} w^2}{r^{n+3}}\right] \geq \frac{2}{r}\left[\frac{D(w,r)}{r^{n+2}} - \frac{2H(w,r)}{r^{n+3}}\right] \geq 0.$$

Proof. The second inequality is due to the Weiss monotonicity formula. Hence, proving the first inequality suffices

$$\frac{\mathrm{d}}{\mathrm{d}r}\left[\frac{\int_{\partial B_r} w^2}{r^{n+3}}\right] = \frac{2}{r}\left[\frac{D(w,r)}{r^{n+2}} - 2\frac{H(w,r)}{r^{n+3}}\right] + \frac{2}{r^{n+3}}\int_{B_r} w\Delta w \,\mathrm{d}x.$$

Note that $w \in C^{1,1}(B_1)$. When $u > 0$, $\Delta w = \Delta u - \Delta v = 0$. If $u = 0$ instead, $\Delta u = 0$ almost everywhere on the set $\{u = 0\}$ and hence $w\Delta w = v\Delta v = v$ almost everywhere on $\{u = 0\}$. Hence $w\Delta w \geq 0$ almost everywhere in B_1. The inequality is verified in either case. □

Remark 5.6. Letting $f(t) = \frac{\int_{\partial B_r} w^2}{r^{n+3}}$, $r = e^t$, $-\infty < t < 0$, then $f_t t(t) \geq 0$ by Lemmas 5.5 and 5.6. This is the reason for the latter being called the convexity lemma.

Theorem 5.4. *For every $x_0 \in \mathrm{sing}\Gamma_u$, the matrix M_{x_0} in Theorem 5.1 is unique.*

Proof. Let $u \in P_1(M)$, and v be one of the blow-ups. Then for λ_0 small enough $|u^{\lambda_0} - v| << 1$. Hence

$$\int_{\partial B_1} w^2 << 1.$$

Moreover, $w = u - v \in C^{1,\alpha}$, $\|w\|_{C^{1,\alpha}(B_1)} << 1$, and as a consequence of Lemmas 5.5 and 5.6 we have

$$0 \leq \frac{D(w,r)}{r^{n+2}} - 2\frac{H(w,r)}{r^{n+3}} \leq D(w,1) - 2H(w,1) << 1$$

and

$$\frac{1}{r^{n+3}}\int_{\partial B_r} w^2 << 1.$$

Hence, the uniqueness of the second order blow-ups follows. □

The regularity theory for singular sets of free boundaries in higher dimensions was also due to L. Caffarelli. It was presented in his survey article (and lectures): The Obstacle Problem, Lezioni Fermiane, [Fermi Lectures] Academia Nazionale dei Lincei, Rome; Scuola Normale Superiore, Pisa, 1998, ii + 54pp. The Obstacle Problem Revisited, J. Fourier Anal. Appl. 4 (1998), no. 4–5, 383–402. The second proof of the the uniqueness of the second order blow ups is taken from the author's paper: On Regularity and Singularity of Free Boundaries in Obstacle Problems, Chin. Ann. Math. 30B (5), 2009, 645–652.

6 Higher regularity of free boundary

6.1 Further regularity of free boundaries

In this section, we will consider the obstacle problem:

$$\begin{cases} \Delta u = a(x) \quad \text{in} \quad \Omega, \\ u = \dfrac{\partial u}{\partial \nu} = 0 \quad \text{on} \quad \Gamma, \end{cases} \tag{6.1}$$

under the following assumptions:

(I) Γ is a C^1-hypersurface,
(II) $u \in C^2(\Omega \cup \Gamma)$.

Other problems one can address are the following:

$$\begin{cases} \Delta u = 0 \quad \text{in} \quad \Omega, \\ u = 0, \quad g(x, \operatorname{grad} u) = 0 \quad \text{on} \quad \Gamma, \end{cases} \tag{6.2}$$

and

$$\begin{cases} \Delta u = 0 \quad \text{in} \quad \Omega, \\ \dfrac{\partial u}{\partial \nu} = 0, \quad |\operatorname{grad} u| = f(x) \quad \text{on} \quad \Gamma, \end{cases} \tag{6.3}$$

or the parabolic problem:

$$\begin{cases} u_t - \Delta u = a(x,t) \quad \text{in} \quad \Omega, \\ u = |\operatorname{grad} u| = 0 \quad \text{on} \quad \Gamma. \end{cases} \tag{6.4}$$

Example 6.1. Consider (6.3), when $n \geq 3$, with $f \equiv 1$. Then:

$$u(x) = x_1, \quad \Gamma = \left\{x^n = \sigma\left(x_2, \cdots, x_{n-1}\right), \quad \sigma \in C^1\right\}$$

is a solution.

The main references for the contents of this section are [11] and [12].

Theorem 6.1. *For the obstacle problem* (6.1) *under the assumptions* (I), (II), *and* $a(x) > 0$, $a \in C^1\left(\overline{\Omega}\right)$, *the following hold:*

(i) *Γ is in $C^{1,\alpha}$ for every $0 < \alpha < 1$,*
(ii) *If $a \in C^{m+\alpha}$, then $\Gamma \in C^{m+1,\alpha}$,*
(iii) *If $a \in C^{\omega}$, then $\Gamma \in C^{\omega}$.*

Here we say functions or surfaces are C^{ω}, it means that they are real-analytic. A more general version of the above theorem under the same assumptions can be formulated as below:

Theorem 6.2. *Let $F\left(x, u, Du, D^2u\right) = 0$ in Ω, where $F \in C^1$ in all variables, and $u = \frac{\partial u}{\partial \nu} = 0$ on Ω. Assume* (I), (II), *as well as $F(0,0,0,0) \neq 0$. Then the following hold:*

(i) Γ *is in* $C^{1,\alpha}$ *for every* $0 < \alpha < 1$,
(ii) *if* $F \in C^{m+\alpha}$, *then* $\Gamma \in C^{m+1+\alpha}$,
(iii) *if* $F \in C^{\omega}$, *then* $\Gamma \in C^{\omega}$.

Before we begin the proof of these theorems, we need to introduce the Hodograph and Legendre transformations, as well as their geometric interpolations.

Definition 6.1 (Hodograph and Legendre Transformations). Given $u \in C^2$ with (u_{ij}) non-singular, the hodograph transformation is a local mapping

$$x \mapsto y = \operatorname{grad} u,$$

and the Legendre transformation is

$$u \mapsto v = \sum_i x_i u_{x_i} - u = xy - u.$$

Definition 6.2 (Geometric interpolation). In (x, y, u) space, consider the graph

$$u(x,y) = \{(x,y,u) \,|\, u = u(x,y)\}$$

or the envelope of its tangent planes, i.e. set up the equation which a plane must satisfy in order to be tangent to the surface. If $\overline{x}, \overline{y}, \overline{u}$ are the running coordinates of a plane whose equation is

$$\overline{u} - \xi\overline{x} - \eta\overline{y} + \omega = 0,$$

then (ξ, η, ω) are called coordinates of this plane. At point $(x, y, u(x,y))$ of the surface, the tangent plane coordinates are

$$\overline{u} - u - (\overline{x} - x)\,u_x - (\overline{y} - y)\,u_y = 0,$$
$$\xi = u_x, \quad \eta = u_y, \quad \omega = xu_x + yu_y - u.$$

We say that the surface is determined if ω is given as a function of ξ, η:

$$\omega = x\xi + y\eta - u.$$

Then we compute

$$\begin{cases} w_\xi = x + \xi\dfrac{\partial x}{\partial \xi} + \eta\dfrac{\partial y}{\partial \xi} - u_x\dfrac{\partial x}{\partial \xi} - u_y\dfrac{\partial y}{\partial \xi} = x, \\ w_\eta = y; \end{cases}$$

$$\omega(\xi,\eta) + u(x,y) = x\xi + y\eta, x = \omega_\xi, \quad y = \omega_\eta, \quad \xi = u_x, \quad \eta = u_y.$$

We have the correspondence

$$(x, y, u, u_x, u_y) \longleftrightarrow (\xi, \eta, \omega, \omega_\xi, \omega_\eta).$$

When

$$u_{xx}u_{yy} - (u_{xy})^2 = \rho > 0,$$

differentiating

$$\begin{cases} u_x = \xi, \\ u_y = \eta, \end{cases}$$

with respect to ξ, η, where

$$\begin{cases} x = x(\xi, \eta), \\ y = y(\xi, \eta), \end{cases}$$

gives the system

$$\begin{pmatrix} u_{xx} & u_{xy} \\ u_{xy} & u_{yy} \end{pmatrix} \begin{pmatrix} \omega_{\xi\xi} & \omega_{\xi\eta} \\ \omega_{\xi\eta} & \omega_{\eta\eta} \end{pmatrix} = \mathrm{I}.$$

Example 6.2. Consider

$$u(x_1, \cdots, x_n) + \omega(\xi_1, \cdots, \xi_n) = x_1\xi_1 + \cdots + x_n\xi_n,$$

where

$$u_{x_i} = \xi_i, \quad \omega_{\xi_i} = x_i, \quad i = 1, \cdots, n.$$

For $F(x, y, u, u_x, u_y, u_{xx}, u_{xy}, u_{yy}) = 0$, one deduces that

$$0 = G = F(\omega_\xi, \omega_\eta, \xi\omega_\xi + \eta\omega_\eta - \omega, \xi, \eta, \rho\omega_{\eta\eta}, -\rho\omega_{\xi\eta}, \rho\omega_{\eta\eta}),$$

we get

$$\rho = \frac{1}{\omega_{\xi\xi}\omega_{\eta\eta} - (\omega_{\xi\eta})^2}.$$

In particular, for the minimal surface equation, one has

$$(1 + \xi^2)\,\omega_{\xi\xi} + 2\xi\eta\omega_{\xi\eta} + (1 + \eta^2)\,\omega_{\eta\eta} = 0.$$

Proof of Theorem. Consider the hodograph transformation:

$$x \mapsto y = (-u_{x_1}, x_2, \cdots, x_n), \quad \text{assuming} \quad u_{x_1x_1}(0) \neq 0;$$

and the corresponding Legendre transformation gives

$$v = u - x_1u_{x_1} = x_1y_1 + u,$$

$$\mathrm{d}v = x_1\mathrm{d}y_1 - u_{x_1}\mathrm{d}x_1 + \mathrm{d}u = x_1\mathrm{d}y_1 + \sum_{i\geq 2} u_{x_i}\mathrm{d}x^i = x_1\mathrm{d}y^1 + \sum_{i\geq 2} u_{x_i}\mathrm{d}y^i.$$

Denoting $u_{x_i} = u_i$, $u_{x_ix_j} = u_{ij}$, $v_{y_i} = v_i$, $v_{y_iy_j} = v_{ij}$:

$$\begin{cases} u_1 = -y_1, \\ u_\alpha = v_\alpha, \quad \alpha \geq 2, \end{cases}$$

and

$$\begin{cases} v_{y_1} = x_1, \\ v_{y_j} = u_{x_j}, \quad j \geq 2. \end{cases}$$

Denote

$$\left(\frac{\partial Z}{\partial Y}\right) = \begin{pmatrix} v_{11} & v_{1\alpha} \\ 0 & I \end{pmatrix}$$

and

$$\left(\frac{\partial Y}{\partial X}\right) = \begin{pmatrix} \frac{1}{v_{11}} & -\frac{v_{1\alpha}}{v_{11}} \\ 0 & I \end{pmatrix}.$$

As $x_1 = v_{y_1}$, $u_{11} = -\frac{1}{v_{11}}$, $u_{1\alpha} = \frac{v_{1\alpha}}{v_{11}}$, $\alpha \geq 2$ and

$$\frac{\partial}{\partial x_\alpha} = \frac{\partial}{\partial y_\alpha} - \frac{v}{v_{11}}\frac{\partial}{\partial y_1}, \quad \frac{\partial}{\partial x_1} = \frac{1}{v_{11}}\frac{\partial}{\partial y_{11}},$$

the equation $\Delta u = a(x)$ is transformed to

$$-\frac{1}{v_{11}} - \frac{1}{v_{11}}\sum_{\alpha\geq 2}(v_{1\alpha})^2 + \sum_{\alpha\geq 2} v_{\alpha\alpha} = \alpha\left(y_2, \cdots, y_n, v_{y_1}\right).$$

Hence, the problem $\Delta u = a(x)$ in Ω with $a \geq 0$ is transformed to

$$\begin{cases} G\left(v_{ij}\right) = a\left(v_1, y_2, \cdots, y_n\right), \\ v|_{\{y_1=0\}} = 0. \end{cases}$$

While the free boundary is flattened out as $\Gamma \mapsto \{y_1 = 0\}$, the domain is transformed also accordingly as $\Omega \mapsto \{y_1 > 0\}$.

Likewise, in the general case, the system

$$\begin{cases} F\left(x, u, Du, D^2u\right) = 0 & \text{in} \quad \Omega, \\ u = \dfrac{\partial u}{\partial \nu} = 0 & \text{on} \quad \Gamma \end{cases}$$

would be transformed to

$$\begin{cases} \tilde{F}\left(y, v, Dv, D^2v\right) = 0 & \text{in} \quad \{y_1 \geq 0\}, \\ v|_{\{y_1=0\}} = 0. \end{cases}$$

At this point we would like to argue that F being elliptic at u implies $\tilde{F}$ being elliptic at v.

Definition 6.3. F is said to be elliptic at $u \in C^2$, if

$$\left(F_{ij}\right) = \left(\frac{\partial F}{\partial u_{ij}}\left(x, u, Du, D^2u\right)\right)$$

is positive definite.

One can check in the special case that

$$-\frac{1}{v_{11}} - \frac{1}{v_{11}}\sum_{\alpha\geq 2}(v_{1\alpha})^2 + \sum_{\alpha\geq 2} v_{\alpha\alpha} = a\left(v_{y_1}, y_2, \cdots, y_n\right)$$

is elliptic. Likewise, a direct calculation yields

$$\tilde{F}_{ij} = \begin{pmatrix} \frac{1+\sum_{\alpha>2}(v_{1\alpha})^2}{(v_{11})^2} & \frac{v_{1\alpha}}{v_{11}} \\ \frac{v_{\alpha 1}}{v_{11}} & \mathrm{I} \end{pmatrix}.$$

In other words,

$$\tilde{F}_{ij}\xi_i\xi_j = \sum_{\alpha\geq 2} \xi_\alpha^2 + 2\sum_{\alpha\geq 2} \frac{v_{1\alpha}}{v_{11}}\xi_\alpha\xi_1 + \frac{1+\sum_{\alpha\geq 2}(v_{1\alpha})^2}{(v_{11})^2}\xi_1^2.$$

Noting that $v \in C^2$, and $v_{y_1} = x_1$, $v_{y_j} = u_{x_j}$, $j \geq 2$ gives $v = -x'u_{x_1} + u$, we get

$$\tilde{F}_{ij}\xi_i\xi_j = \sum_{\alpha\geq 2} \xi_\alpha^2 - (1 = \delta)\sum_{\alpha\geq 2} \xi_\alpha^2 - \frac{1}{1-\delta}\frac{(v_{1\alpha})^2}{v_1 1^2}\xi_1^2 + \frac{1}{v_{11}^2}\xi_1^2 \geq \delta|\xi|^2$$

for $\delta > 0$ suitably small. In general, choosing $\eta_\alpha = \xi_\alpha - \xi_1 v_{1\alpha}$ we obtain

$$\begin{aligned} \tilde{F}_{ij}\xi_i\xi_j &= (F_{11} - 2F_{1\alpha}v_{1\alpha} + F_{\alpha\beta}v_{1\alpha}v_{1\beta})\,\xi_1^2 + 2\,(F_{1\alpha} - F_{\alpha\beta}v_{1\beta})\,\xi_1\xi_\alpha + F_{\alpha\beta}\xi_\alpha\xi_\beta \\ &= F_{ij}\eta_i\eta_j. \end{aligned}$$

Thus, $\tilde{F}$ is elliptic whenever F is elliptic. Hence, through the Legendre transformation we reduced the original problem to

$$\begin{cases} F\,(v_{ij}) = a\,(v_1, y_2, \cdots, y_n)\,, \\ v|_{\{y_1=0\}} = 0 \end{cases}$$

where $v \in C^2$, $a \in C^1$. Hence by standard elliptic regularity we get: $v \in C^{2,\alpha}\left(\overline{D}_+\right)$ for every $0 < \alpha < 1$. Now Γ is the image of $x = X(y)|_{\{y_1=0\}} \in C^{1,\alpha}$, where

$$\left(\frac{\partial X}{\partial Y}\right) = \begin{pmatrix} v_{11} & v_{1\alpha} \\ 0 & \mathrm{I} \end{pmatrix}.$$

Likewise, $a \in C^{m+\alpha}$ implies $v \in C^{m+2+\alpha}$, which in turn gives $\Gamma \in C^{m+1+\alpha}$. The rest of the proof is now standard. □

6.2 Liquid edges

In this section we study the higher regularity of a multi-phase problem via the Hodograph transformation introduced in the preceding section. Consider the system

$$\begin{cases} Mu^1 = Mu^2 = 0 \quad \text{in} \quad \Omega^+, \\ Mu^3 = 0 \quad \text{in} \quad \Omega^-, \quad \text{where} \quad M(u) = \left(\delta_{ij} - \dfrac{u_i u_j}{1+|\nabla u|^2}\right) u_{ij}, \\ u^1 = u^2 = u^3 \quad \text{on} \quad \Gamma, \\ \dfrac{\nabla u^j \nabla u^3 + 1}{\sqrt{1+|\nabla u^j|^2}\sqrt{1+|\nabla u^3|^2}} = \dfrac{1}{2} \quad \text{on} \quad \Gamma, \quad j = 1, 2, \\ \nabla u^1 \neq \nabla u^2 \quad \text{on} \quad \Gamma. \end{cases} \tag{6.5}$$

More generally $\frac{1}{2}$ is replaced by $\cos \mu_j < 1$.

Theorem 6.3. *Suppose that* Γ *is of class* $C^{1,\alpha}$, $u^j \in C^{1,\alpha}(\Omega^+ \cup \Gamma)$, $j = 1, 2$, *and* $u^3 \in C^{1,\alpha}(\Omega^- \cup \Gamma)$, *for some* α, $0 < \alpha \le 1$ *satisfy* (6.5). *If* u^1 *and* u^2 *do not meet at zero angle at* $x = 0$, *i.e.* $\mu_1(0) \ne \mu_2(0)$, *then* Γ *is analytic near* $x = 0$.

Remark 6.1. J. Taylor, J. C. C. Nitsche and others had made some preliminary contributions on the above problem, see [12], [11] and references therein.

We recall the Hodograph transformation. Consider $w \in C^1$, $w_{x_n}(0) > 0$, $y(x) = (x', w(x))$, $x \in B_\epsilon(0)$. Suppose $w|_\Gamma = 0$, $0 \in \Gamma$, $U = y(B_\epsilon \cap \Omega) \subset \{y_n > 0\}$, and $y(\Gamma) \subset \{y_n = 0\}$. Define $\psi(y) = x_n$, $g(y) = (y', \psi(y))$, $\Gamma = \{x_n = \psi(x', 0)\}$. Then after a careful computation, one leads to

$$dx_n = d\psi = \sum_{\alpha \le n-1} [\psi_\alpha dy_\alpha + \psi_n dy_n] = \sum_{\alpha < n} [\psi_\alpha dy_\alpha + \psi_n dw],$$

and

$$\begin{cases} dw = -\sum_{\alpha=1}^{n-1} \dfrac{\psi_\alpha}{\psi_n} dx_\alpha + \dfrac{1}{\psi_n} dx_n, \\ \Delta w = -\sum_{\alpha<n} \left(-\dfrac{\psi_\alpha}{\psi_n}\right)_\alpha + \dfrac{1}{\psi_n} dx_n, \\ w_\alpha = -\dfrac{\psi_\alpha}{\psi_n}, \quad w_n = \dfrac{1}{\psi_n}. \end{cases}$$

Likewise

$$\begin{cases} w_{nn} = \left(\dfrac{1}{\psi_n}\right)_n \dfrac{\partial y_n}{\partial x_n} = -\dfrac{\psi_{nn}}{\psi_n^3}, \\ w_{\alpha\alpha} = \left(-\dfrac{\psi_\alpha}{\psi_n}\right)_\alpha + \left(\dfrac{\psi_\alpha}{\psi_n}\right)_n. \end{cases}$$

Example 6.3. $w \in C^2(\Omega) \cap C^1(\overline{\Omega})$:

$$\begin{cases} \Delta w = a(x, w), \\ w_\nu = f(x) \quad \text{on} \quad \Gamma, \\ w = 0 \quad \text{on} \quad \Gamma, \end{cases}$$

may be transformed into the following:

$$\begin{cases} F\left(D^2\psi, D\psi\right) = a\left(y', \psi, y_n\right), \\ \psi_n - \dfrac{1}{f\left(y', \psi\right)} \sqrt{1 + \sum_{\alpha<n} \psi_\alpha^2} \quad \text{on} \quad \{y_n = 0\} = \Sigma. \end{cases}$$

We leave details to the readers.

Before proceeding with the application of the Hodograph transformation to (6.5), we fix the notation

$$\begin{cases} y = (x', w(x)) \quad x \in \Omega^+, \\ \psi(y) = x_n \end{cases}$$

and

$$\begin{cases} g^+(y) = (y', \psi(y)), & \text{for } \ y \in U \cup \Sigma, \\ g^-(y) = (y', \psi(y) - cy_n), & \text{where } \ C > |\sup \psi_n|, \\ g^-(U) \subset \Omega^-, \quad g^+ = g^- & \text{on } \ y_n = 0. \end{cases}$$

Suppose a, f are analytic near $(x, w) = (0,0)$, $f(0) < 0$, $\nu(0) = -e_n = (0, \cdots, 0, -1)$, $w_{x_n}(0) = a > 0$, $w_\alpha = 0$ at $y = 0$, for $\alpha < n$, $a = -f(0)$. Define

$$L\tilde{\psi} = \frac{\mathrm{d}}{\mathrm{d}\epsilon} F\left(D^2\psi(0) + \epsilon D^2\overline{\psi}, D\psi(0) + \epsilon D\overline{\psi}(0)\right)|_{\epsilon=0}.$$

Computation yields

$$L\tilde{\psi} = -a^3\tilde{\psi}_{nn} - a\sum_{\sigma<n}\tilde{\psi}_{\sigma\sigma} + \text{lower order terms.}$$

This is the linearization of the equation after the transformation. Also define

$$B\overline{\psi} = \frac{\mathrm{d}}{\mathrm{d}\epsilon}\left(\psi_n(0) + \epsilon\overline{\psi}_n(0) - \frac{\sqrt{1 + \sum_\alpha \left(\psi_\alpha(0)_\epsilon \overline{\psi}_\alpha(0)\right)^2}}{f\left(0, \psi(0) + \epsilon\overline{\psi}\right)}\right)\Bigg|_{\epsilon=0}.$$

In other words,

$$B\overline{\psi} = \overline{\psi}_n(0) + g\left(\overline{\psi}(0), \tilde{\psi}_\alpha(0)\right).$$

This is the linearization of the corresponding nonlinear boundary condition.

Now we are ready to apply the above strategy and the Hodograph transformation to (6.5). Let $a_j - \frac{\partial}{\partial x_n}u^j(0)$ for $j = 1, 2$. Assume $a_2 > a_1$ (and hence $u^2 > u^1$) in Ω^+ near $x = 0$. Let $\mu_j = \mu_j(0,0)$. Assume also $\mu_2 \neq \mu_1$ and $\mu_j \neq 0$. $\cos\mu_j = \left(1 + a_j^2\right)^{-1/2}$. Define $w(x) = u^2(x) - u^1(x)$ for $x \in \Omega^+$. Furthermore,

$$\begin{cases} \phi^+(y) = u^1(x) = u^1\left(g^+(y)\right), \\ \phi^-(y) = u^3(x) = u^3\left(g^-(y)\right), \\ \psi(y) = x_n \quad \text{as before.} \end{cases}$$

$$\begin{cases} \nabla w = \left(-\dfrac{\psi_\sigma}{\psi_n}, \dfrac{1}{\psi_n}\right), & x \in \Omega^+, \\ \nabla u^1 = \left(\phi_\alpha^+ - \dfrac{\psi_\alpha}{\psi_n}\phi_n^+, \dfrac{\phi_n^+}{\psi_n}\right), & x \in \Omega^+, \\ \nabla u^2 = \left(\phi_\alpha^+ - \dfrac{\psi_\alpha}{\psi_n}\left(\phi_n^+ + 1\right), \dfrac{1}{\psi_n}\left(\phi_n^+ + 1\right)\right), & x \in \Omega^+, \\ \nabla u^3 = \left(\phi_\alpha^- - \dfrac{\psi_\alpha}{\psi_n - C}\phi_n^-, \dfrac{1}{\psi_n - C}\phi_n^-\right), & x \in \Omega^-. \end{cases}$$

$$\begin{cases} \partial_{x_\alpha} = \partial_{y_\alpha} - \dfrac{\psi_\alpha}{\psi_n}\partial_{y_n}, \quad \partial_{x_n} = \dfrac{1}{\psi_n}\partial_{y_n} & \text{in } \ \Omega^+, \\ \partial_{x_\alpha} = \partial_{y_\alpha} - \dfrac{\psi_\alpha}{\psi_n - C}\partial_{y_n}, \quad \partial_{x_n} = \dfrac{1}{\psi_n - C}\partial_{y_n} & \text{in } \ \Omega^-. \end{cases}$$

The system (6.5) can be expressed as

$$\begin{cases} 0 = Mu^1 = F_1\left(D^2\psi, D\psi, D^2\phi^+, D\phi^+\right), \\ 0 = Mu^2 = F_2\left(D^2\psi, D\psi, D^2\phi^+, D\phi^+\right), \\ 0 = Mu^3 = F_3\left(D^2\psi, D\psi, D^2\phi^-, D\phi^-\right), \end{cases}$$

with boundary conditions

$$\begin{cases} 0 = \Phi_1\left(D\psi, D\phi^+, D\phi^-\right) = \nabla u^1 \nabla u^3 + 1 - \cos\mu_1 \sqrt{1+|\nabla u^1|^2}\sqrt{1+|\nabla u^3|^2}, \\ 0 = \Phi_2\left(D\psi, D\phi^+, D\phi^-\right) = \nabla u^2 \nabla u^3 + 1 - \cos\mu_2 \sqrt{1+|\nabla u^2|^2}\sqrt{1+|\nabla u^3|^2}, \\ 0 = \Phi_3\left(\phi^+, \phi^-\right) = \phi^+ - \phi^- = u^3 - u^1 \quad \text{on } \Sigma. \end{cases}$$

The linearization of the boundary conditions gives

$$\begin{cases} \cos^2\mu_1\left(\overline{\phi_n^+} - a_1\overline{\psi_n}\right) + \dfrac{1}{C\left(a_2-a_1\right)-1}\overline{\phi_n^-} = 0, \\ \cos^2\mu_2\left(\overline{\phi_n^+} - a_2\overline{\psi_n}\right) + \dfrac{1}{C\left(a_2-a_1\right)-1}\overline{\phi_n^-} = 0, \\ \dfrac{a_2}{a_2-a_1}\left(\overline{\phi^+} - a_1\overline{\psi}\right) - \dfrac{a_1}{a_2-a_1}\left(\overline{\phi^+} - a_2\overline{\psi}\right) - \overline{\phi^-} = 0. \end{cases}$$

Similarly, the bulk equations can be also linearized to give the following set of linear equations:

$$\begin{cases} \sum\limits_{\alpha<n}\left(\overline{\phi^+} - a_1\overline{\psi}\right)_{\alpha\alpha} + \left(a_2-a_1\right)^2\cos^2\mu_1\left(\overline{\phi^+} - a_1\overline{\psi}\right)_{nn} = 0, \\ \sum\limits_{\alpha<n}\left(\overline{\phi^+} - a_2\overline{\psi}\right)_{\alpha\alpha} + \left(a_2-a_1\right)^2\cos^2\mu_2\left(\overline{\phi^+} - a_2\overline{\psi}\right)_{nn} = 0, \\ \sum\limits_{\alpha<n}\overline{\phi_{\alpha\alpha}^-} + \left(\dfrac{a_2-a_1}{1-C\left(a_2-a_1\right)}\right)^2\overline{\phi_{nn}^-} = 0. \end{cases}$$

The regularity of solutions of these linear equations with corresponding linear boundary conditions and the corresponding regularity for the nonlinear equations with nonlinear complementary boundary conditions are contained in classical theory of Agmon-Douglis-Norenberg, see [6].

A Elliptic regularity theory — a brief summary

We shall state several classical results from elliptic regularity theory without proof, as we will need them in the notes. We refer to the book by Han and Lin [10], as well as the more comprehensive book by Gilbarg and Trudinger [6] for a more detailed discussion.

Notation A.1. We will need the following notations:

(i) $H_\alpha(u) = \sup_{x,y\in\Omega} \frac{|u(x)-u(y)|}{|x-y|^\alpha}$,
(ii) $\|u\|_0 = \sup_{x\in\Omega} |u(x)|$,
(iii) $\|u\|_\alpha = \|u\|_0 + H_\alpha(u)$,
(iv) $\|u\|_m = \sum_{|\beta|\le m} \|D^\beta u\|_0$,
(v) $\|u\|_{m+\alpha} = \|u\|_m + \sum_{|\beta|=m} H_\alpha(D^\beta u)$,
(vi) $Au = -a_{ij}(x)u_{ij} + b_i(x)u_i + c(x)u$,
(vii) $a_{ij}(x)\xi_i\xi_j \ge \lambda|\xi|^2, \quad \forall x \in \Omega,\ \xi \in \mathbb{R}^n$ (Ellipticity).

Theorem A.1 (Schauder estimate). *Assume that* $\|a_{ij}\|_\alpha + \|b_i\|_\alpha + \|c\|_\alpha \le K$, *and* $\partial\Omega \in C^{2,\alpha}$. *If* $u \in C^2(\bar\Omega)$ *is a solution to*

$$\begin{cases} Au = f & in\ \Omega, \\ u|_{\partial\Omega} = \psi, \end{cases} \tag{A.1}$$

then

$$\|u\|_{2+\alpha} \le C\left[\|f\|_\alpha + \|u\|_0 + \|\psi\|_{2+\alpha}\right], \tag{A.2}$$

where $C = C(\lambda, K, \Omega)$.

Definition A.1. We define

$$\bar H_{j,\alpha}(u) = \sup_{x,y\in\Omega} d_{x,y}^{j+\alpha} \frac{|u(x)-u(y)|}{|x-y|^\alpha},$$

and $\bar H_\alpha = \bar H_{0,\alpha}$, where $d_{x,y} = \min\{d_x, d_y\}$, and $d_x = d(x, \partial\Omega)$. Likewise, we define

$$\overline{\|u\|}_\alpha = \|u\|_0 + \bar H_\alpha(u);$$
$$\overline{\|u\|}_{m+\alpha} = \sum_{\beta\le m} \sup_{x\in\Omega} d_x^{|\beta|} \left|D^\beta u(x)\right| + \sum_{|\beta|=m} \bar H_{m,\alpha}\left(D^\beta u\right).$$

Theorem A.2 (Schauder's interior estimate). *Let* Ω *be a bounded domain with* $\mathrm{diam}(\Omega) = D$, *and suppose that* $\overline{\|a_{ij}\|}_\alpha + \overline{\|b_i\|}_\alpha + \overline{\|c\|}_\alpha \le K$. *If* $u \in L^\infty(\Omega)\cap C^2(\Omega)$ *is a solution to* $Au = f$ *in* Ω, *then for a* $C = C(\lambda, D, K)$ *we have*

$$\overline{\|u\|}_{2+\alpha} \le C\left[\overline{\|f\|}_\alpha + \|u\|_0\right]. \tag{A.3}$$

Theorem A.3 ($W^{2,p}$-estimate). *Let* $\partial\Omega$ *belong to* C^2, $f \in L^p(\Omega)$, *for* $1 < p < \infty$, $\psi \in W^{2,p}(\Omega)$. *Assume*

(i) $|a_{ij}(x) - a_{ij}(y)| \le \omega(|x-y|)$, *where* $\omega(t) \to 0$ *as* $t \to 0$,
(ii) $\sum_{i,j} |a_{ij}| + \sum_i |b_i| + |c| \le K$,
(iii) $\sum_{i,j} a_{ij}\xi_i\xi_j \ge \lambda|\xi|^2$.

If $u \in W^{2,p}$ *is a solution to* (A.1), *then for a* $C = C(\lambda, K, \omega, \Omega)$ *we have*

$$\|u\|_{2,p} \le C\left[\|f\|_p + \|\psi\|_{2,p}\right]. \tag{A.4}$$

Theorem A.4 (Interior $W^{2,p}$-estimate: Agmon-Douglis-Nirenberg, M. Giaquinta). *For all as in Theorem A.3, and $G \subset\subset \Omega$, we have*

$$\|u\|_{2,p}^{G} \leq C(\lambda, K, \omega, \Omega, G)\left(\|f\|_p + \|u\|_p\right). \tag{A.5}$$

And near boundary estimate:

$$\|u\|_{2,p}^{G_1} \leq C\left[\|f\|_p^{\omega} + \|u\|_p^{\omega} + \|\psi\|_{2,p}^{\omega}\right]. \tag{A.6}$$

Here G_1 may be viewed as an intersection of ω with an unit size ball centered at a boundary point of the domain Ω.

Theorem A.5 ($C^{1,\alpha}$-Hölder estimate: C. B. Morrey, S. Campanato). *Consider the operator $A = -\frac{\partial}{\partial_{x_i}}\left[a_{ij}(x)\frac{\partial}{\partial_{x_j}}\right]$, where $a_{ij}(x)\xi_i\xi_j \geq \lambda|\xi|^2$ and $a_{ij} \in C^{\alpha}(\bar{\Omega})$.*

(i) *If u solves (A.1) with $\psi \in C^{1,\alpha}$, $\partial\Omega \in C^{1,\alpha}$, and $f \in L^p(\Omega)$ with $p > n$, then for $\beta = 1 - n/p$, we have*

$$\|u\|_{C^{1+\beta}} \leq C\left[\|f\|_p + \|\psi\|_{1+\beta}\right]. \tag{A.7}$$

(ii) *If u solves (A.1) with $\psi \in C^{1,\alpha}$, $\partial\Omega \in C^{1,\alpha}$, and $f = 0$, then*

$$\|u\|_{C^{1+\alpha}} \leq C\|\psi\|_{1+\alpha}. \tag{A.8}$$

(iii) *If u solves (A.1) with $f \in L^p(\Omega)$, $p > n$, but $a_{ij} \in C^{\alpha}(\Omega)$ only, i.e. not necessarily up to the boundary, then for $\beta = 1 - n/p$, we have the interior estimate*

$$\overline{\|u\|}_{1+\beta} \leq C\left[\|u\|_2 + \|f\|_p\right]. \tag{A.9}$$

(iv) *If u solves (A.1) with $f = 0$, but $a_{ij} \in C^{\alpha}(\Omega)$ only, i.e. not necessarily up to the boundary, then we have the interior estimate*

$$\overline{\|u\|}_{1+\alpha} \leq C\|u\|_2. \tag{A.10}$$

Theorem A.6 (Morrey-Nirenberg-Cordes estimate). *If $\|a_{ij} - \delta_{ij}\|_0 \leq \delta(n,p)$ for δ small, and u is a solution to the problem*

$$\begin{cases} a_{ij}(x)u_{ij}(x) = f(x), & x \in \Omega, \\ u = \psi & \text{on } \partial\Omega, \end{cases}$$

then the following estimate holds:

$$\|u\|_{2,p} \leq C\left[\|f\|_p + \|\psi\|_{2,p}\right]. \tag{A.11}$$

Here the equation is uniformly elliptic, and the constant C will depend on p, n and the ellipticity.

Theorem A.7 (Krylov-Safanov). *For $f \in L^p(\Omega)$, $p > n$, and u solving*

$$\begin{cases} a_{ij}(x)u_{ij}(x) + b_i(x)u_i(x) + c(x)u(x) = f, & x \in \Omega, \\ u = \psi & \text{on } \partial\Omega, \end{cases}$$

we have the estimate

$$\|u\|_{\alpha} \leq C\left(\|f\|_p + \|\psi\|_{\alpha}\right). \tag{A.12}$$

Here again the equation is uniformly elliptic, and the constant α will in general be dependent of p and the dimension n.

Theorem A.8 (Harnack's inequality). *For $u \geq 0$ and $a_{ij}u_{ij} = 0$ in B_2, there holds*

$$\sup_{B_1} u \leq C(n, \lambda) \inf_{B_1} u. \tag{A.13}$$

Theorem A.9 (Krylov-Caffarelli estimate). *For u solving $a_{ij}u_{ij} = 0$ in B_2^+ with $u|_{x_n=0} = 0$, there holds the boundary Harnack estimate $\frac{u}{x_n} \in C^{\alpha}(B_1^+)$.*

Theorem A.10 (De Giorgi-Nash-Moser, Stampacchia). *For u solving the following boundary value problem in divergence form*

$$\begin{cases} (a_{ij}u_j)_i = \operatorname{div} F & \text{in } \Omega, \\ u|_{\partial\Omega} = \psi, \end{cases}$$

we have the estimate

$$\|u\|_{\alpha} \leq C\left(\|F\|_p + \|\psi\|_{\alpha}\right), \tag{A.14}$$

where p is larger than the dimension n, and α is not larger than $1 - p/n$ and a positive constant depending on the dimension and ellipticity constant. It should be noted that the uniform ellipticity are assumed in all above equations.

References

[1] F. J. Almgren. Q valued functions minimizing Dirichlet's integral and the regularity of area minimizing rectifiable currents up to codimension two. Bull. Amer. Math. Soc. (N.S.) **8**(2) (1983), 327–328.

[2] H. Brezis. Functional Analysis, Sobolev Spaces and Partial Differential Equations. Springer Science & Business Media, 2010.

[3] L. A. Caffarelli and S. Salsa. A Geometric Approach to Free Boundary Problems. American Mathematical Soc., 2005.

[4] H. Federer. Geometric Measure Theory. Die Grundlehren der mathematischen Wissenschaften, Band 153, Springer-Verlag New York Inc., New York, 1969.

[5] A. Friedman. Variational Principles and Free-boundary Problems. Courier Dover Publications, 2010.

[6] D. Gilbarg and Neil S. Trudinger. Elliptic Partial Differential Equations of Second Order. Grundlehren der Mathematischen Wissenschaften [Fundamental Principles of Mathematical Sciences], Springer-Verlag, Berlin, 1983.

[7] E. Giusti. Superfici cartesiane di area minima. Rend. Sem. Mat. Fis. Milano **40** (1970), 135–153.

[8] E. Giusti. Minimal Surfaces and Functions of Bounded Variation. Notes on Pure Mathematics **10**. Department of Pure Mathematics, Australian National University, Canberra, Australian National University, 1977.

[9] E. Giusti. On the equation of surfaces of prescribed mean curvature. Existence and uniqueness without boundary conditions. Invent. Math. **46** (2) (1978), 111–137.

[10] Q. Han and F. Lin. Elliptic Partial Differential Equations. Courant Lecture Notes in Mathematics, New York University, Courant Institute of Mathematical Sciences, New York; American Mathematical Society, Providence, RI, 1997.

[11] D. Kinderlehrer and L. Nirenberg. Regularity in free boundary problems. Ann. Scuola Norm. Sup. Pisa Cl. Sci.(4) **4** (2) (1977), 373–391.

[12] D. Kinderlehrer, L. Nirenberg, and J. Spruck. Regularity in elliptic free boundary problems. J. Analyse Math. **34** (1978), 86–119.

[13] D. Kinderlehrer and G. Stampacchia. An Introduction to Variational Inequalities and Their Applications (Classics in Applied Mathematics **31**). Society for Industrial and Applied Mathematics, 2000.

[14] F. Lin. On regularity and singularity of free boundaries in obstacle problems. Chin. Ann. Math. Ser. B **30** (2009), 645–652.

[15] F. Lin. Behaviour of nonparametric solutions and free boundary regularity. Miniconference on Geometry and Partial Differential Equations **2** (Canberra, 1986), Proc. Centre Math. Anal. Austral. Nat. Univ. **12**, 96–116, Austral. Nat. Univ., Canberra, 1987.

[16] F. Lin and X. Yang. Geometric Measure Theory—An Introduction. Advanced Mathematics (Beijing/Boston) **1**, Science Press Beijing, Beijing; International Press, Boston, MA, 2002.

[17] L. Simon. Boundary regularity for solutions of the non-parametric least area problem. Ann. of Math. Second Series **103** (3) (1976), 429–455.

[18] L. Simon. Lectures on Geometric Measure Theory. Proceedings of the Centre for Mathematical Analysis, Australian National University **3**, Australian National University, Centre for Mathematical Analysis, Canberra, 1983.

[19] L. Simon. Theorems on Regularity and Singularity of Energy Minimizing Maps. Lectures in Mathematics ETH Zürich, Based on lecture notes by Norbert Hungerbühler, Birkhäuser Verlag, Basel, 1996.

[20] G. S. Weiss. Partial regularity for weak solutions of an elliptic free boundary problem. Comm. Partial Differential Equations **23** (3–4) (1998), 439–455.

Lectures on the Analysis of Nonlinear Partial Differential Equations Vol. 4
MLM 4, pp. 195–222

The De Giorgi Method for Elliptic and Parabolic Equations and Some Applications

Alexis F. Vasseur*

Abstract

In this lecture notes, we mainly demonstrate the method introduced by De Giorgi to study the regularity of solutions to elliptic equations with rough coefficients.

1 Introduction

This lecture notes are dedicated to the method introduced by De Giorgi to study the regularity of solutions to elliptic equations with rough coefficients. He introduced his technique in 1957 [5] to solve the 19th Hilbert problem. In this work, he showed the regularity of variational solutions to nonlinear elliptic problems. Independently, Nash introduced a similar techniques in 1958 [15]. Subsequently, Moser provided a new formulation of the proof in [14]. Those methods are now usually called De Giorgi-Nash-Moser techniques. The method has been extended to degenerated cases, like the p-Laplacian, first in the elliptic case by Ladyzhenskaya and Uralt'seva [13]. The degenerated parabolic cases were covered later by DiBenedetto [6] (see also DiBenedetto, Gianazza and Vespri [9, 7, 8]). More recently, the method has been extended to integral operators, like the fractional diffusion, in [3, 2] and Kassmann [12] and Felsinger and Kassmann [10]. Further application to fluid mechanics can be found in [1, 16, 11, 4].

In this lecture notes, we will focus on the original formulation of De Giorgi. We will stress on its beautiful geometric interpretation, and how it can be applied in different contexts. We do not try to show the more technical result available, but rather, try to get a simple description of the strength of the method, and the breadth of the possible applications.

*Department of Mathematics, University of Texas at Austin, 1 University Station – C1200, Austin, TX 78712-0257. E-mail: vasseur@math.utexas.edu

The rest of the notes are as follows. in a first section, we introduce the original proof of De Giorgi for elliptic equations. In the second section we investigate the parabolic case. We include a non homogenous case, to illustrate how source terms can be handled. Finally we give an application in the context of compressible fluid mechanics.

2 The original result of De Giorgi

2.1 The result

The 19th Hilbert problem consists in showing the smoothness of local minimizers of convex energy functionals of the form

$$\mathcal{E}(w) = \int_\Omega F(\nabla w)\, dx, \tag{2.1}$$

where F is a smooth function from $\mathbb{R}^N$ to $\mathbb{R}$, and Ω is bounded open set of $\mathbb{R}^N$. This is a generalization of the Dirichlet integral

$$\int_\Omega |\nabla w|^2\, dx.$$

Local minimizers of the Dirichlet integral verify the associate Euler-Lagrange equation which is nothing but the Laplace equation:

$$\Delta w = 0.$$

Solutions of the Laplace equation are known to be C^∞, as it can be easily shown, for instance, by using the mean value formula. We denote $L^2(\Omega)$ the space of square integrable functions on Ω.

Lemma 1. *Let $w \in L^2(\Omega)$ be a solution, in the sense of distributions, to the Laplace equation. For any radially symmetric compactly supported function η with integral 1, and any $x \in \Omega$, we have*

$$w(x) = \int_\Omega \eta(x-y)w(y)\, dy,$$

as long as the support of $\eta(\cdot - x)$ is included in Ω. Especially, for any $n \in \mathbb{N}$ there exists $C_n > 0$ such that for any $x \in \Omega$,

$$|D^n w(x)| \leq C_n \|w\|_{L^2(\Omega)} d(x, \partial\Omega)^{-(n+N/2)},$$

where $d(x, \partial\Omega)$ denotes the distance from x to the boundary Ω.

Proof. This is a very elementary proof. First, for any ball B included in Ω, from the Stokes formula, any solution to the Laplace equation verifies

$$0 = \int_B \Delta w = \int_{\partial B} \partial_r w.$$

Let B_R be the support of η, and define for $0 < t < 1$

$$f(t) = \int_{B_R} \eta(y) w(x + ty)\, dy.$$

Note that, using the polar coordinates

$$f'(t) = \int_{B_R} \eta(y) y \cdot \nabla w(x + ty)\, dy = \int_0^R \left(\eta(r) r \int_{\partial B_r} \partial_r w \right) dr = 0.$$

Moreover, $f(0) = w(x)$, since the integral of η is equal to 1. So, for $t = 1$,

$$w(x) = \int_\Omega \eta(y) w(x + y)\, dy = \int_\Omega \eta(y - x) w(y)\, dy.$$

Consider now a fix function η compactly supported in B_1, and $d = d(x, \partial\Omega)$. The function $\eta_d = d^{-N}\eta(x/d)$ is such that the support of $\eta_d(\cdot - x)$in included in Ω. So

$$\begin{aligned} |D^n w(x)| &= d^{-(N/2+n)} \left| \int d^{-N/2} D^n \eta((y - x)/d) w(y)\, dy \right| \\ &\leq d^{-(N/2+n)} \|D^n \eta\|_{L^2} \|w\|_{L^2}. \end{aligned}$$

□

The Dirichlet case is extremely simple. Surprisingly, the general case is far more difficult, and its resolution involves deep and beautiful mathematics.

First, let us precise the problem. By a local minimizer, it is meant that

$$\mathcal{E}(w) \leq \mathcal{E}(w + \varphi)$$

for any smooth φ compactly supported in Ω. As for the Dirichlet case, such a minimizer w satisfies an Euler-Lagrange equation. We denote $H^1(\Omega)$ the set of functions which are square integrable and such that their derivatives in the sense of distribution are themselves square integrable.

Lemma 2. *Consider F such that $D^2F(p) \leq \Lambda I$, for any $p \in \mathbb{R}^N$, for a fixed $\Lambda > 0$, where I is the identity $N \times N$ matrix. Then, any $w \in H^1(\Omega)$, local minimizer of (2.1), is solution in the sense of distribution to*

$$\operatorname{div}(DF(\nabla w)) = 0, \qquad x \in \Omega, \tag{2.2}$$

where DF is the gradient of the functional $p \to F(p)$.

Proof. This is, again, a very elementary result. For any $\varepsilon > 0$ and smooth function ϕ compactly supported, we have

$$\int_\Omega F(\nabla w + \varepsilon \nabla \phi)\, dx \geq \int_\Omega F(\nabla w)\, dx.$$

Those integrals are finite since F is at most quadratic, and $w \in H^1$. Using a Taylor expansion for F, we find

$$\int_\Omega DF(\nabla w) \cdot \nabla \phi\, dx \geq -\varepsilon \Lambda \int_\Omega |\nabla \phi|^2\, dx.$$

This is true for any $\varepsilon > 0$, so at the limit ε goes to 0

$$\int_\Omega DF(\nabla w) \cdot \nabla \phi \, dx \geq 0.$$

The result is true also for ϕ replaced by $-\phi$. So, finally

$$\int_\Omega DF(\nabla w) \cdot \nabla \phi \, dx = 0.$$

This shows that w is solution to the Euler Lagrange equation is the sense of distribution. □

From now on, we consider only function F which verify

$$\frac{1}{\Lambda} I \leq D^2 F(p) \leq \Lambda I, \qquad \text{for any } p \in \mathbb{R}^N, \tag{2.3}$$

for a fixed $\Lambda > 1$. De Giorgi showed the following answer to the 19th problem of Hilbert:

Theorem 3. *Any local minimizer $w \in H^1(\Omega)$ of (2.1) lies in $C^\infty(\Omega)$.*

Note that the strict convexity is necessary for the result to be true. In one dimension, consider a functional F such that both 1 and -1 are minimum of F. Then $x \to |x|$ is a local minimizer of (2.1), but it is merely Lipschitz.

As long as we have enough regularity, (2.2) can be rewritten in the non divergence form as

$$D^2 F(\nabla w) : D^2 w = 0.$$

For any open set O, we denote $C^\alpha(O)$ the space of function v such that

$$\|v\|_{C^\alpha(O)} = \|v\|_{L^\infty(O)} + \sup_{x,y \in O} \frac{|v(x) - v(y)|}{|x-y|^\alpha}$$

is finite. The strict convexity property of F translates into a strict ellipticity property for the equation on w. Note that if ∇w is C^α, then $D^2 F(\nabla w)$ is also C^α. So (forgetting the dependence on w in $D^2 F(\nabla w)$) the equation can be rewritten as

$$A(x) : D^2 w = 0,$$

for a C^α elliptic matrix A. Then the standard Schauder theory (which was known at the time) ensures that ∇w is $C^{1,\alpha}$. By differentiating the equation we get

$$D^2 F(\nabla w) : D^2 \partial_i w = -D^3 F(\nabla w) \cdot \nabla \partial_i w : D^2 w.$$

Note that the second term is C^α. So, again, forgetting about the dependence of this term with respect to w we can use the Schauder theory on the linear problem

$$A(x) : D^2 \partial_i w = f(x)$$

and find that $w \in C^{3,\alpha}$. Bootstrapping the argument, we get, finally, that $w \in C^\infty(\Omega)$. Note that at this point we have only $\nabla w \in L^2(\Omega)$. We need to fill the gap between L^2 to C^α.

The first idea of De Giorgi is to consider, for every $1 \le i \le N$, the derivative with respect to x_i of (2.2). Denote $u = \partial_i w$, this gives

$$\operatorname{div}(F''(\nabla w)\nabla u) = 0. \tag{2.4}$$

Thanks to (2.3), for every $x \in \Omega$

$$\frac{1}{\Lambda} I \le F''(\nabla w) \le \Lambda I.$$

Forgetting about the dependence of $A(x) = F''(\nabla w)$ on w, it can be rewritten as a classical linear elliptic equation in the divergence form:

$$\operatorname{div}(A(x)\nabla u) = 0, \tag{2.5}$$

with the elliptic condition on A:

$$\frac{1}{\Lambda} I \le A(x) \le \Lambda I, \qquad x \in \Omega. \tag{2.6}$$

De Giorgi showed the following theorem.

Theorem 4. *Let $u \in H^1(\Omega)$ be a weak solution to (2.5) with A verifying (2.6). Then $u \in C^\alpha(\tilde{\Omega})$ for any $\tilde{\Omega} \subset\subset \Omega$, with*

$$\|u\|_{C^\alpha(\tilde{\Omega})} \le C\|u\|_{L^2(\Omega)}.$$

The constant α depends only on Λ and N. The constant C depends on Λ, N, Ω, and $\tilde{\Omega}$.

To apply Theorem 4 to $u = \partial_i w$ and fill the gap by showing that ∇w is C^α, we would need first to show rigorously that indeed $u \in H^1$ (and so $\nabla w \in H^2$) and that it verifies (2.5) in the sense of distribution. This can be shown by standard method reminiscent of the proof that H^1 solutions to an elliptic equation are indeed H^2. For the sake of completeness, we give the proof of this fact at the end of this section.

Theorem 4 is significantly different in spirit with the results on elliptic regularity which existed before. Most of the previous results can be seen as perturbation of the Laplace operator. For instance, when the coefficients are C^α, we can use that locally, when zooming on small ball, the operator is almost constant, and so behaves like the Laplace operator. In the De Giorgi case this is not true anymore. Still, zooming is very important. It is worth noticing right away that the property (2.6) is preserved by the natural scaling of the equation.

From now on, we will denote L any operator $-\operatorname{div}(A(x)\nabla\cdot)$, where A is uniformly elliptic, that verifies (2.6). Note that if a function u verifies $Lu = 0$, then, for any $\lambda \in \mathbb{R}$ and any $\varepsilon > 0$ the function

$$\tilde{u}(y) = \lambda u(x + \varepsilon y) \tag{2.7}$$

is solution of an other equation $\tilde{L}\tilde{u} = 0$ for an operator $\tilde{L}$ which verifies (2.6) for the same value of Λ. This property, based on the linearity of L, will be crucial in the proof.

2.2 Proof of Theorem 4

The proof has two steps. The first step consists in showing that the energy control the supremum bound, and so that the function is indeed bounded. The second step consists in going from L^∞ to the modulus of continuity C^α. It uses a so-called oscillation lemma.

We denote $u_+ = \sup(0, u)$, and for any $r > 0$, B_r stands for the ball of center 0 and radius r. In this section, we show the following lemma.

Lemma 5. *There exists a constant δ depending only on Λ and N such that for any u solution to $Lu = 0$ in B_1, where L verifies (2.6), we have the following property. If*

$$\|u_+\|_{L^2(B_1)} \leq \delta,$$

then

$$\|u_+\|_{L^\infty(B_{1/2})} \leq 1/2.$$

It will be very important later that C does not depend on u. The choice of the size of the balls is not consequential. Changing the size of the ball changes the constant C. But the size of the ball for the L^∞ norm has to be smaller than the ball for the L^2 norm. This is already a regularization result. The solution u can be unbounded on the boundary of B_1. It is regularized inside. This result will be applied later on rescaled functions $\tilde{u}$ as (2.7). As a first easy application of those rescaling techniques, we have the following corollary.

Corollary 6. *Let $u \in L^2(\Omega)$ be a solution to $Lu = 0$ where Ω is a bounded open set, and L verifies (2.6). Then, for any $\tilde{\Omega} \subset\subset \Omega$, $u \in L^\infty(\tilde{\Omega})$.*

Proof. Let d be the distance from $\tilde{\Omega}$ to $\partial\Omega$. For any $x \in \tilde{\Omega}$, consider in B_1

$$\tilde{u}(y) = \delta \frac{d^{N/2}}{\|u\|_{L^2(\Omega)}} u(x + dy).$$

The function $\tilde{u}$ verifies $\tilde{L}\tilde{u} = 0$, where $\tilde{L}$ verifies (2.6), and $\|\tilde{u}\|_{L^2(B_1)} \leq \delta$. So, from Lemma 5,

$$\tilde{u}(x) \leq 1/2.$$

Applying the same result on $-\tilde{u}$ we obtain the bound by below, and

$$|\tilde{u}(x)| \leq 1/2.$$

Hence $\|u\|_{L^\infty(\tilde{\Omega})} \leq d^{-N/2}\|u\|_{L^2(\Omega)}/\delta$. □

We give now the proof of Lemma 5.

Proof. The proof is based on energy. We consider a sequence of level sets of energy on shrinking balls. As mentioned above, we have a regularization effect. So we expect to have a layer close to ∂B_1. In the recursive process, we want to escape from this layer. We consider the family of ball $\tilde{B}_k$ centered at 0 with radius $1/2(1+2^{-k})$. Note that $\tilde{B}_0 = B_1$ and $\tilde{B}_k$ converges to $B_{1/2}$ when k converges to infinity. the sequence of ball $\tilde{B}$ goes in a dyadic way from B_1 to $B_{1/2}$.

We consider in the same way, a family of "energy levels" C_k going in a dyadic way from 0 to 1/2:

$$C_k = 1/2(1-2^{-k}).$$

We define now the sequence of energy above the level set C_k in the ball $\tilde{B}_k$:

$$U_k = \int_{\tilde{B}_k} |u_k(x)|^2 \, dx,$$

where

$$u_k = (u - C_k)_+.$$

The aim is to derive an estimate of the form

$$U_{k+1} \leq C^k U_k^{\beta}, \tag{2.8}$$

for a suitable constant $C > 1$ and $\beta > 1$. It is worth noting that this inequality is *non linear*. This is fundamental. The term C^k converges to infinity when k converges to infinity. But, since $\beta > 1$, if the initial term U_0 (which is the L^2 norm of u_+ in B_1) is small enough, then the nonlinear term will be even smaller, and the sequence U_k will be shown to converge to 0. At the limit this gives

$$\int_{B_{1/2}} (u - 1/2)_+^2 \, dx = 0.$$

This means that in the ball of radius 1/2, the energy above the level 1/2 vanishes. And so, u is smaller than 1/2 in this ball.

The non linear inequality (2.8) is obtained via rather simple tools. We use linear tools as the energy estimate (which is linear since the equation itself is linear) and the Sobolev imbedding. The only nonlinear estimate that we use is the Tchebychev inequality (also known as Markov inequality) which is as follows:

$$|\{x, f(x) \geq R > 0\}| \leq \frac{\|f\|_{L^1}}{R}.$$

To gain the nonlinear estimate, we have to lose in term of level set on f. This explains why we introduce the level sets C_k. We split the proof in several steps.

Step 1. The two first ingredients are the Sobolev inequality, given by

$$\|v\|_{L^p(B_1)} \leq C \|\nabla v\|_{L^2(B_1)}$$

for $p(N) = \frac{2N}{N-2}$, whenever v is supported in B_1, and the energy inequality given by the following lemma.

Lemma 7. (*Energy inequality*) *If u is solution to $Lu = 0$ and $\varphi \in C_0^\infty(B_1)$ then*

$$\int_{B_1} (\nabla[\varphi u_+])^2\,dx \le C\|\nabla\varphi\|_{L^\infty}^2 \int_{B_1\cap \operatorname{supp}\varphi} u_+^2\,dx.$$

If A is symmetric then $C = \Lambda^2$.

Proof. We multiply $Lu = -\mathrm{div}(A(x)\nabla u)$ by $\varphi^2 u_+$ to get

$$\int \nabla^T(\varphi^2 u_+)A\nabla u_+\,dx \le 0.$$

We have to transfer a φ from the left ∇ to the right ∇. Indeed,

$$\begin{aligned}
&\int \nabla^T(\varphi^2 u_+)A\nabla u_+\,dx\\
&= \int \nabla^T(\varphi u_+)A\varphi\nabla u_+\,dx + \int (\varphi u_+)\nabla^T\varphi A\nabla u_+\,dx\\
&= \int \nabla^T(\varphi u_+)A\nabla(\varphi u_+)\,dx - \int \nabla^T(\varphi u_+)A(\nabla\varphi)u_+\,dx\\
&\quad + \int (\varphi u_+)\nabla^T\varphi A\nabla u_+\,dx\\
&= \int \nabla^T(\varphi u_+)A\nabla(\varphi u_+)\,dx - \int \nabla^T(\varphi u_+)(A - A^T)(\nabla\varphi)u_+\,dx\\
&\quad - \int u_+^2\nabla^T\varphi A\nabla\varphi\,dx.
\end{aligned}$$

In the symmetric case $A = A^T$, using (2.6) gives the result with $C = \Lambda^2$. Otherwise we use

$$\begin{aligned}
&\left|\int \nabla^T(\varphi u_+)(A - A^T)(\nabla\varphi)u_+\,dx\right|\\
&\le 2\Lambda\|\nabla(\varphi u_+)\|_{L^2}\|(\nabla\varphi)u_+\|_{L^2}\\
&\le 2\Lambda^{3/2}\left(\int \nabla^T(\varphi u_+)A\nabla(\varphi u_+)\,dx\right)^{1/2}\|(\nabla\varphi)u_+\|_{L^2}\\
&\le \frac{1}{2}\int \nabla^T(\varphi u_+)A\nabla(\varphi u_+)\,dx + 2\Lambda^3\int (\nabla\varphi)^2 u_+^2\,dx.
\end{aligned}$$

And so

$$\begin{aligned}
\int |\nabla(\varphi u_+)|^2\,dx &\le \Lambda\int \nabla^T(\varphi u_+)A\nabla(\varphi u_+)\,dx\\
&\le C\int (\nabla\varphi)^2 u_+^2\,dx.
\end{aligned}$$

□

Step 2. We introduce truncation functions ϕ_k verifying

$$\begin{aligned}
\phi_k &= 1 \qquad \text{in } \tilde{B}_k,\\
&= 0 \qquad \text{in } \tilde{B}_{k-1}^c,
\end{aligned}$$

with

$$|\nabla\phi_k| \le C2^k.$$

Note that

$$U_k \le \int \phi_k^2 u_k^2\, dx.$$

We have also $\mathbf{1}_{\tilde{B}_{k+1}} \le \phi_k$, and $u_{k+1} \le u_k$. The Sobolev inequality with $v = \phi_{k+1}u_{k+1}$, and the energy inequality with $\varphi = \phi_{k+1}$ give

$$\begin{aligned}\left(\int (\phi_{k+1}u_{k+1})^p\, dx\right)^{2/p} &\le C\int |\nabla(\phi_{k+1}u_{k+1})|^2\, dx\\ &\le C2^{2k}\int |\phi_k u_{k+1}|^2\, dx\\ &\le C^k U_k.\end{aligned}$$

Using now the Tchebychev inequality, we get

$$\begin{aligned}U_{k+1} \le \int (\phi_{k+1}u_{k+1})^2\, dx &\le \left(\int (\phi_{k+1}u_{k+1})^p\, dx\right)^{2/p} |\{\phi_{k+1}u_{k+1} > 0\}|^{2/N}\\ &\le C^k U_k |\{\phi_k u_k > 2^{-2k}\}|^{2/N}\\ &= C^k U_k |\{(\phi_k u_k)^2 > 2^{-4k}\}|^{2/N}\\ &\le \frac{C^k}{2^{-8k/N}} U_k^{1+2/N} \le (2^{8/N}C)^k U_k^{1+2/N}.\end{aligned}$$

This gives (2.8) with $\beta = 1 + 2/N$.

Step 3. To show that U_k converges to 0, we compare (2.8) with a geometric series. We want to show that taking U_0 small enough, we have for any k

$$C^k U_k^{\beta-1} \le \frac{1}{(2C)^{\frac{1}{\beta-1}}}. \tag{2.9}$$

Consider k_0 such that

$$\frac{1}{2^{k_0}} \le \frac{1}{(2C)^{\frac{1}{\beta-1}}},$$

and δ small enough such that (2.9) is valid for $k \le k_0$. It is possible to find such that δ since k_0 is fixed. Let us show by induction that (2.9) is still true for any $k \ge k_0$. Fix a $k > k_0$. Assume that it is true for all $j \le k$. Then

$$U_{k+1} \le \frac{1}{(2C)^{\frac{k+1}{\beta-1}}},$$

and

$$C^{k+1}U_{k+1}^{\beta-1} \le 2^{-(k+1)} \le \frac{1}{(2C)^{\frac{1}{\beta-1}}}.$$

Hence we can compare to the geometric sequence and find that U_k converges to 0 when k goes to infinity. Passing into the limit, we find

$$\int_{B_{1/2}} (u-1/2)_+^2\,dx = 0.$$

This implies that $\|u_+\|_{L^\infty} \leq 1/2$. □

2.3 The oscillation lemma

The second step consists in obtaining a so-called oscillation lemma. We denote for any open set D: $\mathrm{osc}_D\, u = \sup_D u - \inf_D u$.

Lemma 8. *Let u be a solution of $Lu = 0$ in B_2 where A verifies (2.6). Then there exists $\lambda(\Lambda, N) < 1$ such that*

$$\mathrm{osc}_{B_{1/2}}\, u \leq \lambda\, \mathrm{osc}_{B_2}\, u\ .$$

This lemma implies C^α regularity of the solutions. Its strength is that it gives a definition that depends only on the L^∞ norm. The proof of C^α regularity follows that way. Consider $u \in H^1(\Omega)$ solution to $Lu = 0$ in Ω, and $\tilde{\Omega} \subset\subset \Omega$. Let $d = d(\tilde{\Omega}, \partial\Omega)$. Take any x_0 in $\tilde{\Omega}$. We introduce the rescaled functions

$$\begin{aligned} \bar{u}_1(y) &= u(x_0 + dy/2), \\ \bar{u}_n(y) &= \bar{u}_{n-1}(y/4). \end{aligned}$$

As before, $\bar{u}_n$ are solutions to (2.5) with diffusion matrices $A_n(y) = A(x_0 + y/4^n)$. Note that A_n verifies (2.6) for the same fixed Λ. We apply recursively Lemma 8 on $\bar{u}_n$. This gives

$$\sup_{|x_0 - x| \leq 4^{-n}} |u(x_0) - u(x)| \leq 2\|u\|_{L^\infty(B_1)}\lambda^n.$$

Note that this estimate does not depend on x_0. Hence u is in $C^\alpha(\tilde{\Omega})$ with

$$\alpha = -\frac{\ln\lambda}{2\ln 2}.$$

We reformulate slightly the oscillation lemma in the following way.

Proposition 9. *Let $v \leq 1$, $Lv = 0$ in B_2. Assume that $|B_1 \cap \{v \leq 0\}| \geq \mu$ ($\mu > 0$). Then $\sup_{B_{1/2}} v \leq 1 - \lambda$, where λ depends only on μ, Λ, and N.*

In other words, if v is a solution of $Lv = 0$, smaller than one in B_2, and is "far from 1" in a set of non trivial measure, it cannot get too close to 1 in $B_{1/2}$.

Let us first show how this leads to Lemma 8. Consider the function

$$v(x) = \frac{2}{\mathrm{osc}_{B_2}\, u}\left(u(x) - \frac{\sup u + \inf u}{2}\right).$$

We have $-1 \le v \le 1$. Assume that v is half of the space smaller than 0 in B_1. Then we can apply Proposition 9 on v which gives that $\operatorname{osc}_{B_{1/2}} v \le 2 - \lambda$. Hence $\operatorname{osc}_{B_{1/2}} u \le (1-\lambda/2) \operatorname{osc}_{B_2} u$. We get the same result if v is half of the space bigger than 0, working with $(-v)$.

To prove Proposition 9, we may first note that if the set

$$|\{v \le 0\} \cap B_1| \ge |B_1| - \delta^2 ,$$

then

$$\|v_+\|_{L^2(B_1)} \le \delta$$

and Lemma 5 imply that $v_+|_{B_{1/2}} \le 1/2$.

So we must bridge the gap between knowing that $|\{v \le 0\}| \ge \frac{1}{2}|B_1|$ and knowing that $|\{v \le 0\}| \ge |B_1| - \delta^2$.

The main tool is the following De Giorgi isoperimetric inequality. It may be considered as a quantitative version of the fact that a function with a jump discontinuity cannot be in H^1.

Lemma 10. *There exists a constant $C_N > 0$ depending only on N such that the following holds true. Consider w such that $\int_{B_1} |\nabla w_+|^2 \, dx \le C_0$. Set*

$$|A| = |\{w \le 0\} \cap B_1|,$$
$$|C| = |\{w \ge 1/2\} \cap B_1|,$$
$$|D| = |\{0 < w < 1/2\} \cap B_1|.$$

Then we have

$$C_0 |D| \ge C_N (|C|\, |A|^{1-\frac{1}{N}})^2 .$$

Proof. Consider $\bar{w} = \sup(0, \inf(w, 1/2))$. Note that $\nabla \bar{w} = \nabla w_+ \mathbf{1}_{\{0 \le w \le 1/2\}}$. For x in A and y in C, we have

$$\begin{aligned} 1/2 = \bar{w}(y) - \bar{w}(x) &= \int_0^1 (y-x) \cdot \nabla \bar{w}(x + t(y-x)) \, dt \\ &\le \int_0^{|y-x|} |\nabla \bar{w}|(x + s e_\sigma) \, ds, \end{aligned}$$

where $e_\sigma = y - x/|y-x|$, and we made the change of variable $s = t|y-x|$. This quantity is smaller than the same integrating along the whole ray coming from x in the direction e_σ in B_1. With a slight abuse of notation (assigning the value 0 to $\nabla \bar{w}$ outside of B_1 for instance), we get

$$1/2 \le \int_0^\infty |\nabla \bar{w}|(x + s e_\sigma) \, ds,$$

integrating this inequality for all $y \in C$, we get

$$\begin{aligned} |C|/2 &\le \int_C \left(\int_0^\infty |\nabla \bar{w}|(x + s e_\sigma) \, ds \right) dy \\ &\le \int_{B_1} \left(\int_0^\infty |\nabla \bar{w}|(x + s e_\sigma) \, ds \right) dy. \end{aligned}$$

Writing the first integral in polar coordinates for $(y-x)$, and noticing that the function does not depend on r we get

$$\begin{aligned} |C|/2 &\leq \int_0^2 r^{N-1} \int_{\mathbb{S}_N} \left(\int_0^\infty |\nabla \bar{w}|(x+se_\sigma)\, ds \right) d\sigma\, dr \\ &\leq \int_{\mathbb{S}_N} \int_0^\infty |\nabla \bar{w}|(x+se_\sigma)\, ds\, d\sigma \\ &= \int_{\mathbb{S}_N} \int_0^\infty s^{N-1} \frac{|\nabla \bar{w}|(x+se_\sigma)}{s^{N-1}}\, ds\, d\sigma. \end{aligned}$$

The last term is nothing but

$$\int_{B_1} \frac{|\nabla \bar{w}|(y)}{|x-y|^{N-1}}\, dy.$$

Integrating $x \in A$ we find

$$|A|\ |C|/2 \leq \int_{B_1} |\nabla \bar{w}|(y) \left(\int_A \frac{dx}{|x-y|^{N-1}} \right) dy.$$

Among all A with the same measure $|A|$ the integral in x is maximized by the ball of radius $|A|^{1/N}$, centered at y

$$\int_A \frac{dx}{|x-y|^{N-1}} \leq |A|^{1/N}.$$

So

$$|A|\ |C|/2 \leq |A|^{1/N} \left(\int_D |\nabla w_+|^2 \right)^{1/2} |D|^{1/2}.$$

Since $\int |\nabla w_+|^2\, dx \leq C_0$ the proof is complete. □

Proof of Proposition 9. We consider the new sequence of truncation

$$w_k = 2^k [v - (1 - 2^{-k})].$$

Note that for any k we have $w_k \leq 1$. So from the energy inequality, we have

$$\int_{B_1} |\nabla (w_k)_+|^2\, dx \leq C_0.$$

We have also $|\{w_k \leq 0\} \cap B_1| \geq \mu$. We apply Lemma 10 recursively on w_k as long as

$$\int_{B_1} (w_{k+1})_+^2\, dx \geq \delta^2.$$

We get

$$|\{w_{k+1} \geq 0\} \cap B_1| = |\{2w_k \geq 1\} \cap B_1| \geq \int_{B_1} (w_{k+1})_+^2\, dx \geq \delta^2.$$

So, from Lemma 10, there exists a constant α which does not depend on k such that

$$|\{0 < w_k < 1/2\} \cap B_1| \geq \alpha.$$

Then

$$|\{w_k \leq 0\} \cap B_1| \geq |\{w_{k-1} \leq 0\} \cap B_1| + \alpha \geq \mu + k\alpha.$$

This clearly fails after a finite number of k. At this k_0 we have for sure that

$$\int_{B_1} (w_{k_0+1})_+^2 \, dx \leq \delta^2.$$

Lemma 5 then implies that $w_{k_0+1} \leq 1/2$ in $B_{1/2}$. Rescaling back to v gives the result with $\lambda = 2^{-(k_0+2)}$. □

2.4 Proof of Theorem 3

To conclude the proof, we need to show that any $w \in H^1(\Omega)$ weak solution to (2.2) is such that $u = \partial_i w \in H^1$ and is a weak solution to (2.4). The proof use elementary techniques of elliptic theory and have nothing to do with the De Giorgi techniques. We present it here for the sake of completeness. It is related to the higher regularity proof of linear elliptic operators in the L^2 theory. In all the section we assume that (2.3) is verified. Recall that a weak solution to (2.2) verifies for every $\phi \in C_c^\infty(\Omega)$

$$\int_\Omega \nabla\phi DF(\nabla w) \, dx = 0.$$

By density, this is still valid for any $\phi \in H_0^1(\Omega)$.

By weak solution to (2.4), we mean that for every $\phi \in C_c^\infty(\Omega)$

$$\int_\Omega \nabla\phi^T D^2F(\nabla w)\nabla u \, dx = 0.$$

Again, by density, this is still true for any $\phi \in H^1(\Omega)$, as long as $u \in H^1(\Omega)$. We need the following proposition.

Proposition 11. *Let $w \in H^1(\Omega)$ be a weak solution to (2.2). Then, for any $\tilde{\Omega} \subset\subset \Omega$, $w \in H^2(\tilde{\Omega})$, and for every i, $u = \partial_i w$ is a weak solution to (2.4).*

As, usual, we will first show a localized version in a fixed ball.

Lemma 12. *Let $w \in H^1(B_3)$ be a weak solution to (2.2). Then, $w \in H^2(B_1)$, and for every i, $u = \partial_i w$ is a weak solution to (2.4) in B_1.*

This lemma implies the proposition in the following way. Denote $d = d(\tilde{\Omega}, \partial\Omega)$, and for any $x \in \tilde{\Omega}$

$$\tilde{w_x}(y) = w\left(x + \frac{d}{3}y\right).$$

The function $\tilde{w}_x \in H^1(B_3)$ and verifies (2.2) in B_3 for $\tilde{F}(y) = F(3y/d)$. Since $\tilde{\Omega}$ is compact, there exists a finite number m of ball $B_{d/3}(x_i)$ such that $\tilde{\Omega} \subset \bigcup_{i=1}^m B_{d/3}(x_i)$, and $w \in H^2(\tilde{\Omega})$. By the theorem of partition of the unity, any function $\phi \in C_c^\infty(\tilde{\Omega})$ can be decomposed as

$$\phi = \sum_{i=1}^m \phi_i$$

where $\phi \in C_c^\infty(B_{d/3}(x_i))$. And so

$$\int_\Omega \nabla\phi^T D^2F(\nabla w)\nabla u\,dx = \sum_{i=1}^m \int_\Omega \nabla\tilde{\phi}_i^T D^2\tilde{F}(\nabla\tilde{w}_{x_i})\nabla\tilde{u}_{x_i}\,dx = 0.$$

We now turn to the proof of the lemma.

Proof. For any $-1 \leq h \leq 1$ we consider the operator defined on $C(\mathbb{R}^N)$

$$T_h\phi(x) = \frac{\phi(x+he_i) - \phi(x)}{h}.$$

For any cut-off function η, non negative, compactly supported in B_2 and equal to 1 in B_1, we can define two operators ηT_h and $T_h(\eta\cdot)$ well defined from $L^2(B_3)$ to $L^2(B_3)$, and from $H^1(B_3)$ to $H_0^1(B_3)$. We have the following properties.

Lemma 13. *For any $\tilde{\phi}, \tilde{\psi} \in L^2(B_3)$, $\phi \in H^1(B_3)$, $\Phi \in [L^2(B_3)]^N$, we have*

$$\int T_h(\eta\tilde{\phi})\tilde{\psi} = -\int \tilde{\phi}\eta T_{-h}\tilde{\psi}, \qquad \nabla[\eta^2 T_h\phi] = \eta^2 T_h[\nabla\phi] + 2\eta[T_h\phi]\nabla\eta, \tag{2.10}$$

$$\eta T_h\{DF(\Phi)\}(x) = \eta T_h\Phi(x)\int_0^1 D^2F(\Phi(x) + t((\Phi(x+he_i) - (\Phi(x)))\,dt, \tag{2.11}$$

$$\|\eta T_h\phi\|_{L^2(B_3)} \leq C_\eta\|\nabla\phi\|_{L^2(B_3)}. \tag{2.12}$$

Moreover, if $\sup_{|h|<1}\|\eta T_h\tilde{\phi}\|_{L^2(B_3)}$ is bounded, then $\partial_i\tilde{\phi} \in L^2(B_1)$ and

$$\|\partial_i\tilde{\phi}\|_{L^2(B_1)} \leq \sup_{|h|<1}\|\eta T_h\tilde{\phi}\|_{L^2(B_3)},$$

$$\partial_i\tilde{\phi} = \lim_{h\to 0}\eta T_h\tilde{\phi},$$

weakly in $L^2(B_1)$.

Proof. The first statement is obtained via a change of variable $y = x + he_i$.

$$\begin{aligned}\int T_h(\eta\tilde{\phi})\tilde{\psi} &= \int \frac{(\eta\tilde{\phi})(x+he_i)\tilde{\psi}(x)}{h}\,dx - \int \frac{(\eta\tilde{\phi})(x)\tilde{\psi}(x)}{h}\,dx \\ &= \int \frac{(\eta\tilde{\phi})(x)\tilde{\psi}(x-he_i)}{h}\,dx - \int \frac{(\eta\tilde{\phi})(x)\tilde{\psi}(x)}{h}\,dx \\ &= -\int \tilde{\phi}\eta T_{-h}\tilde{\psi}.\end{aligned}$$

This means that $-\eta T_{-h}$ is the dual operator of $T_h(\eta\cdot)$ in L^2.

The second equality is just the remark that ∇ and T_h are commuting.

After computation, we find

$$\begin{aligned}&\eta T_h DF(\Phi)(x)\\&=\eta\frac{DF(\Phi(x+he_i))-DF(\Phi(x))}{h}\\&=\frac{\eta[\Phi(x+he_i)-\Phi(x)]}{h}\int_0^1 D^2F(\Phi(x)+t(\Phi(x+he_i)-\Phi(x)))\,dt,\end{aligned}$$

which gives the result.

We have

$$\eta T_h\phi=\eta\int_0^1\partial_i\phi(x+te_i)\,dt.$$

Hence

$$\|\eta T_h\phi\|_{L^2(B_3)}\le\eta_{L^\infty}\|\partial_i\phi\|_{L^2(B_3)}.$$

Consider a sequence of mollifiers for $\varepsilon<1/2$

$$\delta_\varepsilon(x)=\frac{1}{\varepsilon^N}\delta_1(x/\varepsilon),$$

where δ_1 is a smooth nonnegative function compactly supported in B_1 with integral 1. Consider

$$\tilde{\phi}_\varepsilon=\tilde{\phi}*\delta_\varepsilon.$$

Since $\tilde{\phi}_\varepsilon$ is smooth, for $x\in B_2$, $\eta T_h\tilde{\phi}_\varepsilon(x)$ converges to $\eta\partial_i\tilde{\phi}_\varepsilon(x)$ when h goes to 0. But $\eta T_h\tilde{\phi}_\varepsilon=\eta(T_h\tilde{\phi})_\varepsilon$ are uniformly bounded (both with respect to h and ε). Passing into the limit in h, we find that $\partial_i\tilde{\phi}_\varepsilon$ are uniformly bounded in $L^2(B_1)$. In turn, $\partial_i\tilde{\phi}$ is in $L^2(B_1)$ and $T_h\tilde{\phi}$ converges weakly to $\partial_i\tilde{\phi}$. $\square$

Let us fix a cut off function η. We consider the test function

$$\phi=T_{-h}(\eta^2T_hw)\in H_0^1(B_3).$$

we got

$$\begin{aligned}0&=\int\nabla\phi DF(\nabla w)\,dx\\&=-\int\eta\nabla(T_hw)\eta T_hDF(\nabla w)\,dx-2\int\nabla\eta(T_hw)\eta T_hDF(\nabla w)\,dx\\&=-\int\eta\nabla(T_hw)\left[\int_0^1D^2F(\nabla w(x)+t(\nabla w(x+he_i)-\nabla w(x)))\,dt\right]\eta\nabla(T_hw)\,dx\\&\quad-2\int(\nabla\eta)(T_hw)\left[\int_0^1D^2F(\nabla w(x)+t(\nabla w(x+he_i)-\nabla w(x)))\,dt\right]\eta\nabla(T_hw)\,dx.\end{aligned}$$

Using the property (2.3) we get

$$\frac{1}{\Lambda}\int \eta^2|\nabla(T_h w)|^2\,dx \le 2\Lambda\sqrt{\int |\nabla\eta|^2|T_h w|^2\,dx}\sqrt{\int \eta^2|\nabla(T_h w)|^2\,dx}$$

and so

$$\int \eta^2|\nabla(T_h w)|^2\,dx \le 4\Lambda^4\int |\nabla\eta|^2|T_h w|^2\,dx \le C\|w\|^2_{H^1(B_3)}.$$

This gives that $\partial_i w \in H^1(B_1)$. For any test function $\phi \in H^1_0(B_1)$, we consider the test function

$$T_{-h}\phi \in H^1(B_2).$$

We get

$$0 = -\int \nabla\phi T_h DF(\nabla w)\,dx.$$

Since $\partial_i w \in H^1(B_1)$, $T_h DF(\nabla w)$ converges to $\partial_i\{DF(\nabla w)\} = D^2F(\nabla w)\nabla\partial_i w$ weakly in $L^2(B_1)$, and at the limit

$$\int \nabla\phi DF^2(\nabla w)\nabla\partial_i w\,dx = 0,$$

which finishes the proof. □

3 The parabolic case

The parabolic case was first introduced with the techniques of Nash. We present, here, a version with a "De Giorgi packaging". We begin with the homogenous case.

3.1 The homogenous case

Consider the equation

$$\partial_t u - \operatorname{div}(A(t,x)\nabla_x u) = 0 \qquad \text{in } (0,T)\times\Omega, \tag{3.1}$$

where A verifies

$$\frac{1}{\Lambda} \le A(t,x) \le \Lambda I, \qquad \text{for any } (t,x)\in(0,T)\times\Omega, \tag{3.2}$$

but without further assumption on the regularity of A. We have the following theorem.

Theorem 14. *Let $u \in L^\infty(0,T;L^2(\Omega))$, $\nabla u \in L^2((0,T)\times\Omega)$ be a weak solution to (3.1) in $(0,T)\times\Omega$. Then, there exists $\alpha > 0$ such that for any $\tilde\Omega \subset\subset \Omega$, and any $0 < s < T$*

$$u \in C^\alpha((s,T)\times\tilde\Omega).$$

Define $Q_r = (-r, 0) \times B_r$. As in the previous section, we show the proof with going from (Q_2) to $Q_{1/2}$. Theorem 14 can easily be deduced from this case by adequate zooming as in the previous section. We consider the first De Giorgi lemma.

Lemma 15. *There exists $\delta > 0$ such that for any u solution to (3.1) with (3.2) in $(-1,0) \times B_1$: If*

$$\int_{Q_1} |u_+|^2 \, dx \, dt \leq \delta,$$

then

$$u_+ \leq 1/2 \qquad \text{in } Q_{1/2}.$$

Proof. We have now to consider two layers in position and time. We consider the sequence of time

$$T_k = -1/2(1 + 2^{-k}),$$

and the sequence of cylinders

$$\tilde{Q}_k = (T_k, 0) \times \tilde{B}_k,$$

where $\tilde{B}_k$ is defined as before as centered at 0 with radius $1/2(1+2^{-k})$. We define the sequence of energy

$$U_k = \int_{Q_k} |u_k|^2 \, dx \, dt,$$

where as before

$$u_k = (u - C_k)_+,$$

with

$$C_k = 1/2(1 - 2^{-k}).$$

We consider the exact same family of cutoff non negative functions ϕ_k compactly supported in $\tilde{B}_{k+1}$, equal to 1 in $\tilde{B}_k$ and such that $|\nabla \phi_k| \leq C 2^k$.

We obtain the energy inequality in the following way: We multiply the equation by $\phi_{k+1}^2 u_{k+1}$, and integrate in both time and space on $(s,t) \times B_1$ for

$$T_k \leq s \leq T_{k+1} \leq t \leq 0.$$

We find

$$\begin{aligned} &\left(\int \phi_{k+1}^2 |u_{k+1}|^2 \, dx\right)(t) \\ &\leq \left(\int \phi_{k+1}^2 |u_{k+1}|^2 \, dx\right)(s) - \int_s^t \int \nabla(\phi_{k+1}^2 u_{k+1}) A \nabla u \, dx \, d\tau. \end{aligned}$$

Using the energy inequality in x only from the previous section we get

$$\begin{aligned} &\left(\int \phi_{k+1}^2 |u_{k+1}|^2 \, dx\right)(t) + \frac{1}{\Lambda} \int_s^t \int |\nabla(\phi_{k+1} u_{k+1})|^2 \, dx \, d\tau \\ &\leq \left(\int \phi_{k+1}^2 |u_{k+1}|^2 \, dx\right)(s) + \Lambda \int_s^t \int |\nabla \phi_{k+1}|^2 |u_{k+1}|^2 \, dx \, d\tau. \end{aligned}$$

Using the range of s and t, we get

$$\left(\int \phi_{k+1}^2 |u_{k+1}|^2 \, dx\right)(t) + \frac{1}{\Lambda}\int_{T_{k+1}}^t \int |\nabla(\phi_{k+1}u_{k+1})|^2 \, dx \, d\tau$$
$$\leq \left(\int \phi_{k+1}^2 |u_{k+1}|^2 \, dx\right)(s) + \Lambda \int_{T_k}^0 \int |\nabla \phi_{k+1}|^2 |u_{k+1}|^2 \, dx \, d\tau.$$

Taking the mean value in s between T_{k+1} and T_k we get

$$\left(\int \phi_{k+1}^2 |u_{k+1}|^2 \, dx\right)(t) + \frac{1}{\Lambda}\int_{T_{k+1}}^t \int |\nabla(\phi_{k+1}u_{k+1})|^2 \, dx \, d\tau$$
$$\leq 2^{k+2} \int_{T_k}^{T_{k+1}} \int \phi_{k+1}^2 |u_{k+1}|^2 \, dx \, ds + \Lambda \int_{T_k}^0 \int |\nabla \phi_{k+1}|^2 |u_{k+1}|^2 \, dx \, d\tau.$$

Summing the estimate for the worst t, and $t = 0$ we get

$$\mathcal{E}_{k+1} = \sup_{T_{k+1} \leq t \leq 0} \left(\int (\phi_{k+1}u_{k+1})^2 \, dx\right) + \int_{T_{k+1}}^0 \int_{B_1} |\nabla \phi_{k+1} u_{k+1}|^2 \, dx \, dt \leq C^k U_k.$$

The quantity $\mathcal{E}_{k+1}$ controls $\phi_{k+1}u_{k+1}$ in $L^\infty(T_{k+1}, 0; L^2(B_1))$ and, thanks to Sobolev, in the space $L^2(T_{k+1}, 0; L^{2N/(N-2)}(B_1))$. The coefficients $(\infty, 2)$ and $(2, 2N/(N-2))$ verifies both

$$\frac{2}{p} + \frac{N}{q} = \frac{N}{2}.$$

So, by interpolation, $\mathcal{E}_{k+1}$ control the $L^p(L^q)$ norm of any such (p, q) with $p \geq 2$. We consider the one verifying $p = q = 2(2+N)/N$. So

$$\|\phi_{k+1}u_{k+1}\|^2_{L^{2(2+N)/N}(\tilde{Q}_{k+1})} \leq C\mathcal{E}_{k+1}.$$

And so,

$$\begin{aligned}
U_{k+1} &= \int_{\tilde{Q}_{k+1}} |u_{k+1}|^2 \, dx \, dt \\
&= \int_{\tilde{Q}_{k+1}} |u_{k+1}|^2 \mathbf{1}_{\{u_{k+1} > 0\}} \, dx \, dt \\
&= \int_{\tilde{Q}_{k+1}} |u_{k+1}|^2 \mathbf{1}_{\{u_k > 2^{-(k+2)}\}} \, dx \, dt \\
&\leq \mathcal{E}_{k+1} |\{u_k^2 > 2^{-2(k+2)} \cap \tilde{Q}_{k+1}\}|^{\frac{2}{2+N}} \\
&\leq \mathcal{E}_{k+1} \left(2^{2(k+2)} \int_{\tilde{Q}_{k+1}} |u_k|^2 \, dx \, dt\right)^{\frac{2}{2+N}} \\
&\leq C^k U_k^{1+\frac{2}{2+N}}.
\end{aligned}$$

So, a before, there exists δ small enough such that if $U_0 \leq \delta$, we have

$$\lim_{k \to \infty} U_k = 0.$$

This ensures that

$$u_+ \leq 1/2 \qquad \text{on } \tilde{Q}_\infty = (-1/2, 0) \times B_{1/2}. \qquad \square$$

We now turn to the second lemma of De Giorgi. We denote $\tilde{Q} = (-3/2, -1) \times B_1$. We then have the following theorem.

Theorem 16. *There exists a constant* $0 < \lambda < 1$ *such that for any solution* u *of* (3.1) *in* Q_2 *verifying* (3.2), *we have the following:*
If $-1 \leq u \leq 1$ *on* Q_2 *and*

$$|\{u \leq 0\} \cap \tilde{Q}| \geq \frac{|\tilde{Q}|}{2},$$

we have

$$u \leq 1 - \lambda \qquad \text{in } Q_{1/2}.$$

To show this theorem, we need a result similar to the isoperimetric inequality for the parabolic case. We show the following lemma.

Lemma 17. *There exists* $\alpha > 0$ *such that for any* u *solution to* (3.1) *in* Q_2 *verifying* (3.2), *with* $u \leq 1$ *we have the following. Denote*

$$\begin{aligned} |A| &= |\{u \geq 1/2\} \cap Q_1|, \\ |C| &= |\{u \leq 0\} \cap \tilde{Q}|, \\ |D| &= |\{0 < u < 1/2\} \cap (Q_1 \cup \tilde{Q})|. \end{aligned}$$

If $|A| \geq \delta$, $|C| \geq |\tilde{Q}|/2$ *then*

$$|D| \geq \alpha.$$

Proof. We assume that the lemma is wrong. Then we can find a sequence u_k such that the associated quantities

$$\begin{aligned} |A_k| &= |\{u_k \geq 1/2\} \cap Q_1| \geq \delta, \\ |C_k| &= |\{u_k \leq 0\} \cap \tilde{Q}| \geq |\tilde{Q}|/2, \\ |D_k| &= |\{0 < u_k < 1/2\} \cap (Q_1 \cup \tilde{Q})| \leq 1/k. \end{aligned}$$

Consider a cutoff function ϕ in x only, compactly supported in B_2 and equal to 1 in B_1. Consider $\phi^2 v_k$ where $v_k = (u_k)_+$. We have

$$\partial_t \phi^2 (v_k)^2 - \operatorname{div}(\phi^2 A \nabla (v_k)^2) + 2\phi \nabla \phi A \nabla (v_k)^2 + 2\phi^2 \nabla v_k A \nabla v_k = 0. \tag{3.3}$$

Then integrating in x, we have for $-1 < s < t < 0$

$$\left(\int \phi^2 (v_k)^2 \, dx \right)(t) + \frac{1}{\Lambda} \int_s^t |\nabla(\phi v_k)|^2 \leq \left(\int \phi^2 (v_k)^2 \, dx \right)(s) + C(t-s). \tag{3.4}$$

This energy inequality give that ϕv_k is uniformly bounded in $L^2(-2, 0; H^1(B_2))$ (and so is $\phi |v_k|^2$). Hence, from (3.3), $\partial_t v_k$ is uniformly bounded in $L^1(Q_2) +$

$L^\infty(-2,0;W^{-1}(Q_2))$. Using the Aubin lemma, we find that the sequence ϕv_k converges to ϕv up to a subsequence in L^2 when k goes to infinity. Using Tchebychev, for any $\varepsilon > 0$,

$$\lim_{k\to\infty} |\{|v_k - v| \geq \varepsilon\} \cap (-2,0) \times B_1| = 0.$$

Now, if $\varepsilon \leq v \leq 1/2 - \varepsilon$ then either $|v - v_k| \geq \varepsilon$ or $0 < v_k < 1/2$. So

$$\begin{aligned}&|\{\varepsilon \leq v \leq 1/2 - \varepsilon\} \cap Q_1 \cup \tilde{Q}|\\ &\leq |\{|v - v_k| \geq \varepsilon\} \cap Q_1 \cup \tilde{Q}| + |\{0 < v_k < 1/2\} \cap (Q_1 \cup \tilde{Q})|,\end{aligned}$$

which converges to 0. So

$$|\{\varepsilon \leq v \leq 1/2 - \varepsilon\} \cap (Q_1 \cup \tilde{Q})| = 0.$$

This is true for any $\varepsilon > 0$ hence

$$|\{0 < v < 1/2\} \cap (Q_1 \cup \tilde{Q})| = 0.$$

For almost every $t \in (-2,0)$, $\nabla v(t,\cdot) \in L^2(B_1)$. So thanks to the isoperimetric lemma of De Giorgi in the stationary case, for almost every $t \in (-3/2, 0)$ we have either

$$v(t,\cdot) \leq 0 \quad \text{in } B_1, \qquad \text{or } v(t,\cdot) \geq 1/2 \quad \text{in } B_1.$$

In the same way, if $v_k \leq 0$, then either $|v - v_k| \geq \varepsilon$ or $v \leq \varepsilon$. So

$$\begin{aligned}|\tilde{Q}|/2 &\leq |\{v_k \leq 0\} \cap \tilde{Q}| \leq |\{|v - v_k| \geq \varepsilon\} \cap \tilde{Q}| + |\{v \leq \varepsilon\} \cap \tilde{Q}|\\ &\leq |\{v \leq \varepsilon\} \cap \tilde{Q}|.\end{aligned}$$

Passing into the limit $\varepsilon \to 0$, we find

$$|\{v \leq 0\} \cap \tilde{Q}| \geq |\tilde{Q}|/2.$$

So there is some times $-3/2 < s < -1$ such that $v(t,\cdot) \leq 0$ in B_1. Now, consider a nonnegative function in x, $\phi_1 \geq 0$ compactly supported in B_1 of integral 1. the energy inequality ensures that for those time s, and $t > s$

$$\left(\int \phi_1^2 (u_+)^2\, dx\right)(t) \leq \left(\int \phi_1^2 (u_+)^2\, dx\right)(s) + C_0(t-s) \leq C_0(t-s). \tag{3.5}$$

But $\int \phi_1^2 (u_+)^2\, dx$ in all time is either negative or bigger than $1/2$. So it cannot reach the value $1/2$, and finally

$$v(t,x) = 0 \qquad \text{on } Q_1.$$

But, for any $\varepsilon > 0$, if $v_k \geq 1/2$, then either $|v - v_k| \geq \varepsilon$ or $v \geq 1/2 - \varepsilon$. So

$$\delta \leq |\{v_k \geq 1/2\} \cap Q_1| \leq |\{|v - v_k| \geq \varepsilon\} \cap Q_1| + |\{v \geq 1/2 - \varepsilon\} \cap Q_1|.$$

So, passing into the limit first in $k \to \infty$, then $\varepsilon \to 0$, we get

$$|\{v \geq 1/2\} \cap Q_1| \geq \delta.$$

This leads to a contradiction. □

The proof of Theorem 16 and Theorem 14 now follows the lines of the previous section.

3.2 The non homogenous case

For a bounded set Ω, we consider the equation on u

$$\partial_t \rho + \mathrm{div}(\rho V_1) = 0, \tag{3.6}$$

$$\partial_t(\rho u) + \mathrm{div}(\rho u(V_1 + V_2)) + \rho V_3 \cdot \nabla u \\ -\mathrm{div}(A(t,x)\nabla_x u) = f + \mathrm{div}F \qquad \text{in } (0,T)\times\Omega, \tag{3.7}$$

where A and ρ verify

$$\frac{1}{\Lambda} \le \rho(t,x) \le \Lambda, \tag{3.8}$$

$$\frac{1}{\Lambda} I \le A(t,x) \le \Lambda I, \qquad \text{for any } (t,x) \in (0,T)\times\Omega, \tag{3.9}$$

but without further assumption on the regularity of A and ρ. Note that if V_1 is divergence free, and ρ is bounded by below and by above at $t = 0$, then (3.8) is fulfilled for all time and space. The case include the classical parabolic case with drift in the divergence form ($V_1 = 0$, $\rho = 1$, and $V_3 = 0$)

$$\partial_t u + \mathrm{div}(\rho u V_2) + \mathrm{div}(A(t,x)\nabla_x u) = f + \mathrm{div}F,$$

together with the case with the velocity in the non divergence form:

$$\partial_t u + V_3 \cdot \nabla u + \mathrm{div}(A(t,x)\nabla_x u) = f + \mathrm{div}F.$$

We assume that there exists $e > 0$ such that

$$f \in L^{p_f}(0,T;L^{q_f}(\Omega)),\ |V_i|^2 \in L^{p_i}(0,T;L^{q_i}(\Omega)),\ ||F|^2| \in L^{p_F}(0,T;L^{q_F}(\Omega))$$

where (p_f, q_f), (p_i, q_i) and (p_F, q_F) verify

$$1 < p < \infty, \qquad 1 < q < \infty, \qquad \frac{2}{p} + \frac{N}{q} = 2 - e. \tag{3.10}$$

Note that since Ω is bounded, it is not a restriction to consider the same e for all functions. We will denote

$$\|\mathcal{F}\|_{(0,T)\times\Omega} = \sum_{i=1}^{3} \||V_i|^2\|_{L^{p_i}(0,T;L^{q_i}(\Omega))} + \|f\|_{L^{p_F}(0,T;L^{q_f}(\Omega))} + \||F|^2\|_{L^{p_F}(0,T;L^{q_F}(\Omega))}.$$

We will consider only "suitable" solutions which verify for any $K > 0$

$$\partial_t(\rho v_K^2) + \mathrm{div}(\rho v_K^2 V_1 + 2\rho u v_K V_2) + \rho V_3 \cdot \nabla v_K^2 + 2\nabla v_K^T A \nabla v_K \\ \le \mathrm{div}(A\nabla v_K^2) + 2(f + \mathrm{div}F)v_K + 2\rho u V_2 \cdot \nabla v_K, \tag{3.11}$$

where $v_K = (u - K)_+$. This inequality is obtained formally by multiplying Equation (3.7) by $2v_K$ and using Equation (3.6). Indeed,

$$\begin{aligned} v_K\{\partial_t(\rho u) + \mathrm{div}(\rho V_1 u)\} &= \{\partial_t \rho + \mathrm{div}(\rho V_1)\}(u v_K) + \rho\{\partial_t u + V_1 \cdot \nabla u\} v_K \\ &= \rho\left\{\partial_t \frac{v_K^2}{2} + V_1 \cdot \nabla \frac{v_K^2}{2}\right\} \\ &= \partial_t\left(\rho \frac{v_K^2}{2}\right) + \mathrm{div}\left(\rho V_1 \frac{v_K^2}{2}\right). \end{aligned}$$

It is verified, for instance, for any weak solution $u \in L^\infty(0,T;L^2(\Omega))$, $\nabla u \in L^2((0,T)\times\Omega)$ of (3.7), as long as ρ and V_1 are smooth. Even this case is interesting, since the result does not depend on the amount of regularity on those quantities. We have the following theorem.

Theorem 18. *Let $u \in L^\infty(0,T;L^2(\Omega))$, $\nabla u \in L^2((0,T)\times\Omega)$ be a suitable solution to (3.7) in $(0,T)\times\Omega$, where (3.9), (3.8) and (3.10) are verified. Then, there exists $\alpha > 0$ such that for any $\tilde{\Omega} \subset\subset \Omega$, and any $0 < s < T$*

$$u \in C^\alpha((s,T)\times\tilde{\Omega}).$$

Note that for $\varepsilon < 1$,

$$\tilde{u}(s,y) = u(t+\varepsilon^2 s, x+\varepsilon y)$$

is solution to

$$\begin{aligned}
&\partial_s(\tilde{\rho}\tilde{u}) + \mathrm{div}(\tilde{\rho}\tilde{u}(\tilde{V}_1+\tilde{V}_2)) + \tilde{\rho}\tilde{V}_3\cdot\nabla\tilde{u} - \mathrm{div}(\tilde{A}\nabla\tilde{u}) = \tilde{f} + \mathrm{div}\tilde{F},\\
&\partial_t\tilde{\rho} + \mathrm{div}(\tilde{\rho}\tilde{V}_1) = 0,
\end{aligned}$$

with

$$\begin{aligned}
&\tilde{F} = \varepsilon F(t+\varepsilon^2 s, x+\varepsilon y),\\
&\tilde{V}_i = \varepsilon V_i(t+\varepsilon^2 s, x+\varepsilon y),\\
&\tilde{f} = \varepsilon^2 f(t+\varepsilon^2 s, x+\varepsilon y),\\
&\tilde{\rho} = \rho(t+\varepsilon^2 s, x+\varepsilon y),\\
&\tilde{A} = A(t+\varepsilon^2 s, x+\varepsilon y).
\end{aligned}$$

And

$$\begin{aligned}
&\|\tilde{f}\|_{L^{p_1}(L^{q_1})} = \varepsilon^e\|f\|_{L^{p_1}(L^{q_1})},\\
&\||\tilde{V}_i|^2\|_{L^{p_2}(L^{q_2})} = \varepsilon^e\||V_i|^2\|_{L^{p_2}(L^{q_2})},\\
&\||\tilde{F}|^2\|_{L^{p_2}(L^{q_2})} = \varepsilon^e\||F|^2\|_{L^{p_2}(L^{q_2})}.
\end{aligned}$$

Note that $\tilde{\rho}$ and $\tilde{A}$ are verifying (3.8) and (3.9). Moreover, zooming if necessary, we can always assume that the norms of f and V are as small as we need on a fixed cylinder Q.

Note that the equation is not linear anymore. This introduces some new difficulties. The function $\tilde{u}(t,x) = \lambda u(t,x)$ is solution to

$$\partial_t(\rho\tilde{u}) + \mathrm{div}(\rho\tilde{u}(V_1+V_2)) + \rho V_3\cdot\nabla\tilde{u} - \mathrm{div}(A(t,x)\nabla_x\tilde{u}) = \lambda f + \mathrm{div}(\lambda F),$$

where the right-hand side term becomes large when λ increases. This has to be dealt with when engineering the second lemma of De Giorgi.

We first derive the energy inequality in a form that will be useful for both lemmas.

Lemma 19. *Let u be a suitable solution to (3.1), and ϕ be a cut off function in x only, compactly supported in the domain of definition of u. We define $v = (u-K)_+$. Then for every $s < t$ we have*

$$\left(\int \rho\frac{v^2}{2}\phi^2\,dx\right)(t) - \left(\int \rho\frac{v^2}{2}\phi^2\,dx\right)(s) + \frac{1}{2\Lambda}\int_s^t\int |\nabla(\phi v)|^2\,dx\,dt$$
$$\leq C(1+K^2)(\|\mathcal{F}\|_{(t,s)\times\mathrm{supp}\phi}+1)\left(|\{v>0\}\cap\mathrm{supp}\phi|^{\frac{e}{N+2}}\right)$$
$$\cdot|||(|\phi|+|\nabla\phi|)^2(v^2+\mathbf{1}_{\{v>0\}})|||,$$

with

$$|||g||| = \sum_{(p,q)\in\mathcal{I}} \|g\|_{L^p(L^q)},$$

where the sum is taken on

$$\mathcal{I} = \Bigg\{(p,q), \frac{1}{p}+\frac{1}{p'}+\frac{e}{N+2}=1, \frac{1}{q}+\frac{1}{q'}+\frac{e}{N+2}=1,$$
$$(p',q')\in\{(p_i,q_i),(p_F,q_F),(p_f,q_f)\}\Bigg\}.$$

For all the $(p,q)\in\mathcal{I}$ we have

$$0 < \frac{1}{p'}+\frac{e}{N+2} = 1-\frac{N}{2q'}-\frac{Ne}{N+2} < 1,$$
$$0 < \frac{1}{q'}+\frac{e}{N+2} = 1-\frac{2}{N}\left(1-\frac{1}{p'}\right)-\frac{e}{N} < 1.$$

Hence

$$0 < \frac{1}{p} < 1,$$
$$0 < \frac{1}{q} < 1.$$

Moreover,

$$\frac{2}{p}+\frac{N}{q} = N. \tag{3.12}$$

Proof. Multiplying Equation (3.11) by ϕ^2, and integrating in x, and noticing that $vu = v^2 + Kv$, we get

$$\partial_t\int \rho\frac{v^2}{2}\phi^2\,dx + \frac{1}{\Lambda}\int |\nabla(\phi v)|^2\,dx$$
$$\leq \int \Lambda|\nabla\phi|^2 v^2\,dx$$
$$+\int \phi(\nabla\phi)\cdot(V_1+2V_2)\rho v^2\,dx + \int 2K\rho\phi\nabla\phi\cdot V_2 v\,dx$$
$$+\int \rho\phi^2 v\nabla v\cdot(V_2-V_3)\,dx + \int K\rho\phi^2\nabla v\cdot V_2\,dx$$
$$+\int \phi^2 v f\,dx$$

$$-2\int \phi v(\nabla\phi)\cdot F\,dx - \int \phi^2 \nabla v\cdot F\,dx.$$

Using that $\phi\nabla v = \nabla(v\phi) - v\nabla\phi$, $|V_1| \leq 1 + |V_1|^2$, $\rho \leq \Lambda$, $v = v\mathbf{1}_{\{v>0\}}$, and $\nabla v = \nabla v \mathbf{1}_{\{v>0\}}$, we find

$$\begin{aligned}
&\partial_t \int \rho \frac{v^2}{2}\phi^2\,dx + \frac{1}{\Lambda}\int |\nabla(\phi v)|^2\,dx \\
&\leq \frac{1}{2\Lambda}\int |\nabla(\phi v)|^2\,dx \\
&\quad + C\int (\phi + |\nabla\phi|)^2(|V_1|^2 + (1+K^2)|V_2|^2 + |V_3|^2 + |F|^2 + |f|)(v^2 + \mathbf{1}_{\{v>0\}})\,dx.
\end{aligned}$$

Integrating in time, and using Holder gives the result. □

The first lemma is similar to the previous case. Except that it is easier to work with the energy/dissipation of energy instead of $\int |u|^2\,dx\,dt$.

Lemma 20. *There exists $\delta > 0$ such that for any u solution to (3.7) in $(-1,0)\times B_1$: If*

$$|||u_+^2||| \leq \delta,$$

with

$$\|\mathcal{F}\|_{(-1,0)\times B_1} \leq 1,$$

then

$$u_+ \leq 1/2 \qquad \text{in } Q_{1/2}.$$

Proof. We use the same setting as for the homogenous case, except we are working with

$$\mathcal{E}_k = \sup_{T_k \leq t \leq 0}\left(\int (\phi_k u_k)^2\,dx\right) + \int_{T_k}^0\int_{B_1} |\nabla(\phi_k u_k)|^2\,dx\,dt.$$

Following the previous section we have now from the energy inequality

$$\begin{aligned}
\mathcal{E}_{k+1} &\leq C^k \int_{\tilde{Q}_k} |u_k|^2\,dx\,dt \\
&\quad + C^k(1 + \|\mathcal{F}\|_{(-1,0)\times B_1})|||u_k^2 \mathbf{1}_{\{\tilde{Q}_k\}}|||\left(|\{u_{k+1} > 0\}\cap \tilde{Q}_k|^{\frac{\epsilon}{N+2}}\right). \qquad (3.13)
\end{aligned}$$

Thanks to Sobolev, $\mathcal{E}_k$ control the $L^\infty(L^1)$ norm and the $L^1(L^{N/(N-2)})$ norm of $(\phi_k u_k)^2$. By interpolation, it controls all the norms $L^p(L^q)$ with

$$\frac{2}{\bar{p}} + \frac{N}{\bar{q}} = N.$$

Especially all the norms involved in the definition of $|||\cdot|||$. Hence

$$|||\mathbf{1}_{\{\tilde{Q}_k\}} u_k^2||| \leq C\mathcal{E}_k.$$

And

$$\begin{aligned}\mathcal{E}_{k+1} &\leq C^k \mathcal{E}_k |\{u_{k+1} > 0\}|^{2/(N+2)} + C^k \mathcal{E}_k^{1+2/(N+2)} \\ &\leq C^k \left(\mathcal{E}_k^{1+e/(N+2)} + \mathcal{E}_k^{1+2/(N+2)}\right).\end{aligned}$$

Note that $e < 2$. So as long as $\mathcal{E}_k \leq 1$

$$\mathcal{E}_{k+1} \leq 2C^k \mathcal{E}_k^{1+e/(N+2)}.$$

From (3.13), we can have $\mathcal{E}_1$ as small as we want, taking δ small enough. Then, for δ small enough, we have for all k, $\mathcal{E}_k < 1$ and $\mathcal{E}_k$ converges to 0 when $k \to \infty$. □

As before we denote $\tilde{Q} = (-3/2, -1) \times B_1$.

Theorem 21. *There exists $0 < \lambda < 1$, and $\eta > 0$ such that for any suitable solution u of (3.7) with*

$$|||\mathcal{F}|||_{Q_2} \leq \eta :$$

If $u \leq 1$ on Q_2 and

$$|\{u \leq 0\} \cap \tilde{Q}| \geq \frac{|\tilde{Q}|}{2},$$

then $u \leq 1 - \lambda$ in $Q_{1/2}$.

Because the equation is not linear any more, we need the smallness condition η. As before we first show the following lemma.

Lemma 22. *There exists $\alpha > 0$ such that for any u suitable solution to (3.7) with*

$$|||\mathcal{F}|||_{Q_2} \leq 1 :$$

If $|A| \geq \delta$, $|C| > \frac{\tilde{Q}}{2}$ then $|D| \geq \alpha$.

Proof of Theorem 21. The proof goes by contradiction. The energy gives now

$$\begin{aligned}&\left(\int \phi^2 v_k^2\, dx\right)(t) + \frac{1}{2\Lambda}\int_s^t \int |\nabla(\phi v_k)|^2\, dx\, dt \\ &\leq \left(\int \phi^2 v_k^2\, dx\right)(s) + C|t-s|^{e/(N+2)}.\end{aligned}$$

The function ϕv_k is uniformly bounded in $L^2(-2, 0; H^1(B_2))$, and so $\partial_t v_k$ is uniformly bounded in $L^1(Q_2) + L^\infty(-2, 0; W^{-1}(B_2))$. Hence, up to a subsequence, ϕv_k converges in $L^2(Q_2)$ to ϕv. As before, we show that $|D| = 0$. So for almost very time $t \in (-3/2, 0)$, we have either $v(t, \cdot) \leq 0$, or $v(t, \cdot) \geq 1/2$ in B_1. But $|\{v \leq 0\} \cap \tilde{Q}| \geq |\tilde{Q}|/2|$. So, there exists $s_0 \leq -1$ such that $\int \phi^2 v_k^2(s)\, dx = 0$. Consider $s_0 \leq s \leq 0$ the biggest s such that $\int \phi^2 v_k^2(\tau)\, dx = 0$ for $s_0 \leq \tau \leq s$. Assume that $s < 0$. Using the energy inequality, we get that for every $s < t < 0$

$$\begin{aligned}\int \phi^2 v_k^2(t)\, dx &\leq \int \phi^2 v_k^2(s)\, dx + C_0|t-s|^{e/(N+2)} \\ &\leq C_0|t-s|^{e/(N+2)}.\end{aligned}$$

For $|t-s|$ small enough, $\int \phi^2 v_k^2(t)\,dx \leq 1/8$. Hence we cannot have $v(t,\cdot) \geq 1/2$ for those values of t, and from the dichotomy, we have $v(t,\cdot) \leq 0$. Hence $\int \phi^2 v_k^2(t)\,dx = 0$. This contradicts the definition of s. Hence $|A| = 0$, which leads to a contradiction. □

Proof. Consider k^0 such that $k^0\alpha \geq |Q_1 \cup \tilde{Q}|$, where α is defined in the previous lemma, and denote $\eta = \frac{2^{-k^0}}{1+\Lambda}$. We consider, as before, the sequence

$$w_k = 2^k(u - (1 - 2^{-k}))$$

which verifies for any k: $w_k \leq 1$ and $|\{w_k \leq 0\} \cap \tilde{Q}| \geq |\tilde{Q}|/2$, and

$$\partial_t(\rho w_k) - \mathrm{div}(A\nabla w_k) + \mathrm{div}(\rho w_k(V_1+V_2)) + \rho V_3 \cdot \nabla w_k = 2^k f + \mathrm{div}(2^k F - (2^k-1)\rho V_2),$$

and so for every $k \leq k^0$

$$\|\mathcal{F}_k\|_{Q_2} \leq (1+\lambda)2^k \|\mathcal{F}\|_{Q_2} \leq 1.$$

Then, for a fixed $k \leq k_0$,

$$\text{if } \quad |||(w_j)_+^2||| \geq \delta, \qquad \text{for any } \ j \leq k+1,$$

we get

$$\delta \leq |\{w_{j+1} > 0\} \cap Q_1| = |\{w_j > 1/2\} \cap Q_1|.$$

So, from the previous lemma,

$$|\{0 < w_j < 1/2\} \cap (Q_1 \cup \tilde{Q})| \geq \alpha,$$

and

$$|\{w_k < 0\} \cap (Q_1 \cup \tilde{Q})| \geq k\alpha.$$

But this fails for $k = k^0$. So, for sure, there exists a $k \leq k^0$ such that

$$|||(w_{k+1}^2)_+||| \leq \delta,$$

and

$$\|\mathcal{F}_k\|_{Q_2} \leq 1.$$

So from the first lemma, $w_{k+1} \leq 1/2$ in Q_1. This means that $u \leq 1 - 2^{-(2+k^0)}$ in $Q_{1/2}$. This gives the proof of the result with $\lambda = 2^{-(2+k^0)}$. □

This provides the following easy lemma.

Lemma 23. *If* $-1 \leq u \leq 1$ *in* Q_2 *and* $\|\mathcal{F}\|_{Q_2} \leq \eta$ *then*

$$\mathrm{osc}_{Q_{1/2}} \leq 1 - \lambda.$$

We can now prove the main theorem.

Proof. For $d > 0$, we define $\tilde{\Omega}$ the set of points of Ω whose distance from $\partial\Omega$ is greater to d. Consider $\varepsilon \leq \inf((1, 1/||\mathcal{F}||)^{1/e}, d, \sqrt{t_0}/2)$, and for every $(t, x) \in \tilde{\Omega} \times (t_0, T)$ and $s \leq t_0/(2\varepsilon^2)$

$$u_\varepsilon(s, y) = \varepsilon^{(N+1)/2} u(t + \varepsilon^2 s, x + \varepsilon y).$$

The function u_ε is a suitable solution in Q_1 to (3.7) with

$$||\mathcal{F}||_{Q_1} \leq 1.$$

Moreover

$$|||u_\varepsilon^2|||_{Q_1} \leq \varepsilon |||u^2|||_{\tilde{\Omega}}.$$

So, for ε small enough, this quantity will be smaller than δ. Thanks to Lemma 20, for all those $(t, x) \in \tilde{\Omega}$

$$u_\varepsilon \leq 1/2 \qquad \text{in } Q_{1/2}.$$

In the same way we find that u_ε is bounded by below on $Q_{1/2}$. This gives a uniform L^∞ bound of u on $\tilde{\Omega}$.

The C^α regularity is obtained in a similar way. □

References

[1] E. Baer and A. Vasseur. A bound from below on the temperature for the Navier-Stokes-Fourier system. *SIAM J. Math. Anal.*, 45(4): 2046–2063, 2013.

[2] L. Caffarelli, Ch. H. Chan, and A. Vasseur. Regularity theory for parabolic nonlinear integral operators. *J. Amer. Math. Soc.*, 24(3): 849–869, 2011.

[3] L. Caffarelli and A. Vasseur. Drift diffusion equations with fractional diffusion and the quasi-geostrophic equation. *Ann. of Math. (2)*, 171(3): 1903–1930, 2010.

[4] C. Caputo and A. Vasseur. Global regularity of solutions to systems of reaction-diffusion with sub-quadratic growth in any dimension. *Comm. Partial Differential Equations*, 34(10–12): 1228–1250, 2009.

[5] E. De Giorgi. Sulla differenziabilità e l'analiticità delle estremali degli integrali multipli regolari. *Mem. Accad. Sci. Torino. Cl. Sci. Fis. Mat. Nat. (3)*, 3: 25–43, 1957.

[6] E. DiBenedetto. On the local behaviour of solutions of degenerate parabolic equations with measurable coefficients. *Ann. Scuola Norm. Sup. Pisa Cl. Sci. (4)*, 13(3): 487–535, 1986.

[7] E. DiBenedetto, U. Gianazza, and V. Vespri. Forward, backward and elliptic Harnack inequalities for non-negative solutions to certain singular parabolic partial differential equations. *Ann. Sc. Norm. Super. Pisa Cl. Sci. (5)*, 9(2): 385–422, 2010.

[8] E. DiBenedetto, U. Gianazza, and V. Vespri. Harnack type estimates and Hölder continuity for non-negative solutions to certain sub-critically singular parabolic partial differential equations. *Manuscripta Math.*, 131(1–2): 231–245, 2010.

[9] E. DiBenedetto, U. Gianazza, and V. Vespri. *Harnack's Inequality for Degenerate and Singular Parabolic Equations.* Springer Monographs in Mathematics. Springer, New York, 2012.

[10] M. Felsinger and M. Kassmann. Local regularity for parabolic nonlocal operators. *Comm. Partial Differential Equations*, 38(9): 1539–1573, 2013.

[11] Th. Goudon and A. Vasseur. Regularity analysis for systems of reaction-diffusion equations. *Ann. Sci. Éc. Norm. Supér. (4)*, 43(1): 117–142, 2010.

[12] M. Kassmann. A priori estimates for integro-differential operators with measurable kernels. *Calc. Var. Partial Differential Equations*, 34(1): 1–21, 2009.

[13] O. Ladyzhenskaya and N. Uralt'seva. *Linear and Quasilinear Elliptic Equations.* Translated from the Russian by Scripta Technica, Inc. Translation editor: Leon Ehrenpreis. Academic Press, New York-London, 1968.

[14] J. Moser. A new proof of De Giorgi's theorem concerning the regularity problem for elliptic differential equations. *Comm. Pure Appl. Math.*, 13: 457–468, 1960.

[15] J. Nash. Continuity of solutions of parabolic and elliptic equations. *Amer. J. Math.*, 80: 931–954, 1958.

[16] A. Vasseur. Higher derivatives estimate for the 3D Navier-Stokes equation. *Ann. Inst. H. Poincaré Anal. Non Linéaire*, 27(5): 1189–1204, 2010.

Lectures on the Analysis of Nonlinear Partial Differential Equations Vol. 4
MLM 4, pp. 223–269

The 2D Boussinesq Equations with Partial or Fractional Dissipation

Jiahong Wu* †

Abstract

The two-dimensional (2D) incompressible Boussinesq system is an important model in geophysics. Mathematically the 2D Boussinesq equations serve as a lower-dimensional counterpart of the 3D hydrodynamics equations. In fact, the 2D Boussinesq equations retain some key features of the 3D Euler and Navier-Stokes equations such as the vortex stretching mechanism. The global well-posedness problem on the 2D Boussinesq equations with partial or fractional dissipation has attracted considerable attention in the last few years and significant progress has been made. This paper provides an in-depth overview of these new developments and explains the key ideas and approaches in hope that it will be useful for experienced researchers as well as beginners in this field.

1 Introduction

This paper reviews some very recent developments on the global well-posedness issue concerning the incompressible 2D Boussinesq equations with partial or fractional dissipation. In addition, new results that have not been published before are also presented here. There has been very exciting research on the Boussinesq equations and important progress has been made on the well-posedness and related issues. The paper provides a more or less complete view of these recent results and serves as a handy reference for researchers and practitioners on this subject.

The incompressible 2D Boussinesq equations concerned here can be written as

$$\begin{cases} \mathbf{u}_t + \mathbf{u}\cdot\nabla\mathbf{u} = -\nabla p + \nu\Delta\mathbf{u} + \theta\mathbf{e}_2, \\ \nabla\cdot\mathbf{u} = 0, \\ \theta_t + \mathbf{u}\cdot\nabla\theta = \kappa\Delta\theta, \end{cases} \tag{1.1}$$

where $\mathbf{u}$ denotes the 2D velocity field, p the pressure, θ the temperature in the content of thermal convection and the density in the modeling of geophysical fluids,

*Department of Mathematics, Oklahoma State University, 401 Mathematical Sciences, Stillwater, OK 74078, USA. E-mail: jiahong@math.okstate.edu

†Department of Mathematics, Chung-Ang University, Seoul 156-756, Republic of Korea

ν the viscosity, κ the thermal diffusivity, and $\mathbf{e}_2$ is the unit vector in the vertical direction.

The Boussinesq equations model large scale atmospheric and oceanic flows that are responsible for cold fronts and the jet stream (see the books by Gill [35], Pedlosky [59], Majda [51] and others). In addition, the Boussinesq equations also play an important in the study of Rayleigh-Benard convection (see, e.g., the series of papers by Constantin and Doering [23]).

Mathematically the 2D Boussinesq equations serve as a lower-dimensional model of the 3D hydrodynamics equations. In fact, the 2D Boussinesq equations retain some key features of the 3D Euler and Navier-Stokes equations such as the vortex stretching mechanism. The inviscid 2D Boussinesq equations are identical to the Euler equations for the 3D axisymmetric swirling flows. The fundamental issue of whether classical solutions to the 3D Euler and Navier-Stokes equations can develop finite time singularities remains outstandingly open and the study of the 2D Boussinesq equations may shed light on this extremely challenging problem.

In order to model anisotropic Boussinesq flows (the viscosity and the thermal diffusivity are different in the horizontal and vertical directions), (1.1) is slightly generalized to be written as

$$\begin{cases} u_t + uu_x + vu_y = -p_x + \nu_1\, u_{xx} + \nu_2\, u_{yy}, \\ v_t + uv_x + vv_y = -p_y + \nu_1\, v_{xx} + \nu_2\, v_{yy} + \theta, \\ u_x + v_y = 0, \\ \theta_t + u\theta_x + v\theta_y = \kappa_1\, \theta_{xx} + \kappa_2\, \theta_{yy}, \end{cases} \tag{1.2}$$

where u and v are the horizontal and vertical components of $\mathbf{u}$, $\nu_1 \geq 0, \nu_2 \geq 0, \kappa_1 \geq 0$ and $\kappa_2 \geq 0$. (1.2) is called anisotropic Boussinesq equations.

A different type of generalization is to replace the Laplacian by fractional Laplacian, namely

$$\begin{cases} \mathbf{u}_t + \mathbf{u}\cdot\nabla\mathbf{u} = -\nabla p - \nu(-\Delta)^{\alpha}\mathbf{u} + \theta\mathbf{e}_2, \\ \nabla\cdot\mathbf{u} = 0, \\ \theta_t + \mathbf{u}\cdot\nabla\theta = -\kappa(-\Delta)^{\beta}\theta, \end{cases} \tag{1.3}$$

where $0 < \alpha, \beta \leq 1$. This system will be called fractional Boussinesq equations.

We consider the initial-value problems (IVPs) of the equations in (1.2) and in (1.3) with the initial data

$$\mathbf{u}(x,y,0) = \mathbf{u}_0(x,y), \qquad \theta(x,y,0) = \theta_0(x,y).$$

Attention is focused on the global regularity issue: do these IVPs have a global solution for sufficiently smooth data $(\mathbf{u}_0, \theta_0)$?

This problem and related issues have attracted considerable attention in the last few years and significant progress has been made on the 2D Boussinesq equations with partial or fractional dissipation. The goal here is to provide an in-depth review of these recent results with an emphasis on the ideas and techniques. Some very recent new results of the author and his collaborators are also included. For the sake of clarity, we will divide the rest of this paper into nine sections. The second section provides a short derivation of the 2D Boussinesq equations from the

3D rotating Boussinesq equations, which are the most basic model for atmosphere and ocean flows. The third section is concerned with the inviscid Boussinesq equations. The fourth section explains a recent result on the small global solutions of the damped 2D Boussinesq equations. The fifth section describes the work of Chae [15], and Hou and Li [38] on the 2D Boussinesq equations with partial dissipation Δu or $\Delta\theta$. The sixth section details the work of Danchin and Paicu [32] and Larios, Lunasin and Titi [49] on the case of horizontal dissipation. The seventh section outlines the global regularity result for the case when there is only horizontal thermal diffusion. The eighth section presents the work of Cao and Wu [14] on the global well-posedness for the 2D Boussinesq with vertical dissipation and vertical thermal diffusion. The ninth section summarizes recent results on the 2D Boussinesq equations with fractional dissipation and consists of three subsections with each one of them devoted to the subcritical case, the critical case and the supercritical case, respectively. The last section provides open problems with some discussions.

2 Geophysical background

This section explains how the 2D incompressible Boussinesq equations and another related systems of equations, namely the primitive equations can be formally derived from the 3D rotating Boussinesq equations.

Fluid flows in atmosphere and ocean have two distinctive features: rotation and stratification. The simplest model that contains both features is the 3D rotating Boussinesq equation:

$$\begin{cases} \partial_t \mathbf{u} + \mathbf{u}\cdot\nabla\mathbf{u} + f\mathbf{e}_3\times\mathbf{u} = \nu\Delta\mathbf{u} - \frac{1}{\rho_b}\nabla p + \rho g\,\mathbf{e}_3, \\ \nabla\cdot\mathbf{u} = 0, \\ \partial_t\rho + \mathbf{u}\cdot\nabla\rho = \kappa\Delta\rho, \end{cases} \tag{2.1}$$

where $f = 2\Omega\sin\phi$ with Ω being the angular frequency of planetary rotation and ϕ the latitude, ρ_b is a constant for reference density, $\mathbf{e}_3 = (0,0,1)$, and $f\mathbf{e}_3\times\mathbf{u}$ represents the Coriolis forcing. More explicitly, if $\mathbf{u} = (u, v, w)$, then

$$f\mathbf{e}_3\times\mathbf{u} = f\begin{pmatrix} -v \\ u \\ 0 \end{pmatrix},$$

and (2.1) in terms of components becomes

$$\begin{cases} \partial_t u + u\partial_x u + v\partial_y u + w\partial_z u - fv = \nu\Delta u - \frac{1}{\rho_b}\partial_x p, \\ \partial_t v + u\partial_x v + v\partial_y v + w\partial_z v + fu = \nu\Delta v - \frac{1}{\rho_b}\partial_y p, \\ \partial_t w + u\partial_x w + v\partial_y w + w\partial_z w = \nu\Delta w - \frac{1}{\rho_b}\partial_z p + \rho g, \\ \partial_x u + \partial_y v + \partial_z w = 0, \\ \partial_t\rho + u\partial_x\rho + v\partial_y\rho + w\partial_z\rho = \kappa\Delta\rho. \end{cases} \tag{2.2}$$

More details on the 3D rotating Boussinesq equations can be found in [51] (see also [35, 59]).

For atmospheric and oceanic flows in the mid-latitude, the w-equation can be simplified. The terms involving w in the w-equation are small and the w-equation is reduced to

$$\frac{1}{\rho_b}\partial_z p - \rho g = 0. \tag{2.3}$$

This is the so called the hydrostatic balance. (2.3) provides a special solution of (2.2). In fact, $\mathbf{u} = 0$ with p and ρ satisfying (2.3) solves (2.2). The system of equations containing the u-equation, the v-equation in (2.2), the divergence-free condition, the ρ-equation and (2.3), or more explicitly

$$\begin{cases} \partial_t u + u\partial_x u + v\partial_y u + w\partial_z u - fv = \nu\Delta u - \frac{1}{\rho_b}\partial_x p, \\ \partial_t v + u\partial_x v + v\partial_y v + w\partial_z v + fu = \nu\Delta v - \frac{1}{\rho_b}\partial_y p, \\ \frac{1}{\rho_b}\partial_z p - \rho g = 0, \\ \partial_x u + \partial_y v + \partial_z w = 0, \\ \partial_t \rho + u\partial_x \rho + v\partial_y \rho + w\partial_z \rho = \kappa\Delta\rho \end{cases}$$

are called the primitive equations. The primitive equations have been studied by many authors (see, e.g. [12, 35, 50, 51, 59]).

If $f \equiv 0$, (2.2) becomes the 3D Boussinesq equations without rotation. If $f \equiv 0$ and all physical quantities are independent of z, then (2.2) reduces to the 2D Boussinesq equations, which read

$$\begin{cases} \partial_t u + u\partial_x u + v\partial_y u = -\partial_x p + \nu\Delta u, \\ \partial_t v + u\partial_x v + v\partial_y v = -\partial_y p + \nu\Delta v + \theta, \\ \partial_x u + \partial_y v = 0, \\ \partial_t \theta + u\partial_x \theta + v\partial_y \theta = \kappa\Delta\theta. \end{cases}$$

3 The inviscid 2D Boussinesq equations

When the 2D Boussinesq equations do not involve any dissipation, the global regularity problem is extremely difficult and remains open. Even the global existence of weak solutions is unknown. This section presents the local well-posedness result and standard regularity criteria. A discussion on a potential criterion in terms of the vorticity alone is also provided here.

Recall that the inviscid 2D Boussinesq equations are given by

$$\begin{cases} \partial_t \mathbf{u} + (\mathbf{u}\cdot\nabla)\mathbf{u} = -\nabla p + \theta\mathbf{e}_2, \\ \nabla\cdot\mathbf{u} = 0, \\ \partial_t\theta + (\mathbf{u}\cdot\nabla)\theta = 0, \end{cases} \tag{3.1}$$

where $\mathbf{e}_2$ denotes the unit vector in the vertical direction. The issue is whether or not (3.1) has a global solution when supplemented with a reasonably "regular" data,

$$\mathbf{u}(x,0) = \mathbf{u}_0(x), \quad \theta(x,0) = \theta_0(x).$$

The global regularity problem for a general data is currently open. As pointed out in [52], the inviscid Boussinesq can be identified with the 3D axisymmetric Euler with swirl. Recall the 3D axisymmetric Euler equations are given by

$$u_t^\theta + (u^r\partial_r + u^z\partial_z)u^\theta + \frac{u^r u^\theta}{r} = 0, \qquad \frac{\widetilde{D}}{Dt}(ru^\theta) = 0, \quad \frac{\widetilde{D}}{Dt}(ru^\theta)^2 = 0,$$

where $\frac{\widetilde{D}}{Dt} = \partial_t + (u^r\partial_r + u^z\partial_z)$. The swirl component of the vorticity ω^θ satisfies

$$\frac{\widetilde{D}}{Dt}\left(\frac{\omega^\theta}{r}\right) = -\frac{1}{r^4}\partial_z(ru^\theta)^2.$$

While the 2D Boussinesq is given by, with $\frac{D}{Dt} = \partial_t + u\cdot\nabla$,

$$\frac{D}{Dt}\theta = 0, \qquad \frac{D}{Dt}\omega = -\partial_{x_1}\theta.$$

We do have local well-posedness and several regularity criteria (see, e.g., W. E and C. Shu [34], Chae and Nam [19] and Danchin [29]).

Theorem 3.1. *Assume $u_0 \in H^s(\mathbb{R}^2)$ and $\theta_0 \in H^s(\mathbb{R}^2)$ with $s > 2$. Then there is $T = T(\|(\mathbf{u}_0,\theta_0)\|_{H^s}) > 0$ such that the 2D inviscid Boussinesq* (3.1) *has a unique solution $(\mathbf{u},\theta) \in C([0,T];H^s)$. Furthermore, if*

$$\int_0^{T^*} \|\nabla\mathbf{u}\|_{L^\infty} dt < \infty \quad or \quad \int_0^{T^*} \|\nabla\theta\|_{L^\infty} dt < \infty,$$

then the local solution can be extended to $[0,T^*]$.

The local well-posedness can be easily established through the standard ODE theory such as the contraction mapping principle or successive approximation. We remark that $s > 2$ appears to be necessary and it is not clear whether or not the local well-posedness can be obtained in the borderline space H^2. The main reason is that H^1 is not embedded in L^∞. A recent work of Chae and Wu partially addressed this issue (see [21]). Another recent work of Bourgain and Li on the Euler equations hints possible ill-posedness of (3.1) in the borderline space H^2 (see [7]).

In the following we provide the proof for the regularity criteria and start with several simple facts. The first fact is a commutator estimate (see, e.g., Kenig, Ponce and Vega [44]). For any $s \in \mathbb{R}$, we write

$$J^s f = (1-\Delta)^{\frac{s}{2}} f \quad \text{or} \quad \widehat{J^s f(\xi)} = (1+|\xi|^2)^{\frac{s}{2}}\widehat{f}(\xi).$$

Lemma 3.2. *Let $s > 0$ and $1 < p < \infty$. Then*

$$\|J^s(fg)\|_{L^p} \le C\left(\|f\|_{W^{s,p_1}}\|g\|_{L^{p_2}} + \|g\|_{W^{s,p_3}}\|f\|_{L^{p_4}}\right),$$
$$\|J^s(fg) - fJ^s g\|_{L^p} \le C\left(\|f\|_{W^{s,p_1}}\|g\|_{L^{p_2}} + \|g\|_{W^{s-1,p_3}}\|\nabla f\|_{L^{p_4}}\right),$$

where $\frac{1}{p} = \frac{1}{p_1} + \frac{1}{p_2} = \frac{1}{p_3} + \frac{1}{p_4}$ with $p_2, p_4 \in [1,\infty]$ and $p_1, p_3 \in (1,\infty)$, and C's are constants depending on s, p, p_1, p_2, p_3 and p_4. In addition, these inequalities remain valid when $J^s f$ is replaced by $\Lambda^s f$, where

$$\widehat{\Lambda^s f}(\xi) = |\xi|^s \widehat{f}(\xi).$$

The next lemma bounds the L^q-norm of the solution to a simple transport equation. For notational simplicity, we will just use u to denote the velocity field in the rest of this section.

Lemma 3.3. *Let $T > 0$ and let $q \in [1,\infty]$. Assume that $\omega_0 \in L^q(\mathbb{R}^d)$ and $f \in L^1([0,T]; L^q(\mathbb{R}^d))$. Assume that u is Lipschitz and divergence-free. Then the solution ω to the transport equation*

$$\partial_t \omega + u \cdot \nabla \omega = f, \quad \omega(x,0) = \omega_0(x)$$

obeys

$$\|\omega\|_{L^\infty([0,T];L^q)} \le \|\omega_0\|_{L^q} + \int_0^T \|f\|_{L^q} dt.$$

The following lemma is a logarithmic type interpolation inequality.

Lemma 3.4. *Let $\omega \in H^\sigma(\mathbb{R}^2)$ with $\sigma > 1$. Assume $u \in L^2$ satisfies $\omega = \nabla \times u$. Then*

$$\|\nabla u\|_{L^\infty} \le C(1 + \|u\|_{L^2}) + C\|\omega\|_{L^\infty} \log(1 + \|\omega\|_{H^\sigma}),$$

where C's are constants depending on σ only.

Remark 3.5. *Logarithmic type inequalities bound a quantity by another quantity that can almost control the first quantity together with the logarithm of the norm of a higher derivative. In this lemma here ∇u can be written as a singular integral operator on ω, namely $\nabla u = P_2 * \omega$. For any $q \in (1,\infty)$, $\|\nabla u\|_{L^q} \le C\frac{q^2}{q-1}\|\omega\|_{L^q}$ for a constant independent of q. Clearly this bound fails when either $q = 1$ or $q = \infty$. This lemma states that the L^∞-norm can be bounded by $\|\omega\|_{L^\infty}$ times the logarithm of H^σ for any $\sigma > 1$.*

Proof of Lemma 3.4. This lemma can be easily proven via the Littlewood-Paley decomposition and Besov space techniques. The definitions of the Littlewood-Paley decomposition and Besov spaces can be found in many books and papers (say, e.g., [5, 6, 54, 60, 61, 63]). Writing

$$\nabla u = \sum_{j=-1}^{\infty} \Delta_j \nabla u,$$

we have, by the fact that $\|\Delta_j \nabla u\|_{L^\infty} \le C\|\Delta_j \omega\|_{L^\infty}$ for any $j \ge 0$ and Bernstein's inequality,

$$\begin{aligned}\|\nabla u\|_{L^\infty} &\le \|\Delta_{-1}\nabla u\|_{L^\infty} + \sum_{j=0}^{N} C\|\Delta_j\omega\|_{L^\infty} + \sum_{j=N+1}^{\infty} 2^j\|\Delta_j\omega\|_{L^2}\\ &\le C\|u\|_{L^2} + C(N+1)\|\omega\|_{L^\infty} + \sum_{j=N+1}^{\infty} 2^{j(1-\sigma)}2^{\sigma j}\|\Delta_j\omega\|_{L^2}\\ &\le C\|u\|_{L^2} + C(N+1)\|\omega\|_{L^\infty} + \|\omega\|_{H^\sigma}\left(\sum_{j=N+1}^{\infty} 2^{2j(1-\sigma)}\right)^{\frac{1}{2}}\\ &\le C\|u\|_{L^2} + C(N+1)\|\omega\|_{L^\infty} + \|\omega\|_{H^\sigma}2^{(N+1)(1-\sigma)}C(\sigma).\end{aligned}$$

Now taking an integer N such that $\|\omega\|_{H^\sigma}2^{(N+1)(1-\sigma)} \le 1$, we then have $N+1 \le \frac{1}{\sigma-1}\log_2(1+\|\omega\|_{H^\sigma})$. Therefore

$$\|\nabla u\|_{L^\infty} \le C(1+\|u\|_{L^2}) + C\|\omega\|_{L^\infty}\log(1+\|\omega\|_{H^\sigma}).$$

This completes the proof of Lemma 3.4. □

Proof of theorem 3.1. The local existence and uniqueness can be obtained by the Picard type theorems. One can follow the steps leading to the local well-posedness result for the Euler and the Navier-Stokes equations, as in the book of Majda and Bertozzi [52].

Let (u,θ) be a local solution and assume that $\int_0^{T^*}\|\nabla u\|_{L^\infty}dt < \infty$. The goal is to obtain a priori bound for H^s-norm on $[0,T^*]$. To do so, we start with the vorticity formulation

$$\begin{cases}\partial_t\omega + u\cdot\nabla\omega = \partial_x\theta,\\ \partial_t\theta + u\cdot\nabla\theta = 0.\end{cases}$$

Applying J^{s-1} to the vorticity equation, multiplying by $J^{s-1}\omega$ and integrating in space, we obtain, after integration by parts,

$$\frac{1}{2}\frac{d}{dt}\|J^{s-1}\omega\|_{L^2}^2 = \int \partial_x J^{s-1}\theta J^{s-1}\omega - \int J^{s-1}\omega(J^{s-1}\nabla\cdot(u\omega) - u\cdot J^{s-1}\nabla\omega).$$

By Lemma 3.2,

$$\|J^{s-1}\nabla\cdot(u\omega) - u\cdot J^{s-1}\nabla\omega\|_{L^2} \le C\left(\|J^s u\|_{L^2}\|\omega\|_{L^\infty} + \|J^{s-1}\omega\|_{L^2}\|\nabla u\|_{L^\infty}\right).$$

Therefore, by the simple fact $\|J^s u\|_{L^2} = \|J^{s-1}\omega\|_{L^2}$,

$$\frac{d}{dt}\|J^{s-1}\omega\|_{L^2}^2 \le \|J^s\theta\|_{L^2}^2 + \|J^{s-1}\omega\|_{L^2}^2 + C\,\|J^{s-1}\omega\|_{L^2}^2\|\nabla u\|_{L^\infty}.$$

Similarly

$$\frac{d}{dt}\|J^s\theta\|_{L^2}^2 \le C\,\|J^s\theta\|_{L^2}(\|J^s u\|_{L^2}\|\nabla\theta\|_{L^\infty} + \|J^s\theta\|_{L^2}\|\nabla u\|_{L^\infty}).$$

Furthermore, it following from $\partial_t\theta + u\cdot\nabla\theta = 0$ that, for any $q \in [1,\infty]$,

$$\|\nabla\theta\|_{L^q} \le \|\nabla\theta_0\|_{L^q} e^{\int_0^t \|\nabla u\|_{L^\infty}\,d\tau}.$$

In fact, by a standard energy estimate, we have, for any $1 \le q < \infty$,

$$\frac{1}{q}\frac{d}{dt}\|\nabla\theta\|_{L^q}^q \le \|\nabla u\|_{L^\infty}\|\nabla\theta\|_{L^q}^q,$$

which leads to the desired inequality for $q \in (1,\infty)$ and the case $q=\infty$ is obtained by taking $q \to \infty$. Therefore, $Y(t) = \|J^{s-1}\omega\|_{L^2}^2 + \|J^s\theta\|_{L^2}^2$ obeys

$$\frac{d}{dt}Y(t) \le \|\nabla\theta_0\|_{L^\infty} e^{2\int_0^t \|\nabla u\|_{L^\infty}\,d\tau}\, Y(t) + C\,Y(t)(1+\|\nabla u\|_{L^\infty}).$$

By Gronwall's inequality, if $\int_0^{T^*}\|\nabla u\|_{L^\infty}dt < \infty$ then $Y(t) < \infty$ on $[0,T^*]$ or $(u,\theta) \in H^s$.

Now we show that $\int_0^{T^*}\|\nabla\theta\|_{L^\infty}dt < +\infty$ implies that $(u,\theta) \in H^s$ for $t \in [0,T^*]$. According to the estimates above

$$\frac{d}{dt}Y(t) \le CY(t)(\|\nabla\theta\|_{L^\infty} + 1 + \|\nabla u\|_{L^\infty}).$$

By $\partial_t\omega + u\cdot\nabla\omega = \partial_x\theta$ and Lemma 3.3, we have

$$\|\omega(t)\|_{L^\infty} \le \|\omega_0\|_{L^\infty} + \int_0^t \|\nabla\theta\|_{L^\infty}\,d\tau.$$

By Lemma 3.4, for $\sigma > 1$,

$$\|\nabla u\|_{L^\infty} \le C(1+\|u\|_{L^2}) + C\,\|\omega\|_{L^\infty}\log(1+\|\omega\|_{H^\sigma}).$$

Therefore, for $s > 2$,

$$\begin{aligned}\frac{dY(t)}{dt} &\le CY(t)(1+\|\nabla\theta\|_{L^\infty} + \|u\|_{L^2} + \|\omega\|_{L^\infty}\log(1+\|J^{s-1}\omega\|_{L^2}))\\ &\le CY(t)(1+\|\nabla\theta\|_{L^\infty} + \|u\|_{L^2} + \|\omega\|_{L^\infty}\log(1+Y(t))),\end{aligned}$$

which implies immediately that, if $\int_0^{T^*}\|\nabla\theta\|_{L^\infty}dt < \infty$, then $Y(t) < \infty$ on $[0,T^*]$. this completes the proof of Theorem 3.1. □

Finally we make a remark.

Remark 3.6. *For the 3D Euler equation,* $\mathbf{u}_t + \mathbf{u}\cdot\nabla\mathbf{u} = -\nabla p$, $\nabla\cdot\mathbf{u} = 0$, *we have the Beale-Kato-Majda criterion, which says that if*

$$\int_0^T \|\omega(\cdot,t)\|_\infty\,dt < \infty,$$

then the solution remains regular on $[0,T]$. *However, for the inviscid 2D Boussinesq equations, it is unknown if*

$$\int_0^T \|\omega(\cdot,t)\|_\infty \, dt < \infty$$

is enough for the global regularity. This does not appear to be trivial.

What is the difficulty? Following the standard idea, we try the energy estimate. Applying $J^s = (I-\Delta)^{\frac{s}{2}}$ *to* $\partial_t\theta + \mathbf{u}\cdot\nabla\theta = 0$,

$$\partial_t J^s\theta + \mathbf{u}\cdot\nabla J^s\theta = -J^s(u\cdot\nabla\theta) + \mathbf{u}\cdot\nabla J^s\theta,$$

$$\frac{1}{2}\frac{d}{dt}\|J^s\theta\|_{L^2}^2 = \int (-J^s(\mathbf{u}\cdot\nabla\theta) + \mathbf{u}\cdot\nabla J^s\theta)J^s\theta.$$

By the commutator estimate,

$$\|-J^s(u\cdot\nabla\theta) + \mathbf{u}\cdot\nabla J^s\theta\|_{L^2} \le C(\|\nabla\mathbf{u}\|_{L^\infty}\|\theta\|_{H^s} + \|\nabla\theta\|_{L^\infty}\|\mathbf{u}\|_{H^s}).$$

Then,

$$\frac{1}{2}\frac{d}{dt}\|\theta\|_{H^s}^2 \le C\|\nabla\mathbf{u}\|_{L^\infty}\|\theta\|_{H^s}^2 + \|\nabla\theta\|_{L^\infty}\|\mathbf{u}\|_{H^s}\|\theta\|_{H^s}.$$

If $\|\nabla\theta\|_{L^\infty}$ *were not there, then we would combine*

$$\frac{1}{2}\frac{d}{dt}\|\mathbf{u}\|_{H^s}^2 \le C\|\nabla\mathbf{u}\|_{L^\infty}\|\mathbf{u}\|_{H^s}^2 + \|\mathbf{u}\|_{H^s}\|\theta\|_{H^s}$$

with the logarithmic interpolation inequality

$$\|\nabla\mathbf{u}\|_{L^\infty} \le C\|\mathbf{u}\|_{L^2} + C\|\omega\|_{L^\infty}\log(1+\|\mathbf{u}\|_{H^s})$$

to obtain an ODE of the form, $X = \|\mathbf{u}\|_{H^s}^2 + \|\theta\|_{H^s}^2$,

$$\frac{d}{dt}X(t) \le C\|\omega\|_{L^\infty}X(t)\log(1+X(t)) + C\|\mathbf{u}\|_{L^2}.$$

But unfortunately it appears to be difficult to control $\|\nabla\theta\|_{L^\infty}$ *in terms of the vorticity alone. A work of Cao, Constantin, Kiselev and Wu is in progress* [11].

4 Small global solutions to damped Boussinesq equations

Since the the global regularity problem on the inviscid Boussinesq equations appears to be out of reach, a recent work of Adhikari, Cao, Wu and Xu [4] intended to understand this difficult problem by examining how damping affects the regularity of the solutions to the Boussinesq equations. [4] examined the following 2D Boussinesq equations with damping

$$\begin{cases} \partial_t u + (u\cdot\nabla)u + \nu u = -\nabla p + \theta\mathbf{e}_2, & x\in\mathbb{R}^2,\ t>0,\\ \partial_t\theta + (u\cdot\nabla)\theta + \lambda\theta = 0, & x\in\mathbb{R}^2,\ t>0,\\ \nabla\cdot u = 0, & x\in\mathbb{R}^2,\ t>0,\\ u(x,0) = u_0(x), \quad \theta(x,0) = \theta_0(x), & x\in\mathbb{R}^2, \end{cases} \tag{4.1}$$

where $\nu > 0$ and $\lambda > 0$ are real parameters. When there is no confusion, we shall just use u instead of $\mathbf{u}$ for the velocity field.

As we know, the issue of global existence and uniqueness relies crucially on whether or not one can obtain global bounds on the solutions. Thanks to the divergence-free condition $\nabla \cdot u = 0$, global *a priori* bounds for θ in any Lebesgue space L^q and u in L^2 follow directly from simple energy estimates,

$$\|\theta(t)\|_{L^q} \le \|\theta_0\|_{L^q}, \qquad \|u(t)\|_{L^2} \le \|u_0\|_{L^2} + t\,\|\theta_0\|_{L^2}$$

for $1 \le q \le \infty$. However, global bounds for (u, θ) in any Sobolev space, say H^1, can not be easily achieved and the difficulty comes from the vortex stretching term. More precisely, if we resort to the equations of the vorticity ω and $\nabla^\perp\theta$

$$\begin{cases} \partial_t\omega + (u\cdot\nabla)\omega + \nu\omega = \partial_{x_1}\theta, \\ \partial_t(\nabla^\perp\theta) + (u\cdot\nabla)(\nabla^\perp\theta) + \lambda\nabla^\perp\theta = (\nabla^\perp\theta\cdot\nabla)u, \end{cases}$$

we unavoidably have to deal with the "vortex stretching term" $(\nabla^\perp\theta\cdot\nabla)u$, which appears to elude any suitable bound. Here $\nabla^\perp = (-\partial_{x_2}, \partial_{x_1})$. The damping terms are not sufficient to overcome this difficulty. Therefore damping does not appear to make a big difference in dealing with solutions emanating from a general data.

The aim here is at the global existence and uniqueness of small solutions. We remark that, if we consider solutions of (4.1) with initial data in the classical setting, say $(u_0, \theta_0) \in H^s$ with $s > 2$, then it is not difficult to prove the global existence of solutions when we impose a very strong smallness condition such as

$$1 + \|u_0\|_{H^s} + \|\theta_0\|_{H^s} \le C\min\{\nu, \lambda\}, \tag{4.2}$$

where C is a suitable constant independent of ν and λ. In fact, the global regularity follows easily from the local well-posedness in H^s and the global energy inequality for $Y(t) \equiv \|u(t)\|_{H^s} + \|\theta(t)\|_{H^s}$,

$$\frac{d}{dt}Y(t) + \min\{\nu,\lambda\}\,Y(t) \le C(1 + \|u\|_{H^s} + \|\theta\|_{H^s})\,Y(t).$$

However, the smallness assumption (4.2) appears to be too restrict. In particular, it forces ν and λ to be of order 1. It appears that (4.2) can not be easily weakened if we seek solutions in the classical functional setting. This is due to the presence of the forcing term $\theta\mathbf{e}_2$ in the velocity equation and the growth of $\|u(t)\|_{H^s}$ in time. In fact, as shown by Brandolese and Schonbek for the 3D viscous Boussinesq equations, the L^2-norm of u may grow in time when the spatial integral of θ_0 is not zero and when θ does not decay sufficiently fast in time [8]. Our efforts have been devoted to seeking a suitable functional setting so that (4.2) can be relaxed. The right functional space is the homogeneous Besov space $\mathring{B}^1_{\infty,1}$ and our main result can be stated as follows.

Theorem 4.1. *Consider* (4.1) *with* $\nu > 0$ *and* $\lambda > 0$. *Assume that* $(u_0, \theta_0) \in L^2$ *obeys the smallness conditions*

$$\|\nabla u_0\|_{\mathring{B}^0_{\infty,1}} < A_0 \equiv \min\left\{\frac{\nu}{2C_0}, \frac{\lambda}{C_0}\right\} \quad \textit{and} \quad \|\nabla\theta_0\|_{\mathring{B}^0_{\infty,1}} < B_0 \equiv \frac{\nu}{2C_0}\|\nabla u_0\|_{\mathring{B}^0_{\infty,1}} \tag{4.3}$$

for a suitable constant C_0 independent of ν and λ. Then (4.1) *has a unique global solution* (u, θ) *satisfying*

$$(u, \theta) \in L^\infty([0,\infty); L^2), \quad \nabla u, \nabla\theta \in L^\infty([0,\infty); \mathring{B}^0_{\infty,1}).$$

In addition,

$$\sup_{t\geq 0} \|\nabla u(t)\|_{\mathring{B}^0_{\infty,1}} < A_0 \quad \text{and} \quad \sup_{t\geq 0} \|\nabla\theta(t)\|_{\mathring{B}^0_{\infty,1}} < B_0.$$

We remark that the smallness condition (4.3) is weaker than (4.2) in two senses: first, the norm in $\mathring{B}^1_{\infty,1}$ (the smallness for $(\nabla u, \nabla\theta)$ in $\mathring{B}^0_{\infty,1}$ is equivalent to the smallness of (u, θ) in $\mathring{B}^1_{\infty,1}$) is weaker than the norm in H^s due to the embedding $H^s(\mathbb{R}^2) \hookrightarrow \mathring{B}^1_{\infty,1}(\mathbb{R}^2)$ for $s > 2$; and second, (4.3) does not include the factor 1, as in (4.2). $\mathring{B}^1_{\infty,1}$ appears to be a very natural setting if one wants to ensure the uniqueness of the solutions. It may be difficult to further weaken the functional setting.

May it be possible to sharpen the result of Theorem 4.1 by removing one of the damping terms $\nu\, u$ or $\lambda\,\theta$? This problem appears to be extremely challenging. For the 2D Boussinesq equations, it appears that any functional setting guaranteeing the uniqueness of solutions necessarily involves the derivatives of the functions. If $\lambda\,\theta$ is not present, the norm of θ in such a functional setting may grow exponentially (in time) at the rate of $\|\nabla u\|_{L^\infty}$. Consequently the norm of u may grow even when the velocity equation has the damping term νu. It is also clear that, if $\nu\, u$ is missing, then the norm of u is expected to grow. Therefore, when any one of the damping terms is removed, the small data well-posedness problem becomes as difficult as the well-posedness problem for a general initial data.

The detailed proof of Theorem 4.1 can be found in [4].

5 Partial dissipation with either Δu or $\Delta\theta$

If the 2D Boussinesq equations are fully dissipative ((1.1) with $\nu > 0$ and $\kappa > 0$), the global regularity can be easily established (see, e.g., [10]). The proof is almost trivial and similar to that for the 2D Navier-Stokes equations. Therefore, it is natural to examine the 2D Boussinesq equations with partial dissipation. This section reviews the global regularity results on (1.1) with either $\nu > 0$ or $\kappa > 0$. The global regularity issue for these partial dissipation cases was listed by H.K. Moffatt as one of the 21st century PDE problems [56]. The results presented here were obtained by D. Chae [15] and by T. Hou and C. Li [38].

We start with the case when only Δu is present, namely

$$\begin{cases} \partial_t u + u\cdot\nabla u = -\nabla p + \nu\Delta u + \theta \mathbf{e}_2, \\ \nabla\cdot u = 0, \\ \partial_t\theta + u\cdot\nabla\theta = 0. \end{cases} \tag{5.1}$$

The global regularity result can be stated as follows.

Theorem 5.1. *Let* $(u_0, \theta_0) \in W^{2,q}(\mathbb{R}^2)$ *with* $2 < q < \infty$. *Then* (5.1) *has a unique global solution* $(u, \theta) \in C([0,\infty); W^{2,q}(\mathbb{R}^2))$.

Proof. First of all, for any $q \in [1,\infty]$, $\|\theta(t)\|_q \le \|\theta_0\|_q$ and

$$\|u\|_2^2 + \nu \int_0^t \|\nabla u\|_2^2 d\tau \le (\|u_0\|_2 + t\|\theta_0\|_2)^2.$$

In particular, due to $\|\nabla u\|_{L^2} = \|\omega\|_{L^2}$, for any $t > 0$,

$$\nu \int_0^t \|\omega\|_{L^2}^2 \, dt < \infty.$$

For $2 \le q < \infty$, it follows from $\partial_t \omega + u \cdot \nabla \omega = \nu \Delta \omega + \partial_{x_1} \theta$ that

$$\begin{aligned}
&\frac{1}{q}\frac{d}{dt}\|\omega\|_q^q + (q-1)\nu \int |\nabla\omega|^2 |\omega|^{q-2} dx \\
&= \int \partial_{x_1}\theta\omega|\omega|^{q-2} dx \\
&= (q-1)\int \theta\partial_{x_1}\omega|\omega|^{q-2} \\
&= (q-1)\int \theta|\omega|^{\frac{q-2}{2}}\partial_{x_1}\omega|\omega|^{\frac{q-2}{2}} \\
&\le (q-1)\|\nabla\omega|\omega|^{\frac{q-2}{2}}\|_2 \|\theta\|_q \|\omega\|_q^{\frac{q-2}{2}} \\
&\le \frac{\nu(q-1)}{2}\int |\nabla\omega|^2|\omega|^{q-2} dx + (q-1)\frac{1}{2\nu}\|\theta\|_q^2 \|\omega\|_q^{q-2}.
\end{aligned}$$

Therefore,

$$\frac{d}{dt}\|\omega\|_q^q + q(q-1)\nu \int |\nabla\omega|^2|\omega|^{q-2} \le \frac{q(q-1)}{\nu}\|\theta\|_q^2 \|\omega\|_q^{q-2}.$$

In particular,

$$\begin{aligned}
\frac{d}{dt}\|\omega\|_q^2 &\le \frac{2(q-1)}{\nu}\|\theta_0\|_q^2, \\
\|\omega\|_q^2 &\le \|\omega_0\|_q^2 + \frac{2(q-1)}{\nu} t \|\theta_0\|_q^2, \\
\|\omega\|_q &\le \|\omega_0\|_q + \sqrt{\frac{2}{\nu}t(q-1)}\|\theta_0\|_q. \qquad (5.2)
\end{aligned}$$

We thus have obtained a global (in time) bound for $\|\omega\|_q$ with any $2 \le q < \infty$, but this does not lead to a global bound for $\|\omega\|_\infty$. As a consequence,

$$\|u\|_\infty \le C\,\|u\|_2^a \|\nabla u\|_q^{1-a} \le C\,\|u\|_2^a \|\omega\|_q^{1-a} < \infty,$$

where $q \in (2,\infty)$ and $a = \frac{q-2}{2(q-1)}$. We now bound $\|\nabla\omega\|_q$ and $\|\nabla\theta\|_q$.

$$\begin{aligned}
&\frac{1}{q}\frac{d}{dt}\|\nabla\omega\|_q^q + \nu\int |\Delta\omega|^2|\nabla\omega|^{q-2} + \frac{1}{4}(q-2)\nu\int |\nabla|\nabla\omega|^2|^2|\nabla\omega|^{q-4} \\
&= -\int \nabla(u\cdot\nabla\omega)\nabla\omega|\nabla\omega|^{q-2} + \int \partial_{x_1}\nabla\theta\cdot\nabla\omega|\nabla\omega|^{q-2}. \qquad (5.3)
\end{aligned}$$

To see why the dissipation yields the second and third term on the left-hand side, we write the inner product in terms of the components and integrate by parts

$$\begin{aligned}
&\int \nabla\Delta\omega \cdot \nabla\omega|\nabla\omega|^{q-2}\\
&= \int \partial_j\partial_k\partial_k\omega\partial_j\omega|\nabla\omega|^{q-2}\\
&= -\int (\partial_j\partial_k\omega)^2\,|\nabla\omega|^{q-2} - (q-2)\int(\partial_j\partial_k\omega\partial_j\omega)^2|\nabla\omega|^{q-4},
\end{aligned}$$

where repeated indices are summed. To estimate the terms on the right, we integrate by parts,

$$\begin{aligned}
&\int u\cdot\nabla\omega\nabla(\nabla\omega|\nabla\omega|^{q-2})\\
&= \int u\cdot\nabla\omega\left(\Delta\omega|\nabla\omega|^{q-2} + \frac{q-2}{2}\nabla\omega\nabla(|\nabla\omega|^2)|\nabla\omega|^{q-4}\right)\\
&\le \|u\|_\infty\|\nabla\omega\|_q\|\nabla\omega\|_q^{\frac{q}{2}-1}\|\Delta\omega|\nabla\omega|^{\frac{q-2}{2}}\|_2\\
&\quad+\|u\|_\infty\|\nabla\omega\|_q\frac{q-2}{2}\|\nabla\omega\|_q^{\frac{q-2}{2}}\|\nabla|\nabla\omega|^2|\nabla\omega|^{\frac{q-4}{2}}\|_2.
\end{aligned}$$

Then, the entire right hand side of (5.3) can be bounded by

$$\frac{\nu}{2}\|\Delta\omega|\nabla\omega|^{\frac{q-2}{2}}\|_2^2+\frac{4(q-1)}{\nu}(\|u\|_\infty^2\|\nabla\omega\|_q^q+\|\nabla\theta\|_q^q)+\frac{q-2}{2}\nu\|\nabla(|\nabla\omega|^2)|\nabla\omega|^{\frac{q-4}{2}}\|_2^2.$$

For the θ equation, we can get

$$\frac{1}{q}\frac{d}{dt}\|\nabla\theta\|_q^q \le \|\nabla u\|_\infty\|\nabla\theta\|_q^q.$$

Then

$$\frac{1}{q}\frac{d}{dt}(\|\nabla\omega\|_q^q+\|\nabla\theta\|_q^q) \le \frac{4}{\nu}(q-1)\|u\|_\infty^2\|\nabla\omega\|_q^q + \left(\frac{4}{\nu}(q-1)+\|\nabla u\|_\infty\right)\|\nabla\theta\|_q^q.$$

We don't know if $\int_0^t\|\nabla u\|_\infty dt<\infty$. We need a logarithmic type inequality. By lemma 5.2 below,

$$\|\nabla u\|_{L^\infty} \le c(1+\|w\|_{H^1})\left[\log(1+\|\nabla w\|_{L^q})\right]^{\frac{1}{2}}.$$

It then follows that $Y(t) = \|\nabla w\|_{L^q}^q + \|\nabla\theta\|_{L^q}^q$ satisfies

$$\frac{1}{q}\frac{dY(t)}{dt} \le cY(t) + c(1+\|w\|_{H^1})Y(t)\left[\log(1+Y(t))\right]^{\frac{1}{2}}.$$

Since, for any $T>0$,

$$\int_0^T \|w\|_{H^1}\,dt < C(T),$$

we have

$$Y(t) \le C(T).$$

Together with the Gagliardo-Nirenberg inequality

$$\|\nabla u\|_{L^\infty} \le c\|\nabla u\|_{L^2}^a \|\nabla w\|_{L^q}^{1-a} \quad \text{with} \quad a = \frac{q-2}{2(q-1)},$$

we obtain

$$\int_0^T \|\nabla u\|_{L^\infty} dt < \infty,$$

which implies the regularity of the solution. This completes the proof. □

Lemma 5.2. *For $q \in (2,\infty)$, we have*

$$\|f\|_{L^\infty(\mathbb{R}^d)} \le c(1 + \|f\|_{\dot H^{d/2}}) \left[\log\left(1 + \|f\|_{\dot B^{d/2}_{q,\infty}}\right)\right]^{\frac{1}{2}} \tag{5.4}$$

and

$$\|f\|_{L^\infty(\mathbb{R}^d)} \le c\,\|f\|_{L^2} + c\,\|f\|_{\dot B^{d/2}_{2,\infty}} \log\left(1 + \|f\|_{\dot B^{d/2}_{q,\infty}}\right), \tag{5.5}$$

where c's are constants depending on d and q only. In particular, when $d = 2$,

$$\|f\|_{L^\infty} \le c(1 + \|f\|_{\dot H^1}) \left[\log\left(1 + \|\nabla f\|_{L^q}\right)\right]^{\frac{1}{2}}.$$

Proof. We first prove (5.4). Write f as the inhomogeneous Littlewood-Paley decomposition,

$$f = \sum_{j=-1}^{\infty} \triangle_j f.$$

Let $N > 0$ be an integer to be specified later. Then, by Bernstein's inequality,

$$\begin{aligned}
\|f\|_{L^\infty} &\le \|\triangle_{-1} f\|_{L^\infty} + \sum_{j=0}^{N} \|\triangle_j f\|_{L^\infty} + \sum_{j=N+1}^{\infty} \|\triangle_j f\|_{L^\infty} \\
&\le c\|\triangle_{-1} f\|_{L^2} + \sum_{j=0}^{N} 2^{jd(\frac{1}{2}-\frac{1}{\infty})} \|\triangle_j f\|_{L^2} + \sum_{j=N+1}^{\infty} 2^{jd(\frac{1}{q}-\frac{1}{\infty})} \|\triangle_j f\|_{L^q} \\
&\le c\|f\|_{L^2} + (\sum_{j=0}^{N} 1)^{1/2} (\sum_{j=0}^{N} (2^{j\frac{d}{2}} \|\Delta_j f\|_{L^2})^2)^{1/2} + \sum_{j=N+1}^{\infty} 2^{jd(\frac{1}{q}-\frac{1}{\infty})} \|\triangle_j f\|_{L^q} \\
&\le c\|f\|_{L^2} + (N+1)^{1/2} \|f\|_{\dot B^{d/2}_{2,2}} + \sum_{j=N+1}^{\infty} 2^{jd(\frac{1}{q}-\frac{1}{2})} 2^{jd(\frac{1}{2})} \|\triangle_j f\|_{L^q} \\
&= c\|f\|_{L^2} + (N+1)^{1/2} \|f\|_{\dot H^{d/2}} + \|f\|_{\dot B^{d/2}_{q,\infty}} \sum_{j=N+1}^{\infty} 2^{jd(\frac{1}{q}-\frac{1}{2})} \\
&\le c\|f\|_{L^2} + (N+1)^{1/2} \|f\|_{\dot H^{d/2}} + c\|f\|_{\dot B^{d/2}_{q,\infty}} 2^{(N+1)d(\frac{1}{q}-\frac{1}{2})}
\end{aligned}$$

Now taking an integer N such that

$$\|f\|_{\dot{B}^{d/2}_{q,\infty}} 2^{(N+1)d(\frac{1}{q}-\frac{1}{2})} \le 1,$$

we then obtain (5.4). The inequality in (5.5) can be proven by slightly changing the proof above. In fact, if we employ the following bound instead of the one above

$$\sum_{j=0}^{N} 2^{jd(\frac{1}{2}-\frac{1}{\infty})}\|\triangle_j f\|_{L^2} \le (N+1)\sup_j 2^{jd(\frac{1}{2}-\frac{1}{\infty})}\|\triangle_j f\|_{L^2} = (N+1)\|f\|_{\dot{B}^{d/2}_{2,\infty}},$$

then the procedure above leads to (5.5). This completes the proof of Lemma 5.2.

Next, we consider the system with thermal diffusion $\Delta\theta$, namely

$$\begin{cases} \partial_t u + (u\cdot\nabla)u = -\nabla p + \theta \mathbf{e}_2, \\ \partial_t \theta + (u\cdot\nabla)\theta = \kappa\Delta\theta \end{cases}$$

with $\kappa > 0$. This system of equations has also been shown to possess global solutions for sufficiently smooth data. More precisely, we have the following theorem:

Theorem 5.3. *Let* $(\theta_0, u_0) \in W^{2,q}(\mathbb{R}^2), 2 < q < \infty$. *Then the system has a unique global solution*

$$(u,\theta) \in C([0,\infty); W^{2,q}).$$

The following theorem is proven in a very similar way.

Proof. The local existence and uniqueness part is standard. To obtain the global estimate, we reply more on the regularity of θ due to the diffusion in the θ-equation. Clearly, we have the basic bounds, for any $q \in [2,\infty)$,

$$\|\theta\|_{L^q}^q + q(q-1)\kappa \int_0^t \int |\nabla\theta|^2|\theta|^{q-2} dx d\tau = \|\theta_0\|_{L^q}^q,$$

$$\|u\|_{L^2} \le \|u_0\|_{L^2} + t\|\theta_0\|_{L^2}.$$

From the vorticity equation $\partial_t\omega + (u\cdot\nabla)\omega = \partial_{x_1}\theta$, we get

$$\frac{1}{2}\partial_t\|\omega\|_{L^2}^2 = \int (\partial_{x_1}\theta)\omega \le \|\partial_{x_1}\theta\|_{L^2}\|\omega\|_{L^2}.$$

Therefore,

$$\|\omega(T)\|_{L^2} \le \|w_0\|_{L^2} + \int_0^T \|\partial_{x_1}\theta\|_{L^2} dt \le \|\omega_0\|_{L^2} + \left(\int_0^T \|\nabla\theta\|_{L^2}^2 dt\right)^{\frac{1}{2}} \sqrt{T}.$$

Next,

$$\frac{1}{q}\partial_t\|w\|_{L^q}^q = \int (\partial_{x_1}\theta) w|w|^{q-2} dx \le \|\nabla\theta\|_{L^q}\|w\|_{L^q}^{q-1}.$$

Or somply

$$\partial_t\|w\|_{L^q} \le c\|\nabla\theta\|_{L^q}.$$

On the other hand,

$$\frac{1}{q}\partial_t\|\nabla\theta\|_{L^q}^q+\kappa\int|\Delta\theta|^2|\nabla\theta|^{q-2}+\kappa\left(\frac{q-2}{4}\right)\int|\nabla|\nabla\theta|^2|^2|\nabla\theta|^{q-4}dx$$
$$=-\int\nabla((u\cdot\nabla)\theta)\cdot\nabla\theta|\nabla\theta|^{q-2}$$
$$\le C\,\|u\|_{L^\infty}\left[(q-1)\|\nabla\theta\|_{L^q}^{\frac{q}{2}}\|\nabla|\nabla\theta|^2|\nabla\theta|^{\frac{q-4}{2}}\|_{L^2}+\|\nabla\theta\|_{L^q}^{\frac{q}{2}}\|\Delta\theta|\nabla\theta|^{\frac{q-2}{q}}\|_{L^2}\right].$$

Then

$$\frac{1}{q}\partial_t\|\nabla\theta\|_{L^q}^q\le c\frac{(q-1)}{\kappa}\|u\|_{L^\infty}^2\|\nabla\theta\|_{L^q}^q$$

or

$$\partial_t\|\nabla\theta\|_{L^q}\le c(\frac{q-1}{\kappa})\|u\|_{L^\infty}^2\|\nabla\theta\|_{L^q}.$$

Using the Lemma 5.2 again, we get

$$\|u\|_{L^\infty}\le c\|u\|_{L^2}+c(1+\|w\|_{L^2})\sqrt{\log(1+\|\nabla w\|_{L^p})}.$$

Therefore,

$$Z(t)=\|w\|_{L^q}+\|\nabla\theta\|_{L^q}$$

obeys

$$\frac{d}{dt}Z(t)\le c(1+\|w\|_{L^2})Z(t)\log(1+Z(t))+c.$$

This allows us to obtain a global bound for $Z(t)$. This completes the proof of Theorem 5.3.

6 Horizontal dissipation

This section reviews the well-posedness results on the 2D Boussinesq equations with horizontal dissipation from the work of R. Danchin and M. Paicu [32] and from that of Larios, Lunasin and Titi [49]. Consider the IVP for the 2D Boussinesq equations with horizontal dissipation

$$\begin{cases}\partial_t u+(u\cdot\nabla)u=-\nabla p+\nu u_{xx}+\theta\,\mathbf{e}_2,\\ \nabla\cdot u=0,\\ \partial_t\theta+(u\cdot\nabla)\theta=0,\\ u(x,0)=u_0(x),\ \theta(x,0)=\theta_0(x).\end{cases}\tag{6.1}$$

(6.1) has been shown to possess a unique global solution for suitable (u_0,θ_0) and the following theorem combines the main results of [32] with [49].

Theorem 6.1. *Let $u_0\in H^1(\mathbb{R}^2)$ and $\nabla\cdot u_0=0$. Assume $\omega_0=\nabla\times u_0\in\sqrt{L}$, namely*

$$\sup_{q\ge2}\frac{\|\omega_0\|_{L^q}}{\sqrt{q}}<+\infty.$$

Let $\theta_0 \in L^2 \cap L^\infty$. Then the IVP (6.1) has a unique solution (u, θ) satisfying

$$u \in L^\infty_{loc}([0,\infty); H^1), \quad \omega \in L^\infty_{loc}([0,\infty); \sqrt{L}), \quad u_2 \in L^2_{loc}([0,\infty); H^2),$$

$$\theta \in C_b([0,\infty); L^2), \quad \theta \in L^\infty([0,\infty); L^\infty).$$

In addition, if $\theta_0 \in H^s$ with $s \in (1/2, 1)$, then

$$\theta \in L^\infty_{loc}([0,\infty); H^{s-\epsilon}). \tag{6.2}$$

for any $\epsilon > 0$.

Remark 6.2. *The paper of R. Danchin and M. Paicu [32] originally assumed that $\theta_0 \in H^s$ with $s \in (1/2, 1)$ to show the uniqueness. Later Larios, Lunasin and Titi [49] was able to prove the uniqueness without this assumption.*

Remark 6.3. *The regularity result for θ in (6.2) comes from a typical losing type estimate.*

The proof of Theorem 6.1 is divided into several steps, which can be accomplished through the following subsections.

6.1 Global weak solution

This subsection proves the global existence of weak solutions in a very weak functional setting via Friedriches' Method. This method cuts off the high frequencies and thus smooths the functions. The global existence result can be stated as follows.

Theorem 6.4. *Let $\theta_0 \in L^2 \cap L^\infty$ and $u_0 \in H^1$ and $\nabla \cdot u_0 = 0$. Then (6.1) has a global weak solution (u, θ) satisfying $\theta \in L^\infty([0,\infty); L^2 \cap L^\infty)$, $u \in L^\infty_{loc}([0,\infty); H^1)$, $u_2 \in L^2_{loc}([0,\infty); H^2)$.*

Proof. (Friedriches' Method) Let $n \in \mathbb{N}$ and define

$$L^2_n = \{f \in L^2(\mathbb{R}^2) | \operatorname{supp} \widehat{f} \subset B(0,n)\},$$

$$J_n f = (\chi_{B(0,n)} \widehat{f})^\vee,$$

where $\widehat{f}$ and $f^\vee$ denotes the Fourier and the inverse Fourier transforms, respectively, and $\chi_{B(0,n)}$ is the characteristic function on $B(0,n)$. Clearly, $J_n f \in H^\infty = \bigcap_{s \ge 0} H^s$.

Consider the equations

$$\begin{cases} \partial_t \theta + J_n \nabla \cdot (J_n u J_n \theta) = 0, \\ \partial_t u + \mathbb{P} J_n \nabla \cdot (J_n \mathbb{P} u \otimes J_n \mathbb{P} u) = \nu J_n \mathbb{P} \partial_{xx} u + J_n \mathbb{P}(\theta \mathbf{e}_2), \\ u(x,0) = J_n u_0, \quad \theta(x,0) = J_n \theta_0, \end{cases} \tag{6.3}$$

where $\mathbb{P}$ denotes the Leray projection. By the Picard theorem, there exists $T^* > 0$ and a solution $(u, \theta) \in C^1([0,T), L^2_n)$ (of course the solution depends on n, but here we just write (u, θ) for notational convenience). Noticing that $J_n f = f$ if $f \in L^2_n$ and $\mathbb{P}F = F$ if $\nabla \cdot F = 0$, we have

$$\begin{cases} \partial_t \theta + J_n \nabla \cdot (u\theta) = 0, \\ \partial_t u + \mathbb{P} J_n \nabla \cdot (u \otimes u) = \nu \partial_{xx} u + \mathbb{P}(\theta \mathbf{e}_2). \end{cases}$$

By the energy method

$$\|\theta\|_{L^2} \le \|J_n \theta_0\|_2 \le \|\theta_0\|_{L^2},$$

$$\|u\|_{L^2}^2 + 2\nu \int_0^t \|\partial_x u\|_{L^2}^2 dt \le (\|u_0\|_{L^2} + t\|\theta_0\|_{L^2})^2.$$

Taking the curl of the u-equation yields

$$\partial_t \omega + \mathbb{P} J_n (u \cdot \nabla \omega) = \nu \partial_{xx} \omega + \partial_x \theta$$

and thus

$$\begin{aligned} \|\omega\|_2^2 + 2\nu \int_0^t \|\partial_x \omega\|_2^2 d\tau &\le \int_0^t \|\theta_0\|_2 \|\partial_x \omega\|_2 d\tau \\ &\le \nu \int_0^t \|\partial_x \omega\|_2^2 d\tau + \frac{C}{\nu} \int_0^t \|\theta_0\|_2^2 d\tau. \end{aligned}$$

Therefore, $\theta \in L^\infty([0,\infty); L^2)$ and $u \in L^\infty([0,T]; H^1)$ for any $T > 0$. By the Picard Extension Theorem, (θ, u) is global in time and admits bounds that are uniform in n,

$$\theta^{(n)} \in L^\infty([0,\infty); L^2), \qquad u^{(n)} \in L^\infty([0,T]; H^1).$$

In addition, it can be shown that

$$\partial_t \theta^{(n)} \in L^\infty([0,T]; H^{-3/2}), \qquad \partial_t u^{(n)} \in L^\infty([0,T]; H^{-1})$$

Since $L^2 \hookrightarrow H^{-3/2}$ locally and $H^1 \hookrightarrow H^{-1}$ locally, the Aubin-Lions compactness lemma then implies $u^{(n)} \to u$ in H^L_{loc} for any $-1 \le L < 1$ and $\theta^{(n)} \to \theta$ in H^L_{loc} for any $-3/2 \le L < 0$. We can use these convergence to pass the limit in the weak formulation. This completes the proof.

6.2 Local classical solution

This subsection shows that (6.1) has a unique local classical solution.

Theorem 6.5. *Let $(u_0, \theta_0) \in H^s \times H^{s-1}$ with $s > 2$. Then there is $T > 0$ and a unique solution $(u, \theta) \in C([0,T), H^s \times H^{s-1})$ satisfying* (6.1).

Proof. We can either use the mollifier approach as in the book of Majda and Bertozzi [52] or Friedriches' approach above. The crucial part is a local priori bound. Define

$$J^s f = (1-\Delta)^{s/2} f \quad \text{or} \quad \widehat{J^s f}(\xi) = (1+|\xi|^2)^{\frac{s}{2}} \hat{f}(\xi).$$

Then clearly $\|J^s f\|_{L^2} = \|f\|_{H^s}$. It follows from the vorticity equation

$$\partial_t \omega + u \cdot \nabla \omega = \nu \partial_{x_1 x_1} \omega + \partial_{x_1} \theta$$

that

$$\frac{1}{2}\frac{d}{dt}\|J^{s-1}\omega\|_2^2 + \nu\|\partial_{x_1} J^{s-1}\omega\|_2^2 = K_1 + K_2,$$

where

$$K_1 = \int \partial_{x_1} J^{s-1}\theta J^{s-1}\omega,$$
$$K_2 = -\int (J^{s-1}(u\cdot\nabla\omega) - u\cdot\nabla J^{s-1}\omega) J^{s-1}\omega.$$

K_1 and K_2 can be bounded as follows. By integration by parts,

$$|K_1| \le \frac{\nu}{2}\|\partial_{x_1} J^{s-1}\omega\|_2^2 + \frac{1}{2\nu}\|J^{s-1}\theta\|_2^2.$$

Applying the commutator estimates in Lemma 3.2 yields

$$\begin{aligned}|K_2| &\le \|J^{s-1}(u\cdot\nabla\omega) - u\cdot\nabla J^{s-1}w\|_2\|J^{s-1}\omega\|_2\\ &\le C(\|J^s u\|_2\|\omega\|_\infty + \|J^{s-1}\omega\|_2\|\nabla u\|_\infty)\|J^{s-1}\omega\|_2\\ &\le C\,\|\nabla u\|_\infty\|J^{s-1}w\|_2^2.\end{aligned}$$

Similarly, we have

$$\begin{aligned}\frac{1}{2}\frac{d}{dt}\|J^{s-1}\theta\|_2^2 &= -\int (J^{s-1}(u\cdot\nabla\theta) - u\cdot\nabla J^{s-1}\theta)J^{s-1}\theta\\ &\le C\|J^{s-1}\theta\|_2(\|J^s u\|_2\|\theta\|_\infty + \|J^{s-1}\theta\|_2\|\nabla u\|_\infty)\\ &\le C\|\theta\|_\infty(\|J^{s-1}\omega\|_2^2 + \|J^{s-1}\theta\|_2^2) + C\|\nabla u\|_\infty\|J^{s-1}\theta\|_2^2.\end{aligned}$$

Therefore, $Y(t) = \|J^{s-1}\omega\|_2^2 + \|J^{s-1}\theta\|_2^2$ satisfies

$$\frac{d}{dt}Y(t) \le C\,(1 + \|\theta\|_\infty + \|\nabla u\|_\infty)\,Y(t).$$

Since $s > 2$,

$$\|\nabla u\|_\infty \le C\,\|\nabla u\|_{H^{s-1}} \le C\,\|J^{s-1}\omega\|_2$$

and consequently

$$\frac{d}{dt}Y(t) \le C\left(1 + \|\theta\|_\infty + \sqrt{Y(t)}\right)\,Y(t).$$

This inequality implies that $\exists T^*$ such that $Y(t) \le C$ for $t < T^*$. This completes the proof. □

6.3 Global classical solution

This section establishes a global bound for $\|u\|_{H^s}$ and $\|\theta\|_{H^{s-1}}$, which allows us to extend the local solution in the previous subsection to a global one.

Theorem 6.6. *Assume that* $(u_0, \theta_0) \in H^s \times H^{s-1}$ *with* $s > 2$. *Then* (6.1) *has a unique global solution* $(u, \theta) \in C([0, \infty), H^s \times H^{s-1})$.

Proof. It suffices to obtain a global bound for (u, θ) in $H^s \times H^{s-1}$. From the previous proof, $Y(t) = \|J^{s-1}\omega\|_2^2 + \|J^{s-1}\theta\|_2^2$ satisfies

$$\frac{d}{dt} Y(t) \leq C(1 + \|\theta\|_\infty + \|\nabla u\|_\infty) Y(t). \tag{6.4}$$

The trick is still to control $\|\nabla u\|_\infty$ through the following logarithmic interpolation inequality, for $\sigma > 1$,

$$\|f\|_\infty \leq \sup_{q \geq 2} \frac{\|f\|_{L^q}}{\sqrt{q}} \sqrt{\log(1 + \|f\|_{H^\sigma})},$$

as given in Lemma 6.7 below. In particular, for $s > 2$, we have

$$\begin{aligned} \|\nabla u\|_\infty &\leq C \sup_{q \geq 2} \frac{\|\nabla u\|_q}{\sqrt{q}} \sqrt{\log(1 + \|\nabla u\|_{H^{s-1}})} \\ &\leq C \sup_{q \geq 2} \frac{\|\nabla u\|_q}{\sqrt{q}} \sqrt{\log(1 + \|\omega\|_{H^{s-1}})}. \end{aligned} \tag{6.5}$$

The goal is then to show that, for any $T > 0$,

$$\int_0^T \sup_{q \geq 2} \frac{\|\nabla u\|_q}{\sqrt{q}} \, dt < \infty. \tag{6.6}$$

This is accomplished through two steps. First, we have

$$\|\omega(T)\|_2^2 + 2\nu \int_0^T \|\partial_{x_1} \omega\|_2^2 \, dt \leq \frac{2T}{\nu} \|\theta_0\|_2^2$$

and, as in (5.2),

$$\sup_{q \geq 2} \frac{\|\omega\|_q}{\sqrt{q}} \leq \sup_{q \geq 2} \frac{\|\omega_0\|_q}{\sqrt{q}} + \sqrt{\frac{2t}{\nu}} \|\theta_0\|_{L^2 \cap L^\infty}.$$

We caution that the inequality $\|\nabla u\|_q \leq \frac{cq^2}{q-1} \|\omega\|_q$ does not really help. Instead, we use the imbedding inequality, for a constant C independent of q,

$$\sup_{q \geq 2} \frac{\|f\|_q}{\sqrt{q}} \leq C \|f\|_{H^1}$$

and the simple fact that $\|\partial_{x_1} \omega\|_2 = \|\nabla \partial_{x_1} u\|_2$, we have

$$\int_0^T \sup_{q \geq 2} \frac{\|\partial_{x_1} u\|_q}{\sqrt{q}} \, dt \leq C \int_0^T \|\partial_{x_1} \nabla u\|_2 dt = C \int_0^T \|\partial_{x_1} \omega\|_2 dt < \infty.$$

Due to the divergence-free condition $\partial_{x_1} u_1 + \partial_{x_2} u_2 = 0$, we also have

$$\int_0^T \sup_{q\ge 2} \frac{\|\partial_{x_2} u_2\|_q}{\sqrt{q}}\, dt < \infty.$$

In addition, $\partial_{x_2} u_1 = \partial_{x_1} u_2 - \omega$ and thus

$$\int_0^T \sup_{q\ge 2} \frac{\|\partial_{x_2} u_1\|_q}{\sqrt{q}}\, dt \le \int_0^T \sup_{q\ge 2} \frac{\|\partial_{x_1} u_2\|_q}{\sqrt{q}}\, dt + \int_0^T \sup_{q\ge 2} \frac{\|\omega\|_q}{\sqrt{q}}\, dt < \infty.$$

Thus we have proven (6.6). Combining (6.4), (6.5) and (6.6) yields the desired global bound. This completes the proof of Theorem 6.6. □

We now provide the lemma used in the proof of Theorem 6.6.

Lemma 6.7. *For $\sigma > 1$,*

$$\|f\|_\infty \le C(\sigma) \sup_{q\ge 2} \frac{\|f\|_{L^q}}{\sqrt{q}} \sqrt{\log(1 + \|f\|_{H^\sigma})},$$

$$\sup_{q\ge 2} \frac{\|f\|_{L^q}}{\sqrt{q}} \le C\, \|f\|_{H^1},$$

where C's are constants.

6.4 Proof of Theorem 6.1

This subsection outlines the proof of Theorem 6.1.

Proof of Theorem 6.1. It consists of three main steps. The first step proves the existence, the second proves the uniqueness while the last step provides the losing estimates. The first two steps are provided here, but the losing estimates are in [32].

To prove the existence, we first regularize the initial condition. For $\epsilon > 0$, set

$$J_\epsilon f = \rho_\epsilon * f,$$

where $\rho_\epsilon = \frac{1}{\epsilon^2}\rho_0(\frac{x}{\epsilon})$ with $\rho_0 \in C_0^\infty(\mathbb{R}^2)$, $\int_{\mathbb{R}^2} \rho_0(x)\, dx = 1$ and $\rho(x) = 1$ for $|x| \le \frac{1}{2}$. Consider (6.1) with the initial data $u^\epsilon(x,0) = J_\epsilon * u_0$ and $\theta^\epsilon(x,0) = J_\epsilon * \theta_0$. Since, for any $s > 0$,

$$\|u_0^\epsilon\|_{H^s} \le C \frac{1}{\epsilon^{s-1}} \|u_0\|_{H^1}, \quad \|\theta_0^\epsilon\|_{H^s} \le C \frac{1}{\epsilon^s} \|\theta_0\|_{L^2},$$

where C is a constant independent of ϵ. By Theorem 6.6, there exists a unique solution

$$(u^\epsilon, \theta^\epsilon) \in C([0,\infty); H^s \times H^{s-1}).$$

Since $(u^\epsilon, \theta^\epsilon)$ admits a global uniform bound in $H^1 \times L^2$, $(u^\epsilon, \theta^\epsilon) \rightharpoonup (u, \theta)$ in $H^1 \times L^2$ and (u, θ) is a weak solution according to Theorem 6.4.

We now show the uniqueness. R. Danchin and M. Paicu [32] assumed that

$$\theta_0 \in H^s \cap L^\infty \quad \text{with} \quad s \in \left(\frac{1}{2}, 1\right).$$

Larios, Lunasin and Titi [49] obtained the uniqueness by just assuming $\theta_0 \in L^2 \cap L^\infty$. What is presented here is a modified version of Larios, Lunasin and Titi and the essential idea is the Yudovich approach. Let $(u^{(1)}, \theta^{(1)})$ and $(u^{(2)}, \theta^{(2)})$ be two solutions. Then the difference $(\widetilde{u}, \widetilde{\theta})$,

$$\widetilde{u} = u^{(1)} - u^{(2)}, \quad \widetilde{\theta} = \theta^{(1)} - \theta^{(2)},$$

satisfies

$$\begin{cases} \partial_t \widetilde{u} + u^{(1)} \cdot \nabla \widetilde{u} + \widetilde{u} \cdot \nabla u^{(2)} = \partial_{xx} \widetilde{u} - \nabla \widetilde{p} + \widetilde{\theta} \mathbf{e}_2, \\ \partial_t \widetilde{\theta} + u^{(1)} \cdot \nabla \widetilde{\theta} + \widetilde{u} \cdot \nabla \theta^{(2)} = 0. \end{cases}$$

The second equation demands too much regularity on θ. To exchange into a variable with less regularity requirement, we introduce

$$\Delta \xi^{(1)} = \theta^{(1)}, \quad \Delta \xi^{(2)} = \theta^{(2)}, \quad \widetilde{\xi} = \xi^{(1)} - \xi^{(2)}.$$

Clearly, $\widetilde{\xi}$ satisfies

$$\partial_t \Delta \widetilde{\xi} + u^{(1)} \cdot \nabla (\Delta \widetilde{\xi}) + \widetilde{u} \cdot \nabla \Delta \xi^{(2)} = 0.$$

The goal is to show $\widetilde{u} \equiv 0$ and $\widetilde{\xi} \equiv 0$. It follows from simple energy estimates that

$$\frac{1}{2} \frac{d}{dt} \|\widetilde{u}\|_2^2 + \|\partial_x \widetilde{u}\|_2^2 = - \int \widetilde{u} \cdot \nabla u^{(2)} \cdot \widetilde{u} + \int \Delta \widetilde{\xi} \mathbf{e}_2 \cdot \widetilde{u},$$

$$\frac{1}{2} \frac{d}{dt} \|\nabla \widetilde{\xi}\|_2^2 = - \int u^{(1)} \cdot \nabla \Delta \widetilde{\xi}\, \widetilde{\xi} - \int \widetilde{u} \cdot \nabla \Delta \xi^{(2)}\, \widetilde{\xi}.$$

We now estimates the terms on the right.

$$\begin{aligned} \int \Delta \widetilde{\xi}\, \vec{e_2} \cdot u &= \int \Delta \widetilde{\xi}\, \widetilde{u}_2 = \int (\partial_{x_1 x_1} + \partial_{x_2 x_2}) \widetilde{\xi}\, \widetilde{u}_2 \\ &= - \int \partial_{x_1} \widetilde{\xi} \partial_{x_1} \widetilde{u}_2 + \int \partial_{x_2} \widetilde{\xi} \partial_{x_1} \widetilde{u}_1 \\ &\le \frac{1}{2} \|\nabla \widetilde{\xi}\|_2^2 + \frac{1}{2} \|\partial_{x_1} \widetilde{u}\|_2^2. \end{aligned}$$

Integrating by parts and noticing that $\theta^{(2)} = \Delta \widetilde{\xi}^{(2)}$, we have

$$\begin{aligned} - \int \widetilde{u} \cdot \nabla \Delta \xi^{(2)} \widetilde{\xi} &= \int \widetilde{u} \cdot \nabla \widetilde{\xi} \Delta \widetilde{\xi}^{(2)} \\ &\le \frac{1}{2} \|\theta^{(2)}\|_\infty (\|\widetilde{u}\|_2^2 + \|\nabla \widetilde{\xi}\|_2^2). \end{aligned}$$

For notational convenience, we write

$$J_1 = -\int \widetilde{u} \cdot \nabla u^{(2)} \cdot \widetilde{u}, \qquad J_2 = -\int u^{(1)} \cdot \nabla \Delta \widetilde{\xi}\ \widetilde{\xi}.$$

Clearly

$$|J_1| \le \|\widetilde{u}\|_\infty^{\frac{2}{p}} \int |\nabla u^{(2)}||\widetilde{u}|^{2-\frac{2}{p}} \le \|\widetilde{u}\|_\infty^{\frac{2}{p}} \|\nabla u^{(2)}\|_p \|\widetilde{u}\|_2^{2-\frac{2}{p}}.$$

We know that

$$\sup_{p\ge 2} \frac{\|\nabla u^{(2)}\|_p}{p} \le C_0$$

for C_0 independent of p. Also we have $\|\widetilde{u}\|_\infty \le M \equiv C\,\|\widetilde{u}\|_2^a\,\|\nabla \widetilde{u}\|_q^{1-a}$ for any $q > 2$ and $a = \frac{q-2}{2(q-1)}$. Therefore,

$$|J_1| \le C_0\, p\, M^{\frac{2}{p}}\, \|\widetilde{u}\|_2^{2-\frac{2}{p}}.$$

Using the simple fact that a function $f(x) = xA^{\frac{2}{x}}$ has the minimum $2e \log A$, we have

$$|J_1| \le 2C_0\, (\log M - \log \|\widetilde{u}\|_2) \|\widetilde{u}\|_2^2.$$

Integrating by parts yields

$$\begin{aligned} J_2 &= \int u^{(1)} \cdot \nabla \widetilde{\xi}\ \Delta \widetilde{\xi} \\ &= \int u^{(1)} \cdot \nabla \widetilde{\xi}\ \partial_k \partial_k \widetilde{\xi} \\ &= -\int \partial_k u^{(1)} \cdot \nabla \widetilde{\xi}\ \partial_k \widetilde{\xi} - \int u^{(1)} \cdot \nabla \partial_k \widetilde{\xi}\ \partial_k \widetilde{\xi} \\ &= -\int \partial_k u^{(1)} \cdot \nabla \widetilde{\xi}\ \partial_k \widetilde{\xi}. \end{aligned}$$

Therefore,

$$\begin{aligned} |J_2| &\le \int |\nabla u^{(1)}|\ |\nabla \widetilde{\xi}|^2 \\ &\le \|\nabla \widetilde{\xi}\|_\infty^{\frac{2}{p}} \int |\nabla u^{(1)}||\nabla \widetilde{\xi}|^{2-\frac{2}{p}}. \end{aligned}$$

Since $\|\nabla \widetilde{\xi}\|_\infty \le M$,

$$\begin{aligned} |J_2| &\le CpM^{\frac{2}{p}} \|\nabla \widetilde{\xi}\|_2^{2-\frac{2}{p}} \\ &\le C(\log M - \log \|\nabla \widetilde{\xi}\|_2)\ \|\nabla \widetilde{\xi}\|_2^2. \end{aligned}$$

Combining the estimates allows us to conclude that $X(t) = \|\widetilde{u}\|_2^2 + \|\nabla \widetilde{\xi}\|_2^2$ satisfies

$$\frac{d}{dt} X + \|\partial_x \widetilde{u}\|_2^2 \le CX + C\,(\log M - \log X)\, X,$$

where C's are constants.

$$\frac{d}{dt}X - CX \le C(\log M - \log X)X$$
$$\frac{d}{dt}(e^{-Ct}X) \le Ce^{-Ct}(\log M - \log X)X$$
$$X(t) \le e^{Ct}X(0) + C\int_0^t e^{C(t-\tau)}(\log M - \log X)X\,d\tau.$$

Since $X(0) = 0$, we have $X(t) \equiv 0$ for any $t > 0$ by Osgood inequality (see Lemma 6.8 below). This completes the proof of Theorem 6.1. □

Lemma 6.8 (Osgood inequality). *Let $\rho_0 \ge 0$ be a constant and $\alpha(t) \ge 0$ be a continuous function. Assume*

$$\rho(t) \le \rho_0 + \int_0^t \alpha(\tau)w(\rho(\tau))d\tau,$$

where w satisfies

$$\int_1^\infty \frac{1}{w(r)}dr = \infty.$$

Then $\rho_0 = 0$ implies $\rho \equiv 0$, and $\rho_0 > 0$ implies that

$$-\Omega(\rho(t)) + \Omega(\rho_0) \le \int_0^t \alpha(\tau)d\tau,$$

where $\Omega(\rho) = \int_\rho^1 \frac{dr}{w(r)}$.

In particular, the Osgood inequality applies to $w(\rho) = \rho \log \frac{M}{\rho}$ since

$$\int_1^\infty \frac{1}{r(\log M - \log r)}dr = \infty.$$

6.5 Losing regularity estimates for θ

We stated in Theorem 6.1 that, if $\theta_0 \in H^s \cap L^\infty$ with $s \in \left(\frac{1}{2}, 1\right)$, then $\theta \in H^{s-\epsilon}$ for any $\epsilon > 0$. The solution is no longer in the same functional setting as the initial data and the loss of regularity is due to the velocity is not Lipschitz. We state without proof the following theorem (see [32]). Other losing type estimates can be found in [5].

Theorem 6.9. *Consider*

$$\begin{cases} \partial_t \rho + u\cdot\nabla\rho = f, & x \in \mathbb{R}^2, \quad t > 0, \\ \rho(0) = \rho_0, & x \in \mathbb{R}^2 \end{cases}$$

Assume $\rho_0 \in B^s_{2,\infty}(\mathbb{R}^2)$ *with* $-1 < s < 1$, $f \in L^1([0,T]; B^s_{2,\infty})$, $\nabla \cdot u = 0$ *and*

$$\sup_{N\geq 2} \frac{\|\nabla S_N u\|_{L^\infty}}{\sqrt{N}} \leq V(t)$$

for some $V \in L^1(0,T)$. *Then for any* $\epsilon \in (0, \frac{1+s}{2})$,

$$\|\rho\|_{B^{s-\epsilon}_{2,\infty}} \leq C e^{\frac{C_1}{\epsilon}\left(\int_0^t V(\tau)d\tau\right)^2} \left(\|\rho_0\|_{B^s_{2,\infty}} + \|f\|_{L^1_t(B^s_{2,\infty})}\right),$$

where C *and* C_1 *are constants independent of* ϵ.

7 Horizontal thermal diffusion

This section focuses on the 2D Boussinesq equations with only horizontal thermal diffusion

$$\begin{cases} \partial_t u + (u\cdot\nabla)u = -\nabla p + \theta \mathbf{e}_2, \\ \partial_t \theta + (u\cdot\nabla)\theta = \partial_{x_1x_1}\theta, \\ \nabla\cdot u = 0, \\ u(x,0) = u_0(x), \quad \theta(x,0) = \theta_0(x), \end{cases} \tag{7.1}$$

where we have set the coefficient of $\partial_{x_1x_1}\theta$ to be 1, without loss of generality. It has been shown in [32] that (7.1) possesses a unique global solution when (u_0, θ_0) is in a suitable functional setting.

Theorem 7.1. *Assume* $u_0 \in H^1, w_0 = \nabla\times u_0 \in L^\infty, \theta_0 \in H^1 \cap L^\infty, |\partial_{x_1}|^{1+s}\theta_0 \in L^2$ *for* $s \in (0, \frac{1}{2}]$. *Then* (7.1) *has a unique global solution* (u,θ) *satisfying*

$$u \in C([0,\infty); H^1), \qquad w \in L^\infty_{loc}([0,\infty); L^\infty), \qquad \theta \in C([0,\infty); H^1\cap L^\infty),$$
$$|\partial_{x_1}|^{1+s}\theta \in L^\infty_{loc}([0,\infty); L^2), \qquad |\partial_{x_1}|^{2+s}\theta \in L^2_{loc}([0,\infty); L^2).$$

Here $|\partial_{x_1}|^\beta$ with $\beta \in \mathbb{R}$ is defined in terms of its Fourier transform,

$$|\partial_{x_1}|^\beta f(x) = \int e^{ix\cdot\xi}|\xi_1|^\beta \hat{f}(\xi_1,\xi_2)d\xi_1 d\xi_2.$$

Proof. We will not provide the details, but rather briefly outline the major steps. The equations give us for free the global bounds on $\|\theta\|_{L^2\cap L^\infty}$ and $\|u\|_{H^1}$. The first real step is to obtain a global bound for $\|\nabla\theta\|_2$. This can be established by writing the nonlinear term explicitly in terms of the partial derivatives in different directions and fully take advantage of the dissipation in the x_1-direction. The resulting estimate is given by

$$\frac{d}{dt}\|\nabla\theta\|^2_{L^2} + \|\partial_{x_1}\nabla\theta\|^2_{L^2} \leq B(t)\|\nabla\theta\|^2_{L^2},$$

where $B(t)$ is integrable on $[0,\infty)$. Some special consequences of this inequality are

$$\|\omega(\cdot,T)\|_{L^4} \leq C(T), \quad \|u(\cdot,T)\|_{L^\infty} \leq C(T),$$

where $T > 0$ is an arbitrarily fixed and $C(T)$ depends on the initial data and T. In fact, by the Sobolev embedding inequality,

$$\|\partial_{x_1}\theta\|_{L^4} \le C\|\partial_{x_1}\theta\|_{L^2}^{1/2}\|\nabla\partial_{x_1}\theta\|_{L^2}^{1/2},$$

we have

$$\int_0^T \|\partial_{x_1}\theta\|_{L^4}^4\,dt \le C \sup_{t\in[0,T]} \|\partial_{x_1}\theta\|_{L^2}^2 \int_0^T \|\partial_{x_1}\nabla\theta\|_{L^2}^2\,dt < \infty.$$

It then follows from the vorticity equation that

$$\frac{1}{4}\frac{d}{dt}\|\omega\|_{L^4}^4 = \int \partial_{x_1}\theta\omega|\omega|^2dx \le \|\partial_{x_1}\theta\|_{L^4}\|\omega\|_{L^4}^3.$$

Therefore,

$$\|\omega\|_{L^4} \le \|\omega_0\|_{L^4} + \int_0^t \|\partial_{x_1}\theta\|_{L^4}d\tau \le \|\omega_0\|_{L^4} + \left(\int_0^t \|\partial_{x_1}\theta\|_{L^4}^4 d\tau\right)^{1/4} t^{3/4}.$$

Thus,

$$\|u\|_{L^\infty} \le \|u\|_{L^2}^{1/3}\|\nabla u\|_{L^4}^{2/3} \le C\,\|u\|_{L^2}^{1/3}\|\omega\|_{L^4}^{2/3} < \infty.$$

Next we show that, for $s \in (0, 1/2]$,

$$|\partial_{x_1}|^{1+s}\theta \in L^\infty_{loc}([0,\infty); L^2), \quad |\partial_{x_1}|^{2+s}\theta \in L^2_{loc}([0,\infty); L^2). \tag{7.2}$$

Clearly, $|\partial_{x_1}|^{1+s}\theta$ satisfies

$$\partial_t|\partial_{x_1}|^{1+s}\theta - \partial_{x_1x_1}|\partial_{x_1}|^{1+s}\theta = -|\partial_{x_1}|^{1+s}(u\cdot\nabla\theta).$$

Taking the inner product with $|\partial_{x_1}^{1+s}|\theta$, we find

$$\begin{aligned}\frac{1}{2}\frac{d}{dt}\||\partial_{x_1}|^{1+s}\theta\|_{L^2}^2 + \||\partial_{x_1}|^{2+s}\theta\|_{L^2}^2 &= -\int |\partial_{x_1}|^{1+s}(u\cdot\nabla\theta)\,|\partial_{x_1}|^{1+s}\theta\,dx\\ &= -\int |\partial_{x_1}|(u\cdot\nabla\theta)\,|\partial_{x_1}|^{1+2s}\theta\,dx\\ &= \int \frac{\partial_{x_1}}{|\partial_{x_1}|}(\partial_{x_1}u\cdot\nabla\theta + u\cdot\nabla\partial_{x_1}\theta)|\partial_{x_1}|^{1+2s}\theta dx.\end{aligned}$$

Since $s \in (0, 1/2]$,

$$\begin{aligned}\int \frac{\partial_{x_1}}{|\partial_{x_1}|}(u\cdot\nabla\partial_{x_1}\theta)|\partial_{x_1}|^{1+2s}\theta dx &\le \|u\|_{L^\infty}\|\nabla\partial_{x_1}\theta\|_{L^2}\||\partial_{x_1}|^{1+2s}\theta\|_{L^2}\\ &\le \|u\|_{L^\infty}\|\partial_{x_1}\nabla\theta\|_{L^2}^2.\end{aligned}$$

Writing $\partial_{x_1}u\cdot\nabla\theta = \partial_{x_1}u_1\partial_{x_1}\theta + \partial_{x_1}u_2\partial_{x_2}\theta$, we have

$$\begin{aligned}\int \frac{\partial_{x_1}}{|\partial_{x_1}|}(\partial_{x_1}u\cdot\nabla\theta)\,|\partial_{x_1}|^{1+2s}\theta dx &\le \|\omega\|_{L^4}\|\partial_{x_1}\theta\|_{L^4}\||\partial_{x_1}|^{1+2s}\theta\|_{L^2}\\ &\quad + \int \frac{\partial_{x_1}}{|\partial_{x_1}|}\partial_{x_1}u_2\partial_{x_2}\theta|\partial_{x_1}|^{1+2s}\theta.\end{aligned}$$

In addition,

$$\begin{aligned}\int \frac{\partial_{x_1}}{|\partial_{x_1}|}\partial_{x_1}u_2\partial_{x_2}\theta|\partial_{x_1}|^{1+2s}\theta &= -\int \frac{\partial_{x_1}}{|\partial_{x_1}|}u_2\partial_{x_1}\partial_{x_2}\theta|\partial_{x_1}|^{1+2s}\theta dx \\ &\quad -\int \frac{\partial_{x_1}}{|\partial_{x_1}|}|\partial_{x_1}|^s(u_2\partial_{x_2}\theta)|\partial_{x_1}|^{2+s}\theta dx \\ &\leq \|u_2\|_{L^\infty}\|\partial_{x_1}\nabla\theta\|_{L^2}\,\||\partial_{x_1}|^{1+2s}\theta\|_{L^2} \\ &\quad +\||\partial_{x_1}|^s(u_2\partial_{x_2}\theta)\|_{L^2}\||\partial_{x_1}|^{2+s}\theta\|_{L^2} \\ &\leq \|u_2\|_{L^\infty}^2\|\partial_{x_1}\nabla\theta\|_{L^2}^2 + \frac{1}{4}\||\partial_{x_1}|^{2+s}\theta\|_{L^2}^2 \\ &\quad +C\|u\|_{H^1}^2(\|\partial_2\theta\|_{L^2}+\|\partial_1\partial_2\theta\|_{L^2})^2.\end{aligned}$$

Combining the estimates yield, for some $g \in L^1_{loc}([0,\infty))$,

$$\frac{d}{dt}\||\partial_{x_1}|^{1+s}\theta\|_{L^2}^2 + \||\partial_{x_1}|^{2+s}\theta\|_{L^2}^2 \leq g(t).$$

We thus have obtained (7.2). A special consequence is that

$$\omega \in L^\infty_{loc}([0,\infty);L^\infty).$$

This can be obtained by combining

$$\|\omega\|_{L^\infty} \leq \|\omega_0\|_{L^\infty} + \int_0^t \|\partial_{x_1}\theta\|_{L^\infty}d\tau$$

with the simple estimate of the following lemma.

Lemma 7.2. *If* $\frac{1}{s_1}+\frac{1}{s_2}<2,\ s_1>0,\ s_2>0,$ *then*

$$\|f\|_{L^\infty} \leq C(\|f\|_{L^2} + \||\partial_{x_1}|^{s_1}f\|_{L^2} + \||\partial_{x_2}|^{s_2}f\|_{L^2}).$$

Applying this lemma with $s_1 = 1+s,\ s_2 = 1$, we have

$$\|\omega\|_{L^\infty} \leq \|\omega_0\|_{L^\infty} + C\int_0^t (\|\partial_{x_1}\theta\|_{L^2} + \||\partial_{x_1}|^{2+s}\theta\|_{L^2} + \|\partial_{x_1}\partial_{x_2}\theta\|_{L^2})d\tau < \infty.$$

Trivially, interpolating between L^2 and L^∞ yields $\omega \in L^q$ for $q \in [2,\infty)$. More importantly,

$$\nabla u \in L^\infty_{loc}([0,\infty);L) \quad \text{or} \quad \sup_{t\in[0,T]}\sup_{q\geq 2}\frac{\|\nabla u\|_{L^q}}{q} \leq C(T).$$

This completes the part for the existence and regularity part. Finally we show the uniqueness. This is a consequence of the Yudovich type argument. Let $(u^{(1)},\theta^{(1)})$ and $(u^{(2)},\theta^{(2)})$ be two solutions. Then $\tilde{u} = u^{(1)} - u^{(2)}$ and $\tilde{\theta} = \theta^{(1)} - \theta^{(2)}$ satisfy

$$\begin{cases}\partial_t\tilde{u} + u^{(1)}\cdot\nabla\tilde{u} + \tilde{u}\cdot\nabla u^{(2)} = \theta\mathbf{e}_2, \\ \partial_t\tilde{\theta} + u^{(1)}\cdot\nabla\tilde{\theta} + \tilde{u}\cdot\nabla\theta^{(2)} = \partial_{x_1x_1}\tilde{\theta}.\end{cases}$$

The most difficult term we would encounter in the further estimates is

$$\int \tilde{u_2}\, \partial_2 \theta^{(2)}\, \tilde{\theta} dx.$$

One way to handle it is to use the identity

$$I = (I - \partial_{x_1}\partial_{x_2})^{-1} - (I - \partial_{x_1}\partial_{x_2})^{-1}\partial_{x_1}\partial_{x_2}.$$

We shall omit further details. This completes the proof of Theorem 7.1. □

8 Vertical dissipation and vertical thermal diffusion

This section is devoted to the global regularity problem for the 2D Boussinesq equations with vertical dissipation and vertical thermal diffusion. More precisely, we study the global existence and uniqueness of solutions to the IVP for

$$\begin{cases} \partial_t u + u \cdot \nabla u = -\nabla p + \nu \partial_{yy} u + \theta \mathbf{e}_2, \\ \partial_t \theta + u \cdot \nabla \theta = \kappa \partial_{yy} \theta, \\ \nabla \cdot u = 0, \\ u(x,0) = u_0(x), \quad \theta(x,0) = \theta_0(x). \end{cases} \tag{8.1}$$

We remark that there is a geophysical background for this model (see [53]).

The global existence and uniqueness is established in a recent work by Cao and Wu [13]. We should also mention two previous papers [2] and [3] in which partial results have been obtained. The main result can be stated as follows.

Theorem 8.1. *Consider the IVP for the anisotropic Boussinesq equations with vertical dissipation* (8.1). *Let* $\nu > 0$ *and* $\kappa > 0$. *Let* $(u_0, v_0, \theta_0) \in H^2(\mathbb{R}^2)$. *Then, for any* $T > 0$, (8.1) *has a unique classical solution* (u, v, θ) *on* $[0, T]$ *satisfying*

$$(u, v, \theta) \in C([0, T]; H^2(\mathbb{R}^2)).$$

The rest of this section proves this theorem. As we mentioned before, the local existence and uniqueness is not very hard to obtain. Therefore, our effort will be devoted to proving the global *a priori* H^2 bounds for the solution. For the sake of clarity, we should accomplish this big task in several subsections.

8.1 What is the difficulty?

We briefly explain the difficulty and point out a simple fact. The L^2-estimates are easy to get,

$$\|\theta(t)\|_2^2 + 2\kappa \int_0^T \|\partial_y \theta\|_2^2\, dt \le \|\theta_0\|_2^2,$$

$$\|u(t)\|_2^2 + 2\nu \int_0^T \|\partial_y u\|_2^2\, dt \le (\|u_0\|_2 + t\, \|\theta_0\|_2)^2.$$

But the global H^1-bound is hard to prove. Consider the equation for ω, which satisfies

$$\partial_t \omega + u \cdot \nabla \omega = \nu \partial_{yy} \omega + \partial_x \theta. \tag{8.2}$$

As a simple consequence,

$$\frac{1}{2}\frac{d}{dt}\int \omega^2 + \nu \int (\partial_y \omega)^2 = \int \partial_x \theta \omega.$$

In contrast to the horizontal dissipation case, the dissipation in the y-direction does not allow us to hide $\partial_x \theta$. Therefore, if we want to obtain a global bound for $\|\omega\|_2$, then we need to combine it with the estimate of $\nabla \theta$.

$$\frac{1}{2}\frac{d}{dt}\int |\nabla\theta|^2 + \kappa \int |\partial_y \nabla \theta|^2 = -\int \nabla\theta \cdot \nabla u \cdot \nabla \theta. \tag{8.3}$$

To make use of the dissipation in the y-direction, we write

$$\nabla\theta \cdot \nabla u \cdot \nabla\theta = \partial_x u (\partial_x \theta)^2 + \partial_x v \partial_x \theta \partial_y \theta + \partial_y u \partial_x \theta \partial_y \theta + \partial_y v (\partial_y \theta)^2. \tag{8.4}$$

To bound the terms on the right, we need a lemma.

Lemma 8.2. *Assume that $f, g, g_y, h, h_x \in L^2(\mathbb{R}^2)$. Then*

$$\iint |f\, g\, h|\, dxdy \le C\, \|f\|_2\, \|g\|_2^{\frac{1}{2}}\, \|g_y\|_2^{\frac{1}{2}}\, \|h\|_2^{\frac{1}{2}}\, \|h_x\|_2^{\frac{1}{2}}. \tag{8.5}$$

This lemma allows us to bound some of the terms suitably. For example,

$$\begin{aligned}\left|\int \partial_y u \partial_x \theta \partial_y \theta\right| &\le C\, \|\partial_y u\|_2\, \|\partial_x \theta\|_2^{\frac{1}{2}}\, \|\|\partial_{xy}\theta\|_2^{\frac{1}{2}}\, \|\partial_y \theta\|_2^{\frac{1}{2}}\, \|\partial_{xy}\theta\|_2^{\frac{1}{2}} \\ &\le \frac{\kappa}{4}\|\partial_{xy}\theta\|_2^2 + C(\kappa)\, \|\partial_y u\|_2^2\, \|\partial_x\theta\|_2\, \|\partial_y \theta\|_2.\end{aligned}$$

Integrating by parts, we have

$$\begin{aligned}\int \partial_x v \partial_x \theta \partial_y \theta &= -\int \theta\, (\partial_x v \partial_{xy}\theta + \partial_{xy} v \partial_x \theta) \\ &\le \|\theta_0\|_\infty\, \|\partial_x v\|_2\, \|\partial_{xy}\theta\|_2 + \|\theta_0\|_\infty\, \|\partial_x \theta\|_2\, \|\partial_{xy} v\|_2.\end{aligned}$$

However, the term $\int \partial_x u\, (\partial_x \theta)^2$ can NOT be bounded suitably. But if we know

$$\int_0^T \|v(t)\|_{L^\infty}^2\, dt < \infty, \tag{8.6}$$

then we have, after integration by parts,

$$\begin{aligned}\int \partial_x u\, (\partial_x\theta)^2 &= -\int \partial_y v\, (\partial_x \theta)^2 = \int v \partial_x \theta\, \partial_{xy}\theta \\ &\le \frac{\kappa}{4}\|\partial_{xy}\theta\|_2^2 + C(\kappa)\, \|v(t)\|_{L^\infty}^2\, \|\partial_x\theta\|_2^2.\end{aligned}$$

Inserting the estimates above in (8.2) and (8.3), we are able to conclude that, if (8.6) holds, then

$$\|\omega\|_2^2 + \|\nabla\theta\|_2^2 + \nu \int (\partial_y \omega)^2 + \kappa \int |\partial_y \nabla\theta|^2 \le C(T).$$

Unfortunately it appears to be extremely hard to prove (8.6). Therefore, we have to solve this problem through a different route.

8.2 Key ingredients in the proof

This section presents the key ingredients in the proof as well as the proof of Theorem 8.1.

Proposition 8.3. *Assume* $(u_0, v_0, \theta_0) \in H^2$. *Let* (u, v, θ) *be the corresponding classical solution of* (8.1). *Then the quantity*

$$Y(t) = \|\omega\|_{H^1}^2 + \|\theta\|_{H^2}^2 + \|\omega^2 + |\nabla\theta|^2\|_2^2$$

satisfies

$$\begin{aligned}&\frac{d}{dt}Y(t) + \|\omega_y\|_{H^1}^2 + \|\theta_y\|_{H^2}^2 + \int (\omega^2 + |\nabla\theta|^2)\,(\omega^2 + |\nabla\theta_y|^2)\\ &\le C\left(1 + \|\theta_0\|_\infty^2 + \|v\|_\infty^2 + \|u_y\|_2^2 + (1 + \|u\|_2^2)\|v_y\|_2^2\right) Y(t),\end{aligned}$$

where C *is a constant. As a special consequence of this differential inequality, we conclude that, if*

$$\int_0^T \|v(t)\|_{L^\infty}^2\, dt < \infty,$$

then $Y(t) < +\infty$ *on* $[0, T]$.

Proposition 8.4. *Let* $(u_0, v_0, \theta_0) \in H^2(\mathbb{R}^2)$ *and let* (u, v, θ) *be the corresponding classical solution of* (8.1). *Then,*

$$\sup_{r\ge 2} \frac{\|v(t)\|_{L^{2r}}}{\sqrt{r}\log r} \le \sup_{r\ge 2} \frac{\|v_0\|_{L^{2r}}}{\sqrt{r}\log r} + B(t), \tag{8.7}$$

where $B(t)$ *is an explicit integrable function of* $t \in [0, \infty)$ *that depends on* ν, κ *and the initial norm* $\|(u_0, v_0, \theta_0)\|_{H^2}$.

The proof of this proposition is very delicate and provided in another subsection. An exponential bound in r was obtained in [2] and a linear bound in [3], but we really need a bound at this growth level in order for the proof of Theorem 8.1 to work.

Proposition 8.5. *Let* $s > 1$ *and* $f \in H^s(\mathbb{R}^2)$. *Assume that*

$$\sup_{r\ge 2} \frac{\|f\|_r}{\sqrt{r}\log r} < \infty.$$

Then there exists a constant C *depending on* s *only such that*

$$\|f\|_{L^\infty(\mathbb{R}^2)} \le C \sup_{r\ge 2} \frac{\|f\|_r}{\sqrt{r}\log r} \cdot \left[\log(e+\|f\|_{H^s(\mathbb{R}^2)}) \log\log(e+\|f\|_{H^s(\mathbb{R}^2)})\right]^{\frac{1}{2}}.$$

Proof of Theorem 8.1. Applying Proposition 8.4 and using the simple fact that $\|v\|_{H^2}^2 \le \|\omega\|_{H^1}^2 \le Y(t)$, we obtain

$$\frac{d}{dt} Y(t) \le A(t) Y(t) + C\, B^2(t)\, Y(t)\, \log(e+Y(t))\, \log\log(e+Y(t)),$$

where $A(t) = C\,\left(1 + \|\theta_0\|_\infty^2 + \|u_y\|_2^2 + (1+\|u\|_2^2)\|v_y\|_2^2\right)$. An application of Gronwall's inequality then concludes the proof of Theorem 8.1. □

8.3 Proof of Proposition 8.4

Before we provide the real proof, we would like to understand how a bound of this level can be obtained. For this purpose, we make the ansatz

$$\int_0^T \|p\|_\infty^2\, dt < \infty.$$

Then we show that

$$\|v\|_{L^{2r}} \le C\sqrt{r}.$$

Recall the equation for the velocity field

$$\begin{cases} u_t + u u_x + v u_y = -p_x + \nu\, u_{yy}, \\ v_t + u v_x + v v_y = -p_y + \nu\, v_{yy} + \theta. \end{cases} \tag{8.8}$$

Taking the inner product of the second equation in (8.8) with $v\,|v|^{2r-2}$ and integrating by parts, we obtain

$$\begin{aligned} &\frac{1}{2r}\frac{d}{dt}\int |v|^{2r} + \nu(2r-1)\int v_y^2\,|v|^{2r-2} \\ &= (2r-1)\int p\,v_y\,|v|^{2r-2} + \int \theta\, v\,|v|^{2r-2}. \end{aligned}$$

The last term is easy to handle and we focus on the term involving the pressure p.

$$\begin{aligned} &(2r-1)\int p\,v_y\,|v|^{2r-2} \\ &= (2r-1)\int p\,|v|^{r-1}\, v_y\,|v|^{r-1} \\ &\le (2r-1)\,\|p\|_\infty\, \||v|^{r-1}\|_2\, \|v_y|v|^{r-1}\|_2 \\ &\le \frac{\nu(2r-1)}{4}\, \|v_y|v|^{r-1}\|_2^2 + C\,(2r-1)\,\|p\|_\infty^2\, \|v\|_{2r-2}^{2r-2} \\ &\le \frac{\nu(2r-1)}{4}\, \|v_y|v|^{r-1}\|_2^2 + C\,(2r-1)\,\|p\|_\infty^2\, \|v\|_2^{\frac{2}{r-1}}\, \|v\|_{2r}^{2r-2-\frac{2}{r-1}}. \end{aligned}$$

Then, if $\int_0^T \|p\|_\infty^2 \, dt < \infty$,

$$\frac{1}{2r}\frac{d}{dt}\|v\|_{2r}^{2r} \le C\,(2r-1)\,\|p\|_\infty^2\,\|v\|_2^{\frac{2}{r-1}}\,\|v\|_{2r}^{2r-2-\frac{2}{r-1}}$$

would yield $\|v\|_{2r} \le C\sqrt{r}$.

But unfortunately, we do not know if $\int_0^T \|p\|_\infty^2 \, dt < \infty$. What we can show is the following bound.

Proposition 8.6. *Let $(u_0, v_0, \theta_0) \in H^2(\mathbb{R}^2)$ and let (u, v, θ) be the corresponding classical solution of* (8.1). *Then*

$$\|(u(t), v(t))\|_4^4 + \nu \int_0^t \| |(u_y(\tau), v_y(\tau))|\,|(u(\tau), v(\tau))| \|_2^2 \, d\tau \le M_1(t), \qquad (8.9)$$

$$\|p(\cdot, t)\|_2 \le M_2(t), \qquad \int_0^t \|\nabla p(\cdot, \tau)\|_2^2 \, d\tau \le M_3(t), \qquad (8.10)$$

where M_1, M_2 and M_3 are explicit smooth functions of $t \in [0,\infty)$ that depend on ν, κ and the initial norm $\|(u_0, v_0, \theta_0)\|_{H^2}$.

We also need two lemmas.

Lemma 8.7. *Let $f \in H^1(\mathbb{R}^2)$. Let $R > 0$. Denote by $B(0,R)$ the ball centered at zero with radius R and by $\chi_{B(0,R)}$ the characteristic function on $B(0,R)$. Write $f = \overline{f} + \widetilde{f}$ with*

$$\overline{f} = \mathcal{F}^{-1}(\chi_{B(0,R)}\mathcal{F}f) \quad \text{and} \quad \widetilde{f} = \mathcal{F}^{-1}((1-\chi_{B(0,R)})\mathcal{F}f). \qquad (8.11)$$

Then we have the following estimates for $\overline{f}$ and $\widetilde{f}$.

(1) *There exists a pure constant C independent of f and R such that*

$$\|\overline{f}\|_{L^\infty(\mathbb{R}^2)} \le C\sqrt{\log R}\,\|f\|_{H^1(\mathbb{R}^2)}. \qquad (8.12)$$

(2) *For any $2 \le q < \infty$, there is a constant independent of q, R and f such that*

$$\|\widetilde{f}\|_{L^q(\mathbb{R}^2)} \le C\,\frac{q}{R^{\frac{2}{q}}}\,\|\widetilde{f}\|_{H^1(\mathbb{R}^2)} \le C\,\frac{q}{R^{\frac{2}{q}}}\,\|f\|_{H^1(\mathbb{R}^2)} \qquad (8.13)$$

In particular, for $q = 4$,

$$\|\widetilde{f}\|_{L^4(\mathbb{R}^2)} \le \frac{C}{\sqrt{R}}\,\|f\|_{H^1(\mathbb{R}^2)}.$$

Lemma 8.8. *Let $q \in [2,\infty)$. Assume that $f, g, g_y, h_x \in L^2(\mathbb{R}^2)$ and $h \in L^{2(q-1)}(\mathbb{R}^2)$. Then*

$$\iint_{\mathbb{R}^2} |f\,g\,h| \, dxdy \le C\,\|f\|_2\,\|g\|_2^{1-\frac{1}{q}}\,\|g_y\|_2^{\frac{1}{q}}\,\|h\|_{2(q-1)}^{1-\frac{1}{q}}\,\|h_x\|_2^{\frac{1}{q}}, \qquad (8.14)$$

where C is a constant depending on q only. Two special cases of (8.14) *are*

$$\iint |f\,g\,h| \, dxdy \le C\,\|f\|_2\,\|g\|_2^{\frac{2}{3}}\,\|g_y\|_2^{\frac{1}{3}}\,\|h\|_4^{\frac{2}{3}}\,\|h_x\|_2^{\frac{1}{3}} \qquad (8.15)$$

and

$$\iint |f\,g\,h|\,dxdy \le C\,\|f\|_2\,\|g\|_2^{\frac{1}{2}}\,\|g_y\|_2^{\frac{1}{2}}\,\|h\|_2^{\frac{1}{2}}\,\|h_x\|_2^{\frac{1}{2}}. \tag{8.16}$$

Proof of Proposition 8.4.

$$\begin{cases} u_t + uu_x + vu_y = -p_x + \nu\, u_{yy}, \\ v_t + uv_x + vv_y = -p_y + \nu\, v_{yy} + \theta. \end{cases} \tag{8.17}$$

Taking the inner product of the second equation in (8.17) with $v\,|v|^{2r-2}$ and integrating by parts, we obtain

$$\begin{aligned} &\frac{1}{2r}\frac{d}{dt}\int |v|^{2r} + \nu(2r-1)\int v_y^2\,|v|^{2r-2} \\ &= (2r-1)\int p\,v_y\,|v|^{2r-2} + \int \theta\,v\,|v|^{2r-2} \\ &= (2r-1)\int \overline{p}\,v_y\,|v|^{2r-2} + (2r-1)\int \tilde{p}\,v_y\,|v|^{2r-2} + \int \theta\,v\,|v|^{2r-2}. \end{aligned} \tag{8.18}$$

By Hölder's inequality,

$$\int \theta\,v\,|v|^{2r-2} \le \|\theta\|_{2r}\,\|v\|_{2r}^{2r-1}, \tag{8.19}$$

$$\int \overline{p}\,v_y\,|v|^{2r-2} \le \|\overline{p}\|_\infty\,\|v^{r-1}\|_2\,\|v_y v^{r-1}\|_2.$$

Applying Lemma 8.8, we have

$$\int \tilde{p}\,v_y\,|v|^{2r-2} \le C\|\tilde{p}\|_4^{\frac{2}{3}}\,\|\tilde{p}_x\|_2^{\frac{1}{3}}\,\|v^{r-1}\|_2^{\frac{2}{3}}\,\|(r-1)v_y v^{r-2}\|_2^{\frac{1}{3}}\,\|v_y v^{r-1}\|_2.$$

Furthermore, by Hölder's inequality,

$$\begin{aligned} \big\||v|^{r-1}\big\|_2 &= \|v\|_{2(r-1)}^{r-1} \le \|v\|_2^{\frac{1}{r-1}}\,\|v\|_{2r}^{\frac{r(r-2)}{r-1}}, \\ \big\||v|^{r-2}v_y\big\|_2^2 &= \int |v|^{2(r-2)}v_y^2 = \int |v|^{2(r-2)}v_y^{\frac{2(r-2)}{r-1}}\,v_y^{\frac{2}{r-1}} \\ &\le \|v_y\|_2^{\frac{2}{r-1}}\,\big\|v_y\,|v|^{r-1}\big\|_2^{\frac{2(r-2)}{r-1}}. \end{aligned}$$

Therefore,

$$\begin{aligned} \int \overline{p}\,v_y\,|v|^{2r-2} &\le C\,\|\overline{p}\|_\infty\,\|v\|_2^{\frac{1}{r-1}}\,\|v\|_{2r}^{\frac{r(r-2)}{r-1}}\,\|v_y v^{r-1}\|_2, \\ \int \tilde{p}\,v_y\,|v|^{2r-2} &\le C\,(r-1)^{\frac{1}{3}}\,\|\tilde{p}\|_4^{\frac{2}{3}}\,\|\tilde{p}_x\|_2^{\frac{1}{3}}\,\|v\|_2^{\frac{2}{3(r-1)}}\,\|v\|_{2r}^{\frac{2r(r-2)}{3(r-1)}} \\ &\quad\times\|v_y\|_2^{\frac{1}{3(r-1)}}\,\big\|v_y\,|v|^{r-1}\big\|_2^{1+\frac{(r-2)}{3(r-1)}}. \end{aligned}$$

By Young's inequality and Lemma 8.7,

$$\begin{aligned}&(2r-1)\int \overline{p}\, v_y\, |v|^{2r-2}\\ &\le \frac{\nu}{4}(2r-1)\|v_y v^{r-1}\|_2^2 + C(2r-1)(\log R)\, \|p\|_{H^1}^2 \|v\|_2^{\frac{2}{r-1}}\, \|v\|_{2r}^{2r-2-\frac{2}{r-1}}. \qquad (8.20)\end{aligned}$$

By Young's inequality and Lemmas 8.7,

$$\begin{aligned}&(2r-1)\int \tilde{p}\, v_y\, |v|^{2r-2}\\ &\le \frac{\nu}{4}(2r-1)\|v_y v^{r-1}\|_2^2 + C\,(2r-1)(r-1)^{\frac{2r-2}{2r-1}}\\ &\quad\times \|\tilde{p}\|_4^{\frac{4(r-1)}{2r-1}}\, \|\tilde{p}_x\|_2^{\frac{2(r-1)}{2r-1}}\, \|v_y\|_2^{\frac{2}{2r-1}}\, \|v\|_2^{\frac{4}{2r-1}}\, \|v\|_{2r}^{2r-2-\frac{2(r+1)}{2r-1}}\\ &\le \frac{\nu}{4}(2r-1)\|v_y v^{r-1}\|_2^2 + C\,(2r-1)(r-1)^{\frac{2r-2}{2r-1}} R^{-\frac{r-1}{2r-1}}\\ &\quad\times \|p\|_{L^4}^{\frac{2r-2}{2r-1}}\, \|p\|_{H^1}^{\frac{4r-4}{2r-1}}\, \|v_y\|_2^{\frac{2}{2r-1}}\, \|v\|_2^{\frac{4}{2r-1}}\, \|v\|_{2r}^{2r-3-\frac{3}{2r-1}}. \qquad (8.21)\end{aligned}$$

Without loss of generality, we assume $\|v\|_{2r} \ge 1$. Inserting (8.19),(8.20) and (8.21) in (8.18), we have

$$\begin{aligned}&\frac{1}{2r}\frac{d}{dt}\|v\|_{L^{2r}}^{2r} + \frac{\nu}{2}(2r-1)\int v_y^2 |v|^{2r-2}\, dx\\ &\le C(2r-1)(\log R)\, \|p\|_{H^1}^2 \|v\|_2^{\frac{2}{r-1}}\, \|v\|_{2r}^{2r-2}\\ &\quad + C\,(2r-1)(r-1)^{\frac{2r-2}{2r-1}} R^{-\frac{r-1}{2r-1}} \|p\|_{L^4}^{\frac{2r-2}{2r-1}}\, \|p\|_{H^1}^{\frac{4r-4}{2r-1}}\, \|v_y\|_2^{\frac{2}{2r-1}}\, \|v\|_2^{\frac{4}{2r-1}}\, \|v\|_{2r}^{2r-2}\\ &\quad + \|\theta\|_{L^{2r}}\, \|v\|_{L^{2r}}^{2r-1}.\end{aligned}$$

Especially,

$$\begin{aligned}\frac{d}{dt}\|v\|_{L^{2r}}^2 &\le C(2r-1)(\log R)\, \|p\|_{H^1}^2 \|v\|_2^{\frac{2}{r-1}}\\ &\quad + C\,(2r-1)(r-1)^{\frac{2r-2}{2r-1}} R^{-\frac{r-1}{2r-1}} \|p\|_{L^4}^{\frac{2r-2}{2r-1}} \left(\|p\|_{H^1}^2 + \|v_y\|_2^2\right) \|v\|_2^{\frac{4}{2r-1}}\\ &\quad + \|\theta\|_{L^{2r}}^2 + \|v\|_{L^{2r}}^2.\end{aligned}$$

Taking $R = (2r-1)^{\frac{2r-1}{2r-2}}(r-1)^2$, integrating in time and applying Propositions, we obtain

$$\|v(t)\|_{L^{2r}}^2 \le \|v_0\|_{L^{2r}}^2 + B_1(t) r\log r + B_2(t),$$

where B_1 and B_2 are explicit integrable functions. Therefore,

$$\sup_{r\ge 2} \frac{\|v(t)\|_{L^{2r}}^2}{r\log r} \le \sup_{r\ge 2} \frac{\|v_0\|_{L^{2r}}^2}{r\log r} + (B_1(t) + B_2(t)).$$

This completes the proof of Proposition 8.4. □

8.4 Proof of Proposition 8.5

This subsection proves Proposition 8.5.

Proof of Proposition 8.5. By the Littlewood-Paley decomposition, we can write

$$f = S_{N+1}f + \sum_{j=N+1}^{\infty} \Delta_j f,$$

where Δ_j denotes the Fourier localization operator and

$$S_{N+1} = \sum_{j=-1}^{N} \Delta_j.$$

The definitions of Δ_j and S_N are now standard. Therefore,

$$\|f\|_\infty \le \|S_{N+1}f\|_\infty + \sum_{j=N+1}^{\infty} \|\Delta_j f\|_\infty.$$

We denote the terms on the right by I and II. By Bernstein's inequality, for any $q \ge 2$,

$$|I| \le 2^{\frac{2N}{q}} \|S_{N+1}f\|_q \le 2^{\frac{2N}{q}} \|f\|_q.$$

Taking $q = N$, we have

$$|I| \le 4\|f\|_N \le 4\sqrt{N \log N} \sup_{r \ge 2} \frac{\|f\|_r}{\sqrt{r \log r}}.$$

By Bernstein's inequality again, for any $s > 1$,

$$\begin{aligned} |II| &\le \sum_{j=N+1}^{\infty} 2^j \|\Delta_j f\|_2 = \sum_{j=N+1}^{\infty} 2^{-j(s-1)}\, 2^{sj} \|\Delta_j f\|_2 \\ &= C \cdot 2^{-(N+1)(s-1)} \|f\|_{B^s_{2,2}}, \end{aligned}$$

where C is a constant depending on s only. By identifying $B^s_{2,2}$ with H^s, we obtain

$$\|f\|_\infty \le 4\sqrt{N \log N} \sup_{r \ge 2} \frac{\|f\|_r}{\sqrt{r \log r}} + C \cdot 2^{-(N+1)(s-1)} \|f\|_{H^s}.$$

We obtain the desired inequality (8.8) by taking

$$N = \left[\frac{1}{s-1} \log_2(e + \|f\|_{H^s}) \right],$$

where $[a]$ denotes the largest integer less than or equal to a. □

8.5 Open problems

It is currently not clear if the global regularity of Theorem 8.1 still holds if we set either $\nu = 0$ or $\kappa = 0$ in (8.1), even though we can verify that certain parts of the estimates still hold.

9 The 2D Boussinesq with fractional dissipation

This section summarizes the major global well-posedness results for the 2D Boussinesq equations with fractional dissipation and fractional thermal diffusion, namely

$$\begin{cases} u_t + u \cdot \nabla u + \nu \Lambda^\alpha u = -\nabla p + \theta \mathbf{e}_2, \\ \nabla \cdot u = 0, \\ \theta_t + u \cdot \nabla \theta + \kappa \Lambda^\beta \theta = 0, \end{cases} \tag{9.1}$$

where $\Lambda = (-\Delta)^{\frac{1}{2}}$. When $\alpha = \beta = 2$, these equations reduce to the standard 2D Boussinesq equations. One main focus of recent research on the 2D Boussinesq equations is to establish the global regularity for (9.1) with the smallest $\alpha \geq 0$ and $\beta \geq 0$ (see, e.g., [2, 3, 4, 8, 14, 15, 20, 26, 31, 32, 34, 36, 37, 38, 40, 43, 48, 49, 55, 56, 57, 65, 67]).

(9.1) is mathematically significant and the study of the global well-posedness of (9.1) may shed light on the Millennium Prize problem on the 3D Navier-Stokes equations. As we know, the generalized 3D Navier-Stokes equations

$$\begin{cases} \partial_t u + u \cdot \nabla u + \Lambda^\alpha u = -\nabla p, \\ \nabla \cdot u = 0 \end{cases}$$

always possess global classical solutions for $\alpha \geq \frac{5}{2}$ ([42, 62, 64]). One way to understand this Millennium Prize problem is to examine whether the global regularity holds for smaller α. The techniques and tools developed for (9.1) may be useful in dealing with the Navier-Stokes equations. In addition, (9.1) with some special values of α and β may be applicable in modeling physical and geophysical phenomena.

The global regularity problem on (9.1) is not easy when α and β are not sufficiently large. The key obstacle is how to obtain global *a priori* bounds for the Sobolev norms (or equivalent Besov norms) of the solutions. For example, to bound the derivative of the velocity u, or the vorticity $\omega = \nabla \times u$, one resorts to the vorticity equation

$$\partial_t \omega + u \cdot \nabla \omega + \nu \Lambda^\alpha \omega = \partial_1 \theta$$

and immediately realizes that $\alpha \geq 2$ is needed in order to obtain even a global bound for the L^2-norm of ω when no priori information on the derivative of θ is known. Indeed, even the partial dissipation cases $\alpha = 2$ and $\kappa = 0$ or $\beta = 2$ and $\nu = 0$ are not trivial. As described in the previous sections, they have been dealt with by Chae [15] and by Hou and Li [38]. When $\alpha \leq 1$ and $\beta \leq 1$, the situation becomes more difficult and special techniques have to be developed to overcome the difficulty.

As suggested by Jiu, Miao, Wu and Zhang in [40], we classify the parameters α and β into three categories:

(i) the subcritical case, $\alpha + \beta > 1$;
(ii) the critical case, $\alpha + \beta = 1$;
(iii) the supercritical case, $\alpha + \beta < 1$.

One rationale behind this division is that (9.1) in the critical case defined here can indeed be converted into the critical case for the surface quasi-geostrophic type equation, as demonstrated in [40]. It is generally true that the smaller the sum $\alpha + \beta$ is, the more difficult the global regularity problem is. We caution that even the subcritical case may be difficult to handle. In the following we summarize the major results that are currently available on the global well-posedness issue concerning (9.1). For the sake of clarity, we divide into three subsections.

9.1 Results for the subcritical case

Even the subcritical cases appear to be hard and not many results are available. In [26], Constantin and Vicol verified the global regularity for the case

$$\nu > 0, \quad \kappa > 0, \quad \alpha \in (0,2), \quad \beta \in (0,2), \quad \beta > \frac{2}{2+\alpha}.$$

Theorem 9.1. *Consider the Boussinesq equations* (9.1) *with* $\nu > 0$ *and* $\kappa > 0$. *Assume that* $(u_0, \theta_0) \in \mathcal{S}$, *the Schwartz class. If* $\beta > \frac{2}{2+\alpha}$, *then* (9.1) *has a unique global smooth solution.*

Miao and Xue in [55] proved the global existence and uniqueness for (9.1) with $\nu > 0$, $\kappa > 0$ and

$$\alpha \in \left(\frac{6-\sqrt{6}}{4}, 1\right), \quad \beta \in \left(1-\alpha, \min\left\{\frac{7+2\sqrt{6}}{5}\alpha - 2, \frac{\alpha(1-\alpha)}{\sqrt{6}-2\alpha}, 2-2\alpha\right\}\right).$$

A more recent preprint [66] focuses on the ranges of α and β satisfying

$$0 < \alpha < 1, \quad 0 < \beta < 1, \quad \alpha < \beta.$$

The global well-posedness is not trivial and does not follow from any previous work. The major result in [66] can be stated as follows.

Theorem 9.2. *Assume that* $\alpha \in (0,1)$ *and* $\beta \in (0,1)$ *satisfy*

$$\frac{\alpha}{2} + \beta > 1, \quad \beta \geq \frac{2}{3} + \frac{\alpha}{3} \quad \text{and} \quad \beta > \frac{10-5\alpha}{10-4\alpha}.$$

Consider (9.1) *with* (u_0, θ_0) *satisfying* $\nabla \cdot u_0 = 0$ *and*

$$u_0 \in H^1 \cap B^1_{\infty,1} \quad \text{and} \quad \theta_0 \in \mathrm{L}^2 \cap \mathrm{B}^1_{\infty,1}.$$

Then (9.1) *has a unique global solution* (u, θ) *with*

$$u \in L^\infty_{loc}([0,\infty); H^1 \cap B^1_{\infty,1}), \quad \theta_0 \in L^\infty_{loc}([0,\infty); L^2 \cap B^1_{\infty,1}).$$

9.2 Results for the critical case

The global regularity of (9.1) with the critical dissipation

$$0<\alpha\le 1,\quad 0<\beta\le 1,\quad \alpha+\beta=1 \tag{9.2}$$

is an extremely difficult problem. When α and β are in the range in (9.2), standard energy estimates do not yield the global bounds in any Sobolev spaces. Therefore novel ideas are necessary to deal with the 2D Boussinesq equations with such a low dissipation as specified in (9.2).

Hmidi, Keraani and Rousset in [36, 37] were able to establish the global regularity for two special critical cases, namely (9.1) with $\nu>0$, $\kappa=0$ and $\alpha=1$ or (9.1) with $\nu=0$, $\kappa>0$ and $\beta=1$. Their main result for (9.1) with $\nu>0$, $\kappa=0$ and $\alpha=1$ can be stated as follows.

Theorem 9.3. *Let $u_0\in H^1\cap \dot W^{1,q}$ with $q\in[2,\infty)$. Let $\theta_0\in L^2\cap B^0_{\infty,1}$. Then the Boussinesq equations have a unique global solution satisfying*

$$u\in L^\infty_{loc}([0,\infty);H^1\cap \dot W^{1,q})\cap L^1_{loc}([0,\infty);B^1_{\infty,1}),$$

$$\theta\in L^\infty_{loc}([0,\infty);L^2\cap B^0_{\infty,1}).$$

Their main result for (9.1) with $\nu=0$, $\kappa>0$ and $\beta=1$ can be stated as follows.

Theorem 9.4. *Let $u_0\in B^1_{\infty,1}\cap \dot W^{1,q}$ with $q\in[2,\infty)$. Let $\theta_0\in L^q\cap B^0_{\infty,1}$. Then the Boussinesq equations have a unique global solution satisfying*

$$u\in L^\infty_{loc}([0,\infty);B^1_{\infty,1}\cap \dot W^{1,q}),$$

$$\theta\in L^\infty_{loc}([0,\infty);L^q\cap B^0_{\infty,1})\cap \tilde L^1_{loc}([0,\infty);B^1_{q,\infty}).$$

Progress on the general critical case

$$0<\alpha<1,\quad 0<\beta<1,\quad \alpha+\beta=1 \tag{9.3}$$

has been made recently in [40]. Unfortunately, the approaches in [36, 37] do not apply to the general critical case (9.3). When the critical fractional dissipation is split between the velocity equation and the temperature equation in (9.1), the situation becomes very complicated. Magically [40] were able to reduce (9.1) with the general critical case (9.3) into a critical active scalar equation, or more precisely, the generalized critical surface quasi-geostrophic equation. When α is restricted to a suitable range, the global regularity of the critical Boussinesq system (9.1) can be obtained by exploiting the global regularity of the critical scalar equation.

Theorem 9.5. *Let $\alpha_0<\alpha<1$ and $\alpha+\beta=1$, where*

$$\alpha_0=\frac{23-\sqrt{145}}{12}\approx 0.9132.$$

Assume that $u_0 \in B^{\sigma}_{2,1}(\mathbb{R}^2)$ *with* $\sigma \geq \frac{5}{2}$ *and* $\theta_0 \in B^{2}_{2,1}(\mathbb{R}^2)$. *Then the fractional Boussinesq has a unique global solution* (u, θ) *satisfying, for any* $0 < T < \infty$,

$$\begin{aligned} &u \in C([0,T]; B^{\sigma}_{2,1}(\mathbb{R}^2)) \cap L^1([0,T]; B^{\sigma+\alpha}_{2,1}(\mathbb{R}^2)),\\ &\theta \in C([0,T]; B^{2}_{2,1}(\mathbb{R}^2)) \cap L^1([0,T]; B^{2+\beta}_{2,1}(\mathbb{R}^2)). \end{aligned} \tag{9.4}$$

The key component in the proof of Theorem 9.5 is to establish the global *a priori* bounds in the class defined in (9.4). This does not appear to be trivial and the energy methods are not sufficient for this purpose. Although the global bounds for u in $L^\infty([0,T]; L^2)$ and θ in $L^\infty([0,T]; L^q)$ with $q \in [2, \infty]$ can be easily obtained, the global bounds for the derivatives are not evident. The difficulty starts with the global H^1-bound. Multiplying the vorticity equation

$$\partial_t \omega + u \cdot \nabla \omega + \Lambda^\alpha \omega = \partial_1 \theta$$

by ω and integrating in space yields

$$\frac{1}{2}\frac{d}{dt}\|\omega\|^2_{L^2} + \|\Lambda^{\frac{\alpha}{2}}\omega\|^2_{L^2} = \int \partial_1\theta\, \omega.$$

Since we have no control on $\partial_1\theta$, we integrate by parts

$$\int \partial_1 \theta\, \omega = -\int \theta\, \partial_1 \omega \leq \|\theta\|_{L^2}\|\partial_1\omega\|_{L^2}.$$

Then,

$$\frac{1}{2}\frac{d}{dt}\|\omega\|^2_{L^2} + \|\Lambda^{\frac{\alpha}{2}}\omega\|^2_{L^2} \leq \|\theta\|_{L^2}\|\partial_1\omega\|_{L^2}$$

and we need $\frac{\alpha}{2} \geq 1$, or $\alpha \geq 2$ in order to bound the right-hand side by the dissipation.

The novel idea in [40] is to convert the critical 2D Boussinesq problem into a critical problem on the scalar surface quasi-geostrophic (SQG) equation. The global bound is then achieved by taking advantage of the recent progress on the global regularity problem for the critical SQG equation (see, e.g., [9, 17, 18, 22, 24, 25, 26, 27, 28, 45, 46]). The global bounds are obtained through several major steps:

First, we resort to the vorticity equation $\omega = \nabla \times u$,

$$\begin{cases} \partial_t \omega + u \cdot \nabla \omega + \Lambda^\alpha \omega = \partial_1 \theta,\\ u = \nabla^\perp \psi, \quad \Delta \psi = \omega \qquad \text{or} \quad u = \nabla^\perp(-\Delta)^{-1}\omega. \end{cases}$$

Note that "$\partial_1\theta$" is the "vortex stretching" term and is one of the main sources of difficulty. A simple thought is to hide it by introducing a new variable. Writing $\Lambda^\alpha \omega - \partial_1 \theta = \Lambda^\alpha(\omega - \Lambda^{-\alpha}\partial_1\theta)$, we can hide $\partial_1\theta$ by considering

$$G = \omega - \mathcal{R}_\alpha \theta \quad \text{with} \quad \mathcal{R}_\alpha = \Lambda^{-\alpha}\partial_1.$$

Taking the difference of

$$\partial_t \omega + u \cdot \nabla \omega + \Lambda^\alpha G = 0$$

and

$$\partial_t \mathcal{R}_\alpha \theta + u \cdot \nabla \mathcal{R}_\alpha \theta + \Lambda^{\beta-\alpha} \partial_1 \theta = -[\mathcal{R}_\alpha, u \cdot \nabla]\theta,$$

we get

$$\partial_t G + u \cdot \nabla G + \Lambda^\alpha G = [\mathcal{R}_\alpha, u \cdot \nabla]\theta + \Lambda^{\beta-\alpha} \partial_1 \theta.$$

Here we have used the standard commutator notation

$$[\mathcal{R}_\alpha, u \cdot \nabla]\theta = \mathcal{R}_\alpha(u \cdot \nabla \theta) - u \cdot \nabla \mathcal{R}_\alpha \theta.$$

The point is that the commutator term $[\mathcal{R}_\alpha, u \cdot \nabla]\theta$ is less singular than $\partial_1 \theta$ in the vorticity equation.

Second, we show that G is regular. More precisely, we obtain for $\alpha > \alpha_0$,

$$\|\nabla \nabla^\perp (-\Delta)^{-1} G\|_{L^\infty} < \infty. \tag{9.5}$$

This step is lengthy and requires several sub-steps to gradually reach the desired regularity. Details are omitted here and interested readers can consult [40].

Third, we convert the 2D Boussinesq equation into a generalized critical SQG equation. By the Biot-Savart law, $u = \nabla^\perp (-\Delta)^{-1} \omega$,

$$u = \nabla^\perp (-\Delta)^{-1} \omega = \nabla^\perp (-\Delta)^{-1} G + \nabla^\perp (-\Delta)^{-1} \Lambda^{-\alpha} \partial_1 \theta \equiv \widetilde{u} + v.$$

Due to (9.5), $\widetilde{u}$ is Lipschitz. Combining this with the equation for θ, we deduce the generalized critical SQG

$$\begin{cases} \partial_t \theta + u \cdot \nabla \theta + \Lambda^\beta \theta = 0, \\ u = \widetilde{u} + v, \quad v = \nabla^\perp \Lambda^{-3} \Lambda^\beta \, \partial_1 \, \theta. \end{cases} \tag{9.6}$$

For $\alpha + \beta = 1$, (9.6) is seen as a generalized SQG equation of θ and v due to

$$\Lambda^{1-\beta} v = \nabla^\perp \Lambda^{-2} \partial_1 \, \theta.$$

Due to the recent advances on the critical SQG and the generalized critical SQG equations, we can show that the solution (θ, v) of (9.6) is globally regular. Consequently, the critical 2D Boussinesq is globally well-posed.

9.3 Slightly supercritical and supercritical cases

Efforts have also been made to understand the global well-posedness issue concerning the 2D Boussinesq equations that are more singular than the critical case. We present the results in two very recent papers [43, 41]. The first paper [41] establishes the eventual regularity of weak solutions to a supercritical 2D Boussinesq equation while the second paper [43] obtains the global regularity for a slightly supercritical Boussinesq equation.

Very little is known about the supercritical case $\alpha + \beta < 1$. In particular, the global well-posedness of classical solutions of (9.1) with supercritical dissipation remains open. A recent work establishes the eventual regularity of its weak solutions [41]. More precisely, we have the following theorem.

Theorem 9.6. *Consider* (9.1) *with* $\nu > 0$, $\kappa > 0$, $\alpha > \alpha_0$, $\beta > 0$ *and* $\alpha + \beta < 1$, *where*

$$\alpha_0 = \frac{23 - \sqrt{145}}{12} \approx 0.9132.$$

Assume that $(u_0, \theta_0) \in H^s(\mathbb{T}^2)$ *with* $s > 2$, *and* u_0 *and* θ_0 *have zero mean. Let* (u, θ) *be a global weak solution of* (9.1). *Then, there exist* $0 < T_1 \le T_2 < \infty$ *such that* (u, θ) *is actually a classical solution on* $[0, T_1]$ *and on* $[T_2, \infty)$.

This theorem reflects the regularization effect of the dissipation even in the supercritical case. Due to the local well-posedness of classical solutions, it suffices to show the existence of $T_2 > 0$ such that (u, θ) is actually a classical solution on $[T_2, \infty)$. This is not trivial and the energy method does not provide any global bounds controlling the derivatives of (u, θ). The strategy here is to reduce the supercritical 2D Boussinesq equations to a supercritical SQG type equation and then establish the eventual regularity of the supercritical SQG equation. Details can be found in [41].

[43] studied the following generalized vorticity formulation of the 2D Boussinesq equation

$$\begin{cases} \partial_t \omega + u \cdot \nabla \omega = \partial_{x_1} \theta, \\ u = \nabla^\perp \psi, \quad \Delta \psi = \Lambda^\sigma P(\Lambda) \omega, \\ \partial_t \theta + u \cdot \nabla \theta + \Lambda \theta = 0, \end{cases} \tag{9.7}$$

where $\omega = \omega(x, t)$ is a scalar function, $\nabla^\perp = (-\partial_{x_2}, \partial_{x_1})$, and $\sigma \ge 0$ is a real parameter. Here the Fourier multiplier operator $P(\Lambda)$ are defined through the Fourier transform, namely

$$\widehat{P(\Lambda) f}(\xi) = P(|\xi|) \widehat{f}(\xi).$$

When $\sigma = 0$ and $P(\Lambda) = I$, (9.7) reduces to one of the critical 2D Boussinesq vorticity equation

$$\begin{cases} \partial_t \omega + u \cdot \nabla \omega = \partial_{x_1} \theta, \\ u = \nabla^\perp \psi \equiv (-\partial_{x_2}, \partial_{x_1}) \psi, \quad \Delta \psi = \omega, \\ \partial_t \theta + u \cdot \nabla \theta + \Lambda \theta = 0. \end{cases}$$

When either $\sigma > 0$ or $P(\Lambda)$ is a differential operator, the velocity in (9.7) is more singular. We point out that (9.7) does have a corresponding velocity formulation,

$$\begin{cases} \partial_t v + u \cdot \nabla v - \sum_{j=1}^2 u_j \nabla v_j = -\nabla p + \theta \mathbf{e}_2, \\ \nabla \cdot v = 0, \quad u = \Lambda^\sigma P(\Lambda) v, \\ \partial_t \theta + u \cdot \nabla \theta + \Lambda \theta = 0 \end{cases} \tag{9.8}$$

with the corresponding vorticity $\omega = \nabla \times u$.

[43] examines the global (in time) existence and uniqueness of solutions to the initial-value problem (IVP) for (9.7) supplemented with data

$$\omega(x, 0) = \omega_0(x), \quad \theta(x, 0) = \theta_0(x), \quad x \in \mathbb{R}^2. \tag{9.9}$$

It allows a very general class of symbols $P(|\xi|)$ for the operator $P(\Lambda)$, as previously specified in the work of Chae, Constantin and Wu [16, p.36]. More precisely, P is assumed to satisfy the following condition.

Condition 9.7. *The symbol $P(|\xi|)$ assumes the following properties:*

(1) *P is continuous on $\mathbb{R}^2$ and $P \in C^\infty(\mathbb{R}^2 \setminus \{0\})$;*
(2) *P is radially symmetric;*
(3) *$P = P(|\xi|)$ is nondecreasing in $|\xi|$;*
(4) *there exist two constants C and C_0 such that*

$$\sup_{2^{-1}\le|\eta|\le 2} \left|(I-\Delta_\eta)^n\, P(2^j|\eta|)\right| \le C\, P(C_0\, 2^j)$$

for any integer j and $n = 1, 2$.

We remark that (4) in Condition 9.7 is a very natural condition on symbols of Fourier multiplier operators and is similar to the main condition in the Mihlin-Hörmander Multiplier Theorem (see, e.g., [61, p.96]). For notational convenience, we also assume that $P \ge 0$. Some special examples of P are

$$\begin{aligned}
&P(\xi) = \left(\log(1+|\xi|^2)\right)^\gamma \quad \text{with } \gamma \ge 0,\\
&P(\xi) = \left(\log(1+\log(1+|\xi|^2))\right)^\gamma \quad \text{with } \gamma \ge 0,\\
&P(\xi) = |\xi|^\beta \quad \text{with } \beta \ge 0,\\
&P(\xi) = (\log(1+|\xi|^2))^\gamma\, |\xi|^\beta \quad \text{with } \gamma \ge 0 \text{ and } \beta \ge 0.
\end{aligned}$$

The motivation for this study is two fold: first, to extend the work of Chae, Constantin and Wu [16] on the generalized 2D Euler equations to the Euler-Boussinesq system of equations, and second, to explore how far one can go beyond the critical dissipation and still prove the global regularity for the Boussinesq equations with fractional dissipation. The main result states that (9.8) is still globally well-posed when P satisfies two explicit conditions.

Theorem 9.8. *Let $\sigma = 0$. Assume the symbol $P(|\xi|)$ obeys Condition 9.7 and*

$$P(2^k) \le C\,\sqrt{k} \quad \textit{for a constant } C \textit{ and any large integer } k > 0, \tag{9.10}$$

$$\int_1^\infty \frac{1}{r\log(1+r)\,P(r)}\,dr = \infty. \tag{9.11}$$

Let $q > 2$ and let $s > 2$. Consider the IVP (9.7) and (9.9) with $\omega_0 \in B^s_{q,\infty}(\mathbb{R}^2)$ and $\theta_0 \in B^s_{q,\infty}(\mathbb{R}^2)$. Then the IVP (9.7) and (9.9) has a unique global solution (ω, θ) satisfying, for any $T > 0$ and $t \le T$,

$$\omega \in C([0,T]; B^s_{q,\infty}(\mathbb{R}^2)), \quad \theta \in C([0,T]; B^s_{q,\infty}(\mathbb{R}^2) \cap L^1([0,T]; B^{s+1}_{q,\infty}(\mathbb{R}^2)). \tag{9.12}$$

A special consequence is the global well-posedness of (9.7) when P is a "double logarithm", namely

$$P(|\xi|) = (\log(1+\log(1+|\xi|)))^\gamma, \quad \gamma \in [0,1]. \tag{9.13}$$

It is easy to see that $P(|\xi|)$ given by (9.13) verifies Condition 9.7, (9.10) and (9.11). The global well-posedness of (9.7) with $P(|\xi|)$ given by (9.13) is given in the following corollary.

Corollary 9.9. *Let $q > 2$ and let $s > 2$. Let $\omega_0 \in B^s_{q,\infty}(\mathbb{R}^2)$ and $\theta_0 \in B^s_{q,\infty}(\mathbb{R}^2)$. Assume that $\sigma = 0$ and $P(|\xi|)$ is given by (9.13). Then the IVP (9.7) and (9.9) has a unique global solution.*

10 Open problems and discussions

Besides the open problems embedded in the previous sections, we would also like to propose and discuss several problems that may help extend the scope of research on the Boussinesq equations.

The first problem is whether or not the results presented in the previous sections can be extended to the Boussinesq equations for nonhomogeneous flows. More specifically, we consider the 2D nonhomogeneous Boussinesq equations

$$\begin{cases} \partial_t \rho + u\cdot\nabla\rho = 0, \\ \rho(\partial_t u + u\cdot\nabla u) = -\nabla p + \nu_1\,\partial_{x_1x_1}u + \nu_2\,\partial_{x_2x_2}u + \theta\mathbf{e}_2, \\ \rho(\partial_t\theta + u\cdot\nabla\theta) = \kappa_1\,\partial_{x_1x_1}\theta + \kappa_2\,\partial_{x_2x_2}\theta, \\ \nabla\cdot u = 0 \end{cases} \tag{10.1}$$

and study the global well-posedness of (10.1) when some of parameters $\nu_1, \nu_2, \kappa_1, \kappa_2$ are equal to zero. It does not appear to be trivial to extend the results of previous sections to (10.1). The techniques developed by Ping Zhang and his collaborators ([1, 39, 58]) may be useful in dealing with the nonhomogeneous Boussinesq equations.

One could also consider the following 2D nonhomogeneous Boussinesq equations with fractional dissipation

$$\begin{cases} \partial_t \rho + u\cdot\nabla\rho = 0, \\ \rho(\partial_t u + u\cdot\nabla u) + \nu(-\Delta)^\alpha u = -\nabla p + \theta\mathbf{e}_2, \\ \rho(\partial_t\theta + u\cdot\nabla\theta) + \kappa(-\Delta)^\alpha\theta = 0, \\ \nabla\cdot u = 0. \end{cases}$$

The goal would be to establish the global well-posedness for smallest $\alpha \geq 0$ and $\beta \geq 0$. Again it does not appear to be easy to recover the global well-posedness results described in Section 9.

Very little is known on the global well-posedness on the inviscid Boussinesq equations. Even the global existence of weak solutions is currently open. One possible way to resolve this problem would be to apply the approach of De Lellis and Székelydihi ([33]).

Another interesting problem is the time growth (or decay) rates of ω, $\nabla\omega$ and $\nabla\theta$ (in L^2 or L^∞) for the inviscid Boussinesq equations or for the Boussinesq equations with partial or fractional dissipation. Here the spatial domain can be a bounded domain (say a disk), the whole space or a periodic domain. This problem for the inviscid Boussinesq equations are partially motivated by the recent work of Kiselev and Sverak on the 2D Euler equations in a disk ([47]).

Acknowledgements

Wu was partially supported by NSF grant DMS1209153 and by a joint grant with Professor Quansen Jiu from NNSF of China under No.11228102. Wu thanks Professor Ping Zhang for his hospitality.

References

[1] H. Abidi, G. Gui and P. Zhang, On the wellposedness of 3-D inhomogeneous Navier-Stokes equations in the critical spaces, *Arch. Ration. Mech. Anal.* **204** (2012), 189–230.

[2] D. Adhikari, C. Cao and J. Wu, The 2D Boussinesq equations with vertical viscosity and vertical diffusivity, *J. Differential Equations* **249** (2010), 1078–1088.

[3] D. Adhikari, C. Cao and J. Wu, Global regularity results for the 2D Boussinesq equations with vertical dissipation, *J. Differential Equations* **251** (2011), 1637–1655.

[4] D. Adhikari, C. Cao, J. Wu and X. Xu, Small global solutions to the damped two-dimensional Boussinesq equations, *J. Differential Equations* **256** (2014), 3594–3613.

[5] H. Bahouri, J.-Y. Chemin and R. Danchin, *Fourier Analysis and Nonlinear Partial Differential Equations*, Springer, 2011.

[6] J. Bergh and J. Löfström, *Interpolation Spaces, An Introduction*, Springer-Verlag, Berlin-Heidelberg-New York, 1976.

[7] J. Bourgain and D. Li, Strong ill-posedness of the incompressible Euler equation in borderline sobolev spaces, Invent. Math. **201** (2015), 97–157.

[8] L. Brandolese and M. Schonbek, Large time decay and growth for solutions of a viscous Boussinesq system, *Trans. Amer. Math. Soc.* **364** (2012), 5057–5090.

[9] L. Caffarelli and A. Vasseur, Drift diffusion equations with fractional diffusion and the quasi-geostrophic equation, *Ann. Math.* **171** (2010), 1903–1930.

[10] J.R. Cannon and E. DiBenedetto, The initial value problem for the Boussinesq equations with data in L^p, Lecture Notes in Math. **771**, Springer, Berlin, 1980, pp. 129–144.

[11] C. Cao, P. Constantin, A. Kiselev and J. Wu, A regularity criterion in terms of the vorticity for the 2D Boussinesq equations, work in progress.

[12] C. Cao and E. Titi, Global well-posedness of the three-dimensional viscous primitive equations of large scale ocean and atmosphere dynamics, *Ann. Math.* **166** (2007), 245–267.

[13] C. Cao and J. Wu, Global regularity for the 2D MHD equations with mixed

partial dissipation and magnetic diffusion, *Advances in Math.* **226** (2011), 1803–1822.

[14] C. Cao and J. Wu, Global regularity for the 2D anisotropic Boussinesq equations with vertical dissipation, *Arch. Ration. Mech. Anal.* **208** (2013), 985–1004.

[15] D. Chae, Global regularity for the 2D Boussinesq equations with partial viscosity terms, *Adv. Math.* **203** (2006), 497–513.

[16] D. Chae, P. Constantin and J. Wu, Inviscid models generalizing the 2D Euler and the surface quasi-geostrophic equations, *Arch. Ration. Mech. Anal.* **202** (2011), 35–62.

[17] D. Chae, P. Constantin and J. Wu, Dissipative models generalizing the 2D Navier-Stokes and the surface quasi-geostrophic equations, *Indiana Univ. Math. J.* **61** (2012), 1997–2018.

[18] D. Chae, P. Constantin, D. Córdoba, F. Gancedo and J. Wu, Generalized surface quasi-geostrophic equations with singular velocities, *Comm. Pure Appl. Math.* **65** (2012), No. 8, 1037–1066.

[19] D. Chae and H.-S. Nam, Local existence and blow-up criterion for the Boussinesq equations, *Proc. Roy. Soc. Edinburgh Sect. A* **127** (1997), 935–946.

[20] D. Chae and J. Wu, The 2D Boussinesq equations with logarithmically supercritical velocities, *Adv. Math.* **230** (2012), 1618–1645.

[21] D. Chae and J. Wu, Logarithmically regularized inviscid models in borderline Sobolev spaces, *J. Math. Phys.* **53** (2012), No. 11, 115601, 15 pp., a special issue dedicated to Professor Peter Constantin on the occasion of his sixtieth birthday.

[22] P. Constantin, Euler equations, Navier-Stokes equations and turbulence, *Mathematical Foundation of Turbulent Viscous Flows*, 143, Lecture Notes in Math. **1871**, Springer, Berlin, 2006.

[23] P. Constantin and C.R. Doering, Infinite Prandtl number convection, *J. Statistical Physics* **94** (1999), 159–172.

[24] P. Constantin, G. Iyer and J. Wu, Global regularity for a modified critical dissipative quasi-geostrophic equation, *Indiana U. Math. J.* **57** (2008), 2681–2692.

[25] P. Constantin, A. Majda and E. Tabak, Formation of strong fronts in the 2-D quasi-geostrophic thermal active scalar, *Nonlinearity* **7** (1994), 1495–1533.

[26] P. Constantin and V. Vicol, Nonlinear maximum principles for dissipative linear nonlocal operators and applications, *Geom. Funct. Anal.* **22** (2012), No. 5, 1289–1321.

[27] A. Córdoba and D. Córdoba, A maximum princple applied to quasi-geostroohhic equations, *Comm. Math. Phys.* **249** (2004), 511–528.

[28] D. Córdoba and C. Fefferman, Scalars convected by a two-dimensional incompressible flow, *Commun. Pure Appl. Math.* **55** (2002), 255–260.

[29] R. Danchin, Remarks on the lifespan of the solutions to some models of incompressible fluid mechanics, *Proc. Amer. Math. Soc.* **141** (2013), No. 6, 1979–1993.

[30] R. Danchin and M. Paicu, Existence and uniqueness results for the Boussinesq system with data in Lorentz spaces, *Phys. D* **237** (2008), 1444–1460.

[31] R. Danchin and M. Paicu, Global well-posedness issues for the inviscid Boussinesq system with Yudovich's type data, *Comm. Math. Phys.* **290** (2009), 1–14.
[32] R. Danchin and M. Paicu, Global existence results for the anisotropic Boussinesq system in dimension two, *Math. Models Methods Appl. Sci.* **21** (2011), 421–457.
[33] C. De Lellis and L. Székelyhidi, The h-principle and the equations of fluid dynamics, *Bull. Amer. Math. Soc. (N.S.)* **49** (2012), 347–375.
[34] W. E and C. Shu, Samll-scale structures in Boussinesq convection, *Phys. Fluids* **6** (1994), 49–58.
[35] A.E. Gill, *Atmosphere-Ocean Dynamics*, Academic Press (London), 1982.
[36] T. Hmidi, S. Keraani and F. Rousset, Global well-posedness for a Boussinesq-Navier-Stokes system with critical dissipation, *J. Differential Equations* **249** (2010), 2147–2174.
[37] T. Hmidi, S. Keraani and F. Rousset, Global well-posedness for Euler-Boussinesq system with critical dissipation, *Comm. Partial Differential Equations* **36** (2011), 420–445.
[38] T. Hou and C. Li, Global well-posedness of the viscous Boussinesq equations, *Discrete and Cont. Dyn. Syst.* **12** (2005), 1–12.
[39] J. Huang, M. Paicu and P. Zhang, Global solutions to 2-D inhomogeneous Navier-Stokes system with general velocity, *J. Math. Pures Appl.* **100** (2013), No. 6, 806–831.
[40] Q. Jiu, C. Miao, J. Wu and Z. Zhang, The 2D incompressible Boussinesq equations with general critical dissipation, SIAM J. Math. Anal. **46** (2014), 3426–3454.
[41] Q. Jiu, J. Wu and W. Yang, Eventual regularity of the two-dimensional Boussinesq equations with supercritical dissipation, J. Nonlinear Science **25** (2015), 37–58.
[42] N. Katz and N. Pavlović, Finite time blow-up for a dyadic model of the Euler equations, *Trans. Amer. Math. Soc.* **357** (2005), No. 2, 695–708.
[43] D. KC, D. Regmi, L. Tao and J. Wu, The 2D Euler-Boussinesq equations with a singular velocity, *J. Differential Equations* **257** (2014), No.1, 82–108.
[44] C.E. Kenig, G. Ponce, L. Vega, Well-posedness of the initial value problem for the Korteweg-de Vries equation, *J. Amer. Math. Soc.* **4** (1991), 323–347.
[45] A. Kiselev, Nonlocal maximum principles for active scalars, *Adv. Math.* **227** (2011), 1806–1826.
[46] A. Kiselev, F. Nazarov and A. Volberg, Global well-posedness for the critical 2D dissipative quasi-geostrophic equation, *Invent. Math.* **167** (2007), 445–453.
[47] A. Kiselev and V. Sverak, Small scale creation for solutions of the incompressible two dimensional Euler equation, Ann. of Math. **180** (2014), 1205–1220.
[48] M. Lai, R. Pan and K. Zhao, Initial boundary value problem for two-dimensional viscous Boussinesq equations, *Arch. Ration. Mech. Anal.* **199** (2011), 739–760.
[49] A. Larios, E. Lunasin and E.S. Titi, Global well-posedness for the 2D Boussinesq system with anisotropic viscosity and without heat diffusion, *J. Differential Equations* **255** (2013), 2636–2654.
[50] J.-L. Lions, R. Temam and S. Wang, A simple global model for the general

circulation of the atmosphere, *Comm. Pure and Appl. Math.* **50** (1997), 707–752.

[51] A.J. Majda, *Introduction to PDEs and Waves for the Atmosphere and Ocean*, Courant Lecture Notes in Mathematics **9**, AMS/CIMS, 2003.

[52] A.J. Majda and A.L. Bertozzi, *Vorticity and Incompressible Flow*, Cambridge University Press, 2001.

[53] A.J. Majda and M.J. Grote, Model dynamics and vertical collapse in decaying strongly stratified flows, *Phys. Fluids* **9** (1997), 2932–2940.

[54] C. Miao, J. Wu and Z. Zhang, *Littlewood-Paley Theory and Its Applications in Partial Differential Equations of Fluid Dynamics*, Science Press, Beijing, China, 2012 (in Chinese).

[55] C. Miao and L. Xue, On the global well-posedness of a class of Boussinesq-Navier-Stokes systems, *NoDEA Nonlinear Differential Equations Appl.* **18** (2011), 707–735.

[56] H.K. Moffatt, Some remarks on topological fluid mechanics, in: R.L. Ricca (Ed.), *An Introduction to the Geometry and Topology of Fluid Flows*, Kluwer Academic Publishers, Dordrecht, The Netherlands, 2001, pp. 3–10.

[57] K. Ohkitani, Comparison between the Boussinesq and coupled Euler equations in two dimensions, Tosio Kato's Method and Principle for Evolution Equations in Mathematical Physics (Sapporo, 2001), Surikaisekikenkyusho Kokyuroku **1234** (2001), 127–145.

[58] M. Paicu, P. Zhang and Z. Zhang, Global unique solvability of inhomogeneous Navier-Stokes equations with bounded density, *Comm. Partial Differential Equations* **38** (2013), No. 7, 1208–1234.

[59] J. Pedlosky, *Geophysical Fluid Dyanmics*, Springer-Verlag, New York, 1987.

[60] T. Runst and W. Sickel, *Sobolev Spaces of Fractional Order, Nemytskij Operators and Nonlinear Partial Differential Equations*, Walter de Gruyter, Berlin, New York, 1996.

[61] E. Stein, *Singular Integrals and Differentiability Properties of Functions*, Princeton Unviersity Press, Princeton, NJ, 1970.

[62] T. Tao, Global regularity for a logarithmically supercritical hyperdissipative Navier-Stokes equation, *Anal. PDE* **2** (2009), No.3, 361–366.

[63] H. Triebel, *Theory of Function Spaces II*, Birkhauser Verlag, 1992.

[64] J. Wu, Generalized MHD equations, *J. Differential Equations* **195** (2003), No.2, 284–312.

[65] X. Xu, Global regularity of solutions of 2D Boussinesq equations with fractional diffusion, *Nonlinear Analysis: TMA* **72** (2010), 677–681.

[66] W. Yang, Q. Jiu and J. Wu, Global well-posedness for a class of 2D Boussinesq systems with fractional dissipation, J. Differential Equations **257** (2014), 4188–4213.

[67] K. Zhao, 2D inviscid heat conductive Boussinesq equations on a bounded domain, *Michigan Math. J.* **59** (2010), 329–352.

Lectures on the Analysis of Nonlinear Partial Differential Equations Vol. 4
MLM 4, pp. 271–395

Lecture Notes on the Basic Analysis Tools for Critical Dispersive PDEs

Xiaoyi Zhang

Abstract

In these lecture notes, we use spherically symmetric mass critical NLS in sample dimensions as model case, explain how the concentration compactness technique is used to prove scattering for the critical Schrödinger equation. To make the notes self-contained and easily accessible, in the first part of the notes, we introduce the necessary analysis tools. These include discussions on the Fourier transform, Littlewood-Paley theory, Sobolev spaces, and the very important linear profile decomposition. These notes is suitable for advanced undergraduates and graduates. In preparing these notes, we borrowed a great deal of material from [13, 25, 39].

1 Fourier transform

The goal of these lecture notes are to get readers to the frontier of modern analysis on critical dispersive equations. The name "dispersive" comes from the property of such kind of equations: plane waves with different frequencies travel with different velocity, hence as time goes by, solution becomes more spread out. Take 1-d linear Schrödinger equation: $iu_t + \partial_{xx}u = 0$ as an example. Solution with initial data e^{ikx}(k is the frequency, measuring how fast the wave oscillates) is a wave $e^{ik(x-kt)}$ which travels with velocity k. Due to this reason, solution can be better understood in frequency side, hence naturally Fourier analysis is the most convenient tool to study dispersive PDE. The first part of this notes is a brief introduction on frequency analysis and we will start our journey from the very fundamental Fourier transform.

1.1 Fourier transform for L^1 functions

Definition 1.1. *Let $f \in L^1(\mathbb{R}^n)$, we define*

$$\hat{f}(\xi) = \mathcal{F}f(\xi) = \int_{\mathbb{R}^n} e^{-2\pi i x\cdot\xi} f(x)dx$$

as the Fourier transform of f.

Remark 1.2. 1) *The integral on the LHS is convergent due to $f \in L^1$. Normally, we say $\mathbb{R}^n_x$ as the physical space, and $\mathbb{R}^n_\xi$ as the frequency space.*

2) *Fourier transform has several versions depending on where to put the factor 2π. For example, one can also use*

$$\hat{f}(\xi) = \int_{\mathbb{R}^n} e^{-ix\cdot\xi} f(x)dx$$

as the definition (the inverse transform will be changed accordingly). To keep it consistent, we will use the first one throughout the notes.

The following result is direct.

Proposition 1.3. *Let $f, g \in L^1$ and c be a constant. Then $\hat{f}$ and $\hat{g}$ are uniformly continuous. Moreover,*

(1) $\hat{f} + \hat{g} = \widehat{f+g}$, $\widehat{cf} = c\hat{f}$, *for* $g \in L^1$.
(2) *If* $f(x) \geq 0$, *then* $\|f\|_1 = \|\hat{f}\|_\infty = \hat{f}(0)$.
(3) $\widehat{f(\cdot - x_0)}(\xi) = e^{-2\pi i x_0\cdot\xi}\hat{f}(\xi)$.
(4) $\widehat{e^{2\pi i\xi_0\cdot x} f}(\xi) = \hat{f}(\xi - \xi_0)$.
(5) $\widehat{f(\cdot/\lambda)}(\xi) = \lambda^n \hat{f}(\lambda\xi)$.
(6) $\hat{\bar{f}}(\xi) = \overline{\hat{f}(-\xi)}$.
(7) *If* $f \star g \in L^1$, *then* $\widehat{f \star g} = \hat{f}\hat{g}$. *If* $f \cdot g \in L^1$, *then* $\mathcal{F}(fg) = \hat{f} \star \hat{g}$.
(8) *If* $\partial f/\partial x_k \in L^1$, *then* $\widehat{\frac{\partial f}{\partial x_k}} = (2\pi i\xi_k)\hat{f}(\xi)$.
(9) *If* $x_k f(x) \in L^1$, *then* $\widehat{2\pi i x_k f}(\xi) = -\partial/\partial\xi_k \hat{f}(\xi)$.
(10) *(Multiplication formula) Let* $f,\ g \in L^1(\mathbb{R}^n)$. *Then*

$$\int_{\mathbb{R}^n} \hat{f}(\xi)g(\xi)d\xi = \int_{\mathbb{R}^n} f(x)\hat{g}(x)dx.$$

(11) *Fourier transform of a radial function is still a radial function.*

Theorem 1.4 (Riemann-Lebesgue Theorem). *Let $f \in L^1(\mathbb{R}^n)$. Then*

$$\lim_{|\xi|\to\infty} \hat{f}(\xi) = 0.$$

Proof. We first prove this for C_c^∞ functions. Let $\phi \in C_c^\infty(\mathbb{R}^n)$. Then

$$\begin{aligned}
\hat{\phi}(\xi) &= \int_{\mathbb{R}^n} e^{-2\pi ix\cdot\xi}\phi(x)dx \\
&= |\xi|^{-2}|\xi|^2 \int_{\mathbb{R}^n} e^{-2\pi ix\cdot\xi}\phi(x)dx \\
&= -\frac{1}{4\pi^2}|\xi|^{-2}\int_{\mathbb{R}^n} \Delta(e^{-2\pi ix\cdot\xi})\phi(x)dx \\
&= -\frac{1}{4\pi^2}|\xi|^{-2}\int_{\mathbb{R}^n} e^{-2\pi ix\cdot\xi}\Delta\phi(x)dx.
\end{aligned}$$

Hence $|\hat{\phi}(\xi)| \leq |\xi|^{-2}\|\Delta\phi\|_1$ and so $|\hat{\phi}|(\xi) \to 0$ as $|\xi| \to \infty$. The general case follows from a simple density argument. Indeed, as C_c^∞ is dense in L^1, $\forall \varepsilon > 0$, we can take $\phi \in C_c^\infty$ such that $\|f - \phi\|_1 < \frac{\varepsilon}{2}$. Fix this ϕ we can find N large enough such that $|\hat{\phi}(\xi)| < \frac{\varepsilon}{2}$, $\forall |\xi| > N$. Together, we have

$$|\hat{f}(\xi)| \leq |\hat{f}(\xi) - \hat{\phi}(\xi)| + |\hat{\phi}(\xi)| \leq \|f - \phi\|_1 + |\hat{\phi}(\xi)| < \varepsilon.$$

The Theorem is proved. □

In the following we give some well-known explicit examples of Fourier transform.

- (Gaussian) $f(x) = e^{-\pi|x|^2}$. Then $\hat{f}(\xi) = e^{-\pi|\xi|^2}$.
  Indeed, as $f(x)$ is a function of separation of variables, we only need to check the one-dimensional case. From a simple shift of contour and the fact that $\int_{\mathbb{R}} e^{-\pi x^2} dx = 1$, we have

$$\begin{aligned}\hat{f}(\xi) &= \int_{\mathbb{R}} e^{-2\pi i x\cdot\xi - \pi|x|^2} dx = \int_{\mathbb{R}} e^{-\pi|x+i\xi|^2} dx e^{-\pi\xi^2} \\ &= \int_{\mathbb{R}} e^{-\pi x^2} dx e^{-\pi\xi^2} = e^{-\pi\xi^2}.\end{aligned}$$

  Gaussian is an example for function that is left unchanged under the Fourier transform.
- (Poisson Kernel) Let $f(x) = e^{-2\pi|x|}$. Then

$$\hat{f}(\xi) = c_n \frac{1}{(1+|\xi|^2)^{\frac{n+1}{2}}}, \quad c_n = \Gamma\left(\frac{n+1}{2}\right)\pi^{-\frac{n+1}{2}}. \tag{1.1}$$

  $\hat{f}(\xi)$ is called Poisson kernel, it appears in solving the Laplacian equation on half space.

proof The one-dimensional case is easier and can be obtained through a direct computation:

$$\begin{aligned}\hat{f}(\xi) &= \int_{\mathbb{R}} e^{-2\pi|x|} e^{-2\pi i x\xi} dx = 2\mathrm{Re} \int_0^\infty e^{-2\pi\rho} e^{-2\pi i\rho\xi} d\rho \\ &= 2\mathrm{Re} \int_0^\infty e^{-2\pi\rho(1+i\xi)} d\rho = 2\mathrm{Re}\, \frac{1}{2\pi(1+i\xi)} = \frac{1}{\pi(1+\xi^2)}.\end{aligned}$$

A seemingly natural extension for high dimensions is passing to polar-coordinate:

$$\hat{f}(\xi) = \int_{\mathbb{R}^n} e^{-2\pi|x|} e^{-2\pi i x\cdot\xi} dx = \int_0^\infty e^{-2\pi\rho} \int_{S^{n-1}} e^{-2\pi i \pi x\cdot\omega\rho} d\sigma(\omega)\rho^{n-1} d\rho.$$

The integral on the unit sphere can be expressed as the Bessel function and is usually not easy to deal with. Instead, in this kind of problem, there is an often-used technique, namely writing functions into superpositions of Gaussians, whose Fourier transform is already calculated.

Note for $s > 0$, from the following identity (which can be viewed as the inverse formula of (1.1)), we have

$$\begin{aligned} e^{-s} &= \frac{2}{\pi}\int_0^\infty \frac{\cos sx}{1+x^2}dx \\ &= \frac{2}{\pi}\int_0^\infty \cos sx \int_0^\infty e^{-(1+x^2)t}dtdx \\ &= \frac{2}{\pi}\int_0^\infty e^{-t}\int_0^\infty \cos sxe^{-x^2t}dxdt \\ &= \frac{1}{\pi}\int_0^\infty e^{-t}\int_{-\infty}^\infty e^{isx}e^{-x^2t}dxdt \\ &= \frac{1}{\sqrt{\pi}}\int_0^\infty \frac{e^{-t}}{\sqrt{t}}e^{-\frac{s^2}{4t}}dt. \end{aligned}$$

Using this estimate we write

$$\begin{aligned} \hat{f}(\xi) &= \int_{\mathbb{R}^n} e^{-2\pi ix\cdot\xi}e^{-2\pi|x|}dx \\ &= \frac{1}{\sqrt{\pi}}\int_{\mathbb{R}^n} e^{-2\pi ix\cdot\xi}\int_0^\infty \frac{e^{-t}}{\sqrt{t}}e^{-\pi^2|x|^2/t}dtdx \\ &= \frac{1}{\sqrt{\pi}}\int_0^\infty \frac{e^{-t}}{\sqrt{t}}\int_{\mathbb{R}^n} e^{-2\pi ix\cdot\xi}e^{-\pi^2|x|^2/t}dxdt \\ &= \pi^{-\frac{n+1}{2}}\int_0^\infty e^{-t}t^{\frac{n-1}{2}}e^{-t|\xi|^2}dt \\ &= \pi^{-\frac{n+1}{2}}(1+|\xi|^2)^{-\frac{n+1}{2}}\int_0^\infty e^{-s}s^{\frac{n-1}{2}}ds \\ &= \pi^{-\frac{n+1}{2}}\Gamma\left(\frac{n+1}{2}\right)(1+|\xi|^2)^{-\frac{n+1}{2}}. \end{aligned}$$

□

1.2 Fourier inversion formula for L^1 and Fourier transform on L^2

We start from the following useful result.

Lemma 1.5 (Approximation of identity). *Let $\phi \in L^1(\mathbb{R}^n)$ and $\int_{\mathbb{R}^n}\phi(x)dx = 1$. Define*

$$\phi_\epsilon(x) = \epsilon^{-n}\phi\left(\frac{x}{\epsilon}\right).$$

Then $\lim_{\epsilon\to 0}\phi_\epsilon(x) = \delta(x)$ in the distribution sence. Moreover, for any $1 \le p < \infty$,

$$\lim_{\epsilon\to 0}\|\phi_\epsilon \star f - f\|_p = 0.$$

When $p = \infty$, the convergence holds under additional condition that f is uniformly continuous.

Proof. Let $v(x) \in C_c^\infty(\mathbb{R}^n)$, we have

$$\begin{aligned}\int \phi_\epsilon(x)v(x)dx - v(0) &= \int \epsilon^{-n}\phi\left(\frac{x}{\epsilon}\right)v(x)dx - v(0)\\ &= \int_{\mathbb{R}^n} \phi(x)(v(\epsilon x) - v(0))dx \to 0\end{aligned}$$

by dominated convergence. The first convergence is proved.

To prove the convergence in L^p norm, we use Minkowski inequality to write

$$\begin{aligned}\|\phi_\varepsilon * f(x) - f(x)\|_p &= \left\|\int \phi_\varepsilon(y)f(x-y)dy - \int f(x)\phi(y)dy\right\|_p\\ &= \left\|\int \phi(y)(f(x-\varepsilon y) - f(x))dy\right\|_p\\ &\le \int |\phi(y)|\|f(x-\varepsilon y) - f(x)\|_{L^p(dx)}dy\\ &\le \int_{|y|\le r} |\phi(y)|\|f(x-\varepsilon y) - f(x)\|_{L^p(dx)}dy\\ &\quad + 2\int_{|y|>r} |\phi(y)|dy\|f\|_p.\end{aligned}$$

The convergence holds by first taking $\varepsilon \to 0$, then $r \to \infty$. □

Now we are ready to define the inverse Fourier transform

Definition 1.6. *Let $f \in L^1(\mathbb{R}^n)$. We define*

$$\mathcal{F}^{-1}f(x) = \check{f}(x) = \int_{\mathbb{R}^n} e^{2\pi i x\cdot\xi} f(\xi)d\xi.$$

We have the following result.

Theorem 1.7 (Fourier inversion formula). *Let $f \in L^1(\mathbb{R}^n)$, $\hat{f} \in L^1(\mathbb{R}^n)$. Then*

$$\mathcal{F}^{-1}\hat{f}(x) = \mathcal{F}\check{f}(x) = f(x) \quad a.\ e.\ x \in \mathbb{R}^n.$$

Proof. Formally,

$$\begin{aligned}\mathcal{F}^{-1}\hat{f}(x) &= \int_{\mathbb{R}^n} e^{2\pi i x\cdot\xi}\hat{f}(\xi)d\xi\\ &= \int_{\mathbb{R}^n} e^{2\pi i x\cdot\xi}\int_{\mathbb{R}^n} e^{-2\pi i y\cdot\xi} f(y)dyd\xi\\ &= \int_{\mathbb{R}^n}\int_{\mathbb{R}^n} e^{2\pi i(x-y)\xi}d\xi f(y)dy\\ &= \int_{\mathbb{R}^n} \delta(x-y)f(y)dy\\ &= f(x).\end{aligned}$$

To make the argument concrete, we use the approximation of identity. From the dominated convergence, the Fubini's theorem, we write

$$\begin{aligned}\mathcal{F}^{-1}\hat{f}(x) &= \int_{\mathbb{R}^n} e^{2\pi i x\cdot\xi}\hat{f}(\xi)d\xi \\ &= \lim_{t\to 0}\int_{\mathbb{R}^n} e^{2\pi i x\cdot\xi}e^{-t^2\pi|\xi|^2}\hat{f}(\xi)d\xi \\ &= \lim_{t\to 0}\int_{\mathbb{R}^n} e^{2\pi i(x-y)\cdot\xi}e^{-t^2\pi|\xi|^2}d\xi f(y)dy \\ &= \lim_{t\to 0}\int_{\mathbb{R}^n} t^{-n}e^{-\frac{\pi|x-y|^2}{t^2}} f(y)dy \\ &= \lim_{t\to 0} g_t * f(x).\end{aligned}$$

Here $g_t(x) = t^{-n}e^{-\frac{\pi|x|^2}{t^2}}$. A simple observation from the first two lines also shows that the convergence is uniform in x. Therefore,

$$g_t * f(x) \to \mathcal{F}^{-1}\hat{f}(x), \text{ uniform in } x.$$

On the other hand, as it is easy to verify $g_t(x)$ is an approximation to identity we also have

$$\|g_t * f - f\|_1 \to 0, \text{ as } t \to 0.$$

Combining these two facts we obtain

$$\begin{aligned}&m\{|\mathcal{F}^{-1}\hat{f}(x) - f(x)| > \lambda\} \\ &\le m\{|\mathcal{F}^{-1}\hat{f}(x) - g_t * f(x)| > \lambda/2\} + m\{|g_t * f(x) - f(x)| > \lambda/2\} \\ &\le \frac{2}{\lambda}\|g_t * f - f\|_1.\end{aligned}$$

Since t can be chosen arbitrarily small, we have

$$m\{|\mathcal{F}^{-1}\hat{f}(x) - f(x)| > \lambda\} = 0$$

for any $\lambda > 0$. This implies $\mathcal{F}^{-1}\hat{f}(x) = f(x)$ a. e. $x \in \mathbb{R}^n$. □

To extend the Fourier transform to other function space, we need to discuss the Fourier transform on a class of very important functions: the Schwartz functions.

Definition 1.8. *Define $\mathcal{S}(\mathbb{R}^n)$ to be the set of all $C^\infty(\mathbb{R}^n)$ functions such that*

$$\sup_{x\in\mathbb{R}^n} \langle x\rangle^k |\partial_x^\alpha f(x)| < \infty, \ \forall k \ge 0, \ \textit{and multi-index } \alpha \in \mathbb{N}^n.$$

Here, $\langle x\rangle = \sqrt{1+|x|^2}$. The topology of $\mathcal{S}$ can be defined by countable many norms: $\|\langle x\rangle^k \partial_x^\alpha f(x)\|_\infty$ hence $\mathcal{S}$ is a typical Frechét space.

Remark 1.9. *Obviously, $C_c^\infty(\mathbb{R}^n) \subset \mathcal{S}(\mathbb{R}^n) \subset L^p(\mathbb{R}^n)$ for any $1 \le p \le \infty$. $\mathcal{S}$ is a dense subset for $L^p(\mathbb{R}^n)$ when $1 \le p < \infty$ but not in $L^\infty(\mathbb{R}^n)$. A typical example and counter example for Schwartz function is Gaussian: $e^{-|x|^2}$ and $e^{-|x|}$ (why?).*

It is a classical result that both $\mathcal{F}$ and $\mathcal{F}^{-1}$ are continuous map from $\mathcal{S}$ to $\mathcal{S}$. Moreover, one can verify easily that

Lemma 1.10. *Let $f, g \in \mathcal{S}(\mathbb{R}^n)$. Then*

$$\langle \hat{f}, \hat{g} \rangle = \langle f, g \rangle; \; \|\hat{f}\|_2 = \|f\|_2.$$

Here, $\langle \cdot \rangle$ is an L^2 pairing.

Proof. From the definition, we have

$$\begin{aligned}
\langle \hat{f}, \hat{g} \rangle &= \int_{\mathbb{R}^n} \hat{f}(\xi)\bar{\hat{g}}(\xi) d\xi \\
&= \iint_{\mathbb{R}^{2n}} e^{-2\pi i \xi \cdot (x-y)} d\xi f(x)\bar{g}(y) dx dy \\
&= \iint_{\mathbb{R}^{2n}} \delta(x-y) f(x)\bar{g}(y) dx dy \\
&= \int_{\mathbb{R}^n} f(x)\bar{g}(x) dx = \langle f, g \rangle.
\end{aligned}$$

□

Based on this result, one can easily extend the Fourier transform to L^2 space. Indeed, for any $f \in L^2(\mathbb{R}^n)$, we can choose Cauchy sequence $\{f_k\}$ such that $f_k \to f$ in $L^2(\mathbb{R}^n)$. As $\mathcal{F}$ is unitary on $\mathcal{S}$, $\{\hat{f}_k\}$ is also Cauchy in L^2. The limit is then defined as the Fourier transform of f. The results stated in Lemma 1.10 can easily be extended to L^2. We now summarize them here.

Theorem 1.11. *Fourier transform is unitary on $L^2(\mathbb{R}^n)$. More precisely, for any $f, g \in L^2(\mathbb{R}^n)$*

$$\textit{Plancherel:} \quad \langle f, g \rangle = \langle \hat{f}, \hat{g} \rangle = \langle \check{f}, \check{g} \rangle$$

$$\textit{Parseval:} \quad \|f\|_2 = \|\hat{f}\|_2.$$

Remark 1.12. *One can use the following two as exercises for Schwartz function.*

1. *Show that Fourier transform maps $\mathcal{S}(\mathbb{R}^n)$ to $\mathcal{S}(\mathbb{R}^n)$.*
2. *Prove that $\mathcal{S}(\mathbb{R}^n)$ is not dense in $L^\infty(\mathbb{R}^n)$.*

1.3 Fourier transform on other L^p spaces

We first discuss the inversion formula on L^2. We have

Theorem 1.13. *Let $f \in L^2(\mathbb{R}^n)$. Then $\mathcal{F}^{-1}(\hat{f}) = \mathcal{F}(\check{f}) = f(x)$ a. e. $x \in \mathbb{R}^n$.*

Proof. We will need both density argument and the regularization. We first show for any $t \neq 0$

$$\int_{\mathbb{R}^n} e^{2\pi i x \cdot \xi} \hat{f}(\xi) e^{-t^2 \pi |\xi|^2} d\xi = g_t * f(x) \tag{1.2}$$

where $g_t = |t|^{-n} e^{-\frac{\pi |x|^2}{t^2}}$. Indeed, for any $f_j \in \mathcal{S}(\mathbb{R}^n)$, by Fubini Theorem

$$\int_{\mathbb{R}^n} e^{2\pi i x\cdot\xi} \hat{f}_j(\xi) e^{-t^2\pi|\xi|^2} d\xi = \int_{\mathbb{R}^n} e^{2\pi(x-y)\cdot\xi} e^{-t^2\pi|\xi|^2} d\xi f_j(y) dy = g_t * f_j(x). \quad (1.3)$$

Choosing f_j such that $f_j \to f$ in L^2 and using $\|\hat{f}_j - \hat{f}\|_2 \to 0$ in (1.3) we get (1.2).

Second, taking $t \to 0$ in (1.2). From the dominated convergence and Parseval, we know

$$\mathcal{F}^{-1}(\hat{f}(\xi) e^{-t^2\pi|\xi|^2}) \to \mathcal{F}^{-1}\hat{f}$$

in L^2. On the other hand, the approximation of identity says

$$g_t * f \to f$$

in L^2. Therefore

$$f = \mathcal{F}^{-1}\hat{f}$$

as functions in L^2. □

After we have defined the Fourier transform on L^1 and L^2, we can extend it to the L^p space with $1 < p < 2$ following two approaches.

First for any $f \in L^p(\mathbb{R}^n)$, we decompose it into

$$f = f_1 + f_2, \ f_1 = f 1_{\{x; |f(x)>1|\}}, \ f_2 = f 1_{\{x; |f(x)| \le 1\}}.$$

It is easy to see that $f_1 \in L^1$ and $f_2 \in L^2$ thus we can define

$$\hat{f} = \hat{f}_1 + \hat{f}_2 \in L^2 + L^\infty.$$

The alternative way is to view $\mathcal{F}$ as a bounded linear operator on L^1 and L^2

$$\mathcal{F}: \ L^1 \to L^\infty; \ \mathcal{F}: \ L^2 \to L^2$$

with norm 1. By Riesz interpolation, $\mathcal{F}$ can be extended to L^p and we have

Proposition 1.14 (Hausdorff-Young inequality). *For $f \in L^p$ and $1 < p < 2$, $\hat{f} \in L^{p'}$ and*

$$\|\hat{f}\|_{p'} \le \|f\|_p.$$

A direct analysis gives the following

Proposition 1.15 (Convolutions). *Let $f \in L^p(\mathbb{R}^n)$ and $g \in L^q(\mathbb{R}^n)$ with $1 + \frac{1}{r} = \frac{1}{p} + \frac{1}{q}$. Suppose $1 \le p, q, r \le 2$ then we have*

$$\widehat{f * g} = \hat{f} \cdot \hat{g}.$$

For Fourier transform in general L^p space, there is no such analogue, one needs to put this into a general framework which is sketched here. To start with, we define the tempered distribution $\mathcal{S}'(\mathbb{R}^n)$ as the set of linear functionals on Schwartz function class $\mathcal{S}(\mathbb{R}^n)$. Recall $\mathcal{F}$ maps $\mathcal{S}$ to itself continuously, one can extend $\mathcal{F}$ to all tempered distributions through the following

Definition 1.16 (Fourier transform on tempered distributions). *Let $f \in \mathcal{S}'(\mathbb{R}^n)$, we define $\hat{f} \in \mathcal{S}'(\mathbb{R}^n)$ as the Fourier transform of f such that $\forall \phi \in \mathcal{S}(\mathbb{R}^n)$,*

$$(\mathcal{F}f, \phi) = (f, \hat{\phi}).$$

And as such $\mathcal{F}$ is a continuous map from $\mathcal{S}'(\mathbb{R}^n)$ to itself.

Example and counter-examples of $\mathcal{S}'$:

δ-function, polynomials (such as $|x|^k$), $e^{|x|^2}$ are examples distributions; $e^{|x|^2}$ is not a tempered distribution (why?).

At the end of this section, we formally compute the Fourier transform of a typical $\mathcal{S}'$ function. This example is frequently used in the theory of partial differential equations.

Theorem 1.17. *Let $0 < \alpha < n$. Then $c_\alpha \mathcal{F}(|x|^{\alpha-n}) = c_{n-\alpha}|\xi|^{-\alpha}$ and $c_\alpha = \pi^{-\frac{\alpha}{2}}\Gamma(\frac{\alpha}{2})$ holds in the sense of tempered distributions. More precisely, for any $f \in \mathcal{S}(\mathbb{R}^n)$*

$$c_\alpha(|\xi|^{-\alpha}\hat{f}(\xi))\check{}(x) = c_{n-\alpha}\int_{\mathbb{R}^n} |x-y|^{\alpha-n} f(y)dy. \tag{1.4}$$

Proof. We begin with expressing the right hand side of (1.4) into the superposition of Gaussian. This is achieved by using definition of Gamma functions. Recall

$$c_\alpha|\xi|^{-\alpha} = \int_0^\infty e^{-\pi\lambda|\xi|^2}\lambda^{\frac{\alpha}{2}-1}d\lambda.$$

Since $|\xi|^{-\alpha}\hat{f}(\xi)$ is integrable, by Fubini's Theorem

$$\begin{aligned}
c_\alpha(|\xi|^{-\alpha}\hat{f}(\xi))\check{}(x) &= \int_{\mathbb{R}^n} e^{2\pi i x\cdot\xi}|\xi|^{-\alpha}\hat{f}(\xi)d\xi \\
&= \int_0^\infty \int_{\mathbb{R}^n} e^{2\pi i x\cdot\xi}e^{-\pi\lambda|\xi|^2}\hat{f}(\xi)d\xi\lambda^{\frac{\alpha}{2}-1}d\lambda \\
&= \int_0^\infty \lambda^{-\frac{n}{2}}\lambda^{\frac{\alpha}{2}-1}\int_{\mathbb{R}^n} e^{-\pi|x-y|^2/\lambda}f(y)dyd\lambda \\
&= \int_{\mathbb{R}^n}\int_0^\infty \lambda^{-\frac{n-\alpha}{2}-1}e^{-\frac{\pi|x-y|^2}{\lambda}}d\lambda f(y)dy \\
&= c_{n-\alpha}\int_{\mathbb{R}^n}|x-y|^{-n+\alpha}f(y)dy.
\end{aligned}$$

(1.4) is proved. □

1.4 Some discussion on the convergence problem

Fourier inversion formula for L^p functions

$$\mathcal{F}^{-1}\hat{f}(x) = \int_{\mathbb{R}^n} \hat{f}(\xi)e^{2\pi ix\cdot\xi}d\xi.$$

Previously we have proved that if $f \in L^1$, $\hat{f} \in L^1$, then this integral gives f a. e. Same conclusion holds for $f \in L^2$. However, in L^2 case, the integral in ξ does not make sense since $\hat{f}$ is only an L^2 function. To circumvent this problem, our idea was to add a mollifier and consider

$$\lim_{t\to 0}\int_{\mathbb{R}^n} \hat{f}(\xi)e^{-t|\xi|^2\pi}e^{2\pi ix\cdot\xi}d\xi.$$

One can use other mollifiers such as

$$\lim_{t\to 0}\int \hat{f}(\xi)e^{-t\pi|\xi|}e^{2\pi ix\cdot\xi}d\xi$$

or even

$$\lim_{R\to\infty}\int_{|x|\leq R} \hat{f}(\xi)e^{2\pi ix\cdot\xi}d\xi.$$

In all these cases, one can always show the limit, gives rise to f and $\mathcal{F}^{-1}\hat{f}$ at the same time.

So the general question we want to ask is for any $f \in L^p$, how does the limit converge? This kind of question appears in practical problem.

First, consider boundary value problem for Laplacian equation on the half space:

$$\begin{cases} \Delta_{t,x}u = 0, \ (t,x) \in \mathbb{R}^+ \times \mathbb{R}^n, \\ u(0,x) = f(x). \end{cases}$$

Taking Fourier transform in x, we get an second order ODE as follows:

$$(\partial_t^2 - 4\pi^2|\xi|^2)\hat{u}(t,\xi) = 0, \ \hat{u}(0,\xi) = \hat{f}(\xi).$$

Therefore we can write the solution by undoing the Fourier transform:

$$u(t,x) = \mathcal{F}^{-1}(e^{2\pi t|\xi|}\hat{f}(\xi)).$$

It is natural to investigate in which sense $u(t,x)$ approaches to the boundary value as $t \to 0$.

One can also consider the initial value problem for heat equation

$$\frac{\partial u}{\partial t} - \Delta u = 0, \ (t,x) \in R^+ \times \mathbb{R}^n,$$
$$u(0,x) = f(x).$$

Using Fourier transform again, one can write the solution as

$$u(t,x) = \mathcal{F}^{-1}(e^{-4\pi^2 t|\xi|^2}\hat{f}(\xi)).$$

One can ask as $t \to 0$, how the solution $u(t,x)$ approaches its initial value?

In this section, we shall only discuss the convergence problem when the mollifier is a non-smooth cutoff.

Define $S_R(f) = \mathcal{F}^{-1}(\chi_{B(0,R)}\hat{f})$. Then

$$\lim_{R\to\infty} \|S_R f - f\|_p = 0, \ \forall 1 < p < \infty$$

is equivalent to

$$\|S_R f\|_p \leq C_p \|f\|_p.$$

Proof. "$\Rightarrow$" As from scaling argument

$$\|S_1 f\|_p \leq C_p \|f\|_p, \ \forall f \in L^p$$

is equivalent to

$$\|S_R f\|_p \leq C_p \|f\|_p, \ \forall f \in L^p \text{ and } R.$$

It is easy to see the convergence implies the boundedness.

"$\Leftarrow$" It is natural to use smooth cutoff. Let $\phi \in C_c^\infty(B(0,1))$ and $\phi(0) = 1$ (so that $\int_{\mathbb{R}^n} \check{\phi}(x)dx = 1$). Define

$$P_{\leq N} f = \mathcal{F}^{-1}\left(\phi\left(\frac{\xi}{N}\right)\hat{f}(\xi)\right) = N^n \check{\phi}(N\cdot) * f.$$

From the approximation of identity we know

$$\|P_{\leq N} f - f\|_p \to 0, \text{ as } N \to \infty.$$

From triangle we write

$$\begin{aligned}\|S_R f - f\|_p &\leq \|S_R f - S_R P_{\leq N} f\|_p + \|S_R P_{\leq N} f - P_{\leq N} f\|_p \\ &\leq C_p \|f - P_{\leq N} f\|_p + \|S_R P_{\leq N} f - P_{\leq N} f\|_p.\end{aligned}$$

As for $R \geq 4N$, the Fourier support does not overlap, we have $S_R P_{\leq N} f = P_{\leq N} f$, the convergence holds by first letting $R \to \infty$, then $N \to \infty$. □

The L^p boundedness is only true in limited case. In dimension $n = 1$, it can be explicitly computed that

$$\begin{aligned}\mathcal{F}^{-1}\chi_{[-R,R]}(x) &= \int_{-R}^{R} e^{2\pi i x\cdot\xi} d\xi = \left.\frac{e^{2\pi i x\cdot\xi}}{2\pi i x}\right|_{\xi=-R}^{R} \\ &= \frac{\sin(2\pi R x)}{\pi x}.\end{aligned}$$

Hence

$$S_R f = \mathcal{F}^{-1}(\chi_{[-R,R]}\hat{f}(\xi)) = \mathcal{F}^{-1}\chi_{[-R,R]} * f$$
$$= \frac{\sin(2\pi R\cdot)}{\pi\cdot} * f.$$

The function $\frac{\sin(2\pi Rx)}{\pi x}$ is not in L^1, so one can not prove the L^p boundedness just by Young's inequality. We will revisit the problem later after we introduce the Marcinkiewicz interpolation inequality. Nevertheless, in one dimension, the operator S_R is L^p bounded for $1 < p < \infty$.

On the other hand, if we consider the averaging operator, the L^p boundedness is much easier to see. Define

$$\begin{aligned}\sigma_R f(x) &= \frac{1}{R}\int_0^R S_t f(x)dt\\ &= \frac{1}{R}\int_0^R \frac{\sin(2\pi tx)}{\pi x}dt * f\\ &= \frac{1}{R\pi x}\left(\frac{-\cos(2\pi tx)}{2\pi x}\Big|_0^R\right) * f\\ &= \frac{\sin^2(2\pi Rx)}{R(\pi x)^2} * f.\end{aligned}$$

Now as the kernel is L^1 bounded, the L^p boundedness can be proved easily by using the Young's inequality.

In dimension $n \geq 2$, the L^p boundedness holds true when $p = 2$ obviously by Fourier transform. A quite surprising result from C. Fefferman says in other L^p space, S_R is actually not bounded! This is one of his famous work "The multiplier problem for the ball", and was published in Ann. of Math. **94** (1972). Later on, we will see that when the ball is replaced by a rectangle box, the L^p boundedness is always true.

1.5 Localization for Fourier transform

In this lecture, we will understand deeper what the Fourier transform does on a function. In particular we will discuss the issue of localization. To make it easy we only talk about the one-dimensional case.

The localization of a function can be achieved by multiplying a smooth or non-smooth cutoff function. let $\chi_I(t)$ be an indicator function on the interval I and let $\tilde{\chi}_I(t)$ be smoothed out version of χ_I. Then $\chi_I(t)f(t)$ and $\tilde{\chi}_I(t)f(t)$ localize f on and around the interval I. Define

$$\pi_I f = \mathcal{F}^{-1}(\chi_I \hat{f}(\tau)),\ \tilde{\pi}_I f = \mathcal{F}^{-1}(\tilde{\chi}_I(\tau)\hat{f}(\tau)).$$

Then π, $\tilde{\pi}$ localize f in the frequency space. Take $I = [-1, 1]$ as an example, an easy calculation shows

$$\mathcal{F}^{-1}(\chi_{[-1,1]}) = \int_{-1}^{1} e^{-2\pi it\tau}d\tau = \frac{\sin(2\pi s)}{\pi s}$$

and so

$$\pi_I f(t) = \int_{\mathbb{R}} \frac{\sin(2\pi(t-s))}{\pi(t-s)} f(s) ds. \tag{1.5}$$

Suppose f has a compact support Ω, then $\pi_I f$ no longer has compact support. On the other hand, if we look at those t for which

$$\mathrm{dist}(t, \Omega) \gtrsim 1.$$

From (1.5) we have

$$|\pi_I f(t)| \leq \mathrm{dist}(t, \Omega)^{-1} \|f\|_1.$$

Therefore, in frequency space, π_I spread f by a distance $O(1)$ together with a slowly decaying tail.

If we consider the smooth cutoff $\tilde{\pi}_I$, we have

$$\tilde{\pi}_I f = \mathcal{F}^{-1} \chi_{[-1,1]} * f = K * f.$$

Here, after integration by parts j times,

$$K(s) = \int_{\mathbb{R}} e^{2\pi i s\tau} \tilde{\chi}_{[-1,1]}(\tau) d\tau = \int_{\mathbb{R}} \left(\frac{-1}{2\pi i s}\right)^j e^{2\pi i s\tau} \left(\frac{d}{d\tau}\right)^j \tilde{\chi}_{[-1,1]}(\tau) d\tau.$$

Therefore

$$|K(s)| \leq s^{-j}, \ \forall j \in \mathbb{N}.$$

Plugging this estimate we see

$$|\tilde{\pi}_I f(t)| \lesssim \mathrm{dist}(t, \Omega)^{-j} \|f\|_1.$$

We can say that if we use a smooth cutoff, $\tilde{\pi}_I$ spread f by a distance of $O(1)$ together with a rapid decreasing tail.

One may hope by smartly choosing the cutoff function and the non-trivial function f, the tail may vanish so that $\tilde{\pi} f$ still has compact support. The following result says this never happens. To see this, we first need a basic result from complex analysis.

Lemma 1.18. *A nontrivial entire function can only have discrete zero points.*

Proof. We give the proof for the sake of completeness.

Let $f(z)$ be the entire function and $z_n \to z_0$ such that

$$f(z_n) = 0, \quad \text{for } n = 0, 1, \cdots.$$

As $f(z)$ is an entire function, we can write

$$f(z) = \sum_{n=0}^{\infty} a_n (z - z_0)^n.$$

First from $f(z_0) = 0$ we get $a_0 = 0$. Second, we let $z = z_n$ and write

$$\frac{f(z_n) - f(z_0)}{z_n - z_0} = a_1 + \sum_{n=1}^{\infty} a_k(z_n - z_0)^k.$$

Letting $z_n \to z_0$ and noticing $f(z_n) = f(z_0) = 0$ we get $a_1 = 0$. Inductively we can prove

$$a_i = 0, \; i = 1, 2, \cdots,$$

thus $f(z) = 0$. The lemma is proved. □

Lemma 1.19. *There does not exist a non-zero integrable function f such that both f and $\hat{f}$ are compact supported.*

Proof. Suppose f is a non-trivial compact supported function and f is integrable. The integral

$$\hat{f}(t) = \int_{\mathbb{R}} e^{-2\pi i t \tau} f(t) dt$$

is absolutely convergent. By dominated convergence, $\hat{f}$ can be extended to be an entire function

$$\hat{f}(z) = \int_{\mathbb{R}} e^{-2\pi i z t} f(t) dt.$$

Now we can argue by contradiction. Suppose $\hat{f}(t)$ has compact support, in the complex plane, there exists a line for which $\hat{f}(z)$ vanishes on it. From the previous lemma, $f(z) \equiv 0$. A contradiction. □

2 Littlewood-Paley theory

2.1 Littlewood-Paley decomposition

We begin by partition of unity in the frequency space $\mathbb{R}^n_\xi$. Let $\phi(\xi) \in C_c^\infty(\mathbb{R}^n)$ be such that

$$\phi(\xi) = \begin{cases} 1, & |\xi| \leq 1, \\ 0, & |\xi| \geq 2. \end{cases}$$

Define

$$\psi(\xi) = \phi(\xi) - \phi(2\xi),$$
$$\psi_j(\xi) = \psi\left(\frac{\xi}{2^j}\right).$$

Then it's easy to verify that

$$\sum_{j\in\mathbb{Z}} \psi_j(\xi) = 1, \; \forall \xi \neq 0.$$

We define Littlewood-Paley operators as projections on frequency space

$$\begin{aligned} P_k f &= \mathcal{F}^{-1}(\psi_k \hat{f}), \\ P_{\le k} f &= \mathcal{F}^{-1}(\phi_k \hat{f}), \\ P_{>k} f &= (I - P_{\le k}) f. \end{aligned}$$

The purpose of Littlewood-Paley projection is to decompose functions into different frequencies:

$$f = \sum_{k \in \mathbb{Z}} P_k f.$$

The first question we need to understand is when and in what sense the summation converges. Formally we can write

$$\begin{aligned} \sum_{k \in \mathbb{Z}} P_k f &= \lim_{N_1 \to \infty,\ N_2 \to \infty} \sum_{k=-N_1}^{N_2} P_k f \\ &= \lim_{N_2 \to \infty} P_{<N_2} f - \lim_{N_1 \to \infty} P_{<-N_1} f. \end{aligned}$$

The convergence issue is reduced to considering when the first term converges to f and the second term converges to 0.

We have the following result:

Theorem 2.1. *Let $f \in L^p(\mathbb{R}^n)$, $1 \le p \le \infty$. In the case $p = \infty$, we also require f be uniformly continuous. Then*

$$\lim_{k \to \infty} \|P_{\le k} f - f\|_p = 0. \tag{2.1}$$

When $1 < p < \infty$, we have

$$\lim_{k \to \infty} \|P_{\le -k} f\|_p = 0. \tag{2.2}$$

Therefore

$$\lim_{N_1, N_2 \to \infty} \left\| \sum_{k=-N_1}^{N_2} P_k f - f \right\|_p = 0,\ 1 < p < \infty. \tag{2.3}$$

The second convergence fails in L^1 and L^∞ spaces: we have counter examples such that

$$\lim_{N_1 \to \infty} \|P_{\le -N_1} f\|_p \ne 0,\ p = 1,\ \infty.$$

Proof. Recall

$$P_{\le k} f = \mathcal{F}^{-1}\left(\phi\left(\frac{\cdot}{2^k}\right) \hat{f}\right) = 2^{kn} \check{\phi}(2^k \cdot) * f,$$

and for the convolution kernel ϕ,

$$\int \check{\phi} = \phi(0) = 1.$$

From the approximation of identity, we have

$$\lim_{k\to\infty} \|P_{\le k} f - f\|_p = 0$$

when $1 < p < \infty$, $f \in L^p$ and $p = \infty$ and $f \in L^\infty$, f is uniformly continuous. (2.1) is proved.

To see (2.2), we use density argument. Assume $f \in \mathcal{S}(\mathbb{R}^n)$. From

$$P_{\le -k} f(x) = 2^{-kn}\check{\phi}(2^{-k}\cdot) * f(x),$$

and Young's inequality we have

$$\|P_{\le -k} f\|_p \le 2^{-kn} 2^{\frac{kn}{p}} \|\check{\phi}\|_p \|f\|_1 \to 0, \text{ as } k \to \infty.$$

In the case when $p = 1$, the second convergence does not hold for those function which is nontrivial at frequency 0. Indeed, from Fubini's theorem,

$$\begin{aligned}\|P_{\le -k} f\|_1 &\ge 2^{-kn} \int_{\mathbb{R}^n} \check{\phi}(2^{-k}(x-y)) f(y)\,dx dy \\ &= \int \check{\phi}(x)dx \int f(y)dy \\ &= \phi(0)\hat{f}(0) = \hat{f}(0) = \int f(x)dx.\end{aligned}$$

Therefore for those functions with $\int f(x)dx > 0$, the convergence (2.2) do not hold. The counter-example for L^∞ can be simply taken as $f = 1$. □

We now introduce the fundamental Bernstein inequalities for the Littlewood-Paley operators.

Proposition 2.2 (Bernstein inequality). *Let* $1 \le p \le q \le \infty$, $s \ge 0$ *and* $f \in L^p(\mathbb{R}^n)$. *Then*

$$\begin{aligned}\|P_{\le k} f\|_q &\lesssim 2^{kn(\frac{1}{p}-\frac{1}{q})}\|f\|_p, \\ \|P_k f\|_q &\lesssim 2^{kn(\frac{1}{p}-\frac{1}{q})}\|f\|_p, \\ \||\nabla|^s P_{\le k} f\|_p &\lesssim 2^{ks}\|f\|_p, \\ \|P_k f\|_p &\sim 2^{-ks}\||\nabla|^s P_k f\|_p, \\ \|P_{>k} f\|_p &\lesssim 2^{-ks}\||\nabla|^s f\|_p.\end{aligned}$$

Proof. From the definition,

$$P_{\le k} f(x) = 2^{kn}\check{\phi}(2^k\cdot) * f(x).$$

Using Young's inequality, we have

$$\begin{aligned}\|P_{\leq k}f\|_q &\leq 2^{kn}\|f\|_p\|\check{\phi}(2^k\cdot)\|_{(1+\frac{1}{q}-\frac{1}{p})^{-1}}\\ &= 2^{kn}2^{-kn(1+\frac{1}{q}-\frac{1}{p})}\|\check{\phi}\|_{\frac{pq}{pq+p-q}}\|f\|_q \lesssim 2^{kn(\frac{1}{p}-\frac{1}{q})}\|f\|_p.\end{aligned}$$

The second inequality can be proved similarly.

For the third one, we start from the formulation

$$\begin{aligned}|\nabla|^s P_{\leq k}f &= \mathcal{F}^{-1}\left(|2\pi\xi|^s\phi\left(\frac{\xi}{2^k}\right)\right) * f\\ &= 2^{ks}2^{kn}|2\pi|^s\mathcal{F}^{-1}(|\xi|^s\phi)(2^k\xi) * f.\end{aligned}$$

Then the inequality follows quickly from the Young's inequality:

$$\begin{aligned}\||\nabla|^s P_{\leq k}f\|_p &\lesssim 2^{ks}\|\mathcal{F}^{-1}(|\xi|^s\phi)\|_1\|f\|_p\\ &\lesssim 2^{ks}\|f\|_p.\end{aligned}$$

The second inequality is then proved, so is the "$\lesssim$" part of the fourth inequality. The "$\gtrsim$" can be dealt with similarly if we write

$$P_k f = |\nabla|^{-s}\tilde{P}_k\,|\nabla|^s P_k f$$

and show the operator norm $\||\nabla|^{-s}P_k\|_{L^p\to L^p} \lesssim 2^{-k}$. Here $\tilde{P}_k$ is the fattened P_k, the Fourier support of $\tilde{P}_k$ is 1 on the Fourier support of P_k. The natural choice is

$$\tilde{P}_k = P_{k-1} + P_k + P_{k+1}.$$

We skip the detailed computation. Finally, we can use the fourth inequality to prove the last one by dyadic decomposition. As the case $s = 0$ is trivial we assume $s > 0$. We have

$$\begin{aligned}\|P_{>k}f\|_p &\leq \sum_{j>k}\|P_j f\|_p \lesssim \sum_{j>k}2^{-js}\||\nabla|^s f\|_p\\ &\lesssim 2^{-ks}\||\nabla|^s f\|_p.\end{aligned}$$ □

Remark 2.3. *In Bernstein estimate, the* 1, 2, 3, 5 *items can be slightly improved by keeping the same frequency cutoff. For example, for the first estimate we can write as*

$$\|P_k f\|_q \lesssim 2^{kn(\frac{1}{p}-\frac{1}{q})}.$$

This can be done by introducing the fattened operators. For example, we can write

$$P_k = \tilde{P}_k P_k;\ P_{\leq k} = P_{\leq k}P_{\leq k+2};\ P_{>k} = P_{>k}P_{>k-2}.$$

It is obvious that same properties for the fattened operator can be proved similarly.

2.2 An application: Sobolev embedding

Sobolev space is a very important tool in the study of PDEs. Here we give a simple example showing how the Littlewood-Paley decomposition can be used to prove embedding results between Sobolev spaces. More detailed discussion on Sobolev spaces and embedding results can be found in later chapters.

For $1 \le p < \infty$, we define Sobolev space $H^{1,p}(\mathbb{R}^n)$ to be the completion of $\mathcal{S}(\mathbb{R}^n)$ with respect the norm

$$\|f\|_{H^{1,p}} = \|f\|_p + \|\nabla f\|_p.$$

We have the non-sharp Sobolev embedding

Lemma 2.4. *Let $1 \le p < q \le \infty$ be such that*

$$1 - \frac{n}{p} > -\frac{n}{q}.$$

Then we have $H^{1,p}(\mathbb{R}^n) \subset L^q(\mathbb{R}^n)$. Moreover, for any $f \in H^{1,p}$,

$$\|f\|_q \lesssim_{p,q,n} \|f\|_{H^{1,p}}.$$

Proof. From density argument, it suffices to prove the estimate for $f \in \mathcal{S}(\mathbb{R}^n)$. We decompose f into dyadic pieces and use Bernstein to get

$$\begin{aligned}
\|f\|_q &\le \sum_{k\in\mathbb{Z}} \|P_k f\|_q \\
&\le \sum_{k=-\infty}^{0} \|P_k f\|_q + \sum_{k=1}^{\infty} \|P_k f\|_q \\
&\lesssim \sum_{k=-\infty}^{0} 2^{kn(\frac{1}{p}-\frac{1}{q})} \|f\|_p + \sum_{k=1}^{\infty} 2^{kn(\frac{1}{p}-\frac{1}{q})} \|P_k f\|_p \\
&\lesssim \|f\|_p + \sum_{k=1}^{\infty} 2^{kn(\frac{1}{p}-\frac{1}{q})-1} \|\nabla f\|_p \\
&\lesssim \|f\|_p + \|\nabla f\|_p.
\end{aligned}$$

The estimate is proved. □

For $1 \le p < \infty$, we define the homogeneous Sobolev space $\dot{H}^{1,p}(\mathbb{R}^n)$ as the completion of $\mathcal{S}(R^n)$ under the norm $\|f\|_{\dot{H}^{1,p}} = \|\nabla f\|_p$. We have the following sharp embedding:

Lemma 2.5 (Sharp Sobolev embedding). *Let $1 < p < q < \infty$ and $1 - \frac{n}{p} = -\frac{n}{q}$. Then $\dot{H}^{1,p}(\mathbb{R}^n) \subset L^q(\mathbb{R}^n)$. Moreover for any $f \in \dot{H}^{1,p}$,*

$$\|f\|_q \lesssim_{p,q,n} \|\nabla f\|_p. \tag{2.4}$$

Proof. In Fourier side,

$$\hat{f}(\xi) = \frac{\xi}{|\xi|^2} \cdot \xi \hat{f}(\xi) = \mathcal{F}(\Delta^{-1}\nabla \cdot \nabla f).$$

Therefore we have

$$f(x) = \Delta^{-1}\partial_j \ (\partial_j f).$$

To show (2.4), it suffices to prove

$$\|\Delta^{-1}\partial_j g\|_q \lesssim \|g\|_p, \ \forall g \in L^p(\mathbb{R}^n), \ \text{and } j = 1, 2, \cdots, n.$$

By Marcinkiewicz interpolation (in the Appendix), we only need to show for $1 < p < q < \infty$ and $1 - \frac{n}{p} = -\frac{n}{q}$ that

$$\|\Delta^{-1}\partial_j g\|_{q,\infty} \lesssim \|g\|_p.$$

By scaling we can assume $\|g\|_p = 1$.

We first look at each frequency of g to get an inspiration.

From Bernstein inequality

$$\|\Delta^{-1}\partial_j P_k g\|_\infty \lesssim 2^{-k} 2^{k\frac{n}{p}} \leq C 2^{k(\frac{n}{p}-1)}.$$

Therefore if we choose $\lambda > C2^{k(\frac{n}{p}-1)}$,

$$m\{x : \ |\Delta^{-1}\partial_j P_k g(x)| > \lambda\} = 0.$$

This puts an automatic assumption on λ: $\lambda \lesssim 2^{k(\frac{n}{p}-1)}$, which is equivalent to

$$2^k \gtrsim \lambda^{1-\frac{n}{p}}.$$

Using this we can estimate

$$\begin{aligned} m\{x : |\Delta^{-1}\partial_j P_k g(x)| > \lambda\} &\leq \lambda^{-p} \|\Delta^{-1}\partial_j P_k g\|_p^p \\ &\lesssim 2^{-pk}\lambda^{-p} \\ &\lesssim \lambda^{-p-\frac{p}{\frac{n}{p}-1}} = \lambda^{-q}. \end{aligned}$$

Therefore

$$\|\Delta^{-1}\partial_j P_k g\|_{q,\infty} \lesssim 1.$$

For the whole function, we decompose at frequency 2^{k_0} and follow the same idea.

$$\begin{aligned} &m\{x : |\Delta^{-1}\partial_j g(x)| > \lambda\} \\ &\leq m\{x : |\Delta^{-1}\partial_j P_{>k_0} g(x)| > \lambda/2\} + m\{x : |\Delta^{-1}\partial_j P_{\leq k_0} g(x)| > \lambda/2\}. \end{aligned}$$

As for the low frequency

$$\|\Delta^{-1}\partial_j P_{\le k_0} g\|_\infty \le C2^{k_0(\frac{n}{p}-1)},$$

if we take k_0 such that

$$\lambda = C2^{k_0(\frac{n}{q}-1)}$$

for some large constant C, we have

$$\|\Delta^{-1}\partial_j P_{\le k_0} g\|_\infty < \frac{\lambda}{2}.$$

Thus only the high frequencies are left, we can use again the Chebyshev's inequality to estimate

$$\begin{aligned} m\{x: \ |\Delta^{-1}\partial_j g(x) > \lambda\} &\le \lambda^{-p}\|\Delta^{-1}\partial_j P_{>k_0} g\|_p^p \\ &\lesssim \lambda^{-p}2^{-pk_0} \\ &= \lambda^{-q}. \end{aligned}$$

We finally have

$$\|\Delta^{-1}\partial_j g\|_{q,\infty} \lesssim 1.$$

The lemma is proved. □

Remark 2.6. *One can use the same method introduced in this part, prove the general Sobolev inequality:*

Let $1 < p < q < \infty$, $s > 0$ be such that $s - \frac{n}{p} = -\frac{n}{q}$. Let $f \in \mathcal{S}(\mathbb{R}^n)$. Prove

$$\|f\|_q \lesssim \||\nabla|^s f\|_p.$$

The above inequality implies:

$$\dot{H}^{s,p}(\mathbb{R}^n) \subset L^q(\mathbb{R}^n).$$

Here, $\dot{H}^{s,p}$ refers to the general Bessel potential space, it can be defined as the completion of all Schwartz function with respect to the norm $\|\cdot\|_{\dot{H}^{s,p}}$. As it coincides with classical Sobolev space when s is positive integer, we will make no difference between the two spaces. See later part for detailed discussion.

2.3 Littlewood-Paley theorem

Let $1 < p < \infty$, $f \in L^p$. Then we have

$$f = \sum_{k\in\mathbb{Z}} P_k f.$$

From triangle we have the cheap inequality

$$\sup_{k\in\mathbb{Z}} \|P_k f\|_p \le \|f\|_p \le \sum_{k\in\mathbb{Z}} \|P_k f\|_p.$$

However, if we take advantage of the cancelation between high and low frequencies, the estimate can be improved significantly. The most obvious case is $p = 2$ because of Parseval. We have

$$\begin{aligned}
\|f\|_2^2 &= \int_{\mathbb{R}^n} |\hat{f}(\xi)|^2 d\xi \\
&= \sum_{k\in\mathbb{Z}} \int_{2^k \le |\xi| \le 2^{k+1}} |\hat{f}(\xi)|^2 d\xi \\
&\le \sum_{k\in\mathbb{Z}} \int |\widehat{P_k f} + \widehat{P_{k+1} f}|^2 d\xi \\
&\lesssim \sum_{k\in\mathbb{Z}} \int |P_k f|^2 dx \\
&= \|(\sum_{k\in\mathbb{Z}} |P_k f|^2)^{\frac{1}{2}}\|_2^2.
\end{aligned}$$

On the other hand,

$$\begin{aligned}
\|(\sum_{k\in\mathbb{Z}} |P_k f|^2)^{\frac{1}{2}}\| &= \sum_{k\in\mathbb{Z}} \int |P_k f|^2 dx \\
&= \sum_{k\in\mathbb{Z}} \int |\widehat{P_k f}(\xi)|^2 d\xi \\
&= \sum_{k\in\mathbb{Z}} \int \psi^2\left(\frac{\xi}{2^k}\right) |\hat{f}(\xi)|^2 d\xi \\
&\lesssim \sum_{k\in\mathbb{Z}} \int_{2^{k-1} \le |\xi| \le 2^{k+1}} |\hat{f}(\xi)|^2 d\xi \\
&\lesssim \sum_{k\in\mathbb{Z}} \int_{2^{k-1} \le |\xi| \le 2^k} |\hat{f}(\xi)|^2 d\xi + \sum_{k\in\mathbb{Z}} \int_{2^k \le |\xi| \le 2^{k+1}} |\hat{f}(\xi)|^2 d\xi \\
&\lesssim \|f\|_2^2.
\end{aligned}$$

To summarize, we have

$$\|f\|_2 \sim \|(\sum_{k\in\mathbb{Z}} |P_k f|^2)^{\frac{1}{2}}\|_2.$$

In the L^p space, the cancelation is not easy to see, but the similar estimate holds and this is an important result from Littlewood-Paley.

Theorem 2.7 (Littlewood-Paley theorem). *Denote*

$$Sf = \Big(\sum_{k\in\mathbb{Z}} |P_k f|^2\Big)^{\frac{1}{2}},$$

as the square function of f. Then for $1 < p < \infty$,

$$\|Sf\|_p \sim_p \|f\|_p.$$

Littlewood-Paley theorem can be proved by using Rademacher functions which are introduced as follows.

Define $r_0(t)$ be a period function function on $\mathbb{R}$ with period 1. On $[0, 1)$, it is defined as

$$r_0(t) = \begin{cases} 1, & 0 \le t < \frac{1}{2}, \\ -1, & \frac{1}{2} \le t < 1. \end{cases}$$

Define

$$r_m(t) = r_0(2^m t).$$

Then r_m has the period $\frac{1}{2^m}$. The following property is easily seen.

Proposition 2.8 (Orthogonality for Rademacher functions). *Let $0 \le m_1 < m_2 < \cdots < m_k < \infty$ be integers. Then*

$$\int_0^1 r_{m_1}(t) r_{m_2}(t) \cdots r_{m_k}(t)\,dt = 0.$$

Proof. First we note that

$$\int_{\frac{k}{2^m}}^{\frac{k+1}{2^m}} r_m(t)\,dt = 0. \tag{2.5}$$

Now we partition

$$[0, 1] = \bigcup_{j=0}^{2^{m_k}-1} \left[\frac{j}{2^{m_k}}, \frac{j+1}{2^{m_k}}\right]$$

and write $\int_0^1$ into 2^{m_k} pieces. Note on each subinterval $[\frac{j}{2^{m_k}}, \frac{j+1}{2^{m_k}}]$, $r_{m_l}(t)$ with $m_l > m_k$ remains constant, the results follows immediately from (2.5). □

The square function estimate relies crucially on

Theorem 2.9. *Let $1 < p < \infty$. Let $\{a_m\}$ be an l^2 sequence. Then for*

$$F(t) = \sum_{m=0}^{\infty} a_m r_m(t),$$

we have

$$\|F\|_{L_t^p([0,1])} \sim \|F\|_{L_t^2([0,1])} = \left(\sum_{m=0}^{\infty} a_m^2\right)^{\frac{1}{2}}.$$

Proof. We first show the second part. From the orthogonality of Radamache functions, we have

$$\begin{aligned}\|F(t)\|^2_{L^2_t([0,1])} &= \int_0^1 F^2(t)dt = \int_0^1 \sum_{m=0}^{\infty} a_m r_m(t) \sum_{n=0}^{\infty} a_n r_n(t)dt \\ &= \sum_{m,n=0}^{\infty} a_m a_n \int_0^1 r_m(t) r_n(t)dt \\ &= \sum_{m=0}^{\infty} a_m^2.\end{aligned}$$

Generally,

$$\begin{aligned}\|F(t)\|^4_{L^4_t([0,1])} &= \int_0^1 F^4(t)dt \\ &= \sum_{0\le m_1,m_2,m_3,m_4<\infty} a_{m_1}a_{m_2}a_{m_3}a_{m_4} \int_0^1 r_{m_1}(t)r_{m_2}(t)r_{m_3}(t)r_{m_4}(t)dt.\end{aligned}$$

From the orthogonality, the only nonvanishing case is the four integers are pairwise equal. Therefore

$$\|F\|^4_{L^4_t([0,1])} = C_2 \sum_{0\le m,l<\infty} a_m^2 a_l^2 = C_2 \|F\|^4_{L^2_t([0,1])}.$$

Similarly, for any integer k, the same argument yields

$$\|F\|_{L^{2k}_t([0,1])} \le C_k \|F\|_{L^2_t([0,1])}.$$

For any $p > 2$, we can find k such that $2 < p < 2k$. From this and the interpolation we have

$$\|F\|_{L^p_t([0,1])} \le C_p \|F\|_{L^2_t([0,1])}.$$

Using Hölder, one also has $\|F\|_{L^2_t([0,1])} \le \|F\|_{L^p_t([0,1])}$, therefore for $p \ge 2$,

$$\|F\|_{L^p_t([0,1])} \sim_p \|F\|_{L^2_t([0,1])} = \left(\sum_{m=0}^{\infty} a_m^2\right)^{\frac{1}{2}}.$$

Now we look at the case $1 \le p < 2$. From interpolation

$$\begin{aligned}\|F\|_{L^2_t([0,1])} &\le \|F\|^{\theta}_{L^p_t([0,1])} \|F\|^{1-\theta}_{L^4_t([0,1])} \\ &\le C_p \|F\|^{\theta}_{L^p_t([0,1])} \|F\|^{1-\theta}_{L^p_t([0,1])}.\end{aligned}$$

We get

$$\|F\|_{L^2_t([0,1])} \lesssim_p \|F\|_{L^p_t([0,1])}.$$

Hölder inequality implies $\|F\|_{L_t^p([0,1])} \leq \|F\|_{L_t^2([0,1])}$ so again we have

$$\|F\|_{L_t^p([0,1])} \sim_p \|F\|_{L_t^2([0,1])} = \Big(\sum_{m=0}^{\infty} a_m^2\Big)^{\frac{1}{2}}.$$

The Theorem is proved. □

To complete the proof, we need to resort to a classical result known as multiplier theorem which determines the L^p boundedness of the operator directly from its "multiplier" on the Fourier side.

Theorem 2.10 (Mikhlin-Hörmander multiplier theorem). *Let $1 < p < \infty$ and $k > \frac{n}{2}+1$ be an integer number. The symbol $m(\xi) \in C^k(\mathbb{R}^n)$ satisfies the condition*

$$\|m\|_\infty \leq B,$$
$$|\partial_\xi^j m(\xi)| \leq B|\xi|^{-j} \quad j \leq k.$$

Then the Fourier multiplier defined by

$$\widehat{Tf}(\xi) = m(\xi)\hat{f}(\xi)$$

is L^p bounded:

$$\|Tf\|_{L^p(\mathbb{R}^n)} \lesssim B\|f\|_{L^p(\mathbb{R}^n)}.$$

Now we are ready to give the proof of Theorem 2.7.

Proof of square function estimate

Proof. We first prove the first half estimate

$$\|Sf\|_p \lesssim \|f\|_p.$$

Using Theorem 2.9 we can write

$$\begin{aligned}(\sum_{k\in\mathbb{Z}} |P_k f|^2)^{\frac{1}{2}} &\leq (\sum_{m=0}^{\infty} |P_m f|^2)^{\frac{1}{2}} + (\sum_{m=0}^{\infty} |P_{-m} f|^2)^{\frac{1}{2}} \\ &\lesssim \|\sum_{m=0}^{\infty} P_m f\ r_m(t)\|_{L_t^p([0,1])} + \|\sum_{m=0}^{\infty} P_{-m} f\ r_m(t)\|_{L_t^p([0,1])}.\end{aligned}$$

Therefore

$$\begin{aligned}\|(\sum_{k\in\mathbb{Z}} |P_k f|^2)^{\frac{1}{2}}\|_{L_x^p} &\lesssim \|\sum_{m=0}^{\infty} P_m f\ r_m(t)\|_{L_{t,x}^p([0,1]\times\mathbb{R}^n)} \\ &\quad + \|\sum_{m=0}^{\infty} P_{-m} f\ r_m(t)\|_{L_{t,x}^p([0,1]\times\mathbb{R}^n)} \\ &\leq \sup_{t\in[0,1]} \Big(\|\sum_{m=0}^{\infty} P_m f\ r_m(t)\|_{L_x^p} + \|\sum_{m=0}^{\infty} P_{-m} f\ r_m(t)\|_{L_x^p}\Big).\end{aligned}$$

Define

$$T_t f(x) = \sum_{m=0}^{\infty} P_m f \, r_m(t), \quad T_t' f(x) = \sum_{m=0}^{\infty} P_{-m} f \, r_m(t).$$

We need to show T_t, T_t' are bounded operator from L^p to L^p, with the bound independent of t.

We only consider T_t as the other one can be dealt with similarly. Define the symbol

$$m_t(\xi) = \sum_{m=0}^{\infty} \psi\Big(\frac{\xi}{2^m}\Big) r_m(t).$$

Obviously, T_t is a Fourier multiplier with symbol $m_t(\xi)$. As supp$\psi \subset \{\xi, \frac{1}{2} \leq |\xi| \leq 2\}$, for every $\xi \in \mathbb{R}^n$, there are at most two terms in the summation that survive. Therefore we have

$$|m_t(\xi)| \leq 2,$$
$$|\partial_\xi^\alpha m_t(\xi)| \lesssim_\psi \frac{1}{2^{m|\alpha|}} \chi_{|\xi| \sim 2^m} \lesssim \frac{1}{|\xi|^\alpha}, \ \forall \alpha.$$

The L^p boundedness comes directly from the multiplier theorem.

From the proof we also see that the L^p boundedness still holds true if we replace P_k by $\tilde{P}_k$, where $\tilde{P}_k$ are variants of P_k, several acceptable examples of variants are

$$\tilde{P}_k = \sum_{l=-k_0}^{k_1} P_{k+l},$$
$$\tilde{P}_k = |\nabla|^s 2^{-ks} P_k.$$

Now we use this to show the second half of Littlewood-Paley theorem can be obtained by duality. Indeed,

$$\begin{aligned} \|f\|_p &= \sup_{\|g\|_{p'}=1} \int_{R^n} f(x) g(x) dx \\ &= \sup_{\|g\|_{p'}=1} \int_{\mathbb{R}^n} \hat{f}(\xi) \hat{g}(\xi) d\xi \\ &= \sup_{\|g\|_{p'}=1} \sum_{k \in \mathbb{Z}} \int_{\mathbb{R}^n} \psi\Big(\frac{\xi}{2^k}\Big) \hat{f}(\xi) \hat{g}(\xi) d\xi. \end{aligned}$$

Since $\sum_{j=k-1}^{k+1} \psi(\frac{\xi}{2^j}) = 1$ on supp $\psi(\frac{\xi}{2^k})$, writing $\tilde{P}_k = P_{k-1} + P_k + P_{k+1}$ we continue the estimate

$$\begin{aligned} \|f\|_p &= \sup_{\|g\|_{p'}=1} \sum_{k \in \mathbb{Z}} \int_{\mathbb{R}^n} P_k f \tilde{P}_k g d\xi \\ &\leq \sup_{\|g\|_{p'=1}} \|(\sum_{k \in \mathbb{Z}} |P_k f|^2)^{\frac{1}{2}}\|_p \|(\sum_{k \in \mathbb{Z}} |\tilde{P}_k g|^2)^{\frac{1}{2}}\|_{p'} \\ &\lesssim \|Sf\|_p \|g\|_{p'} \leq \|Sf\|_p. \end{aligned}$$

The proof is completed. □

It is easy to see that the square function estimate still holds if P_k is replaced by some other operator with similar properties as P_k. As a consequence of the Littlewood-Paley theorem, we have

Corollary 2.11. *Let $\tilde{P}_k$ be an analogue operator to P_k, then both of following two estimates are true:*

$$\Big\|\Big(\sum_{k\in\mathbb{Z}}|\tilde{P}_k f|^2\Big)^{\frac12}\Big\|_p \lesssim \|f\|_p, \quad \forall f\in L^p, 1<p<\infty. \tag{2.6}$$

$$\Big\|\sum_{k\in\mathbb{Z}}\tilde{P}_k f_k\Big\|_p \lesssim \Big\|\Big(\sum_{k\in\mathbb{Z}}|f_k|^2\Big)^{\frac12}\Big\|_p, \quad \forall\ l^2 \text{ sequence } f_k,\ 1<p<\infty. \tag{2.7}$$

Proof. It suffices to show the two estimates are equivalent as (2.6) is already proved.

Suppose (2.6) holds, we have

$$\begin{aligned}
\Big\|\sum_{k\in\mathbb{Z}}\tilde{P}_k f_k\Big\|_p &= \sup_{\|g\|_{p'}=1}\sum_{k\in\mathbb{Z}}\int_{\mathbb{R}^n}\tilde{P}_k f_k g\,dx \\
&= \sup_{\|g\|_{p'}=1}\sum_{k\in\mathbb{Z}}\int_{\mathbb{R}^n} f_k\tilde{P}_k g\,dx \\
&\le \sup_{\|g\|_{p'}=1}\Big\|\Big(\sum_{k\in\mathbb{Z}}|f_k|^2\Big)^{\frac12}\Big\|_p\Big\|\Big(\sum_{k\in\mathbb{Z}}|\tilde{P}_k g|^2\Big)^{\frac12}\Big\|_{p'} \\
&\lesssim \Big\|\Big(\sum_{k\in\mathbb{Z}}|f_k|^2\Big)^{\frac12}\Big\|_p.
\end{aligned}$$

This proves (2.7).

On the other hand, suppose (2.7) is true, we have

$$\begin{aligned}
\Big\|\Big(\sum_{k\in\mathbb{Z}}|\tilde{P}_k f|^2\Big)^{\frac12}\Big\|_p &= \sup_{\|g_k\|_{L^{p'}l^2}=1}\sum_{k\in\mathbb{Z}}\int_{\mathbb{R}^n}\tilde{P}_k f g_k\,dx \\
&\le \sup_{\|g_k\|_{L^{p'}l^2}=1}\int_{\mathbb{R}^n} f\sum_{k\in\mathbb{Z}}\tilde{P}_k g_k\,dx \\
&\le \|f\|_p\Big\|\sum_{k\in\mathbb{Z}}\tilde{P}_k g_k\Big\|_{p'} \\
&\lesssim \|f\|_p\Big\|\Big(\sum_{k\in\mathbb{Z}}|g_k|^2\Big)^{\frac12}\Big\|_{p'} \\
&\le \|f\|_p.
\end{aligned}$$

(2.6) is verified. □

The more original way to prove the Littlewood-Paley theorem is to use the Calderón-Zygmund theory for the singular operator. We define

Definition 2.12. *A Calderón-Zygmund operator T is a linear operator on complex-valued functions which is bounded from $L^2(\mathbb{R}^n)$ to $L^2(\mathbb{R}^n)$ and which can be written into the following way*

$$Tf(x) = \lim_{\varepsilon \to 0} \int_{|x-y|>\varepsilon} K(x,y)f(y)dy = P.V. \int_{\mathbb{R}^n} K(x,y)f(y)dy$$

for some kernel K with the bounds

$$|K(x,y)| \lesssim \frac{1}{|x-y|^n},$$
$$|\nabla K(x,y)| \lesssim \frac{1}{|x-y|^{n+1}}$$

for all $x \neq y$.

Example. Hilbert transform

$$\mathcal{H}f(x) = P.\ V. \int_{\mathbb{R}^n} \frac{1}{x-y} f(y)dy$$

is a typical example of C-Z operator. The L^2 boundedness is a direct consequence of

$$\widehat{\mathcal{H}f}(\xi) = -isgn(\xi).$$

Remark 2.13. *The C-Z operator theory can be found in any high level harmonic analysis textbook, for instance, [38]. One can refer it for more details. We remark here the general C-Z operator can deal with functions with vector value. For instance, denote X, Y as two Banach spaces, we can assume $f : \mathbb{R}^n \to X$. For each $x, y \in \mathbb{R}^n$, $K(x,y)$ is a linear operator maps X to Y. The integral converges in Y in the principle value sense. $|K(x,y)|$ or $|\nabla K(x,y)|$ can be understood as the operator norm.*

All C-Z operators satisfy the following

Theorem 2.14. *For $1 < p < \infty$, C-Z operator is bounded from $L^p(\mathbb{R}^n)$ to $L^p(\mathbb{R}^n)$.*

Proof of Littlewood-Paley theorem using C-Z operator theorem

Proof. We first prove

$$\|Sf\|_p \lesssim \|f\|_p.$$

One can view S induces an operator from $f : \mathbb{R}^n \to \mathbb{C}$ to $Tf : \mathbb{R}^n \to l^2$ and $Sf(x) = \|Tf(x)\|_{l^2}$. Here l^2 is a space containing sequences $\{a_l\}$ such that

$$\sum_{l=-\infty}^{\infty} |a_l|^2 < \infty.$$

We can write

$$\begin{aligned} Tf(x) &= \{P_k f(x)\}_{k=-\infty}^{\infty} \\ &= \left\{ \int_{\mathbb{R}^n} 2^{kn}\check{\psi}(2^k(x-y))f(y)dy \right\}_{k=-\infty}^{\infty} \\ &:= \int_{\mathbb{R}^n} K(x,y)f(y)dy,\ K(x,y) = \{2^{kn}\check{\psi}(2^k(x-y))\}_{k=-\infty}^{\infty}. \end{aligned}$$

The fact that T is an L^2 bounded operator has been verified at the beginning of this section. We only need to check the bound on the kernel $K(x,y)$. Indeed, as $\check{\psi}$ is a Schwartz function,

$$\begin{aligned} |K(x,y)|^2 = \|K(x,y)\|_{l^2} &= \sum_{k=-\infty}^{\infty} 2^{2kn}|\check{\psi}(2^k(x-y))|^2 \\ &= \sum_{k:|2^k(x-y)|\le 1} 2^{2kn} + \sum_{k:|2^k(x-y)|>1} 2^{kn}(2^k|x-y|)^{-10n} \\ &\lesssim |x-y|^{-n}. \end{aligned}$$

The similar computation gives

$$|\nabla K(x,y)| \lesssim |x-y|^{-n-1}.$$

Using C-Z theorem, T is bounded from $L^p(\mathbb{R}^n;\mathbb{C})$ to $L^p(\mathbb{R}^n;l^2)$, which is the same by saying

$$\|Sf\|_p \lesssim \|f\|_p.$$

The same argument can also prove that

$$\left\|\left(\sum |\tilde{P}_k f|^2\right)^{\frac{1}{2}}\right\|_p \lesssim \|f\|_p$$

for fattened $\tilde{P}_k = P_{k-1} + P_k + P_{k+1}$. A dual statement of it gives

$$\|f\|_p \lesssim \|Sf\|_p. \qquad \square$$

Remark 2.15. *As an exercise, one can also use the same technique to prove the multiplier theorem. Assume T is an operator with symbol $m(\xi)$, we can write*

$$\begin{aligned} \widehat{Tf}(\xi) &= m(\xi)\hat{f}(\xi) \\ &= \sum_{k\in\mathbb{Z}} m(\xi)\psi\left(\frac{\xi}{2^k}\right)\hat{f}(\xi). \end{aligned}$$

Hence

$$Tf(x) = \sum_{k\in\mathbb{Z}} \widetilde{\tilde{P}_k} f$$

where $\widetilde{\widetilde{P_k}}$ *is a variant of* P_k: $\widetilde{\widetilde{P_k}} = \mathcal{F}^{-1}(m(\xi)\psi(\frac{\xi}{2^k}))*$. *Now for any* $g \in L^{p'}$ *we estimate the integral by splitting into dyadic frequencies:*

$$\begin{aligned}&\int_{\mathbb{R}^n} Tf(x)g(x)dx\\ &= \sum_{k\in\mathbb{Z}} \int_{\mathbb{R}^n} \widetilde{\widetilde{P_k}} f \cdot g dx\\ &= \sum_{k\in\mathbb{Z}} \int_{\mathbb{R}^n} \widetilde{\widetilde{P_k}} f \cdot \tilde{P}_k g dx.\end{aligned}$$

Here $\tilde{P}_k = P_{k-1} + P_k + P_{k+1}$ *as* $\tilde{P}_k = I$ *if the frequency is supported at* $\mathrm{supp}\hat{P}_k$. *We continue to estimate the integral by using Hölder inequality*

$$\begin{aligned}\int_{\mathbb{R}^n} Tf(x)g(x)dx &\leq \|(\sum_{k\in\mathbb{Z}} |\widetilde{\widetilde{P_k}} f|^2)^{\frac{1}{2}}\|_p \|(\sum_{k\in\mathbb{Z}} |\widetilde{\widetilde{P_k}} f|^2)^{\frac{1}{2}}\|_{p'}\\ &\lesssim \|(\sum_{k\in\mathbb{Z}} |\widetilde{\widetilde{P_k}} f|^2)^{\frac{1}{2}}\|_p \|g\|_{p'}.\end{aligned}$$

Taking supremum in those g *such that* $\|g\|_{p'} = 1$ *we obtain*

$$\|Tf\|_p \lesssim \|(\sum_{k\in\mathbb{Z}} |\widetilde{\widetilde{P_k}} f|^2)^{\frac{1}{2}}\|_p.$$

The L^p *boundedness for* T *is reduced to showing a "pseudo" square function estimate*

$$\|(\sum_{k\in\mathbb{Z}} |\widetilde{\widetilde{P_k}} f|^2)^{\frac{1}{2}}\|_p \lesssim \|f\|_p.$$

Again, this can be done by introducing a C-Z operator with kernel

$$\left\{\mathcal{F}^{-1}\left(m(\xi)\psi\left(\frac{\xi}{2^k}\right)(x-y)\right)\right\}_{k\in\mathbb{Z}}.$$

We will leave the verification of being a C-Z operator to interested readers. We need to point out that the restriction on the smoothness is not optimal in this method (need $C^k, k > n$*).*

Remark 2.16. *Before leaving this section, we note that the Littlewood-Paley theorem can actually be proved to be true for the following examples of variants of* P_k.

1. $\tilde{P}_k$ *are fattened* P_k. $\tilde{P}_k = \sum_{j=k-k_0}^{k+k_1} P_j f$ *for some fixed number* k_0, k_1.
2. $\tilde{P}_k = 2^{-k}\nabla P_k$.
3. $\tilde{P}_k = 2^{ks}|\nabla|^{-s} P_k$.
4. *Assume* $m(\xi)$ *satisfy* $m(\xi) \in C^\infty(\mathbb{R}^n - \{0\})$ *and*

$$|\partial_\xi^\alpha m(\xi)| \lesssim |\xi|^{-|\alpha|}, \ \forall \text{ multi-index } \alpha.$$

Define $\tilde{P}_k f = \mathcal{F}^{-1}(m(\xi)\psi(\frac{\xi}{2^k}))$.

3 Sobolev space

Sobolev spaces are effective tools to study PDEs because it appropriately measure distances between solutions. There are comprehensive study on the theory of Sobolev space we will not be able to cover all of them in this lecture. Instead, we will only focus on the Sobolev space defined on whole space $\mathbb{R}^n$ and we will describe them mainly from the Fourier side.

Classical Sobolev space $W^{k,p}(\mathbb{R}^n)$ for integer number $k \geq 0$, $1 \leq p \leq \infty$ are defined as subspace of Lebesgue space $L^p(\mathbb{R}^n)$. It contains all functions in $L^p(\mathbb{R}^n)$ for which all derivatives, up to order k, belong to L^p. When equipped with the norm

$$\|f\|_{W^{k,p}} = \sum_{\alpha \in \mathbb{N}^n,\ 0 \leq |\alpha| \leq k} \|\partial^\alpha f\|_p.$$

$W^{k,p}$ is a Banach space and when $p = 2$, $W^{k,2}$ becomes a Hilbert space.

We can also define Sobolev spaces involving fractional derivatives. Such space is called "Bessel-Potential space".

We first define the nonlocal fractional derivative. For $s \in \mathbb{R}$, we define $|\nabla|^s$ and $\langle \nabla \rangle^s$ as follows:

$$|\nabla|^s f = \mathcal{F}^{-1}(2\pi|\xi|)^s \hat{f}(\xi)),$$
$$\langle \nabla \rangle^s f = \mathcal{F}^{-1}(\langle 2\pi|\xi| \rangle^s \hat{f}(\xi)).$$

Here, for a number a, $\langle a \rangle = (1 + |a|^2)^{\frac{1}{2}}$.

Remark 3.1. *In the case $s \in (-n, 0)$, $|\nabla|^s$ can be written as*

$$\begin{aligned} |\nabla|^s f(x) &= \mathcal{F}^{-1}((2\pi|\xi|)^s) * f \\ &= c_{n,s} \int_{\mathbb{R}^n} \frac{1}{|x-y|^{n+s}} f(y) dy. \end{aligned}$$

In the case $1 < p < \infty$ and k is a positive integer, ∇^k is equivalent to the non-local operator $|\nabla|^k$ in L^p space. For example, we can show that

$$\|\nabla f\|_p \lesssim \||\nabla| f\|_p,$$
$$\||\nabla f\|_p \lesssim \|\nabla f\|_p.$$

Indeed, writing

$$\nabla = \frac{\nabla}{|\nabla|} |\nabla|,$$
$$|\nabla| = \frac{\nabla}{|\nabla|} \cdot \nabla,$$

the L^p boundedness is reduced to showing the L^p boundedness of Riesz transform (or Hilbert transform if dimension=1), which is guaranteed by Mikhlin-Hörmander multiplier theorem. The case $k \geq 2$ can be considered similarly, the verification will be left to the interested readers.

Due to their coincidence in this general case, for the rest part of this notes, we will still name them all as "Sobolev space". However, the two definitions are indeed not the same in the endpoint cases $p = 1$ and $p = \infty$. So we adopt a different notation H.

After all these discussion, we are ready to define general Sobolev space.

Definition 3.2. *Let $1 \leq p \leq \infty$ and $s \in \mathbb{R}$, we define the inhomogeneous Sobolev space $H^{s,p}(\mathbb{R}^n)$ to be the set of all tempered distributions such that*

$$\|f\|_{H^{s,p}} = \|\langle\nabla\rangle^s f\|_p < \infty.$$

For all choices of s and p, $H^{s,p}(\mathbb{R}^n)$ are Banach space. In the case $1 \leq p < \infty$, Schwartz functions are dense in $H^{s,p}$.

We also define homogeneous Sobolev space $\dot{H}^{s,p}(\mathbb{R}^n)$ to be the set of all tempered distributions such that

$$\|f\|_{\dot{H}^{s,p}} = \||\nabla|^s f\|_p < \infty.$$

Again, $\dot{H}^{s,p}(\mathbb{R}^n)$ are Banach space and for $1 \leq p < \infty$, Schwartz function are dense in $\dot{H}^{s,p}$.

When $p = 2$, we usually omit p and write the space directly as H^s and $\dot{H}^s$.

Using Littlewood-Paley theory, we give the following characterization

Theorem 3.3. *Let $1 < p < \infty$. For any $f \in H^{s,p}(\mathbb{R}^n)$ we have*

$$\|f\|_{H^{s,p}} \sim \left\| \left(\sum_{k\in\mathbb{Z}} \langle 2^k\rangle^{2s} |P_k f|^2 \right)^{\frac{1}{2}} \right\|_p.$$

Proof. We first prove $\gtrsim$. By writing

$$\begin{aligned} \langle 2^k\rangle^s P_k f &= \langle 2^k\rangle^s \langle\nabla\rangle^{-s} P_k \, \langle\nabla\rangle^s f \\ &= \tilde{P}_k \langle\nabla\rangle^s f \end{aligned}$$

for $\tilde{P}_k = \langle 2^k\rangle^s \langle\nabla\rangle^{-s}$, we can directly use (2.6) to get

$$\begin{aligned} \Big\|\Big(\sum_{k\in\mathbb{Z}} \langle 2^k\rangle^{2s} |P_k f|^2\Big)^{\frac{1}{2}}\Big\|_p &= \Big\|\Big(\sum_{k\in\mathbb{Z}} |\tilde{P}_k \langle\nabla\rangle^s f|^2\Big)^{\frac{1}{2}}\Big\|_p \\ &\lesssim \|\langle\nabla\rangle^s f\|_p. \end{aligned}$$

Next we show the $\lesssim$ part. We write

$$\begin{aligned} \|f\|_{H^{s,p}} = \|\langle\nabla\rangle^s f\|_p &= \Big\|\sum_{k\in\mathbb{Z}} P_k \langle\nabla\rangle^s f\Big\|_p \\ &= \Big\|\sum_{k\in\mathbb{Z}} (P_{k-1} + P_k + P_{k+1}) \langle 2^k\rangle^{-s} \langle\nabla\rangle^s \cdot \langle 2^k\rangle^s P_k f\Big\|_p. \end{aligned}$$

Applying (2.7) to $\tilde{P}_k = (P_{k-1} + P_k + P_{k+1})\langle 2^k\rangle^{-s}\langle\nabla\rangle^s$ and $f_k = \langle 2^k\rangle^s P_k f$ we immediately get

$$\|f\|_{H^{s,p}} \lesssim \|(\sum_{k\in\mathbb{Z}} |\langle 2^k\rangle^s P_k f|^2)^{\frac{1}{2}}\|_p.$$

The theorem is proved. □

Several quick observations:

1. For $s \geq 0$, $1 < p < \infty$,

$$\|f\|_{H^{s,p}} \sim \|f\|_p + \||\nabla|^s f\|_p.$$

As an example, we show the equivalence when $k = 1$. The $\gtrsim$ part is easy to show by Mihlin-Hörmander multiplier theorem. Now we show the $\lesssim$ part.

We have

$$\begin{aligned}
\|f\|_{H^{1,p}} &= \|\langle\nabla\rangle f\|_p \\
&= \|(\sum_{k\in\mathbb{Z}} \langle 2^k\rangle^2 |P_k f|^2)^{\frac{1}{2}}\|_p. \\
&\leq \|(\sum_{k\in\mathbb{Z}} |P_k f|^2)^{\frac{1}{2}}\|_p + \|\sum_{k\in\mathbb{Z}} |2^k|^2 |P_k f|^2\|_p \\
&\sim \|(\sum_{k\in\mathbb{Z}} |P_k f|^2)^{\frac{1}{2}}\|_p + \|(\sum_{k\in\mathbb{Z}} |2^k||P_k f|^2)^{\frac{1}{2}}\|_p \\
&\sim \|f\|_p + \||\nabla| f\|_p.
\end{aligned}$$

The last thing is to show $\||\nabla| f\|_p \sim \|\nabla f\|_p$. Indeed, by writing

$$\begin{aligned}
|\nabla| f &= \frac{\nabla\cdot}{|\nabla|}\nabla f, \\
\nabla f &= \frac{\nabla}{|\nabla|}|\nabla f|.
\end{aligned}$$

The equivalence of two norms is reduced to showing both operators $\frac{\nabla\cdot}{|\nabla|}$ and $\frac{\nabla}{|\nabla|}$ are L^p bounded. This is obvious from the L^p boundedness of Riesz transform.

2. Let $s \geq 0$, then

$$\|f\|_{W^{s,p}} \sim \|P_{\leq 0} f\|_p + \|P_{>0} f\|_{\dot{W}^{s,p}}. \tag{3.1}$$

4 Estimates in Sobolev spaces

In this section, we give several important estimates in the Sobolev spaces. Most of them will be proved using Littlewood-Paley theory.

4.1 Interpolations

In Lebesgue space, the following interpolation holds as an application of Hölder inequality. For $p_1 < q < p_2$ and

$$\frac{1}{q} = \frac{\theta}{p_1} + \frac{1-\theta}{p_2},$$

we have

$$\|f\|_q \leq \|f\|_{p_1}^{\theta} \|f\|_{p_2}^{1-\theta}.$$

Similar estimate holds in Sobolev spaces. We start from a simple example:

Lemma 4.1. *For $f \in H^1(\mathbb{R}^2)$, we have*

$$\|f\|_4 \lesssim \|f\|_2^{\frac{1}{2}} \|\nabla f\|_2^{\frac{1}{2}}.$$

Proof. We first note that the estimate is scale invariant, moreover the power can be uniquely determined from scaling argument.

The estimate is obvious if we consider single frequency. From interpolation in Legesgue space and Bernstein inequality we have

$$\begin{aligned}\|P_k f\|_4 &\leq \|P_k f\|_2^{\frac{1}{2}} \|P_k f\|_\infty^{\frac{1}{2}} \\ &\lesssim \|P_k f\|_2^{\frac{1}{2}} (2^k \|P_k f\|_2)^{\frac{1}{2}} \\ &\lesssim \|P_k f\|_2^{\frac{1}{2}} \|\nabla P_k f\|_2^{\frac{1}{2}}.\end{aligned}$$

For general f, we use square function estimate to write

$$\begin{aligned}\|f\|_4^2 &\sim \|(\sum_{k\in\mathbb{Z}} |P_k f|^2)^{\frac{1}{2}}\|_4^2 \\ &\leq \|\sum_{k\in\mathbb{Z}} |P_k f|^2\|_2 \leq \sum_{k\in\mathbb{Z}} \||P_k f|^2\|_2 \\ &= \sum_{k\in\mathbb{Z}} \|P_k f\|_4^2 \\ &\lesssim \sum_{k\in\mathbb{Z}} \|P_k f\|_2 \|\nabla P_k f\|_2 \\ &\leq (\sum_{k\in\mathbb{Z}} \|P_k f\|_2^2)^{\frac{1}{2}} (\sum_{k\in\mathbb{Z}} \|\nabla P_k f\|_2^2)^{\frac{1}{2}} \\ &\lesssim \|f\|_2 \|\nabla f\|_2.\end{aligned}$$

□

The general interpolation inequality is written as follows:

Theorem 4.2. *Let $1 < p, p_1, p_2 < \infty$ and $0 < \theta < 1$. Let $0 \leq \sigma \leq s\theta$ and*

$$\sigma - \frac{n}{p} = (1-\theta)\left(-\frac{n}{p_1}\right) + \theta\left(s - \frac{n}{p_2}\right).$$

Then we have

$$\|f\|_{\dot{H}^{\sigma,p}} \lesssim \|f\|_{p_1}^{1-\theta} \|f\|_{\dot{H}^{s,p_2}}^{\theta}.$$

Proof. We first prove the case when

$$\sigma = s\theta.$$

In this case $\frac{1}{p} = \frac{1-\theta}{p_1} + \frac{\theta}{p_2}$. We write

$$\begin{aligned}
\|f\|_{\dot H^{\sigma,p}} &= \||\nabla|^\sigma f\|_p \\
&= \|(\sum_{k\in\mathbb{Z}} 2^{2k\sigma}|P_k f|^2)^{\frac12}\|_p \\
&= \|2^{k\sigma}|P_k f|\|_{L^p l^2} \\
&= \||P_k f|^{1-\theta}\cdot 2^{k\sigma}|P_k f|^\theta\|_{L^p l^2} \\
&\le \||P_k f|^{1-\theta}\|_{L^{\frac{p_1}{1-\theta}} l^{\frac{2}{1-\theta}}} \|2^{ks\theta}|P_k f|^\theta\|_{L^{\frac{p_2}{\theta}} l^{\frac{2}{\theta}}} \\
&\le \|P_k f\|_{L^{p_1} l^2}^{1-\theta} \|2^{ks} P_k f\|_{L^{p_2} l^2}^{\theta} \\
&\lesssim \|f\|_{p_1}^{1-\theta} \|f\|_{\dot W^{s,p}}^{\theta}.
\end{aligned}$$

For the general case, we need to use Sobolev embedding first. Let r be determined by

$$s - \frac{n}{r} = \sigma - \frac{n}{p}.$$

From Sobolev embedding and the interpolation we just proved, we have

$$\|f\|_{\sigma,p} \lesssim \|f\|_{\dot W^{s\theta,r}} \lesssim \|f\|_{p_1}^{1-\theta}\|f\|_{\dot W^{s,p_2}}^{\theta}.$$

The proof is finished. □

4.2 Applications to the sharp Sobolev embedding

Previously we give the proof of sharp Sobolev embedding by using Marcinkiewicz interpolation. Here reprove one such embedding directly using the square function estimates.

Lemma 4.3. *In three dimensions, we have*

$$\|f\|_{L^3(\mathbb{R}^3)} \lesssim \|f\|_{\dot H^{\frac12}(\mathbb{R}^3)}.$$

Proof. Using Littlewood-Paley theorem, we write

$$\begin{aligned}
\|f\|_3^3 &= \int_{\mathbb{R}^3} (\sum_{k\in\mathbb{Z}} |P_k f|^2)^{\frac32} dx \\
&= \int_{\mathbb{R}^3} \sum_{k_1\in\mathbb{Z}} |P_{k_1} f|^2 (\sum_{k_2\in\mathbb{Z}} |P_{k_2} f|^2)^{\frac12} dx \\
&\lesssim \sum_{k_1>k_2} \int_{\mathbb{R}^3} |P_{k_1} f|^2 |P_{k_2} f| dx + \sum_{k_1\le k_2} \int_{\mathbb{R}^3} |P_{k_1} f|^2 |P_{k_2} f| dx \\
&= I + II.
\end{aligned}$$

We first control I. Using Hölder, Bernstein inequality we have

$$
\begin{aligned}
I &\lesssim \sum_{k_1>k_2} \|P_{k_1} f\|_2^2 \|P_{k_2} f\|_\infty \\
&\lesssim \sum_{k_1>k_2} 2^{\frac{3k_2}{2}} \|P_{k_1} f\|_2^2 \|P_{k_2} f\|_2 \\
&\lesssim \sum_{k_1>k_2} \|P_{k_1} f\|_2^2 2^{k_2} \||\nabla|^{\frac{1}{2}} P_{k_2} f\|_2 \\
&\lesssim \|f\|_{\dot{H}^{\frac{1}{2}}} \sum_{k_1>k_2} 2^{k_2} \|P_{k_1} f\|_2^2 \\
&\lesssim \|f\|_{\dot{H}^{\frac{1}{2}}} \sum_{k_1 \in \mathbb{Z}} 2^{k_1} \|P_{k_1} f\|_2^2 \\
&\lesssim \|f\|^3_{\dot{H}^{\frac{1}{2}}}.
\end{aligned}
$$

Now we control II.

$$
\begin{aligned}
II &\leq \sum_{k_1 \leq k_2} \int_{\mathbb{R}^3} |P_{k_1} f|^2 |P_{k_2} f| dx \\
&\leq \sum_{k_1 \leq k_2} \|P_{k_1} f\|_\infty \|P_{k_1} f\|_2 \|P_{k_2} f\|_2 \\
&\leq \sum_{k_1 \leq k_2} 2^{\frac{3k_1}{2}} \|P_{k_1} f\|_2 \|P_{k_1} f\|_2 \|P_{k_2} f\|_2 \\
&\leq \sum_{k_1 \leq k_2} 2^{k_1} \|P_{k_1} f\|_2 \|P_{k_2} f\|_2 \||\nabla|^{\frac{1}{2}} f\|_2 \\
&\leq \sum_{k_1 \leq k_2} \frac{2^{\frac{k_1}{2}}}{2^{\frac{k_2}{2}}} \||\nabla|^{\frac{1}{2}} P_{k_1} f\|_2 \||\nabla|^{\frac{1}{2}} P_{k_2} f\|_2 \||\nabla|^{\frac{1}{2}} f\|_2 \\
&\leq \sum_{k_1 \leq k_2} \left(\frac{2^{k_1/2}}{2^{k_2/2}} \|P_{k_1} |\nabla|^{\frac{1}{2}} f\|_2^2 + \frac{2^{k_1/2}}{2^{k_2/2}} \|P_{k_2} |\nabla|^{\frac{1}{2}} f\|_2^2 \right) \||\nabla|^{\frac{1}{2}} f\|_2 \\
&\lesssim \sum_{k \in \mathbb{Z}} \|P_k |\nabla|^{\frac{1}{2}} f\|_2^2 \||\nabla|^{\frac{1}{2}} f\|_2 \\
&\lesssim \|f\|^3_{\dot{H}^{\frac{1}{2}}}.
\end{aligned}
$$

Putting the two pieces together, we prove Lemma 4.3. □

4.3 Product estimate

In this part, we show some examples of using square function estimates to estimate product of two functions in Sobolev spaces. We start with a simple example.

Lemma 4.4. *Let $s \geq 0$ and $f, g \in L^\infty \cap H^s$. Then*

$$
\|fg\|_{H^s} \lesssim \|f\|_\infty \|g\|_{H^s} + \|f\|_{H^s} \|g\|_\infty.
$$

Proof. Without loss of generality we only consider the case $s > 0$ and estimating the norm $\|fg\|_{\dot{H}^s}$. If s is an integer number, we may follow the classical way to prove such inequality. That involves using product rule for derivatives and interpolation for treating the mixed terms. However, using the square function estimate, the argument can be much simpler.

We start from a simple fact. If $\hat{f}$ is supported on the set A, $\hat{g}$ is supported on the set B, then the frequency support of $f \cdot g$ is $A + B$. Therefore by decomposing

$$f \cdot g = f_{>k-2}g + f_{\leq k-2}g_{>k-2} + f_{\leq k-2}g_{\leq k-2},$$

we have

$$P_k(f \cdot g) = P_k(f_{>k-2}g) + P_k(f_{\leq k-2}g_{>k-2}).$$

As we see that in the above decomposition, at least one function is in high frequency. Plugging this estimate and throwing away the bounded operator P_k we have

$$\begin{aligned}\|f \cdot g\|_{\dot{H}^s}^2 &= \sum_{k \in \mathbb{Z}} 2^{2ks} \Big(\|f_{>k-2}g\|_2^2 + \|g_{>k-2}f_{\leq k-2}\|_2^2 \Big) \\ &\leq \sum_{k \in \mathbb{Z}} 2^{ks} \|f_{>k-2}\|_2^2 \|g\|_\infty \\ &\quad + \sum_{k \in \mathbb{Z}} 2^{ks} \|g_{>k-2}\|_2^2 \|f\|_\infty.\end{aligned}$$

The two terms can be treated similarly. We only consider the first one. Note that if $f_{>k-2}$ is replaced by $f_{\sim k}$, the estimate follows directly from definition. Treating the high frequency counterpart involves one more step of estimate using discrete Young's inequality. Specifically we have

$$\begin{aligned}\sum_{k \in \mathbb{Z}} 2^{ks} \|f_{>k-2}\|_2^2 &= \sum_{k \in \mathbb{Z}} 2^{ks} \sum_{j > k-2} \|P_j f\|_2^2 \\ &= \sum_{k \in \mathbb{Z}} \sum_{j \in \mathbb{Z}} \chi_{k-j<2} 2^{2(k-j)s} 2^{2js} \|P_j f\|_2^2 \\ &= \left\| \chi_{l<2} 2^{2ls} \star 2^{2js} \|P_j f\|_2^2 \right\|_{l^1} \\ &\leq \|\chi_{l<2} 2^{2ls}\|_{l^1} \left\| 2^{2js} \|P_j f\|_2^2 \right\|_{l^1} \\ &\lesssim \|f\|_{\dot{H}^s}^2.\end{aligned}$$

In the last line we have used the fact $s > 0$. Plugging this estimate and using the similar estimate to treat the second term we proved Lemma 4.4. □

We can prove more general product estimates. It is worthwhile to point out that in the recent work of Bourgain-Li [30], they are able to deal with the endpoint case when $p = \infty$.

Theorem 4.5. *Let $s \geq 0$. Let $1 < p, p_1, p_2, p_3, p_4 < \infty$ and $\frac{1}{p} = \frac{1}{p_1} + \frac{1}{p_2} = \frac{1}{p_3} + \frac{1}{p_4}$ or $1 < p < \infty$, $p_2 = p_3 = \infty$. We have*

$$\||\nabla|^s(fg)\|_p \lesssim \||\nabla|^s f\|_{p_1}\|g\|_{p_2} + \|f\|_{p_3}\||\nabla|^s g\|_{p_4}.$$

Proof. Using square function estimates we have

$$\begin{aligned}\||\nabla|^s(fg)\|_p &= \left\|\left(\sum_{k\in\mathbb{Z}} 2^{2ks}|P_k(fg)|^2\right)^{\frac{1}{2}}\right\|_p = \|2^{ks}P_k(fg)\|_{L^p l^2} \\ &\leq \|2^{ks}P_k(f_{>k-2}g)\|_{L^p l^2} + \|2^{ks}P_k(f_{\leq k-2}g_{>k-2})\|_{L^p l^2} \\ &= I + II.\end{aligned}$$

Here again we used the decomposition

$$P_k(fg) = P_k(f_{>k-2}g) + P_k(f_{\leq k-2}g_{>k-2}).$$

To estimate I, we first notice that we can throw away the operator P_k by using the boundedness of it. Then we have

$$\begin{aligned}I &\lesssim \|2^{ks}f_{>k-2}g\|_{L^p l^2} \\ &\lesssim \|2^{ks}f_{>k-2}\|_{L^{p_1}l^2}\|g\|_{L^{p_2}} \\ &\lesssim \left\|\sum_{j>k-2} 2^{(k-j)s}2^{js}P_j f\right\|_{L^{p_1}l^2}\|g\|_{L^{p_2}} \\ &\lesssim \left\|\sum_{j\in\mathbb{Z}} 2^{(k-j)s}\chi_{k-j<2}2^{js}P_j f\right\|_{L^{p_1}l^2}\|g\|_{p_2} \\ &\lesssim \|2^{js}P_j f\|_{L^{p_1}l^2}\|g\|_{p_2} \\ &\lesssim \||\nabla|^s f\|_{p_1}\|g\|_{p_2}.\end{aligned}$$

We can run the same argument for II if we use the fact

$$|P_{\leq k}f(x)| \lesssim Mf(x). \tag{4.1}$$

and the L^p $(1 < p \leq \infty)$ boundedness for maximal functions. For the sake of convenience, we recall the definition of maximal function

$$Mf(x) = \sup_{r>0}\frac{1}{vol(B(x,r))}\int_{B(x,r)} f(y)dy.$$

To prove (4.1), we write

$$\begin{aligned}P_{\leq k}f(x) &= 2^{kn}\int_{\mathbb{R}^n}\check{\phi}(2^k(x-y))f(y)dy \\ &= 2^{kn}\int_{2^k|x-y|\leq 1}\check{\phi}(2^k(x-y))f(y)dy \\ &\quad + \sum_{R>1,\ R\ dyadic} 2^{kn}\int_{R\leq 2^k|x-y|<2R}\check{\phi}(2^k(x-y))f(y)dy.\end{aligned}$$

Therefore, we have

$$\begin{aligned}|P_{\leq k}f(x)| &\leq Mf(x) + \sum_{R>1,\ R\ dyadic} R^{-10n}2^{kn}\int_{|x-y|\sim\frac{R}{2^k}}|f(y)|dy\\ &\lesssim Mf(x) + \sum_{R>1,\ R\ dyadic} R^{-9n}Mf(x)\\ &\lesssim Mf(x).\end{aligned}$$

□

4.4 Fractional chain rule for general nonlinear functions

Let $F(z)$ be a power type function $F(z) = |z|^p z$ for some $p > 0$. If p is not an even integer, the frequency support of $F(u)$ is no longer separable, as a consequence, one can not localize u into high frequency even if $F(u)$ is localized in high frequencies. Despite the technical complexity, we still have

Theorem 4.6. *Let $1 < p < \infty$ and $\frac{1}{p_1} + \frac{1}{p_2} = 1$. Let $F(z) \in C^1$. Then for $0 \leq s \leq 1$,*

$$\|F(u)\|_{\dot{H}^{s,p}} \lesssim \|O(F'(u))\|_{p_1}\||\nabla|^s u\|_{p_2}.$$

Remark 4.7. *The term $O(F'(u))$ is a simple writing for functions that behave similarly with the first derivative of F evaluated at u or certain frequencies of u. To control frequencies of u, we can use the pointwise estimate like $|P_k u(x)| \lesssim Mu(x)$.*

From the Fundamental Theorem of Calculus, if F is a real function we use the formula

$$F(s) = F(s_0) + \int_0^1 F'(\theta s + (1-\theta)s_0)d\theta(s - s_0),$$

and simply record it as $F(s) = F(s_0) + O(F'(s))(s - s_0)$. If F is a complex function, we use the formula

$$\begin{aligned}F(z) = F(z_0) &+ \int_0^1 \frac{\partial F}{\partial z}(\theta z + (1-\theta)z_0)d\theta(z - z_0)\\ &+ \int_0^1 \frac{\partial F}{\partial \bar{z}}(\theta z + (1-\theta)z_0)d\theta(\bar{z} - \bar{z}_0).\end{aligned}$$

For simplicity, we also write simply as $F(z) = F(z_0) + O(F'(z))(z - z_0)$.

Now we give the proof of this theorem.

Proof. We assume $0 < s < 1$ as the two endpoint cases can be proved easily. Using the decomposition

$$F(u) = F(u_{\leq k}) + O[F'(u)]u_{>k},$$

we write

$$\begin{aligned}
\|F(u)\|_{\dot{H}^{s,p}} &= \|2^{ks}P_kF(u)\|_{L^pl^2} \\
&\lesssim \|2^{ks}P_kF(u_{\le k})\|_{L^pl^2} + \|2^{ks}O(F'(u))u_{>k}\|_{L^pl^2} \\
&\lesssim \|O(F'(u))\|_{p_1}\|2^{ks}u_{>k}\|_{L^{p_2}l^2} + \|2^{ks}P_kF(u_{\le k})\|_{L^pl^2} \\
&\lesssim \|O(F'(u))\|_{p_1}\|u\|_{\dot{H}^{s,p_2}} + \|2^{ks}P_kF(u_{\le k})\|_{L^pl^2}.
\end{aligned}$$

To estimate the second part, we need to take advantage that all function u are localized in low frequencies. Denote $\tilde{P}_k = |\Delta|^{-1}\nabla P_k 2^k$,

$$\begin{aligned}
\|2^{ks}P_kF(u_{\le k})\|_{L^pl^2} &\lesssim \|2^{ks}|\Delta|^{-1}\nabla \cdot P_k\nabla F(u_{\le k})\|_{L^pl^2} \\
&\lesssim \|2^{k(s-1)}\tilde{P}_k \cdot \nabla u_{\le k}F'(u_{\le k})\|_{L^pl^2} \\
&\lesssim \|2^{k(s-1)}\tilde{P}_k\nabla u_{\le k}\|_{L^{p_2}l^2}\|F'(u_{\le k})\|_{p_1} \\
&\lesssim \|2^{k(s-1)}\nabla u_{\le k}\|_{L^{p_2}l^2}\|O(F'(u))\|_{p_1} \\
&\lesssim \|O(F'(u))\|_{p_1}\|u\|_{\dot{H}^{s,p_2}}.
\end{aligned}$$

Note that here we have used the boundedness of $\tilde{P}_k$ and

$$\begin{aligned}
\|2^{k(s-1)}\nabla u_{\le k}\|_{L^{p_2}l^2} &\lesssim \|2^{k(s-1)}\nabla P_k u\|_{L^{p_2}l^2} \\
&\lesssim \|2^{ks}P_k u\|_{L^{p_2}l^2},
\end{aligned}$$

which can be proved similarly by using Young's inequality. Finally, putting the two pieces together, we prove the estimate. □

4.5 Trace theorem

An $H^{s,p}(\mathbb{R}^n)$ function, when restricted to $(n-1)$-dimensional subspace, will lose $\frac{1}{p}$ derivative. The result of this kind is called trace theorem. The non-sharp version of the trace theorem can be proved simply by Sobolev embedding.

For example, let f be a Schwartz function on $R^n_x = R_{x_1}R^{n-1}_{x'}$, we then have

$$\begin{aligned}
\|f\|_{L^\infty_{x_1}H^{s-\frac{1}{p},p}_{x'}(\mathbb{R}\times\mathbb{R}^{n-1})} &\le \|\langle\nabla_{x'}\rangle^{s-\frac{1}{p}}f\|_{L^p_{x'}L^\infty_{x_1}} \\
&\lesssim \|\langle\nabla_{x_1}\rangle^{\frac{1}{p}+\varepsilon}\langle\nabla_{x'}\rangle^{s-\frac{1}{p}}\|_{L^p_x} \\
&\lesssim \|\langle\nabla_x\rangle^{s+\varepsilon}f\|_{L^p_x}.
\end{aligned}$$

The proof of sharp version is much more complicated. Here we just show a special example.

Let $s > \frac{1}{2}$ and f be a Schwartz function. By changing of the coordinate if necessary, we only need to prove

$$\|f(0,x')\|_{H^{s-\frac{1}{2}}(\mathbb{R}^{n-1})} \lesssim \|f\|_{H^s}.$$

Indeed, by using Fourier inversion formula we write

$$\mathcal{F}_{x'}f(0,\xi') = \int_{\mathbb{R}} \mathcal{F}_x f(\xi_1,\xi')d\xi_1.$$

Then

$$\begin{aligned}\langle\xi'\rangle^{s-\frac{1}{2}}\mathcal{F}_{x'}f(0,\xi') &= \int_{\mathbb{R}}\mathcal{F}_x f(\xi_1,\xi')\langle\xi'\rangle^{s-\frac{1}{2}}d\xi_1\\ &= \int_{\mathbb{R}}\mathcal{F}_x f(\xi_1,\xi')\langle\xi\rangle^{s}\frac{\langle\xi'\rangle^{s-\frac{1}{2}}}{\langle\xi\rangle^{s}}d\xi_1\\ &\lesssim \Big(\int_{\mathbb{R}}|\mathcal{F}_x f(\xi_1,\xi')|^2\langle\xi\rangle^{2s}d\xi_1\Big)^{\frac{1}{2}}\Big(\int_{\mathbb{R}}\frac{\langle\xi'\rangle^{2(s-\frac{1}{2})}}{\langle\xi\rangle^{2s}}d\xi_1\Big)^{\frac{1}{2}}\\ &\lesssim \Big(\int_{\mathbb{R}}|\mathcal{F}_x f(\xi_1,\xi')|^2\langle\xi\rangle^{2s}d\xi_1\Big)^{\frac{1}{2}}.\end{aligned}$$

Taking $L^2_{\xi'}$ norm on both sides gives the desired estimate.

4.6 Hardy's inequality

The general Hardy's inequality says for any Schwartz function f and $1 < p < \infty$ and $0 < s < \frac{n}{p}$, that

$$\Big\|\frac{f}{|x|^s}\Big\|_p \lesssim \||\nabla|^s f\|_p.$$

The proof of the Hardy's inequality is a classical result and can be found in many places. Again, we will only show the proof of a special case by using the square function estimate.

For $f \in \mathcal{S}(\mathbb{R}^n)$ and $0 < s < \frac{n}{2}$, we will show

$$\|\frac{f}{|x|^s}\|_2 \lesssim \||\nabla|^s f\|_2.$$

Indeed, partition into dyadic pieces we have

$$\begin{aligned}\Big\|\frac{f}{|x|^s}\Big\|_2^2 &= \sum_{j\in\mathbb{Z}} 2^{-2js}\|f\|^2_{L^2(2^j\le|x|<2^{j+1})}\\ &\lesssim \sum_{j\in\mathbb{Z}} 2^{-2js}\|\chi_{\sim 2^j} f\|_2^2\\ &\lesssim \sum_{j\in\mathbb{Z}} 2^{-2js}\Big(\sum_{k\in\mathbb{Z}}\|\chi_{\sim 2^j}P_k f\|_2\Big)^2.\end{aligned}$$

Notice that

$$\begin{aligned}\|\chi_{\sim 2^j}P_k f\|_2 &\lesssim 2^{jn/2}2^{kn/2}\|P_k f\|_2\\ &\lesssim 2^{jn/2}2^{kn/2}2^{-sk}\||\nabla|^s P_k f\|_2,\\ \|\chi_{\sim 2^j}P_k f\|_2 &\lesssim 2^{-sk}\||\nabla|^s P_k f\|_2.\end{aligned}$$

We have

$$\begin{aligned}\left\|\frac{f}{|x|}\right\|_2^2 &\lesssim \sum_{j\in\mathbb{Z}}\Big(\sum_{k\in\mathbb{Z}}\min(2^{(j+k)\frac{n}{2}}2^{-(j+k)s}, 2^{-(j+k)s})\||\nabla|^s P_k f\|_2\Big)^2 \\ &= \|\min(2^{m\frac{n}{2}-ms}, 2^{-ms})\|_{l^1}^2\||\nabla|^s P_k f\|_{l^2}^2 \\ &\lesssim \||\nabla|^s f\|_2^2.\end{aligned}$$

4.7 Other function spaces

Besides the Sobolev spaces, other spaces are also used to study PDEs. For more comprehensive studies, one refers to see [1]. Here we only introduce some examples.

For $s \in \mathbb{R}$, $1 \le p, q \le \infty$, we define $B^s_{p,q}$ as the Besov space which includes all functions that are finite when measured by the norm

$$\|f\|_{B^s_{p,q}} = \|\langle 2^k\rangle^s P_k f\|_{l^q L^p}.$$

It is easily seen that for Besov spaces, the following embedding holds:

$$B^s_{p,1} \subset B^s_{p,q} \subset B^s_{p,\infty}.$$

For the same set of index, we can define the Trieble-Lizorkin space $F^s_{p,q}$ equipped with the norm:

$$\|f\|_{F^s_{p,q}} = \|\langle 2^k\rangle^s P_k f\|_{L^p l^q}.$$

Comparing with the norm in Sobolev spaces:

$$\|f\|_{H^{s,p}} = \|\langle 2^k\rangle^s P_k f\|_{L^p l^2}.$$

It is easily seen that

$$\begin{aligned} &F^s_{p,q} \subset H^{s,p}, \quad \text{when } q \le 2; \\ &H^{s,p} \subset F^s_{p,q}, \quad \text{when } q \ge 2; \\ &H^{s,p} \subset B^s_{p,p}, \quad \text{when } p \ge 2; \\ &B^s_{p,p} \subset H^{s,p}, \quad \text{when } p \le 2. \end{aligned}$$

In all definitions above, we can do a direct extension, to define the homogeneous counterparts.

5 Linear profile decomposition (Part 1)

Note: Starting from this section, we will use a slightly different notation. We use P_N, $N \in 2^{\mathbb{Z}}$ to denote the L-P operator around dyadic frequency N. $P_{\le N}$, $P_{>N}$ can be interpreted similarly. We should point out that part of the material in Section 5 and Section 6 are borrowed from the Clay Lecture notes [25] by R. Killip and M. Visan.

In finite dimensional space, we know all closed bounded sets are compact. Such property is obviously not true for infinite dimensional space. For example, the orthogonal basis $\{e^{ikx}\}_{k\in\mathbb{Z}}$ for the space $L^2([0,2\pi])$ is uniformly bounded, but there is no convergence subsequence in it. In this section, we will discuss in more details how the compactness fails for a bounded sequence in several typical Sobolev spaces.

Take homogeneous Sobolev space $\dot{H}^1(\mathbb{R}^n)$ for example. Like discussed before, $\dot{H}^1(\mathbb{R}^n)$ is Hilbert space and its norm can be given by the inner product:

$$\langle f, g\rangle_{\dot{H}^1} = \int_{\mathbb{R}^n} \nabla f \cdot \nabla \bar{g} dx.$$

As $\mathbb{R}^n$ is scaling and translation invariant, $\dot{H}^1(\mathbb{R}^n)$ is also scaling and translation invariant. Next we will construct several examples of bounded sets, whose compactness are lost due exactly to these symmetries.

(1) Take $f \in \dot{H}^1(\mathbb{R}^n) \neq 0$, we define

$$f_\lambda(x) = \lambda^{1-\frac{n}{2}} f\left(\frac{x}{\lambda}\right).$$

It is easy to check

$$\|f_\lambda\|_{\dot{H}^1} = \|f\|_{\dot{H}^1}.$$

We will show as $\lambda \to 0$ or $\lambda \to \infty$, f_λ converge weakly to 0. Hence $\{f_\lambda\}$ is not compact.

To see this, we use density argument. Since $\mathbb{C}_c^\infty$ is dense in $\dot{H}^1(\mathbb{R}^n), \forall \varepsilon > 0$, we take $\psi \in \mathbb{C}_c^\infty(\mathbb{R}^n)$ such that $\|f - \psi\|_{\dot{H}^1(\mathbb{R}^n)} \leq \varepsilon$. Therefore $\forall \varphi \in \mathbb{C}_c^\infty$

$$\begin{aligned}|\langle f_\lambda, \varphi\rangle_{\dot{H}^1}| &\leq |\langle f_\lambda - \psi_\lambda, \varphi\rangle_{\dot{H}^1}| + |\langle \psi_\lambda, \varphi\rangle_{\dot{H}^1}| \\ &\leq \|f_\lambda - \psi_\lambda\|_{\dot{H}^1}\|\varphi\|_{\dot{H}^1} + \min(\|\psi_\lambda\|_2\|\varphi\|_{\dot{H}^2}, \|\psi_\lambda\|_{\dot{H}^2}\|\varphi\|_2) \\ &\leq \varepsilon\|\psi\|_{\dot{H}^1} + \min(\lambda\|\psi\|_2\|\varphi\|_{\dot{H}^2}, \lambda^{-1}\|\psi\|_{\dot{H}^2}\|\varphi\|_2).\end{aligned}$$

Take $\lambda \to +\infty$ or 0, the last term converges to 0. As ε is taken arbitrarily, we finally show

$$\lim_{\lambda\to 0,\infty} |\langle f_\lambda, \varphi\rangle_{\dot{H}^1}| = 0, \text{as desired.}$$

Therefore one possible way to lose compactness is scaling.

(2) Take $f \in \dot{H}^1(\mathbb{R}^n)$, one can easily verify that $\{f(\cdot + x_n)\}$ for $|x_n| \to \infty$ converges weakly to 0 with universal $\dot{H}^1(\mathbb{R}^n)$-*norm*. Thus translation is also a way to lose compactness.

(3) Now let us discuss another "unusual" way to lose compactness: the dispersion. As our course is mainly related with Schrödinger equation, we will use Schrödinger propagator to "spread" a given function.

Let $e^{it\Delta}$ be the Schrödinger propagator, $e^{it\Delta}f$ represents solution of linear Schrödinger equation

$$iu_t + \Delta u = 0,$$

with initial data f. Suppose $f \in \dot{H}^1(\mathbb{R}^n)$, we consider a sequence of function $\{e^{it_n\Delta}f\}_{n=1}^{\infty}$ with $t_n \to \infty$. As $e^{it\Delta}$ is unitary operator on $\dot{H}^1(\mathbb{R}^n)$, we have

$$\|e^{it_n\Delta}f\|_{\dot{H}^1} = \|f\|_{\dot{H}^1}.$$

Therefore, the noncompactness of $e^{it_n\Delta}f$ can by proved by showing again,

$$e^{it_n\Delta}f \rightharpoonup 0, \text{ weakly in } \dot{H}^1(\mathbb{R}^n).$$

Indeed, taking $\varphi \in \mathbb{C}_c^{\infty}(\mathbb{R}^n)$, we define $F(t) = \langle e^{it\Delta}f, \varphi\rangle_{\dot{H}^1}$. Note first $F(t)$ is uniformly Lipschitz as

$$\left|\frac{d}{dt}F(t)\right| \leq \|f\|_{\dot{H}^1}\|\varphi\|_{\dot{H}^3}.$$

Second, by Strichartz estimate (which will be introduced shortly),

$$\|F(t)\|_{L^{\frac{2(n+2)}{n-2}}([T,\infty))} \leq \|e^{it\Delta}f\|_{L_{t,x}^{\frac{2(n+2)}{n-2}}([T,\infty))} \|\Delta\varphi\|_{L_x^{\frac{2(n+2)}{n+6}}} \to 0.$$

The weak convergence follows immediately from the two estimates.

Now Let us discuss in more details the effect of dispersion. Assume first that $f(x)$ is a simple plane wave with frequency ξ_0

$$f(x) = e^{ix\cdot\xi_0}.$$

Easy to verify

$$e^{it\Delta}f(x) = e^{ix\cdot\xi_0 - it|\xi_0|^2} = e^{i(x-\xi_0 t)\cdot\xi_0} = f(x - \xi_0 t).$$

That is, what Schrödinger propagator does is to translate the plane wave by the amount $\xi_0 t$, ξ_0 can be regarded as the velocity that the wave propagates under the Schrodinger propagator.

More generally, for a wave packet with frequency ξ_0 (i.e., a collection of plane waves with dominating frequency ξ_0),

$$f(x) = \int e^{ix\cdot\xi}\chi_{|\xi-\xi_0|<\lambda}d\xi, \quad \lambda \ll 1,$$

we have

$$e^{it\Delta}f(x) = \int e^{i(x\cdot\xi - t|\xi|^2)}\chi_{|\xi-\xi_0|<\lambda}d\xi.$$

As ξ_0 is dominating, we do a Taylor expansion for the phase function around ξ_0

$$\Phi(\xi) = xt - t\xi^2 = \Phi(\xi_0) + (x - 2t\xi_0)(\xi - \xi_0) + O(|\xi - \xi_0|^2).$$

Roughly

$$\begin{aligned} e^{it\Delta}f(x) &\approx \int e^{i(x-2t\xi_0)\xi}\chi_{|\xi-\xi_0|<\lambda}d\xi \; e^{i(\Phi(\xi_0)-\xi_0(x-2t\xi_0))} \\ &\sim f(x-2\xi_0 t). \end{aligned}$$

Therefore, roughly speaking $e^{it\Delta}$ is nothing else but a "frequency-dependant" translation.

There is another source of losing compactness which we can not observe in $\dot{H}^1$ space: the oscillation.

(4) Consider $L^2(\mathbb{R}^n)$. Let $|\xi_n| \to \infty$ as $n \to \infty$. Then the sequence $\{e^{ix\cdot\xi_n}f(x)\}_{n=1}^{\infty}$ is not compact in L^2 simply because

$$\mathcal{F}(e^{2\pi ix\cdot\xi_n}f(x)) = \hat{f}(\xi-\xi_n)$$

and Plancherel.

In the above discussion we analyze three possibilities of losing compactness: dilation, translation and oscillation. In the following example, we will show that if we consider the bounded set in slightly weaker topology, the compactness can be recovered by precluding these symmetries. Results of these kind are called "linear profile decomposition". We will give several examples, starting with the one with 0 symmetries.

5.1 0 symmetry—Compactness

Let $\Omega \subset \mathbb{R}^2$ be a compact regime. From Sobolev embedding, we know

$$H^1(\Omega) \hookrightarrow L^4(\Omega).$$

As the scaling of L^4 is sandwiched between L^2 and $\dot{H}^1$, (i.e. the L^4 invariant scaling transformation $f \to f_\lambda = \lambda^{-\frac{1}{2}}f(x/\lambda)$ will make either L^2 or $\dot{H}^1$ norm explode as $\lambda \to 0$ or ∞), one can not construct non-compact sequence by using scaling transformation. Due to the loss of translation invariance, there is no non-compact sequence constructed only by translation either. Therefore a natural conclusion is the embedding is compact, which is the result stated below.

Lemma 5.1. *Let Ω be compact in $\mathbb{R}^2$. Then $H^1(\Omega) \hookrightarrow L^4(\Omega)$ is compact.*

Proof. Let $\{f_n\}_{n=1}^{\infty}$ be a bounded sequence in $H^1(\Omega)$, out goal is to show $\{f_n\}_{n=1}^{\infty}$ has a convergent subsequence in $L^4(\Omega)$. By density argument, we can always assume $f_n \in C_c^{\infty}(\Omega)$.

We will resort to a compactness criterion, for example, the well known Arzala-Ascoli theorem for continuous functions.[1]

[1]Arzela-Ascoli: Let Ω be a compact domain in $\mathbb{R}^n$. If $\{g_n\}_{n=1}^{\infty} \subset C(\Omega)$ is uniformly bounded and equi-continuous then $\{g_n\}_{n=1}^{\infty}$ is precompact in $C(\Omega)$.

To make $\{f_n\}$ uniformly bounded and equi-continuous we need to smoothout the functions, we will use convolution with another C_c^∞ function to keep the finiteness of the support.

Now take $\varphi \in C_c^\infty(B(0,1))$ be such that $0 \le \varphi \le 1$ and $\int \varphi = 1$. Define $\varphi_\varepsilon = \varepsilon^{-2}\varphi(\frac{x}{\varepsilon})$. Then it is easy to verify

(1) $f_n * \varphi_\epsilon \in C_c^\infty(\Omega)$.
(2) $\|f_n * \varphi_\varepsilon\|_\infty \le \|\varphi_\varepsilon\|_2 \|f_n\|_2 \le \varepsilon^{-1}\|f_n\|_2$.
(3) $\|\nabla f_n * \varphi_\varepsilon\|_\infty \le \|\nabla \varphi_\varepsilon\|_2 \|f_n\|_2 \le \varepsilon^{-2}\|f_n\|_2$.
(4)

$$\begin{aligned}
\|f_n - f_n * \varphi_\varepsilon\|_{L_x^2} &= \left\| f_n(x) - \int_{\mathbb{R}^2} \varepsilon^{-2}\varphi\left(\frac{y}{\varepsilon}\right) f_n(x-y)dy \right\|_{L_x^2} \\
&\le \left\| f_n(x) - \int_{\mathbb{R}^2} f_n(x-\varepsilon y)\varphi(y)dy \right\|_{L_x^2} \\
&= \left\| \int_{\mathbb{R}^2} (f_n(x-\varepsilon y) - f_n(x))\varphi(y)dy \right\|_{L_x^2} \\
&\lesssim \varepsilon \|\nabla f_n\|_2 \|\varphi\|_1 \lesssim \varepsilon.
\end{aligned}$$

(5) $\|\nabla(f_n - f_n * \varphi_\varepsilon)\|_{L_x^2} \lesssim 1$.
(6) $\|f_n - f_n * \varphi_\varepsilon\|_{L^4} \lesssim \varepsilon^{\frac{1}{2}}$.

Let $\varepsilon = 1$, from (1) (2) (3) and the interpolation $\|f\|_4 \le \|f\|_\infty^{\frac{1}{2}} \|f\|_2^{\frac{1}{2}}$ we can extract a subsequence such that $f_{1j} * \varphi_1$ is Cauchy in L^4. In particular

$$\begin{aligned}
&\|f_{1j} * \varphi_1 - f_{1k} * \varphi_1\|_4 \le 1, \\
&\|f_{1j} * (1 - \varphi_1)\|_4 \lesssim 1.
\end{aligned}$$

Let $\varepsilon = \frac{1}{2}$, we consider the sequence $\{f_{1j}\}$. Again, we can extract a subsequence

$$f_{21}, f_{22}, \cdots, f_{2j}, \cdots$$

for which

$$\begin{aligned}
&\|f_{2j} * \varphi_{\frac{1}{2}} - f_{2k} * \varphi_{\frac{1}{2}}\|_4 \le \frac{1}{2}, \\
&\|f_{2j} * (1 - \varphi_{\frac{1}{2}})\|_4 \lesssim \sqrt{\frac{1}{2}}.
\end{aligned}$$

Inductively, for $\varepsilon = \frac{1}{2^n}$, we can have the subsequence $\{f_{nj}\}_{j=1}^\infty$

$$\begin{aligned}
&\|f_{nj} * \varphi_{\frac{1}{2^n}} - f_{nk} * \varphi_{\frac{1}{2^n}}\|_4 \le \frac{1}{2^n}, \\
&\|f_{nj} * (1 - \varphi_{\frac{1}{2^n}})\|_4 \lesssim 2^{-\frac{n}{2}}.
\end{aligned}$$

Finally as usual we take the diagonal sequence $\{f_{nn}\}_{n=1}^\infty$. We will show $\{f_{nn}\}$ is

Cauchy in L^4. Indeed, $\forall\ m \geq n$, we estimate

$$\begin{aligned}\|f_{nn} - f_{mm}\|_4 &\leq \|(f_{nn} - f_{mm}) * \varphi_{\frac{1}{2^n}}\|_4 \\ &\quad + \|f_{nn} * (1 - \varphi_{\frac{1}{2^n}})\|_4 + \|f_{mm} * (1 - \varphi_{\frac{1}{2^n}})\|_4 \\ &\lesssim 2^{-\frac{n}{2}}.\end{aligned}$$

The lemma is proved. □

5.2 Linear profile decomposition relative to $H^1(\mathbb{R}^2) \hookrightarrow L^4(\mathbb{R}^2)$

In this part, we investigate the compactness of the Sobolev embedding $H^1 \hookrightarrow L^4$ in the whole space $\mathbb{R}^2$. Due to the incomparability of the scaling of the two spaces, the only enemy of losing compactness is the spatial translation. Indeed, for $|x_n| \to \infty$, $f \in H^1(\mathbb{R}^2)$, $\{f(\cdot + x_n)\}_{n=1}^{\infty}$ is not compact in L^4. Based on this example, we can construct another counterexample, the two bubbles run away from each other.

Take $\varphi_1, \varphi_2 \in C_c^\infty(\mathbb{R}^2), \varphi_1 \geq 0,\ \varphi_2 \geq 0,\ x_n^1,\ x_n^2 \in \mathbb{R}^2$ such that

$$\lim_{n\to\infty} |x_n^1 - x_n^2| = \infty.$$

We consider

$$f_n(x) = \varphi^1(x - x_n^1) + \varphi^2(x - x_n^2).$$

Then $\{f_n(x)\}_{n=1}^{\infty}$ is not precompact in $L^4(\mathbb{R}^2)$.

Indeed, on the one hand

$$\begin{aligned}\|f_n\|_4^4 &= \int_{\mathbb{R}^2} |f_n(x)|^4 dx \\ &= \|\varphi^1\|_4^4 + \|\varphi^2\|_4^4 + 4\int (\varphi^1(x - x_n^1))^3 \varphi^2(x - x_n^2) dx \\ &\quad + 6\int (\varphi^1(x - x_n^1))^2 (\varphi^2(x - x_n^2))^2 dx \\ &\quad + 4\int \varphi^1(x - x_n^1)(\varphi^2(x - x_n^2))^3 dx.\end{aligned}$$

The last three terms are all 0 as n sufficiently large, for example,

$$\begin{aligned}\int (\varphi^1(x - x_n^1))^3 \varphi^2(x - x_n^2) dx &= \int (\varphi^1(x))^3 \varphi^2(x + x_n^1 - x_n^2) dx \\ &\equiv 0 \text{ as } n \to \infty.\end{aligned}$$

Therefore,

$$\|f_n\|_4^4 = \|\varphi^1\|_4^4 + \|\varphi^2\|_4^4,$$

as n sufficiently large. On the other hand, passing to subsequence, f_n can only has three possibilities of weak convergence

$$\begin{aligned}&x_n^1 \to x_\infty^1, f_n \rightharpoonup \varphi^1(x - x_\infty^1), \\ &x_n^2 \rightharpoonup x_\infty^2, f_n \rightharpoonup \varphi^2(x - x_\infty^2), \\ &|x_n^1|, |x_n^2| \to \infty, f_n \rightharpoonup 0.\end{aligned}$$

In any of the above case,

$$\|w - \lim_{n\to\infty} f_n\|_4^4 < \|\varphi^1\|_4^4 + \|\varphi^2\|_4^4.$$

Therefore, the weak convergence can not be strong.

Similarly, one can construct non-compact sequence, each of the function is summations of finite (infinite) bubbles, and these bubbles run away from each other as $n \to \infty$.

The next theorem asserts that this is the only possible case for any bounded sequence. We have the following linear profile decomposition.

Theorem 5.2. *Let $\{f_n\}_{n=1}^{\infty}$ be bounded in $H^1(\mathbb{R}^2)$. Then there exists a subsequence (which we still denote as f_n), $J^* \in \{0, 1, 2, \cdots, n, \cdots\} \cup \{\infty\}$, $\{\varphi^j\}_{j=1}^{J^*} \subset H^1(\mathbb{R}^2)$, $\{x_n^j\}_{j=1}^{J^*}$, such that $\forall\ 0 \le J < J^*$,*

- $f_n = \sum_{j=1}^{J} \varphi^j(x - x_n^j) + w_n^J$.
- $\forall j \neq k, |x_n^j - x_n^k| \to \infty$ *as* $n \to \infty$.
- *Decoupling:*

$$\lim_{n\to\infty} (\|f_n\|_{H^1}^2 - \sum_{j=1}^{J} \|\varphi^j\|_{H^1}^2 - \|w_n^J\|_{H^1}^2) = 0.$$

- $w_n^J(x + x_n^j) \rightharpoonup 0$ *weakly in* $H^1(\mathbb{R}^2)$ *for* $1 \le j \le J$.
- $\lim_{J\to J^*} \overline{\lim}_{n\to\infty} \|w_n^J\|_4 = 0$.

The proof of the theorem is repeatedly using the next result which tells you how to take the first bubble.

Lemma 5.3. *Let $\{f_n\}_{n=1}^{\infty}$ satisfy*

$$\lim_{n\to\infty} \|f_n\|_{H^1} = A,$$
$$\lim_{n\to\infty} \|f_n\|_4 = \varepsilon > 0.$$

Then $\exists\ x_n \in \mathbb{R}^2$ and $\varphi^1 \in H^1(\mathbb{R}^2)$ such that passing to subsequence, we can write

$$f_n(x) = \varphi^1(x - x_n) + w_n(x),$$
$$w_n(x + x_n) \rightharpoonup 0 \text{ weakly in } H^1(\mathbb{R}^2),$$
$$\|w_n\|_4^4 \le \varepsilon^4(1 - c\varepsilon^4/A^4),$$
$$\lim_{n\to\infty} (\|f_n\|_{H^1}^2 - \|\varphi^1\|_{H^1}^2 - \|w_n\|_{H^1}^2) = 0.$$

Proof. The location x_n is where f_n is "big". The best norm to observe x_n is L^∞. However, since f_n is not necessarily in L^∞, we need to cut into frequencies. Choose C sufficiently large, from Bernstein we have

$$\begin{aligned}\|P_{>C} f_n\|_4 &\le C^{-\frac{1}{2}} \|\nabla P_{>C} f_n\|_2 \\ &\le C^{-\frac{1}{2}} A \le \frac{\varepsilon}{2}\end{aligned}$$

by taking $C = 4(\frac{A}{\varepsilon})^2$. Therefore for the low frequencies, we have

$$\begin{aligned}\frac{\varepsilon}{2} \le \|P_{\le C} f_n\|_4 &\le \|P_{\le C} f_n\|_\infty^{\frac12} \|P_{\le C} f_n\|_2^{\frac12}\\ &\le A^{\frac12} \|P_{\le C} f_n\|_\infty^{\frac12},\end{aligned}$$

and

$$\|P_{\le C} f_n\|_\infty \ge \frac{\varepsilon^2}{4A}.$$

From this we can find $x_n \in \mathbb{R}^2$ such that

$$|P_{\le C} f_n(x_n)| \ge \frac{\varepsilon^2}{4A}$$

i.e,

$$\frac{\varepsilon^2}{4A} \le |P_{\le C} f_n(x_n)| = \left| \int C^2 \check{\phi}(Cy) f_n(x_n + y) dy \right|. \tag{5.1}$$

Now as $\{f_n(x_n + x)\}_{n=1}^\infty$ is bounded in H^1, passing to a subsequence, it converges weakly to φ^1, i.e.,

$$\varphi^1 = w - \lim_{n\to\infty} f_n(x_n + x).$$

Continuing from (5.1), we get

$$\begin{aligned}\frac{\varepsilon^2}{4A} &\le \left| \int C^2 \check{\phi}(Cy) f_n(x_n + y) dy \right|\\ &\to |\int C^2 \check{\phi}(Cy) \varphi^1(y) dy|\\ &\le \|C^2 \check{\phi}(C\cdot)\|_2 \|\varphi^1\|_2\\ &\lesssim \|\varphi^1\|_2.\end{aligned}$$

Thus,

$$\|\varphi^1\|_{H^1} \ge \|\varphi^1\|_2 \ge c\frac{\varepsilon^4}{A^3}.$$

Similarly,

$$\|\varphi^1\|_4 \ge c\frac{\varepsilon^4}{A^3}.$$

Define

$$w_n(x) = f_n(x) - \varphi^1(x - x_n),$$

we easily see that

$$\begin{aligned} w_n(x + x_n) &\rightharpoonup 0 \text{ weakly in } H^1(\mathbb{R}^2),\\ \|f_n\|_{H^1}^2 &= \|f_n - \varphi^1(\cdot - x_n) + \varphi^1(\cdot - x_n)\|_{H^1}^2\\ &= \|f_n - \varphi^1(\cdot - x_n)\|_{H^1}^2 + \|\varphi^1\|_{H^1}^2 + 2\langle w_n(\cdot + x_n), \varphi^1\rangle_{H^1}.\end{aligned}$$

Thus,

$$\|f_n\|_{H^1}^2 - \|f_n - \varphi^1(\cdot - x_n)\|_{H^1}^2 + \|\varphi^1\|_{H^1}^2 = o_n(1).$$

We finally show that by moving away the bubble, the L_x^4-norm will decrease by non-trivial amount. This will guarantee that after countably many times, the process of taking bubble will be finished and the error converges to 0 in L^4.

The estimate we will use is

$$\lim_{n\to\infty} \int_{\mathbb{R}^2} |f_n|^4 - |f_n - \varphi(\cdot - x_n)|^4 - |\varphi|^4 dx = 0. \tag{5.2}$$

And we will discuss the proof separately. Finally we note that with this estimate, we immediately have

$$\begin{aligned}\|w_n\|_4^4 &\le \varepsilon^4 - c\frac{\varepsilon^{16}}{A^{12}}\\ &\le \varepsilon^4\left(1 - c\frac{\varepsilon^{12}}{A^{12}}\right).\end{aligned}$$

Lemma 5.3 is proved. □

We now prove (5.2) in a general setting.

Lemma 5.4 (Refined-Fatou). *Let g_n be bounded in $L^p(\mathbb{R}^n), 1 < p < \infty$. Let g_n converge almost everywhere to g. Then*

$$\lim_{n\to\infty} \int_{\mathbb{R}^n} |g_n|^p - |g_n - g|^p - |g|^p dx = 0.$$

Remark 5.5. *To apply this lemma to* (5.2), *we first note that by translation invariant,* (5.2) *is equivalent to*

$$\int_{\mathbb{R}^n} |f_n(x + x_n)|^4 - |f_n(x + x_n) - \varphi|^4 - |\varphi|^4 dx \to 0.$$

We can apply Lemma 5.4 *with $p = 4$, $g_n(x) = f_n(x + x_n)$. The fact that passing to subsequence $g_n \to \varphi$ a.e. comes from* (1) *$g_n \rightharpoonup \varphi$ weakly in H^1,* (2) *$H_{loc}^1(\mathbb{R}^2) \hookrightarrow L_{loc}^4(\mathbb{R}^2)$ is compact embedding.*

Proof of Lemma 5.4. From the elementary inequality, for $1 < p < \infty$,

$$||a + b|^p - |a|^p - |b|^p| \le c_p(|a|^{p-1}|b| + |a||b|^{p-1})$$

we estimate

$$\left|\int |g_n|^p - |g_n - g|^p - |g|^p dx\right| \le c_p \int |g_n - g|^{p-1}|g|dx + \int |g_n - g||g|^{p-1}dx.$$

As $g \in L^p$, $|g|^{p-1} \in L^{\frac{p}{p-1}}$, by density argument, to prove convergence, it is enough to show for any $\psi \in C_c^\infty(\mathbb{R}^d)$ that

$$\lim_{n\to\infty} \int |g_n - g|^{p-1}\psi dx = 0,$$

and

$$\lim_{n\to\infty}\int |g_n - g|\psi dx = 0. \tag{5.3}$$

We only prove (5.3), since the other one can be discussed similarly. Denote $h_n = |g_n - g|$. We have $h_n \in L^p$ and $h_n \to 0$ a.e., thus

$$\begin{aligned}\lim_{n\to\infty}\int h_n\psi dx &= \lim_{n\to\infty}\int h_n 1_{|h_n|\le N}\psi dx + \lim_{n\to\infty}\int h_n 1_{|h_n|>N}\psi dx\\ &\le \lim_{n\to\infty}\|h_n\|_p mes\{|h_n|>N\}^{\frac{1}{p'}}\|\psi\|_\infty\\ &\le \lim_{n\to\infty}\|h_n\|_p\left(\frac{\|h_n\|_p^p}{N^p}\right)^{\frac{1}{p'}}\|\psi\|_\infty \lesssim N^{-\frac{p}{p'}}.\end{aligned}$$

Here for the first term we have used dominating convergence. Finally as N is taken arbitrarily, we proved (5.3), hence Lemma 5.4. □

Finally we use Lemma 5.3 to illustrate the proof of Theorem 5.2. One can complete the proof by mimicking the proof in the coming Section 5.3.

Proof. Let $\{f_n\}$ be a bounded sequence in $H^1(\mathbb{R}^2)$. If

$$\liminf_{n\to\infty}\|f_n\|_4 = 0, \tag{5.4}$$

then we simply take $f_n := w_n^0, J^* = 0$, no bubble need to be taken, all the conclusions hold automatically. Otherwise we can apply Lemma 5.3 to w_n^0 and passing subsequence we get φ^1 and x_n^1 such that

$$\begin{aligned}f_n &= w_n^0,\\ w_n^0 &= \varphi^1(x - x_n^1) + w_n^1.\end{aligned}$$

We repeat the same process J times, we have

$$\begin{aligned}w_n^1 &= \varphi^2(x - x_n^2) + w_n^2,\\ &\cdots\cdots\cdots\cdots\\ w_n^{J-1} &= \varphi^J(x - x_n^J) + w_n^J.\end{aligned}$$

If for some finite J, $\liminf_{n\to\infty}\|w_n^J\|_4 = 0$, we stop and set $J^* = J$, otherwise we keep on doing and set $J^* = \infty$. Note each of the sequence is taken as subsequence from the previous one. If the process terminates at $J^* < \infty$, we simply take the last one, otherwise we take diagonal indices as explained before. The only thing that needs to be seen is

$$|x_n^j - x_n^k| \to \infty \text{ as } n\to\infty,\ \forall j\ne k.$$

Take for example $j = 1, k = 2$. From the construction we know

$$\begin{aligned}w_n^1(x + x_n^1) &\rightharpoonup 0 \text{ in } H^1,\\ w_n^2(x + x_n^2) &\rightharpoonup 0 \text{ in } H^1.\end{aligned}$$

Suppose by contradiction that $|x_n^1 - x_n^2| \to x_0$ and $|x_0| < \infty$, we have

$$\begin{aligned}0 \leftharpoondown\ & w_n^2(x + x_n^2) \\ &= w_n^1(x + x_n^2) - \varphi^2(x) \\ &= w_n^1(x + x_n^1 + x_n^2 - x_n^1) - \varphi^2(x) \\ &\rightharpoonup 0 - \varphi^2(x).\end{aligned}$$

A contradiction! The lemma is proved. □

5.3 Linear profile decomposition relative to $\dot{H}^1(\mathbb{R}^3) \hookrightarrow L^6(\mathbb{R}^3)$

Consider the Sobolev embedding

$$\dot{H}^1(\mathbb{R}^3) \hookrightarrow L^6(\mathbb{R}^3).$$

As the two spaces has the same scaling, the loss of compactness can be caused by translation and scaling

$$f \to f_\lambda(x) = \lambda^{-\frac{1}{2}} f\left(\frac{x}{\lambda}\right).$$

The corresponding linear profile decomposition is stated as follows

Theorem 5.6 ($\dot{H}^1(\mathbb{R}^3) \hookrightarrow L^6(\mathbb{R}^3)$ linear profile decomposition[19]). *Let $\{f_n\}_{n=1}^\infty$ be bounded sequence in $\dot{H}^1(\mathbb{R}^3)$. Then there exists a subsequence (which is still denoted by $\{f_n\}_{n=1}^\infty$), $J^* \in \{0, 1, 2, \cdots\} \cup \{\infty\}$, linear profiles $\{\varphi^j\}_{j=1}^{J^*} \subset \dot{H}^1$, parameter $\{(x_n^j, \lambda_n^j)\}_{j=1}^{J^*} \subset \mathbb{R}^3 \times \mathbb{R}^+$ such that for any $0 \le J < J^*$, we can write*

- $f_n = \sum_{j=1}^J g_n^j \varphi^j + w_n^J$, $w_n^J \in \dot{H}^1$, $g_n^j \varphi^j = (\lambda_n^j)^{-\frac{1}{2}} \varphi^j(\frac{x - x_n^j}{\lambda_n^j})$.
- *Orthogonality condition: $\forall j \ne k$*

$$\lim_{n\to\infty} \left(\frac{\lambda_n^j}{\lambda_n^k} + \frac{\lambda_n^k}{\lambda_n^j} + \frac{1}{\lambda_n^j \lambda_n^k} |x_n^j - x_n^k|^2 \right) = \infty.$$

- *Decoupling:*

$$\lim_{n\to\infty} (\|f_n\|_{\dot{H}^1}^2 - \sum_{j=1}^J \|\varphi^j\|_{\dot{H}^1}^2 - \|w_n^J\|_{\dot{H}^1}^2) = 0,$$

$$\lim_{n\to\infty} (\|f_n\|_{L^6}^6 - \sum_{j=1}^J \|\varphi^j\|_{L^6}^6 - \|w_n^J\|_{L^6}^6) = 0.$$

- $(g_n^j)^{-1} w_n^J \rightharpoonup 0$ *weakly in* $\dot{H}^1, \forall\ 1 \le j \le J$.
- $\lim_{J\to J^*} \overline{\lim}_{n\to\infty} \|w_n^J\|_6 = 0$.

Again, to prove this result, one has to be clear on how to take the first bubble. The previous discussion has shown how to find location of the bubbles. The scaling of the bubble corresponds to its frequency. We begin with the following crucial inequality.

Lemma 5.7 (Generalized Gagliardo-Nirenberg inequality). *Let $f \in \dot{H}^1(\mathbb{R}^3)$. Then*

$$\|f\|_6 \lesssim \|f\|_{\dot{H}^1}^{\frac{1}{3}} \big(\sup_{N \in 2^{\mathbb{Z}}} N^{-\frac{1}{2}} \|P_N f\|_\infty\big)^{\frac{2}{3}}.$$

Remark 5.8. 1. *If there is a bounded sequence in $\dot{H}^1$ with non-trivial L^6-norm, one can use this estimate to find a frequency N that is responsible for the non-triviality of the L^6-norm. The inverse of the frequency is the scaling we are looking for.*

2. *The last one is indeed the Besov-norm*

$$\|f\|_{\dot{B}^{-\frac{1}{2}}_{\infty,\infty}} = \sup_{N \in 2^{\mathbb{Z}}} \|P_N f\|_\infty N^{-\frac{1}{2}}.$$

General Besov-norm can be defined as follows:

$$\|f\|_{\dot{B}^s_{p,q}} = \Big(\sum_{N \in 2^{\mathbb{Z}}} (N^s \|P_N f\|_p)^q \Big)^{\frac{1}{q}}.$$

The scaling of such space can still be counted as $s - \frac{n}{p}$. (Why?)

Proof of Lemma 5.7. From the square function estimate

$$\|f\|_6 \approx \Big\| \big(\sum_{N \in 2^{\mathbb{Z}}} |P_N f|^2 \big)^{\frac{1}{2}} \Big\|_6,$$

we have

$$\begin{aligned}
\|f\|_6^6 &\approx \int_{\mathbb{R}^3} \Big(\sum_{N \in 2^{\mathbb{Z}}} |P_N f|^2 \Big)^3 dx \\
&\lesssim \sum_{N_1 \ge N_2 \ge N_3} \int_{\mathbb{R}^3} |P_{N_1} f|^2 |P_{N_2} f|^2 |P_{N_3} f|^2 dx \\
&\lesssim \sum_{N_1 \ge N_2 \ge N_3} \|P_{N_1} f\|_2^2 \|P_{N_2} f\|_\infty^2 \|P_{N_3} f\|_\infty^2 \\
&\lesssim \sum_{N_1 \ge N_2 \ge N_3} \|P_{N_1} f\|_2^2 N_2 N_3 \big(\sum_{N \in 2^{\mathbb{Z}}} N^{-\frac{1}{2}} \|P_N f\|_\infty \big)^4 \\
&\lesssim \sum_{N_1 \in 2^{\mathbb{Z}}} N_1^2 \|P_{N_1} f\|_2^2 \big(\sup_{N \in 2^{\mathbb{Z}}} N^{-\frac{1}{2}} \|P_N f\|_\infty \big)^4 \\
&\lesssim \|f\|_{\dot{H}^1}^2 \|f\|_{\dot{B}^{-\frac{1}{2}}_{\infty,\infty}}^4 .
\end{aligned}$$

Taking 1/6 power on both sides gives the desired result. □

Lemma 5.9 (Taking the first bubble). *Let $\{f_n\}_{n=1}^\infty$ be a bounded sequence in $\dot{H}^1(\mathbb{R}^3)$ satisfying*

$$\lim_{n \to \infty} \|f_n\|_{\dot{H}^1} = A,$$
$$\lim_{n \to \infty} \|f_n\|_6 = \varepsilon.$$

Then there exists a subsequence (which we still denote as f_n), linear profile $\varphi \in \dot{H}^1, (x_n, \lambda_n) \in \mathbb{R}^3 \times \mathbb{R}^+$, $w_n \in \dot{H}^1(\mathbb{R}^3)$ such that

$$f_n(x) = g_n\varphi(x) + w_n(x)$$

with the properties that

- $g_n^{-1} w_n \rightharpoonup 0$ *weakly in* $\dot{H}^1$,
- $\|w_n\|_6 \leq \varepsilon(1 - c(\frac{\varepsilon}{A})^3)$, $\|w_n\|_{\dot{H}^1} \leq A(1 - c(\frac{\varepsilon}{A})^3)$,
- $\lim_{n\to\infty}(\|f_n\|_{\dot{H}^1}^2 - \|\varphi\|_{\dot{H}^1}^2 - \|w_n\|_{\dot{H}^1}^2) = 0$,
- $\lim_{n\to\infty}(\|f_n\|_6^6 - \|\varphi\|_6^6 - \|w_n\|_6^6) = 0$.

Here, $g_n f(x) = \lambda_n^{-\frac{1}{2}} f(\frac{x - x_n}{\lambda_n})$, *and*

$$g_n^{-1} f(x) = \lambda_n^{\frac{1}{2}} f(x_n + \lambda_n x).$$

Proof. Applying generalized G-N inequality Lemma 5.7 to f_n we obtain

$$\varepsilon \lesssim A^{\frac{1}{3}} \Big(\sum_N \|P_N f\|_\infty N^{-\frac{1}{2}}\Big)^{\frac{2}{3}}.$$

Thus there exist x_n and $N_n \in 2^{\mathbb{Z}}$ such that

$$|P_{N_n} f_n(x_n)| N_n^{-\frac{1}{2}} \gtrsim \left(\frac{\varepsilon^3}{A}\right)^{\frac{1}{2}}.$$

Recall also

$$\begin{aligned} & N_n^{-\frac{1}{2}} |P_{N_n} f_n(x_n)| \\ &= N_n^{-\frac{1}{2}} \left| \int N_n^3 \psi(N_n y) f_n(x_n - y) dy \right| \\ &= N_n^{-\frac{1}{2}} \left| \int \psi(y) f_n(x_n + N_n^{-1} y) dy \right|. \end{aligned} \tag{5.5}$$

Let $\lambda_n = N_n^{-1}$, we consider the sequence of functions $g_n^{-1} f_n(y) = f_n(x_n + N_n^{-1} y) N_n^{-\frac{1}{2}}$. From the weak compactness of $\dot{H}^1(\mathbb{R}^3)$, there exists a subsequence and $\varphi \in \dot{H}^1$ such that

$$g_n^{-1} f_n \rightharpoonup \varphi \text{ weakly in } \dot{H}^1(\mathbb{R}^3).$$

Moreover, (5.5) implies (taking $n \to \infty$)

$$\begin{aligned} \left(\frac{\varepsilon^3}{A}\right)^{\frac{1}{2}} &\lesssim \left| \int \psi(y) g_n^{-1} f_n(y) dy \right| \\ &= \left| \int \frac{\nabla}{\Delta} \psi(y) \cdot \nabla g_n^{-1} f_n(y) dy \right| \\ &\to \left| \int \frac{\nabla}{\Delta} \psi(y) \nabla \varphi(y) dy \right| \\ &\lesssim \| |\nabla|^{-1} \psi \|_2 \|\nabla \varphi\|_2. \end{aligned}$$

Therefore, $\|\nabla\varphi\|_2 \gtrsim (\frac{\varepsilon^3}{A})^{\frac{1}{2}}$.

Similarly, as $\dot{H}^1 \hookrightarrow L^6$, we know $g_n^{-1}f_n \rightharpoonup \varphi$ weakly in L^6, and we have

$$\left(\frac{\varepsilon^3}{A}\right)^{\frac{1}{2}} \lesssim \|\psi\|_{\frac{6}{5}}\|\varphi\|_6 \Rightarrow \|\varphi\|_6 \gtrsim \left(\frac{\varepsilon^3}{A}\right)^{\frac{1}{2}}. \tag{5.6}$$

Now define

$$w_n(x) = f_n(x) - g_n\varphi(x).$$

We can write

$$f_n(x) = g_n\varphi(x) + w_n(x).$$

We check the four items as follows

- Simply by definition, $g_n^{-1}w_n = g_n^{-1}f_n - \varphi \rightharpoonup 0$ in $\dot{H}^1$.
- $\dot{H}^1$-decoupling is a simple consequence of the above fact. Indeed, as

$$g_n^{-1}f_n = \varphi + g_n^{-1}w_n,$$
$$\|g_n^{-1}f_n\|_{\dot{H}^1}^2 = \|\varphi\|_{\dot{H}^1}^2 + \|g_n^{-1}w_n\|_{\dot{H}^1}^2 + 2\int \nabla\varphi \cdot \nabla g_n^{-1}w_n dx.$$

Undoing the symmetries we get

$$\lim_{n\to\infty}(\|f_n\|_{\dot{H}^1}^2 - \|\varphi\|_{\dot{H}^1}^2 - \|w_n\|_{\dot{H}^1}^2) = 0, \text{ as desired.}$$

- L^6-*decoupling* and the lower bound (5.6) implies the decreaseness of $\|w_n\|_6$

$$\|w_n\|_6^6 \le \varepsilon^6 - c\left(\frac{\varepsilon^3}{A}\right)^3 \le \varepsilon^6\left(1 - c\left(\frac{\varepsilon}{A}\right)^3\right),$$
$$\|w_n\|_{\dot{H}^1}^2 \le A^2 - c\frac{\varepsilon^3}{A} = A^2\left(1 - c\left(\frac{\varepsilon}{A}\right)^3\right).$$

- Finally we only need to show the decoupling in L^6. Again we will use the Refined-Fatou. First, we note that $g_n^{-1}f_n$ is uniformly bounded in L^6. Second as $g_n^{-1}f_n \in H_{loc}^1$, from the compactness of the embedding $H_{loc}^1 \hookrightarrow L_{loc}^4$, by passing to a subsequence, $g_n^{-1}f_n \to \varphi$ a.e. From Refined Fatou (applied to $g_n^{-1}f_n$ and φ in L^6) we know

$$\lim_{n\to\infty}(\|g_n^{-1}f_n\|_6^6 - \|g_n^{-1}f_n - \varphi\|_6^6 - \|\varphi\|_6^6) = 0.$$

Undoing the scaling again, we obtain

$$\lim_{n\to\infty}(\|f_n\|_6^6 - \|\varphi\|_6^6 - \|w_n\|_6^6) = 0.$$

The Lemma is proved. □

Next we give the proof of Theorem 5.6.

Proof. We will proceed inductively and extract one bubble at a time. To start with, we set

$$w_n^0 := f_n.$$

If passing to a subsequence

$$\lim_{n\to\infty} \|w_n^0\|_6 = 0.$$

Then we stop and set $J^* = 0$ otherwise we can use Lemma 5.9 to extract the first bubble. Repeating this process J steps. Now suppose we have the decomposition up to the level $J \geq 0$:

$$f_n = \sum_{j=1}^{J} g_n^j \phi^j + w_n^J$$

with the stated properties:

$$\begin{cases} (g_n^J)^{-1} w_n^J \rightharpoonup 0 \text{ weakly in } \dot{H}^1, \\ \|f_n\|_{\dot{H}^1}^2 - \sum_{j=1}^J \|\phi^j\|_{\dot{H}^1}^2 - \|w_n^J\|_{\dot{H}^1}^2 = o_n(1), \\ \|f_n\|_{L^6}^6 - \sum_{j=1}^J \|\phi^j\|_{L^6}^6 - \|w_n^J\|_{L^6}^6 = o_n(1). \end{cases} \tag{5.7}$$

Passing to a subsequence if necessary, we set

$$A_J = \lim_{n\to\infty} \|w_n^J\|_{\dot{H}^1},$$
$$\varepsilon_J = \lim_{n\to\infty} \|w_n^J\|_{L^6}.$$

If $\varepsilon_J = 0$ then we stop. If not, we use the Lemma to obtain

$$w_n^J = g_n^{J+1}\phi^{J+1} + w_n^{J+1}.$$

Here

$$(g_n^{J+1})^{-1} w_n^J \rightharpoonup \phi^{J+1} \text{ weakly in } \dot{H}^1,$$
$$\lim_{n\to\infty} (\|w_n^J\|_{\dot{H}^1}^2 - \|\phi^{J+1}\|_{\dot{H}^1}^2 - \|w_n^{J+1}\|_{\dot{H}^1}^2) = 0.$$

The second equality implies, from the inductive assumption that

$$\lim_{n\to\infty} \Big(\|f_n\|_{\dot{H}^1}^2 - \sum_{j=1}^{J+1} \|\phi^j\|_{\dot{H}^1}^2 - \|w_n^{J+1}\|_{\dot{H}^1}^2\Big) = 0.$$

By the same token we have

$$\lim_{n\to\infty} (\|f_n\|_{L^6}^6 - \sum_{j=1}^{J+1} \|\phi^j\|_{L^6}^6 - \|w_n^{J+1}\|_{L^6}^6) = 0.$$

Moreover,

$$(A_{J+1})^2 := \lim_{n\to\infty} \|w_n^{J+1}\|_{\dot{H}^1}^2 \leq A_J^2 \left(1 - c\left(\frac{\varepsilon_J}{A_J}\right)^3\right) \leq A_J^2,$$

$$(\varepsilon_{J+1})^6 \leq \lim_{n\to\infty} \|w_n^{J+1}\|_{L^6}^6 \leq \varepsilon_J^6 \left(1 - c\left(\frac{\varepsilon_J}{A_J}\right)^3\right).$$

Therefore, the L^6-*norm* is decreasing.

If $\varepsilon_{J+1} = 0$, then we stop and set $J^* = J + 1$ otherwise we repeat the procedure again. If the procedure does not terminate at finite steps, we set $J^* = \infty$. In that case,

$$\varepsilon_J \to 0 \text{ as } J \to \infty$$

which proves,[2]

$$\lim_{J\to J^*} \overline{\lim_{n\to\infty}} \|w_n^J\|_{L^6} = 0. \tag{5.8}$$

To close the proof, one only needs to verify the orthogonality condition. We argue by contradiction, suppose the orthogonality condition fails for some pair (j, k). Without loss of generality, we assume (j, k) is the first such pair in the sense that for all pairs $(j, l), j < l < k$, the orthogonality condition holds true. Passing to a subsequence we can assume

$$\frac{\lambda_n^j}{\lambda_n^k} \to \lambda_0 \in (0, \infty), \quad \frac{1}{\sqrt{\lambda_n^j \lambda_n^k}}(x_n^j - x_n^k) \to x_0.$$

From the inductive assumption, we know

$$\begin{aligned} w_n^j &= g_n^{j+1}\phi^{j+1} + w_n^{j+1} \\ &= g_n^{j+1}\phi^{j+1} + g_n^{j+2}\phi^{j+2} + w_n^{j+2} \\ &= \sum_{l=j+1}^{k-1} g_n^l \phi^l + w_n^{k-1}, \\ w_n^{k-1} &= w_n^j - \sum_{l=j+1}^{k-1} g_n^l \phi^l. \end{aligned}$$

From this and the definition of ϕ^k, we have

$$\begin{aligned} \phi^k &= w - \lim_{n\to\infty} (g_n^k)^{-1} w_n^{k-1} \\ &= w - \lim_{n\to\infty} (g_n^k)^{-1} w_n^j - \sum_{l=j+1}^{k-1} w - \lim_{n\to\infty} (g_n^k)^{-1} g_n^l \phi^l \\ &= w - \lim_{n\to\infty} (g_n^k)^{-1} g_n^j (g_n^j)^{-1} w_n^j - \sum_{l=j+1}^{k-1} w - \lim_{n\to\infty} (g_n^k)^{-1} g_n^j (g_n^j)^{-1} g_n^l \phi^l \\ &= 0. \end{aligned}$$

The convergence in the last equality follows from the next two claims.

(1) (j, l) is an orthogonal pair. Then we have

$$(g_n^j)^{-1} g_n^l \phi^l \rightharpoonup 0 \text{ weakly in } \dot{H}^1.$$

[2]Although in this process, the $\dot{H}^1$-*norm* is also decreasing the error is not small in $\dot{H}^1$. An example: $f_n = e^{it_n\Delta}\phi, t_n \to \infty$.

Indeed, from density argument, we can replace ϕ^l by ψ, a $C_c^\infty(\mathbb{R}^3)$ function, we are left to estimating, $\forall\ h \in C_c^\infty(\mathbb{R}^3)$

$$
\begin{aligned}
&|\langle (g_n^j)^{-1} g_n^l \psi, h\rangle_{\dot H^1}| \\
=&\left|\left\langle \left(\frac{\lambda_n^j}{\lambda_n^l}\right)^{\frac12} \psi\left(\frac{\lambda_n^j x + x_n^j - x_n^l}{\lambda_n^l}\right), h\right\rangle_{\dot H^1}\right| \\
\le& \min\left(\|h\|_{\dot H^2}\|\psi\|_2 \left(\frac{\lambda_n^j}{\lambda_n^l}\right)^{-1},\ \|\psi\|_{\dot H^2}\|h\|_2\left(\frac{\lambda_n^j}{\lambda_n^l}\right)\right).
\end{aligned}
$$

Suppose in the orthogonal condition,

$$\frac{\lambda_n^j}{\lambda_n^l} + \frac{\lambda_n^l}{\lambda_n^j} \longrightarrow \infty.$$

The above estimate implies that

$$\lim_{n\to\infty} |\langle (g_n^j)^{-1} g_n^l \psi, h\rangle_{\dot H^1}| = 0.$$

Thus we can assume

$$\frac{\lambda_n^j}{\lambda_n^l} \to \lambda_0 \in (0,\infty) \text{ and } \frac{1}{\sqrt{\lambda_n^j\lambda_n^l}}|x_n^j - x_n^l| \to \infty.$$

In this case

$$\left|\frac{x_n^j - x_n^l}{\lambda_n^l}\right| = \left|\frac{x_n^j - x_n^l}{\sqrt{\lambda_n^j\lambda_n^l}}\sqrt{\frac{\lambda_n^j}{\lambda_n^l}}\right| \to \sqrt{\lambda_0}\frac{|x_n^j - x_n^l|}{\sqrt{\lambda_n^j\lambda_n^l}} \to \infty.$$

For n sufficiently large, the spatial supports of the two functions do not overlap. Therefore $\langle (g_n^j)^{-1} g_n^l \psi, h\rangle_{\dot H^1} \equiv 0$.

(2) (j,k) is non-orthogonal pair. Then for any $\psi \in \dot H^1$

$$
\begin{aligned}
(g_n^k)^{-1} g_n^j \psi(x) &= \left(\frac{\lambda_n^k}{\lambda_n^j}\right)^{\frac12} \psi\left(\frac{\lambda_n^k}{\lambda_n^j}x + \frac{x_n^k - x_n^j}{\sqrt{\lambda_n^k\lambda_n^j}}\sqrt{\frac{\lambda_n^k}{\lambda_n^j}}\right) \\
&\to (\lambda_0)^{-\frac12}\psi(\lambda_0^{-1}x - x_0\sqrt{\lambda_0^{-1}}).
\end{aligned}
$$

The theorem is proved. □

6 Linear profile decomposition (Part 2)

Our next goal is to establish the (Schrödinger) flow-dependent linear profile decomposition. This is the out-come of investigating the loss of compactness of Strichartz estimate. Such results were heavily referred in the study of critical problem, especially the mass and energy critical NLS.

To start with, we introduce Strichartz estimate for linear Schrödinger operator.

6.1 Strichartz estimate for linear Schrödinger equation

Definition 6.1. *Let $u_0 \in \mathcal{S}'(\mathbb{R}^d)$, we use*

$$e^{it\Delta}u_0 = \mathcal{F}(e^{-i4\pi^2 t|\xi|^2}\hat{u}_0(\xi))$$

to denote the distribution solution of linear Schrödinger equation

$$iu_t + \Delta u = 0, u(0,x) = u_0(x).$$

Obviously, $e^{it\Delta}$ is an isometry on L^2 and any Hilbert-type Sobolev space

$$\|e^{it\Delta}u_0\|_{H^s} = \|u_0\|_{H^s}, s \in \mathbb{R}.$$

Remark 6.2. *In quantum physics, $|u(t,x)|^2$ is interpreted as the probability density of a micro-particle appearing at (t,x). Therefore a natural conservation law is*

$$\int_{\mathbb{R}^d} |u(t,x)|^2 dx$$

which corresponds to the total probability density. As the solution u has been scaled out, the constant is not necessarily 1.

The fundamental estimate we will rely on is the following:

Lemma 6.3 ($L^1 - L^\infty$ dispersive estimate).

$$\|e^{it\Delta}f\|_{L_x^\infty(\mathbb{R}^d)} \leq |t|^{-\frac{d}{2}}\|f\|_{L_x^1(\mathbb{R}^d)}. \tag{6.1}$$

Proof. This follows trivially from the explicit formula

$$e^{it\Delta}f = (4\pi it)^{-\frac{d}{2}}\int_{\mathbb{R}^d} e^{\frac{i|x-y|^2}{4t}}f(y)dy.$$ □

Interpolating (6.1) with L^2 conservation one obtains

$$\|e^{it\Delta}f\|_p \leq |t|^{-d(\frac{1}{2}-\frac{1}{p})}\|f\|_{p'}, \ 2 \leq p \leq \infty. \tag{6.2}$$

We are now ready to introduce the Strichartz estimate.

Lemma 6.4. *(p,q) is said to be admissible, if*

$$\frac{2}{p} + \frac{d}{q} = \frac{d}{2}, \ 2 \leq p, q \leq \infty, \ (p,q,d) \neq (2,\infty,2).$$

Then solution of linear Schrödinger equation

$$iu_t + \Delta u = f, \ u(0,x) = u_0$$

which, by Duhamel formula can be written as

$$u(t) = e^{it\Delta}u_0 - i\int_0^t e^{i(t-s)\Delta}f(s)ds$$

has the following estimates

$$\|u\|_{L_t^pL_x^q(I\times\mathbb{R}^d)} \lesssim \|u_0\|_2 + \|f\|_{L_t^{\tilde{p}'}L_x^{\tilde{q}'}(I\times\mathbb{R}^d)}$$

for any time interval $I, 0 \in I$ *and any admissible pairs* (p,q), $(\tilde{p},\tilde{q})$. *The estimate for the forbidden pair* $(2,\infty)$ *in* $d=2$ *is still true in the spherically symmetric case* [40].

Proof. Here we only provide the proof in the non-endpoint case, i.e. $p,\tilde{p}\neq 2$. The endpoint case remained open for quite a long time and was finally settled by Keel-Tao [21]. We start by considering the estimate

$$\|e^{it\Delta}u_0\|_{p,q} \lesssim \|u_0\|_2. \tag{6.3}$$

Here, $\|\cdot\|_{p,q}$ is a simple writing of the spacetime norm $\|\cdot\|_{L_t^pL_x^q(\mathbb{R}\times\mathbb{R}^d)}$. As by equivalent description of the Lebesgue norm:

$$\|F\|_{p,q} = \sup_{\|g\|_{p',q'}=1} \langle F, g\rangle_{t,x},$$

and

$$\begin{aligned}\langle e^{it\Delta}u_0, g\rangle_{t,x} &= \langle u_0, \int_{\mathbb{R}} e^{-it\Delta}g(t)dt\rangle_x \\ &\leq \|u_0\|_2\Big\|\int_{\mathbb{R}} e^{-it\Delta}g(t)dt\Big\|_2.\end{aligned}$$

It suffices to estimate

$$\Big\|\int_{\mathbb{R}} e^{-it\Delta}g(t)dt\Big\|_2 \lesssim \|g\|_{p',q'}. \tag{6.4}$$

On the other hand, using pairing we have

$$\begin{aligned}\left\|\int_{\mathbb{R}} e^{-it\Delta}g(t)dt\right\|_2^2 &= \left\langle \int_{\mathbb{R}} e^{-it\Delta}g(t)dt, \int_{\mathbb{R}} e^{-is\Delta}g(s)ds\right\rangle_x \\ &= \left\langle g(t), \int_{\mathbb{R}} e^{i(t-s)\Delta}g(s)ds\right\rangle_{t,x} \\ &\lesssim \|g\|_{p',q'}\left\|\int_{\mathbb{R}} e^{i(t-s)\Delta}g(s)ds\right\|_{p,q} \\ &\lesssim \|g\|_{p',q'}^2\end{aligned}$$

provided we prove

$$\left\|\int_{\mathbb{R}} e^{i(t-s)\Delta}g(s)ds\right\|_{p,q} \lesssim \|g\|_{p',q'}. \tag{6.5}$$

(6.5) is a direct consequence of the dispersive estimate and Hardy-Littlewood-Sobolev inequality: we have

$$\begin{aligned}LHS\ (6.5) &\lesssim \left\|\int_{\mathbb{R}} |t-s|^{-d(\frac{1}{2}-\frac{1}{q})}\|g(s)\|_{q'}ds\right\|_{L_t^p} \\ &\lesssim \|g\|_{p',q'}.\end{aligned}$$

(6.3) is proved. Finally by Christ-Kislev Lemma [9] to prove

$$\left\| \int_0^t e^{i(t-s)\Delta} f(s)ds \right\|_{p,q} \lesssim \|f\|_{\tilde{p}',\tilde{q}'}$$

with $p > 2$, $\tilde{p}' < 2$, it suffices to show

$$\left\| \int_{\mathbb{R}} e^{i(t-s)\Delta} f(s)ds \right\|_{p,q} \lesssim \|f\|_{\tilde{p}',\tilde{q}'}.$$

However, this follows easily from (6.3) and (6.4). The actual version in Lemma 6.4 follows by introducing cutoff function in time. The non-endpoint Strichartz estimate is proved. □

We give a remark regarding the Strichartz estimate.

Remark 6.5. *The scaling explanation of admissible pairs. Consider the linear Schrödinger equation*

$$iu_t + \Delta u = f.$$

Obviously,

$$u_\lambda(t,x) = u(\lambda^2 t, \lambda x)$$

solves the equation:

$$(i\partial_t + \Delta)u_\lambda(t,x) = \lambda^2 f_\lambda(t,x).$$

Strichartz estimate for solution u_λ is given by

$$\|u_\lambda\|_{q,r} \lesssim \|u_\lambda(0)\|_2 + \|\lambda^2 f_\lambda\|_{\tilde{q}',\tilde{r}'}.$$

Undoing the scaling one obtains

$$\lambda^{-\frac{2}{q}-\frac{d}{r}} \|u\|_{q,r} \lesssim \lambda^{-\frac{d}{2}} \|u_0\|_2 + \lambda^{2-(\frac{2}{\tilde{q}'}+\frac{d}{\tilde{r}'})} \|f\|_{\tilde{q}',\tilde{r}'}$$

which implies

$$\frac{2}{q} + \frac{d}{r} = \frac{d}{2} = \left(\frac{2}{\tilde{q}'} + \frac{d}{\tilde{r}'}\right) - 2 = -\frac{2}{\tilde{q}} - \frac{d}{\tilde{r}} + d$$

i.e,

$$\frac{2}{q} + \frac{d}{r} = \frac{2}{\tilde{q}} + \frac{d}{\tilde{r}} = \frac{d}{2}.$$

It is also easy to see that the inhomogeneous estimate (i.e. $u_0 = 0$) is expected to hold for larger range of indices. One can refer to [17] for more detailed discussion.

Remark 6.6. *Several often used admissible pairs:*

$$d = 2,\ (\infty, 2),\ (4,4),\ (2,\infty)\ (\textit{only in the radial case}).$$
$$d = 1,\ (\infty, 2),\ (6,6),\ (4,\infty).$$
$$d \geq 3,\ (\infty, 2),\ (2(d+2)/d,\ 2(d+2)/d),\ (2, 2d/(d-2)).$$

As "∇" operator compute with the linear propagator $e^{it\Delta}$, *we have*

$$\|\nabla e^{it\Delta}u_0\|_{q,r} \lesssim \|u_0\|_{\dot{H}^1}.$$

This together with the Sobolev embedding give rise to a different set of space-time estimate.

Example: $d = 3$, $(10, \frac{30}{13})$ is an admissible pair,

$$\|e^{it\Delta}u_0\|_{L^{10}_{t,x}} \lesssim \|e^{it\Delta}u_0\|_{L^{10}_t \dot{H}^{1,\frac{30}{13}}} \lesssim \|u_0\|_{\dot{H}^1}. \tag{6.6}$$

The above estimate shows $e^{it\Delta}$ is a bounded linear operator from $\dot{H}^1$ to $L^{10}_{t,x}$. Obviously this operator is not compact. The failure of the compactness is caused by the following three noncompact groups:

Translation:

$$u_n(x) = u_0(x + x_n),\ |x_n| \to \infty,$$
$$e^{it\Delta}u_n(x) = (e^{it\Delta}u_0(\cdot + x_n))(x) = e^{it\Delta}u_0(x + x_n).$$

Scaling:

$$u_\lambda(x) = \lambda^{-\frac{1}{2}}u_0\left(\frac{x}{\lambda}\right),\ \lambda \to 0 \text{ or } \infty,$$
$$e^{it\Delta}u_\lambda = e^{it\Delta}\left(\lambda^{-\frac{1}{2}}u_0\left(\frac{\cdot}{\lambda}\right)\right) = \lambda^{-\frac{1}{2}}(e^{i\frac{t}{\lambda^2}}u_0)\left(\frac{x}{\lambda}\right).$$

Translation in time

$$u_n(x) \to e^{it_n\Delta}u_0(x),\ t_n \to \infty,$$
$$e^{it\Delta}u_n(x) = e^{i(t+t_n)\Delta}u_0(x).$$

There is one more symmetry in L^2-level Strichartz estimate:

$$\|e^{it\Delta}u_0\|_{L^4_{t,x}(\mathbb{R}\times\mathbb{R}^2)} \lesssim \|u_0\|_2. \tag{6.7}$$

In dimension d, the index of the spacetime norm should be $\frac{2(d+2)}{d}$. That symmetry corresponds to oscillation for L^2 functions.

$$\begin{cases} u_n(x) = e^{ix\cdot\xi_n}u_0(x), \\ e^{it\Delta}u_n(x) = e^{-it|\xi_n|^2}e^{ix\cdot\xi_n}(e^{it\Delta}u_0)(x - 2t\xi_n). \end{cases} \tag{6.8}$$

The second transformation is also known as Galilean boost. To show (6.8) introduce a non-compact group, we argue as follows.

Let $|\xi_n| \to \infty$, the second line in (6.8) implies

$$\|e^{it\Delta}(e^{ix\cdot\xi_n}u_0)\|_{L^4_{t,x}(\mathbb{R}\times\mathbb{R}^2)} = \|e^{it\Delta}u_0\|_{L^4_{t,x}(\mathbb{R}\times\mathbb{R}^2)}.$$

On the other hand, we will show the weak convergence in $L^4_{t,x}(\mathbb{R}\times\mathbb{R}^2)$. To this end, we take $\psi, h \in C_c^\infty(\mathbb{R}\times\mathbb{R}^2)$ and consider

$$\langle e^{it\Delta}(e^{ix\cdot\xi_n}u_0), h\rangle_{(t,x)} = \left\langle e^{ix\cdot\xi_n}u_0, \int_{\mathbb{R}} e^{-it\Delta}h(t)dt\right\rangle_x .$$

Since $\| \int_{\mathbb{R}} e^{it\Delta}h(t)dt\|_2 \lesssim \|h\|_{L^{\frac{4}{3}}_{t,x}(\mathbb{R}\times\mathbb{R}^2)}$ by dual Strichartz estimate, the convergence to 0 follows immediately from the density argument and the fact $|\xi_n| \to \infty$.

Obviously, due to the additional Galilean boost, the linear profile decomposition in L^2 is much more involved than that in $\dot{H}^1$. Next we will discuss the flow dependent linear profile decomposition in this two levels and we start from the simpler $\dot{H}^1$ case.

6.2 Schrödinger flow dependent linear profile decomposition in $\dot{H}^1$

We have the following result.

Theorem 6.7 ($e^{it\Delta} : \dot{H}^1 \to L^{10}_{t,x}$ linear profile decomposition in $d = 3$). *Let $\{f_n\}_{n=1}^\infty$ be a bounded sequence in $\dot{H}^1$. Then there exists a subsequence (which we still denote as $\{f_n\}_{n=1}^\infty$), $J^* \in \{0,1,2,\cdots\}\cup\{\infty\}$, such that $\forall\ 0 \le J \le J^*$, the following hold true:*

1. $f_n = \sum_{j=1}^J g_n^j \varphi^j + w_n^J$ *with* $w_n^J \in \dot{H}^1$ *and* $g_n^j\varphi^j(x) = (\lambda_n^j)^{-\frac{1}{2}}(e^{it_n^j\Delta}\varphi^j)(\frac{x-x_n^j}{\lambda_n^j})$.
2. $\lim_{J\to J^*}\limsup_{n\to\infty}\|e^{it\Delta}w_n^J\|_{L^{10}_{t,x}} = 0$, $(g_n^J)^{-1}w_n^J \rightharpoonup 0$ *weakly in* $\dot{H}^1$.
3. $\forall j \ne k$, g_n^j *and* g_n^k *are orthogonal in the sense that*

$$\lim_{n\to\infty}\left(\frac{\lambda_n^j}{\lambda_n^k} + \frac{\lambda_n^k}{\lambda_n^j} + \frac{1}{\lambda_n^j\lambda_n^k}(|x_n^j - x_n^k|^2 + |(\lambda_n^j)^2 t_n^j - (\lambda_n^k)^2 t_n^k|)\right) = \infty.$$

4. *Decoupling*

$$\lim_{n\to\infty}\Big(\|f_n\|_{\dot{H}^1}^2 - \sum_{j=1}^J \|\varphi^j\|_{\dot{H}^1}^2 - \|w_n^J\|_{\dot{H}^1}^2\Big) = 0,$$

$$\lim_{n\to\infty}\Big(\|f_n\|_6^6 - \sum_{j=1}^J \|\varphi^j\|_6^6 - \|w_n^J\|_6^6\Big) = 0,$$

$$\lim_{n\to\infty}\Big(\|e^{it\Delta}f_n\|_{10,10}^{10} - \sum_{j=1}^J \|e^{it\Delta}\varphi^j\|_{10,10}^{10} - \|e^{it\Delta}w_n^J\|_{10,10}^{10}\Big) = 0.$$

Remark 6.8. *This result was first obtained by Keraani in [23]. Kenig-Merle first realized the full strength of this result and applied it in their ground breaking work on the focusing energy critical NLS [22]. The argument of the proof, known as "concentration compactness" is very much inspiring for later work on critical dispersive equations.*

We start with the following very important estimate which plays the same role as Lemma 5.7.

Lemma 6.9 (Generalized Strichartz inequality). *Let $f \in \dot{H}^1(\mathbb{R}^3)$. Then we have*

$$\|e^{it\Delta}f\|_{10,10} \lesssim \|f\|_{\dot{H}^1}^{\frac{1}{3}}\big(\sup_{N\in 2^{\mathbb{Z}}} N^{-\frac{1}{2}}\|e^{it\Delta}P_N f\|_{\infty,\infty}\big)^{\frac{2}{3}}. \tag{6.9}$$

Proof. Applying square function estimate we have

$$\begin{aligned}
&\|e^{it\Delta}f\|_{10,10}^{10}\\
\approx\ & \|(\sum_{N\in 2^{\mathbb{Z}}}|e^{it\Delta}P_N f|^2)^{\frac{1}{2}}\|_{10,10}^{10}\\
=\ & \int_{\mathbb{R}\times\mathbb{R}^3}\big(\sum_{N\in 2^{\mathbb{Z}}}|e^{it\Delta}P_N f|^2\big)^5 dxdt,\\
\lesssim\ & \sum_{N_1\geq N_2\geq\cdots\geq N_5}\int_{\mathbb{R}\times\mathbb{R}^3}|e^{it\Delta}P_{N_1}f|^2|e^{it\Delta}P_{N_2}f|^2\cdots|e^{it\Delta}P_{N_5}f|^2dxdt\\
\lesssim\ & \sum_{N_1\geq N_2\geq\cdots\geq N_5}\|e^{it\Delta}P_{N_1}f\|_{\frac{10}{3},\frac{10}{3}}^2\|e^{it\Delta}P_{N_2}f\|_{\frac{10}{3},\frac{10}{3}}^{\frac{4}{3}}\\
&\cdot\|e^{it\Delta}P_{N_2}f\|_{\infty,\infty}^{\frac{2}{3}}\|e^{it\Delta}P_{N_3}f\|_{\infty,\infty}^2\cdots\|e^{it\Delta}P_{N_5}f\|_{\infty,\infty}^2\\
\lesssim\ & \sum_{N_1\geq N_2\geq\cdots\geq N_5}\|P_{N_1}f\|_2^2\|P_{N_2}f\|_2^{\frac{4}{3}}N_2^{\frac{1}{3}}N_3N_4N_5\big(\sup_{N\in 2^{\mathbb{Z}}}N^{-\frac{1}{2}}\|e^{it\Delta}P_Nf\|_{\infty,\infty}\big)^{\frac{20}{3}}\\
\lesssim\ & \sum_{N_1\geq N_2}N_2^{\frac{10}{3}}\|P_{N_1}f\|_2^2\|P_{N_2}f\|_2^{\frac{4}{3}}\cdot\big(\sup_{N\in 2^{\mathbb{Z}}}N^{-\frac{1}{2}}\|e^{it\Delta}P_Nf\|_{\infty,\infty}\big)^{\frac{20}{3}}\\
\lesssim\ & \big(\sum_{N_1\geq N_2}N_2^2\|P_{N_1}f\|_2^2\big)\|f\|_{\dot{H}^1}^{\frac{4}{3}}\big(\sup_{N\in 2^{\mathbb{Z}}}N^{-\frac{1}{2}}\|e^{it\Delta}P_Nf\|_{\infty,\infty}\big)^{\frac{20}{3}}\\
\lesssim\ & \|f\|_{\dot{H}^1}^{\frac{10}{3}}\big(\sup_{N\in 2^{\mathbb{Z}}}N^{-\frac{1}{2}}\|e^{it\Delta}P_Nf\|_{\infty,\infty}\big)^{\frac{20}{3}}.
\end{aligned}$$

(6.9) follows immediately. The Lemma is proved. □

Based on this crucial estimate, we have the following result which tells how to take the first bubble.

Lemma 6.10 (Inverse Strichartz inequality). *Let $\{f_n\}_{n=1}^{\infty}$ be a bounded sequence in $\dot{H}^1(\mathbb{R}^3)$ such that*

$$\lim_{n\to\infty}\|f_n\|_{\dot{H}^1} = A,\ \lim_{n\to\infty}\|e^{it\Delta}f_n\|_{10,10} = \varepsilon > 0.$$

Then there exists a subsequence in n, $(\lambda_n, x_n, t_n) \in \mathbb{R}^+\times\mathbb{R}^3\times\mathbb{R}$ and $\varphi \in \dot{H}^1(\mathbb{R}^3)$ such that

1. $f_n = g_n\varphi + w_n$,
2. $g_n^{-1}w_n \rightharpoonup 0$ *weakly in* $\dot{H}^1$,

3. *Decoupling:*

$$\lim_{n\to\infty}(\|f_n\|_{\dot H^1}^2 - \|\varphi\|_{\dot H^1}^2 - \|w_n\|_{\dot H^1}^2) = 0,$$
$$\lim_{n\to\infty}(\|e^{it\Delta}f_n\|_{10,10}^{10} - \|e^{it\Delta}\varphi\|_{10,10}^{10} - \|e^{it\Delta}w_n\|_{10,10}^{10}) = 0.$$

In particular

$$\|e^{it\Delta}w_n\|_{10,10}^{10} \le \varepsilon^{10}\Big(1 - c\Big(\frac{\varepsilon}{A}\Big)^5\Big).$$

Proof. Applying Lemma 6.9 to f_n we obtain

$$\varepsilon \le A^{\frac13} \sup_{N\in 2^{\mathbb{Z}}} (N^{-\frac12}\|e^{it\Delta}P_N f_n\|_{\infty,\infty})^{\frac23}.$$

Thus there exist $N_n \in 2^{\mathbb{Z}}, s_n, x_n$ such that

$$N_n^{-\frac12}|(e^{is_n\Delta}P_{N_n}f_n)(x_n)| \gtrsim \Big(\frac{\varepsilon}{A^{\frac13}}\Big)^{\frac32}. \tag{6.10}$$

We first simplify the right hand side of (6.10), with the goal of putting all the symmetries on f.

First it is not hard to verify

$$P_{N_n}g(x) = P_1\Big(g\Big(\frac{\cdot}{N_n}\Big)\Big)(N_n x).$$

From this we can write

$$\begin{aligned} e^{is_n\Delta}P_{N_n}f_n(x_n) &= (P_{N_n}e^{is_n\Delta}f_n)(x_n) \\ &= P_1\Big[(e^{is_n\Delta}f_n)\Big(\frac{\cdot}{N_n}\Big)\Big](N_n x_n) \\ &= P_1\Big[(e^{is_n\Delta}f_n)\Big(\frac{\cdot}{N_n} + x_n\Big)\Big](0). \end{aligned}$$

This together with (6.10) gives

$$\begin{aligned} \Big(\frac{\varepsilon}{A^{\frac13}}\Big)^{\frac32} &\le N_n^{-\frac12}|e^{is_n\Delta}P_{N_n}f_n(x_n)| \\ &= \Big|P_1\Big[N_n^{-\frac12}(e^{is_n\Delta}f_n)\Big(\frac{\cdot}{N_n} + x_n\Big)\Big](0)\Big| \\ &= \Big|P_1\Big[e^{is_nN_n^2\Delta}\Big(N_n^{-\frac12}f_n\Big(\frac{\cdot}{N_n} + x_n\Big)\Big)\Big](0)\Big|. \end{aligned} \tag{6.11}$$

Defining

$$t_n = -s_nN_n^2,\ \lambda_n = N_n^{-1},$$

we can write

$$e^{is_n N_n^2 \Delta}\left(N_n^{-\frac{1}{2}} f_n\left(\frac{\cdot}{N_n} + x_n\right)\right) = e^{-it_n\Delta}(\lambda_n^{\frac{1}{2}} f_n(\lambda_n \cdot + x_n)) = g_n^{-1} f_n.$$

As g_n^{-1} preserve the $\dot{H}^1$-norm, $\|g_n^{-1} f_n\|_{\dot{H}^1} \to A$, passing to a subsequence, we can extract a weak limit

$$\lim_{n\to\infty} g_n^{-1} f_n = \varphi \text{ weakly in } \dot{H}^1. \tag{6.12}$$

On the other hand, (6.11) implies

$$\begin{aligned}\left(\frac{\varepsilon}{A^{\frac{1}{3}}}\right)^{\frac{3}{2}} &\lesssim \left|\int g_n^{-1} f_n(x)\check{\psi}(x)dx\right| \\ &\to \left|\int \varphi(x)\check{\psi}(x)dx\right| \\ &\le \|\varphi\|_6 \|\check{\psi}\|_{\frac{6}{5}}.\end{aligned} \tag{6.13}$$

Hence,

$$\|\varphi\|_{\dot{H}^1} \gtrsim \|\varphi\|_6 \gtrsim \left(\frac{\varepsilon}{A^{\frac{1}{3}}}\right)^{\frac{3}{2}}.$$

Moreover, from (6.13), for $\forall\ t \in \mathbb{R}$, we have

$$\begin{aligned}\left(\frac{\varepsilon}{A^{\frac{1}{3}}}\right)^{\frac{3}{2}} &\lesssim |\langle\varphi, \check{\psi}\rangle| = |\langle e^{it\Delta}\varphi, e^{it\Delta}\check{\psi}\rangle| \\ &\lesssim \|e^{it\Delta}\varphi\|_{L_x^{10}} \|e^{it\Delta}\check{\psi}\|_{L_x^{\frac{10}{9}}}.\end{aligned} \tag{6.14}$$

From Mihlin-Hömander multiplier theorem, it is easy to verify that

$$\|e^{it\Delta}\check{\psi}\|_{L_x^{\frac{10}{9}}} \lesssim_{|t|} \|\check{\psi}\|_{L_x^{\frac{10}{9}}}.$$

For $|t| \le 1$, this estimate is uniform in t. Integrating (6.14) over $|t| \le 1$, we get

$$\left(\frac{\varepsilon}{A^{\frac{1}{3}}}\right) \lesssim \|e^{it\Delta}\varphi\|_{L_{t,x}^{10}([-1,1]\times\mathbb{R}^3)} \lesssim \|e^{it\Delta}\varphi\|_{10,10}. \tag{6.15}$$

Finally, we prove the decoupling in $\dot{H}^1$, L^6 and $L_{t,x}^{10}$. We define

$$w_n = f_n - g_n\varphi$$

i.e.

$$f_n = g_n\varphi + w_n, \text{ or } g_n^{-1} f_n = \varphi + g_n^{-1} w_n. \tag{6.16}$$

The $\dot{H}^1$ decoupling is a direct consequence of weak convergence $g_n^{-1} w_n \rightharpoonup 0$ in $\dot{H}^1$. The L^6-*decoupling* is an application of refined Fatou (as explained earlier). The $L_{t,x}^{10}$ decoupling is a bit involved. From the fact

$$|\partial_t|^{\frac{1}{2}} e^{it\Delta} f \sim |\nabla| e^{it\Delta} f$$

we know

$$\{e^{it\Delta} g_n^{-1} f_n\}_{n=1}^{\infty}$$

is uniformly bounded in $\dot{H}^{\frac{1}{2}}_{t,x}$. Thus by Rellich theorem, on any compact set in $\mathbb{R} \times \mathbb{R}^3$, the embedding into $L^2_{t,x}$ is compact, therefore passing to a subsequence if necessary,

$$e^{it\Delta} g_n^{-1} f_n \to e^{it\Delta}\varphi, \text{ a.e.}$$

On the other hand

$$\begin{aligned}\|e^{it\Delta} g_n^{-1} f_n\|_{10,10} &= \|g_n^{-1} e^{it\lambda_n^2 \Delta} f_n\|_{10,10} \\ &= \|e^{it\Delta} f_n\|_{10,10} \lesssim \|f_n\|_{\dot{H}^1} \lesssim A.\end{aligned}$$

Applying refined Fatou again yields the decoupling in $L^{10}_{t,x}$. □

We now sketch the proof of Theorem 6.7.

Proof. The proof of Theorem 6.7 is an inductive application of inverse Strichartz inequality Lemma 6.10 as explained in the proof of Theorem 5.6. The only part that needs attention is proving the orthogonality for distinct pairs. Therefore we will only give the crucial ingredients and leave the details to interested readers.

The proof of orthogonality relies on the following two statements:

Claim 1: If g_n^j and g_n^l are not orthogonal, then $(g_n^j)^{-1} g_n^l$ is compact group in $\dot{H}^1$.

Claim 2: If g_n^j and g_n^l are orthogonal, then as $n \to \infty$,

$$(g_n^j)^{-1} g_n^l f \rightharpoonup 0, \text{ weakly in } \dot{H}^1. \tag{6.17}$$

We only prove Claim 2, which by density, can be reduced to showing for $\forall \varphi, \psi \in C_c^\infty(\mathbb{R}^3)$, that

$$\langle (g_n^j)^{-1} g_n^l \varphi, \psi \rangle_{\dot{H}^1} \to 0 \text{ as } n \to \infty. \tag{6.18}$$

Here from definition we can write

$$(g_n^j)^{-1} g_n^l \varphi(x) = \left(\frac{\lambda_n^j}{\lambda_n^l}\right)^{\frac{1}{2}} \left[e^{i(t_n^l - t_n^j (\frac{\lambda_n^j}{\lambda_n^l})^2)\Delta} \varphi\right] \left(\frac{\lambda_n^j x + x_n^j - x_n^l}{\lambda_n^l}\right).$$

First we assume

$$\frac{\lambda_n^j}{\lambda_n^l} + \frac{\lambda_n^l}{\lambda_n^j} \to \infty \text{ as } n \to \infty.$$

Then

$$\begin{aligned}|RHS(6.18)| &\le \min(\|\Delta\psi\|_2 \|(g_n^j)^{-1} g_n^l \varphi\|_2,\ \|\psi\|_2 \|\Delta (g_n^j)^{-1} g_n^l \varphi\|_2) \\ &\lesssim \min\left(\|\Delta\psi\|_2 \|\varphi\|_2 \frac{\lambda_n^l}{\lambda_n^j}, \|\psi\|_2 \|\Delta\varphi\|_2 \frac{\lambda_n^j}{\lambda_n^l}\right) \\ &\to 0 \text{ as } n \to \infty.\end{aligned}$$

Thus we can assume

$$\frac{\lambda_n^j}{\lambda_n^l} \to \lambda_0 \in (0, \infty).$$

If $\frac{1}{\lambda_n^j \lambda_n^l}|(\lambda_n^j)^2 t_n^j - (\lambda_n^l) t_n^l| \to \infty$, Then

$$\left| t_n^l - t_n^j \left(\frac{\lambda_n^j}{\lambda_n^l}\right)^2 \right| = \left| \frac{(\lambda_n^l)^2 t_n^l - (\lambda_n^j)^2 t_n^j}{\lambda_n^j \lambda_n^l} \frac{\lambda_n^j}{\lambda_n^l} \right| \to \infty,$$

$$\begin{aligned} |RHS(6.18)| &\lesssim \|(g_n^j)^{-1} g_n^l \varphi\|_6 \|\Delta\psi\|_{\frac{6}{5}} \\ &\lesssim \|e^{i\left(t_n^l - t_n^j (\frac{\lambda_n^j}{\lambda_n^l})^2\right)\Delta} \varphi\|_6 \|\Delta\psi\|_{\frac{6}{5}} \\ &\lesssim \left| t_n^l - t_n^j \left(\frac{\lambda_n^j}{\lambda_n^l}\right)^2 \right|^{-1} \|\varphi\|_{\frac{6}{5}} \|\Delta\psi\|_{\frac{6}{5}} \to 0. \end{aligned}$$

The only scenario left is

$$\begin{cases} \frac{\lambda_n^j}{\lambda_n^l} \to \lambda_0, \text{ and } \frac{1}{\lambda_n^j \lambda_n^l}|x_n^j - x_n^l|^2 \to \infty, \\ \frac{1}{\lambda_n^j \lambda_n^l}\left((\lambda_n^l)^2 t_n^l - (\lambda_n^j)^2 t_n^j\right) \to t_0. \end{cases} \tag{6.19}$$

(6.18) is then reduced to showing

$$\left\langle \lambda_0^{\frac{1}{2}} e^{i\lambda_0 t_0 \Delta} \varphi\left(\lambda_0 x + \frac{x_n^j - x_n^l}{\lambda_n^l}\right), \psi \right\rangle_{\dot{H}^1} \to 0 \tag{6.20}$$

under the condition (6.19). As $|\frac{x_n^j - x_n^l}{\lambda_n^l}|^2 = \frac{|x_n^j - x_n^l|^2}{\lambda_n^j \lambda_n^l} \frac{\lambda_n^j}{\lambda_n^l} \to \infty$, (6.20) follows easily from the non-overlapping of the spatial supports. □

6.3 Schrödinger flow dependent linear profile decomposition in L^2—The basics

The linear profile decomposition associated with the L^2-level Strichartz inequality is much more complicated due to the additional degree of freedom. In this lecture, to minimize the technical complexity, we will only discuss the proof with the radial assumption.

Theorem 6.11 (Linear profile decomposition related to $\|e^{it\Delta}u_0\|_{L_{t,x}^{\frac{2(d+2)}{d}}(\mathbb{R}\times\mathbb{R}^d)} \lesssim \|u_0\|_2$). *Let $\{f_n\}_{n=1}^\infty$ be a radial bounded sequence in $L^2(\mathbb{R}^d)$. Then there exist a subsequence, $J^* \in \{0, 1, \cdots\} \cup \{\infty\}$, $\{\varphi^j\}_{j=1}^{J^*} \subset L^2$ and $\{(\lambda_n^j, t_n^j)\}_{j=1}^{J^*} \subset \mathbb{R}^+ \times \mathbb{R}$ such that $\forall\ 0 \le J \le J^*$,*

1. $f_n = \sum_{j=1}^J g_n^j \varphi^j + w_n^J$, $g_n^j \varphi^j(x) = (\lambda_n^j)^{-\frac{d}{2}}(e^{it_n^j\Delta}\varphi^j)(\frac{x}{\lambda_n^j})$.
2. $w_n^J \in L^2$ *and*

$$\begin{aligned} &(g_n^J)^{-1} w_n^J \rightharpoonup 0 \text{ weakly in } L^2, \\ &\lim_{J\to J^*} \overline{\lim_{n\to\infty}} \|e^{it\Delta} w_n^J\|_{\frac{2(d+2)}{d}, \frac{2(d+2)}{d}} = 0. \end{aligned}$$

3. *Orthogonality condition:*

$$\lim_{n\to\infty}\left(\frac{\lambda_n^j}{\lambda_n^k}+\frac{\lambda_n^k}{\lambda_n^j}+\frac{1}{\lambda_n^j\lambda_n^k}|(\lambda_n^j)^2t_n^j-(\lambda_n^k)^2t_n^k|\right)=\infty, \forall j\neq k.$$

4. *Decoupling:*

$$\lim_{n\to\infty}(\|f_n\|_2^2-\sum_{j=1}^{J}\|\varphi^j\|_2^2-\|w_n^J\|_2^2)=0.$$

With the radial assumption, one does not need to consider how to localize the function in both physical and frequency spaces. However, the difficulty still remains as the L^2 level Strichartz estimate is the one with the lowest regularity, one can not interpolate like in the $\dot{H}^1$-*case*. The crucial step then, is to establish a break scaling estimate, which is direct in the radial case. We start with the following smoothing estimate (see [12, 36, 46] for general discussion on smoothing estimate).

Lemma 6.12 (Radial Smoothing). *Let $f\in L^2(\mathbb{R}^d)$ be spherically symmetric. Then*

$$\||x|^{\frac{d-1}{2}}|\nabla|^{\frac{1}{2}}e^{it\Delta}f\|_{L_t^2(\mathbb{R})}\lesssim\|f\|_{L_x^2(\mathbb{R}^d)} \tag{6.21}$$

uniformly in x.

Remark 6.13. *This estimate is not the required break-scaling estimate, as the norm on the LHS scales the same way as L^2. The reason why it is called smoothing estimate is that, normally, the linear flow does not gain smoothness in L^2 space. This estimate says for a.e. t, the linear flow is smoother.*

Proof. The proof is an application of Plancherel in time. We start with the following fact

$$\left\|\int_0^\infty e^{-it\rho^2}h(\rho)d\rho\right\|_{L_t^2}^2=\frac{1}{2}\int_0^\infty|h(\rho)|^2\frac{d\rho}{\rho}.$$

Indeed,

$$\begin{aligned}\left\|\int_0^\infty e^{-it\rho^2}h(\rho)d\rho\right\|_{L_t^2}&=\left\|\int_0^\infty e^{-it\eta}h(\sqrt{\eta})\frac{1}{2}\eta^{-\frac{1}{2}}d\eta\right\|_{L_t^2}\\&=\frac{1}{2}\left(\int_0^\infty|h(\sqrt{\eta})\eta^{-\frac{1}{2}}|^2d\eta\right)^{\frac{1}{2}}\\&=\frac{\sqrt{2}}{2}\left(\int_0^\infty|h(\rho)|^2\frac{d\rho}{\rho}\right)^{\frac{1}{2}}.\end{aligned}$$

Now

$$\begin{aligned}\||x|^{\frac{d-1}{2}}|\nabla|^{\frac{1}{2}}e^{it\Delta}f\|_{L_t^2}^2 &= \left\||x|^{\frac{d-1}{2}}\int_{\mathbb{R}^d} e^{ix\cdot\xi-it|\xi|^2}\hat{f}(\xi)|\xi|^{\frac{1}{2}}d\xi\right\|_{L_t^2}^2 \\ &= \left\||x|^{\frac{d-1}{2}}\int_0^\infty e^{-it\rho^2}\int_{S^{d-1}} e^{i|x|\rho w_1}d\sigma(w)\hat{f}(\rho)\rho^{d-\frac{1}{2}}d\rho\right\|_{L_t^2}^2 \\ &= \frac{1}{2}|x|^{d-1}\int_0^\infty\left(\int_{S^{d-1}} e^{i|x|\rho w_1}d\sigma(w)\right)^2|\hat{f}(\rho)|^2\rho^{2d-1}\frac{d\rho}{\rho} \\ &\lesssim \int_0^\infty |\hat{f}(\rho)|^2\rho^{d-1}d\rho \approx \|f\|_2^2.\end{aligned}$$

□

Now we introduce the refined version of the above estimate.

Lemma 6.14 (Frequency localized smoothing estimate). *Let $f\in L^2(\mathbb{R}^d)$ be radial and $\frac{1}{2}<\alpha<\frac{d}{2}$. Then*

$$\||x|^{-\alpha}P_{\sim 1}e^{it\Delta}f\|_{L^2_{t,x}(\mathbb{R}\times\mathbb{R}^d)}\lesssim\|f\|_{L^2(\mathbb{R}^d)}.$$

Remark 6.15. *If there is no frequency projection, the correct power of α should be 1. The fact that α can go below 1 render this estimate breaks the scaling.*

Proof.

$$|x|^{-\alpha}P_{\sim 1}e^{it\Delta}f(x)=|x|^{-\alpha}\int_{\mathbb{R}^d} e^{ix\cdot\xi-it|\xi|^2}\hat{f}(\xi)\psi(\xi)d\xi,$$

here $\psi(\xi)$ is a smooth cutoff on the annulus $|\xi|\sim 1$. Following the same principle as the last lemma, we write,

$$\begin{aligned}&\||x|^{-\alpha}P_{\sim 1}e^{it\Delta}f\|_{L^2_{t,x}(\mathbb{R}\times\mathbb{R}^d)}^2 \\ &= \left\||x|^{-\alpha}\int_0^\infty e^{-it\rho^2}\int_{S^{d-1}} e^{i|x|\rho w_1}d\sigma(\omega)\hat{f}(\rho)\psi(\rho)\rho^{d-1}d\rho\right\|_{L^2_{x,t}}^2 \\ &\approx \int_{\mathbb{R}^d}|x|^{-2\alpha}\int_0^\infty\left(\int_{S^{d-1}} e^{i|x|\rho w_1}d\sigma(\omega)\right)^2|\hat{f}(\rho)|^2\psi^2(\rho)\rho^{2(d-1)}\frac{d\rho}{\rho}dx \\ &\lesssim \int_0^\infty|\hat{f}(\rho)|^2\rho^{2(d-1)}\psi^2(\rho)\int_{\mathbb{R}^d}|x|^{-2\alpha}\langle|x|\rho\rangle^{-(d-1)}dx\frac{d\rho}{\rho}.\end{aligned}$$

For $\frac{1}{2}<\alpha<\frac{d}{2}$, the integration in x is bounded. Note also that $\rho\sim 1$, we immediately have

$$\||x|^{-\alpha}P_{\sim 1}e^{it\Delta}f\|_{L^2_{t,x}(\mathbb{R}\times\mathbb{R}^d)}^2\lesssim\int_0^\infty|\hat{f}(\rho)|^2\rho^{d-1}d\rho\approx\|f\|_2^2.$$

□

To remove the negative power in $|x|$, we need another piece of estimate, an estimate with positive power of $|x|$. Interpolating between these two gives the correct estimate. Here we used the form appeared in [27, 28].

Lemma 6.16 (Weighted radial Strichartz). *Let* $F(t,x) \in L_{t,x}^{\frac{2(d+2)}{d+4}}(\mathbb{R}\times\mathbb{R}^d)$, $u_0 \in L^2(\mathbb{R}^d)$ *be radial. Then*

$$u(t) = e^{it\Delta}u_0 - i\int_0^t e^{i(t-s)\Delta}F(s)ds$$

satisfies

$$\||x|^{\frac{2(d-1)}{q}}u\|_{L_t^q L_x^{\frac{2q}{q-4}}} \lesssim \|u_0\|_{L_x^2} + \|F\|_{L_{t,x}^{\frac{2(d+2)}{d+4}}(\mathbb{R}\times\mathbb{R}^d)}$$

for all $4 \le q \le \infty$.

Proof. We first prove the homogeneous estimate

$$\||x|^{\frac{2(d-1)}{q}}e^{it\Delta}u_0\|_{q,\frac{2q}{q-4}} \lesssim \|u_0\|_2. \tag{6.22}$$

In the case $q=\infty$, this is the usual Strichartz

$$\|e^{it\Delta}u_0\|_{\infty,2} \lesssim \|u_0\|_2.$$

We only need to prove the other endpoint case:

$$\||x|^{\frac{d-1}{2}}e^{it\Delta}u_0\|_{4,\infty} \lesssim \|u_0\|_2. \tag{6.23}$$

(6.22) then follows by interpolating between these two.

From duality, (6.23) is equivalent to show

$$\left\|\int_{\mathbb{R}} e^{-it\Delta}|x|^{\frac{d-1}{2}}f(s)ds\right\|_2 \lesssim \|f\|_{\frac{4}{3},1}. \tag{6.24}$$

The

$$\begin{aligned}(LHS(6.24))^2 &= \left\langle \int_{\mathbb{R}} e^{-it\Delta}|x|^{\frac{d-1}{2}}f(t)dt, \int_{\mathbb{R}} e^{-is\Delta}|x|^{\frac{d-1}{2}}f(s)ds\right\rangle_x \\ &= \int_{\mathbb{R}}\left\langle f(t), \int_{\mathbb{R}}|x|^{\frac{d-1}{2}}e^{i(t-s)\Delta}|x|^{\frac{d-1}{2}}ds\right\rangle_x dt \\ &\lesssim \|f\|_{\frac{4}{3},1}\||x|^{\frac{d-1}{2}}\int_{\mathbb{R}} e^{i(t-s)\Delta}|x|^{\frac{d-1}{2}}f(s)ds\|_{4,\infty}.\end{aligned}$$

To prove (6.24), it suffices to show

$$\left\||x|^{\frac{d-1}{2}}\int_{\mathbb{R}} e^{i(t-s)\Delta}|x|^{\frac{d-1}{2}}f(s)ds\right\|_{4,\infty} \lesssim \|f\|_{\frac{4}{3},1}$$

which holds provided the weighted dispersive estimate

$$\||x|^{\frac{d-1}{2}}e^{it\Delta}|x|^{\frac{d-1}{2}}\varphi\|_{L_x^\infty} \lesssim |t|^{-\frac{1}{2}}\|\varphi\|_{L_x^1} \tag{6.25}$$

for spherically symmetric function φ. From fundamental solution of Schrödinger propagator, we can write

$$\begin{aligned} &|x|^{\frac{d-1}{2}} e^{it\Delta} |x|^{\frac{d-1}{2}} \varphi \\ &\sim |t|^{-\frac{d}{2}} |x|^{\frac{d-1}{2}} \int e^{\frac{i|x-y|^2}{4t}} |y|^{\frac{d-1}{2}} \varphi(y) dy \\ &\sim |t|^{-\frac{d}{2}} |x|^{\frac{d-1}{2}} \int_{\mathbb{R}^d} e^{\frac{i(|x|^2+|y|^2)}{4t} - \frac{2ix\cdot y}{4t}} |y|^{\frac{d-1}{2}} \varphi(y) dy \\ &\sim |t|^{-\frac{d}{2}} |x|^{\frac{d-1}{2}} \int_0^\infty e^{\frac{i(|x|^2+|y|^2)}{4t}} \int_{S^{n-1}} e^{-\frac{i|x|\rho w}{2t}} d\sigma(w) \varphi(\rho) \rho^{\frac{d-1}{2}} \rho^{d-1} d\rho. \end{aligned} \tag{6.26}$$

Using the stationary phase estimate:

$$\left| \int_{S^{d-1}} e^{-\frac{i|x|\rho w}{2t}} d\sigma(w) \right| \lesssim \left(\frac{|x|\rho}{t} \right)^{-\frac{d-1}{2}},$$

we continue estimate (6.26) as

$$RHS(6.25) \lesssim |t|^{-\frac{1}{2}} \int_0^\infty |\varphi(\rho)| \rho^{d-1} d\rho \lesssim |t|^{-\frac{1}{2}} \|\varphi\|_{L^1_x}$$

as desired.

Once (6.22) is proved, the estimate for inhomogeneous part comes from Christ-Kislev Lemma and (6.22). The Lemma is proved. □

As a consequence, we have

Corollary 6.17. *Let $u_0 \in L^2(\mathbb{R}^d)$ be spherically symmetric. Then*

$$\| |x|^{\frac{d-1}{3}} e^{it\Delta} u_0 \|_{6,6} \lesssim \|u_0\|_2.$$

With both of the two crucial estimates Lemma 6.14 and Corollary 6.17, we will be able to estimate the linear flow in $L^q_{t,x}$ space, with q below the borderline $\frac{2(d+2)}{d}$. Again, to keep the presentation with minimal complexity, we will choose to work in $d = 2$ and give a complete proof of linear profile decomposition in this dimension. The argument can be generalized to all dimensions without difficulty. And we will leave the extension to the interested readers.

In $d = 2$, the local smoothing implies

$$\| |x|^{-\alpha} P_{\sim 1} e^{it\Delta} u_0 \|_{2,2} \lesssim \|u_0\|_2.$$

This together with weighted Strichartz

$$\| |x|^{\frac{1}{3}} e^{it\Delta} u_0 \|_{6,6} \lesssim \|u_0\|_2$$

implies that

$$\|P_{\sim 1} e^{it\Delta} u_0\|_{L_{t,x}^{\frac{2(1+3\alpha)}{\alpha+1}}} \lesssim \|u_0\|_2.$$

$\frac{2(1+3\alpha)}{\alpha+1}$ is increasing in α, the smallest index is $\frac{10}{3}$, corresponding to taking $\alpha = \frac{1}{2}$, when $\frac{2(1+3\alpha)}{\alpha+1} > 4$, the estimate can be proved by usual Strichartz and Bernstein, the most remarkable point is that this estimate falls below the admissible scaling. To summarize, we have

Lemma 6.18. *$u_0 \in L^2_{rad}(\mathbb{R}^2)$, $\frac{10}{3} < q \leq \infty$, then*

$$\|P_{\sim 1} e^{it\Delta} u_0\|_{q,q} \lesssim \|u_0\|_2,$$

equiavilently by scaling, $\forall N \in 2^{\mathbb{Z}}$,

$$\|P_{\sim N} e^{it\Delta} u_0\|_{q,q} \lesssim N^{1-\frac{4}{q}} \|u_0\|_2. \tag{6.27}$$

In particular,

$$\|P_{\sim N} e^{it\Delta} u_0\|_{\frac{11}{3},\frac{11}{3}} \lesssim N^{-\frac{1}{11}} \|u_0\|_2. \tag{6.28}$$

Now with the estimate (6.28), we are ready to prove the following generalized Strichartz inequality.

Lemma 6.19. *Let $f \in L^2_{rad}(\mathbb{R}^2)$. Then*

$$\|e^{it\Delta} f\|_{L^4_{t,x}} \lesssim \|f\|_2^{\frac{3}{4}} \big(\sup_{N \in 2^{\mathbb{Z}}} \|e^{it\Delta} P_N f\|_{L^{\frac{22}{5}}_{t,x}} N^{-\frac{1}{11}} \big)^{\frac{1}{4}}. \tag{6.29}$$

Proof. Using the square function estimate, we write

$$\begin{aligned}
&\|e^{it\Delta} f\|^4_{L^4_{t,x}} \\
&\lesssim \int_{\mathbb{R}\times\mathbb{R}^2} \Big(\sum_{N \in 2^{\mathbb{Z}}} |e^{it\Delta} P_N f|^2 \Big)^2 dxdt \\
&\lesssim \sum_{N_1 \geq N_2} \int_{\mathbb{R}\times\mathbb{R}^2} |e^{it\Delta} P_{N_1} f|^2 |e^{it\Delta} P_{N_2} f|^2 dxdt \\
&\lesssim \sum_{N_1 \geq N_2} \|e^{it\Delta} P_{N_1} f\|_{\frac{11}{3},\frac{11}{3}} \|e^{it\Delta} P_{N_1} f\|_{4,4} \|e^{it\Delta} P_{N_2} f\|_{4,4} \|e^{it\Delta} P_{N_2} f\|_{\frac{22}{5},\frac{22}{5}} \\
&\lesssim \sum_{N_1 \geq N_2} N_1^{-\frac{1}{11}} \|f\|_2 \|P_{N_1} f\|_2 \|P_{N_2} f\|_2 N_2^{\frac{1}{11}} \sup_{N \in 2^{\mathbb{Z}}} N^{-\frac{1}{11}} \|e^{it\Delta} P_N f\|_{\frac{22}{5},\frac{22}{5}} \\
&\lesssim \sum_{N_1 \geq N_2} \Big(\frac{N_2}{N_1}\Big)^{\frac{1}{11}} \|P_{N_1} f\|_2 \|P_{N_2} f\|_2 \|f\|_2 \sup_{N \in 2^{\mathbb{Z}}} N^{-\frac{1}{11}} \|e^{it\Delta} P_N f\|_{\frac{22}{5},\frac{22}{5}} \\
&\lesssim \sum_{N_1 \geq N_2} \Big(\frac{N_2}{N_1}\Big)^{\frac{1}{11}} (\|P_{N_1} f\|_2^2 + \|P_{N_2} f\|_2^2) \|f\|_2 \sup_{N \in 2^{\mathbb{Z}}} N^{-\frac{1}{11}} \|e^{it\Delta} P_N f\|_{\frac{22}{5},\frac{22}{5}} \\
&\lesssim \|f\|_2^3 \sup_{N \in 2^{\mathbb{Z}}} N^{-\frac{1}{11}} \|e^{it\Delta} P_N f\|_{\frac{22}{5},\frac{22}{5}}.
\end{aligned}$$

Power $\frac{1}{4}$ on both sides yields the final result. □

Using interpolation and usual Strichartz inequality, one controls

$$\begin{aligned}\|e^{it\Delta}P_N f\|_{L^{\frac{22}{5}}_{t,x}} &\lesssim \|e^{it\Delta}P_N f\|^{\frac{10}{11}}_{L^4_{t,x}} \|e^{it\Delta}P_N f\|^{\frac{1}{11}}_{L^\infty_{t,x}} \\ &\lesssim \|f\|_2^{\frac{10}{11}} \|e^{it\Delta}P_N f\|^{\frac{1}{11}}_{L^\infty_{t,x}}.\end{aligned}$$

Thus (6.29) can be improved as

$$\|e^{it\Delta}f\|_{L^4_{t,x}} \lesssim \|f\|_2^{\frac{43}{44}} \big(\sup_{N\in 2^{\mathbb{Z}}} \|e^{it\Delta}P_N f\|_{L^\infty_{t,x}} N^{-1}\big)^{\frac{1}{44}}. \tag{6.30}$$

Based on (6.30), we have the following

Lemma 6.20 (Inverse Strichartz inequality). *Let $\{f_n\}_{n=1}^\infty \subset L^2_{rad}(\mathbb{R}^2)$ be bounded sequence satisfying*

$$\begin{aligned}&\lim_{n\to\infty} \|f_n\|_2 = A,\\ &\lim_{n\to\infty} \|e^{it\Delta}f_n\|_{4,4} = \varepsilon > 0.\end{aligned}$$

Then there exists a subsequence, $\{(t_n,\lambda_n)\}_{n=1}^\infty \subset \mathbb{R}\times\mathbb{R}^+, \varphi \in L^2$ such that

1. *Decomposition: there exists $w_n \in L^2$ such that*

$$\begin{aligned}&f_n(x) = g_n\varphi(x) + w_n(x),\\ &g_n\varphi(x) = \lambda_n^{-1}(e^{it\Delta}\varphi)\Big(\frac{x}{\lambda_n}\Big),\\ &g_n^{-1}w_n \rightharpoonup 0 \text{ weakly in } L^2(\mathbb{R}^2).\end{aligned}$$

2. *Norm decreasing: there exist constant c, C such that*

$$\begin{aligned}&\|w_n\|_2^2 \le A^2\left(1 - c\left(\frac{\varepsilon}{A}\right)^C\right),\\ &\|e^{it\Delta}w_n\|_{4,4}^4 \le \varepsilon^4\left(1 - c\left(\frac{\varepsilon}{A}\right)^C\right).\end{aligned}$$

3. *Norm decoupling:*

$$\begin{aligned}&\lim_{n\to\infty}(\|f_n\|_2^2 - \|\varphi\|_2^2 - \|w_n\|_2^2) = 0,\\ &\lim_{n\to\infty}(\|e^{it\Delta}f_n\|_{4,4}^4 - \|e^{it\Delta}\varphi\|_{4,4}^4 - \|e^{it\Delta}w_n\|_{4,4}^4) = 0.\end{aligned}$$

Proof. Applying (6.30) to f_n, we obtain

$$\varepsilon \lesssim A^{\frac{43}{44}}\big(\sup_{N\in 2^{\mathbb{Z}}} \|e^{it\Delta}P_N f_n\|_{\infty,\infty} N^{-1}\big)^{\frac{1}{44}}.$$

Therefore there exist $s_n \in \mathbb{R}, N_n \in 2^{\mathbb{Z}}, x_n \in \mathbb{R}^2$ such that

$$N_n^{-1}|e^{is_n\Delta}P_{N_n}f_n(x_n)| \gtrsim \frac{\varepsilon^{44}}{A^{43}} \tag{6.31}$$

or

$$N_n^{-1}\left|P_1\left(e^{is_n\Delta}f_n\left(\frac{\cdot}{N_n}\right)\right)(N_nx_n)\right| \gtrsim \frac{\varepsilon^{44}}{A^{43}}. \qquad (6.32)$$

Here comes a very important step, we will show $|N_nx_n| \lesssim C(A,\epsilon)$ uniformly. To see this, we begin with the decay property for radial functions: for $f \in H^1_{rad}(\mathbb{R}^d)$ [3],

$$|x|^{\frac{d-1}{2}}|f(x)| \lesssim \|f\|_2^{\frac{1}{2}}\|\nabla f\|_2^{\frac{1}{2}}. \qquad (6.33)$$

Applying (6.33) to $e^{is_n\Delta}P_{N_n}f_n$, we obtain

$$|x_n|^{\frac{1}{2}}|(e^{is_n\Delta}P_{N_n}f_n)(x_n)| \lesssim AN_n^{\frac{1}{2}} \Leftrightarrow |x_nN_n|^{\frac{1}{2}}|N_n^{-1}(e^{is_n\Delta}P_{N_n}f)(x_n)| \lesssim A.$$

This together with (6.31) implies

$$|x_nN_n|^{\frac{1}{2}} \cdot \frac{\varepsilon^{44}}{A^{43}} \lesssim A \Rightarrow |x_nN_n| \le C(A,\epsilon).$$

Due to the uniform boundedness of $|x_nN_n|$, we can assume, passing to a subsequence that

$$N_nx_n \to x_0.$$

On the other hand, from (6.32) we have

$$\frac{\varepsilon^{44}}{A^{43}} \lesssim |P_1(g_n^{-1}f_n)(N_nx_n)| = \left|\int \psi(N_nx_n+y)g_n^{-1}f_n(y)dy\right|.$$

As $g_n^{-1}f_n$ is uniform bounded in L^2, we have for some $\varphi \in L^2_{rad}$,

$$\lim_{n\to\infty} g_n^{-1}f_n \rightharpoonup \varphi, \quad \text{weakly in } L^2.$$

Therefore,

$$\frac{\varepsilon^{44}}{A^{43}} \lesssim \left|\int \psi(x_0+y)\varphi(y)dy\right| \lesssim \|\psi\|_2\|\varphi\|_2.$$

This implies

$$\|\varphi\|_2 \gtrsim \frac{\varepsilon^{44}}{A^{43}}.$$

Now we can write

$$f_n = g_n\varphi + w_n.$$

The remaining properties follow exactly the same argument as the proof in the $\dot{H}^1$-*level* linear profile decomposition, we omit the steps. □

[3] Indeed, (6.33) is proved simply by integration by parts,

$$r^{d-1}f^2(r) = r^{d-1}\int_r^\infty \frac{d}{d\rho}f^2(\rho)d\rho \lesssim r^{d-1}\int_r^\infty |f(\rho)f'(\rho)|d\rho$$

$$\lesssim \left(\int_r^\infty |f(\rho)|^2\rho^{d-1}d\rho\right)^{\frac{1}{2}}\left(\int_r^\infty |f'(\rho)|^2\rho^{d-1}d\rho\right)^{\frac{1}{2}} \lesssim \|f\|_2^{\frac{1}{2}}\|\nabla f\|_2^{\frac{1}{2}}.$$

Finally, we record the general L^2 linear profile decomposition.

Theorem 6.21. [2, 5, 34] *Let f_n be a bounded sequence in $L^2(\mathbb{R}^d)$. Then there exist a subsequence (which we still denote as f_n), $J^* \in \{0, 1, 2, \cdots\} \cup \{\infty\}$, $\{\phi^j\}_{j=1}^{J^*} \in L^2$, parameters $\{(\lambda_n^j, t_n^j, x_n^j, \xi_n^j)\}_{j=1}^{J^*} \subset \mathbb{R}^+ \times \mathbb{R} \times \mathbb{R}^d \times \mathbb{R}^d$ such that for each $1 \leq J \leq J^*$, $J < \infty$, we have*

- *Decomposition: there exists $w_n^J \in L^2$ such that*

$$f_n = \sum_{j=1}^{J} g_n^j \phi^j + w_n^J,$$

$$g_n^j \phi^j = \lambda_n^j e^{i\xi_n^j \cdot x} e^{it_n^j \Delta} \phi^j \left(\frac{x - x_n^j}{\lambda_n^j} \right).$$

- *Decoupling:*

$$\lim_{n \to \infty} \left(\|f_n\|_2^2 - \sum_{j=1}^{J} \|\phi^j\|_2^2 - \|w_n^J\|_2^2 \right) = 0.$$

- *Vanishing in $L^{\frac{2(d+2)}{d}}$:*

$$\lim_{J \to J^*} \limsup_{n \to \infty} \|e^{it\Delta} w_n^J\|_{L_{t,x}^{\frac{2(d+2)}{d}}} = 0.$$

- *Orthogonality: for each $j \neq k$, g_n^j and g_n^k are orthogonal in the sense that*

$$\lim_{n \to \infty} \left(\frac{\lambda_n^j}{\lambda_n^k} + \frac{\lambda_n^k}{\lambda_n^j} + \lambda_n^j \lambda_n^k |\xi_n^j - \xi_n^k|^2 + \frac{|t_n^j (\lambda_n^j)^2 - t_n^k (\lambda_n^k)^2|}{\lambda_n^j \lambda_n^k} + \frac{|x_n^j - x_n^k - 2t_n^j (\lambda_n^j)^2 (\xi_n^j - \xi_n^k)|^2}{\lambda_n^j \lambda_n^k} \right) = \infty.$$

7 A review on local and global theory of mass critical nonlinear Schrödinger equation

In the past lectures, we briefly introduce several examples of linear profile decomposition. In the next few sections, we will give an example, to show how this technique can be used in critical nonlinear Schrödinger equations. To keep everything in the simplest form, we will use mass critical NLS in sample dimensions with the additional radial assumption as example. To start with, we go through the local and global theory for this problem with general data. And we will discuss $3d$ as a representative case. All the other dimensions can be treated similarly.

We consider the initial value problem

$$\begin{cases} iu_t + \Delta u = \pm |u|^{\frac{4}{3}} u, \\ u(0, x) = u_0(x). \end{cases} \tag{7.1}$$

Here, $u(t, x)$ is a complex valued function on $\mathbb{R} \times \mathbb{R}^3$. In "+" case, the nonlinearity helps the dispersive Δ, making the solution more spread out, and is referred to as "defocusing"; In the "–" case, the nonlinearity is against with the Δ, making the solution more concentrated, thus is called "focusing". The initial data $u_0 \in H^s(\mathbb{R}^3)$, the reason why we choose this space is because the linear propagator $e^{it\Delta}$ generates a group only in Hilbert type Sobolev space H^s.

We understand the solution of (7.1) in the following sense:

Definition 7.1 (Strong solution). *We call $u(t,x)$ a strong solution of* (7.1) *if it lies in certain Banach space (depending on the choice of s and will be specified in specific settings) and solves the following integration equation in distribution sense for each t,*

$$u(t) = e^{it\Delta}u_0 \mp i\int_0^t e^{i(t-s)\Delta}|u|^{\frac{4}{3}}u(s)ds. \tag{7.2}$$

The problems of interest for solutions of (7.1) are the following:

1. Given $u_0 \in H^s$, is there a unique (local in time) solution to (7.1)?
2. Does the solution depend continuously on the initial data?
3. Is the solution global in time? If it is, what is the asymptotic behavior of the solution as $|t| \to \infty$? If it is not, how well one can understand the exact dynamic of finite blowup solutions?

Since the linear solution is well-defined, the classical approach of obtaining the local solution is "perturbative", namely, viewing the nonlinear equation as a small perturbation of the linear one. This could happen in short time or when u_0 is small.

We now introduce two such cases:

Theorem 7.2. *Let $u_0 \in H^1$, then there exists $T = T(\|u_0\|_{H^1}) \approx \|u_0\|_{H^1}^{-c}$, s.t.* (7.1) *has a unique solution*

$$u(t,x) \in C_t^0 H^1 \cap L_t^{\frac{10}{3}} H_x^{1,\frac{10}{3}}([0,T]\times\mathbb{R}^3).$$

Proof. Define a solution map

$$\Phi(u) := e^{it\Delta}u_0 \mp i\int_0^t e^{i(t-s)\Delta}|u|^{\frac{4}{3}}u(s)ds.$$

The goal is to show Φ is contractive on a closed ball in certain Banach space. From Strichartz estimate, we have

$$\begin{aligned}\|\Phi(u)\|_{X_T} \le C\|u_0\|_{H^1} + C(&\||u|^{\frac{4}{3}}u\|_{L_{t,x}^{\frac{10}{7}}([0,T]\times\mathbb{R}^3)} \\ &+\|\nabla(|u|^{\frac{4}{3}}u)\|_{L_{t,x}^{\frac{10}{7}}([0,T]\times\mathbb{R}^3)}).\end{aligned}$$

Here $X_T = L_t^\infty H_x^1 \cap L_t^{\frac{10}{3}} L_x^{\frac{10}{3}}([0,T]\times\mathbb{R}^3)$. Using Hölder inequality, we continue to estimate

$$\begin{aligned}\|\Phi(u)\|_{X_T} &\le C\|u_0\|_{H^1} + C\|u\|^{\frac{4}{3}}_{L_{t,x}^{\frac{10}{3}}}\|u\|_{L_t^{\frac{10}{3}}H^{1,\frac{10}{3}}} \\ &\le C\|u_0\|_{H^1} + CT^{\frac{2}{5}}\|u\|^{\frac{4}{3}}_{L_t^\infty H^1}\|u\|_{X_T} \\ &\le C\|u_0\|_{H^1} + CT^{\frac{2}{5}}\|u\|^{\frac{7}{3}}_{X_T}.\end{aligned}$$

By the same token,

$$\|\Phi(u) - \Phi(v)\|_{X_T} \leq CT^{\frac{2}{5}}(\|u\|_{X_T}^{\frac{4}{3}} + \|v\|_{X_T}^{\frac{4}{3}})\|u - v\|_{X_T}.$$

Now choose T such that

$$CT^{\frac{2}{5}}(2C\|u_0\|_{H^1})^{\frac{4}{3}} \leq \frac{1}{2} \Leftrightarrow T \leq c\|u_0\|_{H^1}^{-\frac{3}{4}}$$

for a sufficiently small c, we can verify for such T, Φ is a contraction mapping from $B(0, 2C\|u_0\|_{H^1}) \subset X_T$ to itself. Therefore, there exists a unique solution in $B(0, 2C\|u_0\|_{H^1})$. The uniqueness on a big set X_T can also be proved similarly, as well as the continuity in H^1. The theorem is then proved. □

Remark 7.3 (Maximal lifespan solution). *This local result constructs a solution on the time interval $[0, T]$, in particular, $u(T)$ is a well defined H^1 function. One can repeat this process using $u(T)$ as an initial data to extend the solution beyond T. If the iteration terminates at time T^*, we say u is a maximal lifespan solution on $[0, T^*)$. The case $T^* = \infty$ corresponds to a global (forward in time) solution.*

Remark 7.4 (Blowup criterion). *From this local result, we see that for this problem, solution can be extended as long as H^1-norm of the solution is a priori bounded. This leads to the following blowup criterion:*

Let u be a maximal lifespan solution to (7.1) on $[0, T^*)$. Then either $T^* = \infty$ or $T^* < \infty$, $\|u(t)\|_{H^1} \to \infty$ as $t \to T^*$.

We will compare this result with the next where we construct the solution only in L^2. The dependence of the lifespan on the initial data will be different.

Theorem 7.5. *Let $u_0 \in L^2(\mathbb{R}^3)$. Then there exist an absolute constant $\eta_0 > 0$ and $T > 0$ such that whenever*

$$\|e^{it\Delta}u_0\|_{L_{t,x}^{\frac{10}{3}}([0,T]\times\mathbb{R}^3)} \leq \eta_0$$

(7.1) has a unique solution

$$u(t, x) \in C_t^0 L^2 \cap L_{t,x}^{\frac{10}{3}}([0, T] \times \mathbb{R}^3) := X_T.$$

In particular, if $\|u_0\|_2 \ll 1$, $T = \infty$, u is global forward in time and

$$\|u\|_{L_{t,x}^{\frac{10}{3}}([0,\infty))} < \infty. \tag{7.3}$$

This global estimate implies u scatters forward in time: there exists $u_+ \in L^2$ such that

$$\|u(t) - e^{it\Delta}u_+\|_2 \to 0, \ \text{as } t \to \infty. \tag{7.4}$$

The same statement can be made backward in time.

Proof. Let Φ still be the solution map. From Strichartz and Hölder inequality, we have

$$\begin{aligned}\|\Phi(u)\|_{L_t^\infty L_x^2} &\le \|u_0\|_2 + C\||u|^{\frac{4}{3}}u\|_{L_{t,x}^{\frac{10}{3}}} \\ &\le \|u_0\|_2 + C\|u\|^{\frac{7}{3}}_{L_{t,x}^{\frac{10}{3}}}, \\ \|\Phi(u)\|_{L_{t,x}^{\frac{10}{3}}} &\le \eta_0 + C\|u\|^{\frac{7}{3}}_{L_{t,x}^{\frac{10}{3}}}.\end{aligned}$$

Define closed set

$$\begin{aligned}B = \{&\|u\|_{L_t^\infty L_x^2([0,T]\times\mathbb{R}^3)} \le \|u_0\|_2 + \eta_0, \\ &\|u\|_{L_{t,x}^{\frac{10}{3}}([0,T]\times\mathbb{R}^3)} \le 2\eta_0\} \subset X_T.\end{aligned}$$

Then the two estimates show that for $u, v \in B$ and sufficiently small η_0,

$$\begin{aligned}\|\Phi(u)\|_{L_t^\infty L_x^2([0,T]\times\mathbb{R}^3)} &\le \|u_0\|_2 + C(2\eta_0)^3 \le \|u_0\|_2 + \eta_0, \\ \|\Phi(u)\|_{L_{t,x}^{\frac{10}{3}}([0,T]\times\mathbb{R}^3)} &\le \eta_0 + C(2\eta_0)^3 \le 2\eta_0, \\ \|\Phi(u) - \Phi(v)\|_{X_T} &\le \frac{1}{2}\|u - v\|_{X_T}.\end{aligned}$$

A fixed point argument yields the unique solution.

Suppose $\|u_0\|_2 \ll 1$, from Strichartz, $\|e^{it\Delta}u_0\|_{L_{t,x}^{\frac{10}{3}}[0,\infty)} \le \eta_0$, u is a forward global solution. Next we show scattering by first deriving u_+ through a formal argument. As

$$u(t) = e^{it\Delta}u_0 \mp i\int_0^t e^{i(t-s)\Delta}|u|^{\frac{4}{3}}u(s)ds,$$

we have

$$e^{-it\Delta}u(t) = u_0 \mp i\int_0^t e^{-is\Delta}|u|^{\frac{4}{3}}u(s)ds.$$

Therefore, formally

$$u_+ = u_0 \mp i\int_0^\infty e^{-is\Delta}|u|^{\frac{4}{3}}u(s)ds. \tag{7.5}$$

Indeed, by Strichartz, it is easy to show $\int_0^\infty e^{-is\Delta}|u|^{\frac{4}{3}}u(s)ds$ converges in L^2, thus the RHS defines an L^2-function. Now it suffices only to show (7.4) using (7.3). Indeed from Strichartz again,

$$\begin{aligned}\|u(t) - e^{it\Delta}u_+\|_{L^2} &= \left\|\int_t^\infty e^{-is\Delta}|u|^{\frac{4}{3}}u(s)ds\right\|_{L^2} \\ &\lesssim \|u\|^{\frac{7}{3}}_{L_{t,x}^{\frac{10}{3}}([t,\infty))} \to 0 \text{ as } t \to \infty.\end{aligned}$$

□

Remark 7.6. *In this result, the lifespan is defined in this "unusual manner": T is taken such that*

$$\|e^{it\Delta}u_0\|_{L^{\frac{10}{3}}_{t,x}([0,T]\times\mathbb{R}^3)} \leq \eta_0.$$

Now if

$$u_{0\lambda} = \lambda^{-\frac{3}{2}}u_0\left(\frac{x}{\lambda}\right), \quad \textit{hence } \|u_{0\lambda}\|_2 = \|u_0\|_2,$$

we can easily verify that

$$e^{it\Delta}u_{0\lambda} = \lambda^{-\frac{3}{2}}(e^{i\frac{t}{\lambda^2}\Delta}u_0)\left(\frac{x}{\lambda}\right),$$
$$\|e^{it\Delta}u_{0\lambda}\|_{L^{\frac{10}{3}}_{t,x}([0,\lambda^2T]\times\mathbb{R}^3)} = \|e^{it\Delta}u_0\|_{L^{\frac{10}{3}}_{t,x}([0,T]\times\mathbb{R}^3)} \leq \eta_0.$$

This shows that data with the same L^2 norm could have different lifespan because of their different profiles. Therefore the lifespan depends not only on the norm, but also on the profile of the data!

We even have specific examples.

Let $R(x)$ be the solution of the elliptic equation

$$\Delta R - R + |R|^{\frac{4}{3}}R = 0. \tag{7.6}$$

From the standard elliptic theory, there are infinite many solutions to (7.6). From every R, one can construct two solutions of NLS with totally different behaviors:

(1) $e^{it}R(x)$,

(2) $(1-t)^{-\frac{3}{2}}e^{-i\frac{|x|^2}{4(1-t)}+i\frac{t}{1-t}}R(\frac{x}{1-t})$.

Both of the two solutions are connected via pseudo-conformal transformation which is invariant for (7.1) (see [4, 39] for instant). The two examples have the same L^2-norm. However, the first one is a global solution, the second one blows up at $t = 1$.

We state the exact blowup criterion as follows:

Corollary 7.7 (Blowup criterion with L^2-data). *Let $u : [0, T^*) \times \mathbb{R}^3 \to \mathbb{C}$ be a maximal lifespan solution of NLS that is constructed in Theorem 7.5. Then either $T^* = \infty$, or $T^* < \infty$, $\|u\|_{L^{\frac{10}{3}}_{t,x}([0,T^*)\times\mathbb{R}^3)} = \infty$.*

Proof. We only need to prove if $T^* < \infty$ and

$$\|u\|_{L^{\frac{10}{3}}_{t,x}([0,T^*)\times\mathbb{R}^2)} < \infty. \tag{7.7}$$

Then we can extend our solution beyond T^*, which contradicts the maximality of T^*.

Let T be a time before T^* to be determined. From Duhamel formula, we can write

$$u(t) = e^{i(t-T)\Delta}u(T) \mp i\int_T^t e^{i(t-s)\Delta}|u|^{\frac{4}{3}}u(s)ds.$$

From Strichartz and triangle inequality we can write

$$\|e^{i(t-T)\Delta}u(T)\|_{L^{\frac{10}{3}}_{t,x}([T,T^*))} \le \|u\|_{L^{\frac{10}{3}}_{t,x}([T,T^*))} + C\|u\|^{\frac{7}{3}}_{L^{\frac{10}{3}}_{t,x}([T,T^*))}.$$

Thanks to the condition (7.7), we can take T sufficiently close to T^* such that

$$\|e^{i(t-T)\Delta}u(T)\|_{L^{\frac{10}{3}}_{t,x}([T,T^*))} \le \frac{\eta_0}{2}. \tag{7.8}$$

Here η_0 is the required constant in Theorem 7.5. Now fix this T. Note by Strichartz

$$\|e^{i(t-T)\Delta}u(T)\|_{L^{\frac{10}{3}}_{t,x}(\mathbb{R}\times\mathbb{R}^3)} \le C\|u(T)\|_2,$$

we can take δ small enough such that

$$\|e^{i(t-T)\Delta}u(T)\|_{L^{\frac{10}{3}}_{t,x}([T^*,T^*+\delta])} \le \frac{\eta_0}{2}. \tag{7.9}$$

Combining the two estimates (7.8) and (7.9) we obtain

$$\|e^{i(t-T)\Delta}u(T)\|_{L^{\frac{10}{3}}_{t,x}([T,T*+\delta))} \le \eta_0.$$

Local theory just proved guarantee that u can be extended to $[T, T*+\delta)$ which contradicts to the fact that T^* is maximal. □

The last two theorems show different dependance of the lifespan on the initial data. In Theorem 7.5, the lifespan also depends on profile of the data, thus is more "nonlinear". A general way to recognize these different behaviors is through scaling heuristic which we explain here.

Consider the powertype semilinear Schrödinger equation:

$$\begin{cases} iu_t + \Delta u = |u|^p u, \ p > 0, \\ u(0,x) = u_0(x) \in \dot{H}^s(\mathbb{R}^d). \end{cases} \tag{7.10}$$

The equation enjoys the scaling symmetry: If $u(t,x)$ is a solution, then so is

$$u_\lambda(t,x) = \lambda^{-\frac{2}{p}} u\left(\frac{t}{\lambda^2}, \frac{x}{\lambda}\right),$$

whose data is given by

$$u_{0\lambda}(x) = \lambda^{-\frac{2}{p}} u_0\left(\frac{x}{\lambda}\right).$$

The critical index s_c is such that the $\dot{H}^{s_c}$ norm of the initial data is invariant under the scaling. It is easy to find

$$s_c = \frac{d}{2} - \frac{2}{p}.$$

The problem (7.10) is subcritical when $s > s_c$[4], supercritical when $s < s_c$ and is critical when $s = s_c$.

Suppose u is a maximal lifespan solution on $[0, T^*)$ with initial data u_0. Then

$$u_\lambda : [0, \lambda^2 T^*) \times \mathbb{R}^d \to \mathbb{C} \text{ with } u_\lambda(0) = u_{0\lambda}, \text{ and } \|u_{0\lambda}\|_{\dot{H}^s} = \lambda^{s_c - s}.$$

For the subcritical problem, shrinking the size of data ($\lambda \to \infty$) yields a solution with longer lifespan, solution behaves in a linear manner. For the subcritical problem, this scaling heuristic indicates that small solution blows up sooner than the big solution, which shows the problem is highly nonlinear. As matter of fact, inspired by the scaling argument, one can prove that the supercritical problem is illposed in the sense that the data solution map is not continuous (or uniform continuous) [8].

In the past few decades, there have been comprehensive study on the subcritical and critical problems, with several main breakthroughs made in recent years on the critical problems. There are several references from which one can find reviews on this. In this notes, we will make our goal straight by skipping that part and jump over directly to the representative critical problems— for which the criticality coincides with the conservation laws:

$$\text{Mass conservation: } M(u(t)) = \int_{\mathbb{R}^d} |u(t,x)|^2 dx,$$

$$\text{Energy conservation: } E(u(t)) = \frac{1}{2}\int_{\mathbb{R}^d} |\nabla u|^2 dx \pm \frac{1}{p+2}\int |u|^{p+2} dx.$$

The case $s_c = 0$ or $p = \frac{4}{d}$ is called mass critical, while the case $s_c = 1$, $p = \frac{4}{d-2}$ corresponds to energy critical. To write it up, we have

$$\begin{cases} iu_t + \Delta u = \pm |u|^{\frac{4}{d}} u \\ u(0,x) = u_0(x) \in L^2 \end{cases} \rightsquigarrow \text{Mass critical,} \tag{7.11}$$

$$\begin{cases} iu_t + \Delta u = \pm |u|^{\frac{4}{d-2}} u \\ u(0,x) = u_0(x) \in H^1 \end{cases} \rightsquigarrow \text{Energy critical.} \tag{7.12}$$

The lecture notes [25] use 3d energy critical NLS as an example, showing how the linear profile decomposition can be used to prove the scattering for solutions. In the next few chapters, we will do the discussion for mass critical NLS.

Before leaving this chapter, we recall the stability result for the mass critical NLS, which is part of the local theory.

Lemma 7.8 (Stability). *Let $\tilde{u}$ be a near solution on I containing 0 in the sense that*

$$\begin{cases} i\tilde{u}_t + \Delta \tilde{u} = |\tilde{u}|^{\frac{4}{d}} \tilde{u} + e, \\ \tilde{u}(0,x) = \tilde{u}_0(x) \end{cases}$$

[4]In the actual problem, initial data is assumed to belong to the inhomogeneous space H^s instead of the homogeneous one.

and satisfy

$$\|\tilde{u}\|_{L_{t,x}^{\frac{2(d+2)}{d}}(I\times\mathbb{R}^d)} \le L, \quad \|\tilde{u}_0\|_2 \le M,$$
$$\|e\|_{L_{t,x}^{\frac{2(d+2)}{d}}(I\times\mathbb{R}^d)} \le \varepsilon.$$

Let $u_0 \in L^2$ be such that

$$\|e^{it\Delta}(u_0 - \tilde{u}_0)\|_{L_{t,x}^{\frac{2(d+2)}{d}}} \le \varepsilon,$$
$$\|u_0 - \tilde{u}_0\|_2 \le M'.$$

Then there exists $\epsilon_0 > 0$ such that when $\epsilon \le \epsilon_0$, NLS (7.11) *has a unique solution with data u_0 and*

$$\begin{cases} \|u - \tilde{u}\|_{L_{t,x}^{\frac{2(d+2)}{d}}} \le C(L, M, M')\varepsilon, \\ \|u\|_{L_t^q L_x^r} \le C(L, M, M')M', \ \forall\ (q,r) \text{ admissible.} \end{cases} \tag{7.13}$$

Proof. From blowup criterion, it suffices to prove (7.13) by assuming u already exists on I. The first step is to establish the short-time statement.

Step 1: Assume $L = \delta \ll 1$.

Let $v = \tilde{u} - u$. Then

$$\begin{cases} iv_t + \Delta v = |\tilde{u}|^{\frac{4}{d}}\tilde{u} - |u|^{\frac{4}{d}}u + e, \\ v(0) = \tilde{u}_0 - u_0. \end{cases}$$

From Duhamel formula and Strichartz estimate,

$$\begin{aligned} \|v\|_{L_{t,x}^{\frac{2(d+2)}{d}}} &\lesssim \|e^{it\Delta}(\tilde{u}_0 - u_0)\|_{L_{t,x}^{\frac{2(d+2)}{d}}} + \|e\|_{L_{t,x}^{\frac{2(d+2)}{d+4}}} + \||\tilde{u}|^{\frac{4}{d}}\tilde{u} - |u|^{\frac{4}{d}}u\|_{L_{t,x}^{\frac{2(d+2)}{d+4}}} \\ &\lesssim 2\varepsilon + (\|v\|^{\frac{4}{d}}_{L_{t,x}^{\frac{2(d+2)}{d}}} + \|\tilde{u}\|^{\frac{4}{d}}_{L_{t,x}^{\frac{2(d+2)}{d}}})\|v\|_{L_{t,x}^{\frac{2(d+2)}{d}}} \\ &\lesssim 2\varepsilon + \delta^{\frac{4}{d}}\|v\|_{L_{t,x}^{\frac{2(d+2)}{d}}} + \|v\|^{1+\frac{4}{d}}_{L_{t,x}^{\frac{2(d+2)}{d}}}. \end{aligned}$$

Take δ small (but a universal small constant), a continuity argument[5] shows that

$$\|v\|_{L_{t,x}^{\frac{2(d+2)}{d}}(I\times\mathbb{R}^d)} \le C\varepsilon.$$

All other estimates in (7.13) follow from Strichartz inequality.

---

[5]Continuity argument:
Let $X(s)$ be a quantity continuous in s and $X(0) \le \varepsilon$. Suppose on $[0,T]$, $X(s)$ satisfies

$$X(s) \le \varepsilon + C_1 X(s)^\alpha, \ \alpha > 1.$$

Then there exists $\varepsilon_0(C_1, \alpha)$ such that when $\varepsilon < \varepsilon_0$, $\|X(s)\|_{L_s^\infty[0,T]} \le 2\varepsilon$.

To proceed to the next step we will prove the preservation of the smallness condition. Assume $I = [0, T]$ from Duhamel we write

$$v(T) = e^{iT\Delta}v(0) - i\int_0^T e^{i(T-s)\Delta}(|\tilde{u}|^{\frac{4}{d}}\tilde{u} - |u|^{\frac{4}{d}}u + e)(s)ds,$$

$$e^{i(t-T)\Delta}v(T) = e^{it\Delta}v(0) - i\int_0^T e^{i(t-s)\Delta}(|\tilde{u}|^{\frac{4}{d}}\tilde{u} - |u|^{\frac{4}{d}}u + e)(s)ds.$$

Hence by triangle and Strichartz

$$\begin{aligned}\|e^{i(t-T)\Delta}v(T)\|_{L_{t,x}^{\frac{2(d+2)}{d}}} &\le \|e^{it\Delta}v(0)\|_{L_{t,x}^{\frac{2(d+2)}{d}}} + C\||\tilde{u}|^{\frac{4}{d}}\tilde{u} - |u|^{\frac{4}{d}}u + e\|_{L_{t,x}^{\frac{2(d+2)}{d+4}}} \\ &\le \varepsilon + C\delta^{\frac{4}{d}}\|v\|_{L_{t,x}^{\frac{2(d+2)}{d}}} + C\|v\|^{1+\frac{4}{d}}_{L_{t,x}^{\frac{2(d+2)}{d}}} \\ &= \varepsilon + C\delta^{\frac{4}{d}}\varepsilon + C(C\varepsilon)^{1+\frac{4}{d}}. \\ &\le 2\varepsilon.\end{aligned}$$

Step 2: Partition the interval $I = \bigcup_{j=1}^{J} I_j$ such that $\frac{1}{2}\delta \le \|\tilde{u}\|_{L_{t,x}^{\frac{2(d+2)}{d}}} \le \delta$. Then $J = C(\frac{L}{\delta})^{\frac{2(d+2)}{d}}$. Choose small constant ε_0 which incorporates the increment or the error in each iteration. Inductively applying the result from Step 1 for $\varepsilon \le \varepsilon_0$ yields the desired result. □

8 Global theory for mass critical NLS: introduction and preliminaries

In the defocusing case, all solutions should exist globally and scatter. In the focusing case, scattering holds as long as

$$M(u) < M(Q).$$

Here, Q is the *ground state* solution for (7.6), which is defined as the unique, positive, radial solution for (7.6). From the variational principle, Q also serves as the extremizer of sharp Gagliardo-Nirenberg inequality. We have

Lemma 8.1. *Let $f \in H^1$. Then*

$$\frac{d}{2(d+2)}\|f\|^{\frac{2(d+2)}{d}}_{\frac{2(d+2)}{d}} \le \left(\frac{M(f)}{M(Q)}\right)^{\frac{4}{d}}\frac{1}{2}\|\nabla f\|_2^2.$$

Moreover, if the equality holds, there exists θ_0, x_0, λ_0 such that

$$f(x) = ce^{i\theta_0}\lambda_0^{-1}Q\Big(\frac{x - x_0}{\lambda_0}\Big).$$

The argument that supports the criticality of Q is as follows. Suppose for solution u, $M(u) < M(Q)$. Then from the sharp inequality, we have

$$E(u) = \frac{1}{2}\|\nabla u\|_2^2 - \frac{d}{2(d+2)}\|u\|_{\frac{2(d+2)}{d}}^{\frac{2(d+2)}{d}}$$

$$\geq \frac{1}{2}\Big(1 - \Big(\frac{M(u)}{M(Q)}\Big)^{\frac{4}{d}}\Big)\|\nabla u\|_2^2 \geq c\|\nabla u\|_2^2.$$

Therefore, $\|u(t)\|_{H^1} \leq C(E, M)$. From here, the iteration on the local result immediately gives the global solution. This was an earlier result of Weinstein [48].

The conjectured result is the scattering in critical space L^2.

Conjecture 8.2. *Consider mass critical NLS* (7.11). *Then in the defocusing case, solution exists globally and scatters. In the focusing case, scattering holds if* $M(u) < M(Q)$.

Until recently, this conjecture is completely settled, in the radial case by the earlier work of Tao-Visan-Zhang [45, 44], Killip-Tao-Visan [27], Killip-Visan-Zhang [28] and in the nonradial case by a series work of B. Dodson [13, 14, 15, 16]. The remaining part of the notes will have a heavy citation on those work (which we will no longer specify).

Now we are ready to discuss the proof of this Scattering Conjecture. The proof is a proof by contradiction, roughly it contains two parts.

Part one: Extracting the minimal mass blowup solutions[6] if assuming the scattering conjecture fails.

Part two: Killing the minimal mass blowup solutions by employing PDE techniques.

We first discuss part one. As the argument does not depend on dimensions and sign of the nonlinearity so heavily, we will present the defocusing case in $d = 2$ for simplicity. The general cases need minor modification which we will skip.

We argue by contradiction. Define

$$L(M) = \sup\{S_I(u); u : I \times \mathbb{R}^2 \to \mathbb{C} \ s.t \ M(u) \leq M, u \text{ radial}\}.$$

Here

$$S_I(u) = \int_I |u(t,x)|^4 dx dt \tag{8.1}$$

is called the *scattering size* of the solution on I. The supremum is taken over all solutions which are radially symmetric with *mass* $\leq M$.

[6]Here the blowup is the blowup of the scale invariant spacetime integral $\int_{\mathbb{R}\times\mathbb{R}^d} |u(t,x)|^{\frac{2(d+2)}{d}} dt dx$, whose finiteness implies the scattering of the solution (see Theorem 7.5). Solution does not blow up at finite time, as matter of fact, it could be a global solution like $e^{it}R$.

From the classical local theory, $L(M) \leq CM$ for sufficiently small M. From the stability, $L(M)$ is left continuous in M. Therefore the failure of the conjecture is equivalent to the existence of the critical mass $M_c > 0$ such that

$$L(M) < \infty, \ \forall\, M < M_c,$$
$$L(M) = \infty, \ M \geq M_c.$$

The fact that $L(M_c) = \infty$ gives us a "minimizing sequence" in the following sense: there exist $u_n : I_n \times \mathbb{R}^2 \to \mathbb{C}$ for which $S(u_n) \to \infty$ and $M(u_n) \leq M_c$, $M(u_n) \to M_c$. Next we prove a crucial result for this sequence: this sequence is precompact in L^2 by modulo symmetries of the equation. In the radial case, there is only one symmetry that needs to be considered: the scaling.

Proposition 8.3 (Palais-Smale condition modulo scaling). *Let $u_n : I_n \times \mathbb{R}^2 \to \mathbb{C}$ be sequence of radial solutions to defocusing cubic NLS, which satisfy*

$$M(u_n) \leq M_c, \ M(u_n) \to M_c,$$
$$\lim_{n\to\infty} S_{\geq t_n}(u_n) = \lim_{n\to\infty} S_{\leq t_n}(u_n) = +\infty.$$

Then there exists $N_n \in 2^{\mathbb{Z}}$ such that the set $\{N_n^{-1} u_n(t_n, \frac{x}{N_n})\}$ is precompact in L^2.

Proof. By time translation invariant, we can assume $t_n = 0$, thus we have

$$\lim_{n\to\infty} S_{\geq 0}(u_n) = \lim_{n\to\infty} S_{\leq 0}(u_n) = +\infty.$$

Applying the L^2 linear profile decomposition to $u_n(0)$, we obtain

$$u_n(0) = \sum_{j=1}^{J} (\lambda_n^j)^{-1} (e^{it_n^j \Delta} \varphi^j) \left(\frac{x}{\lambda_n^j} \right) + w_n^J \tag{8.2}$$

with the stated properties.

We first make several simplifications.

First by passing to a subsequence and using the usual diagonal procedure, we can assume for each j, t_n^j has a limit $t_n^j \to \pm\infty$ or $t_n^j \to t_0^j \in (-\infty, \infty)$. In the case when t_n^j has finite limit t_0^j, by redefining φ^j as $e^{it_0^j \Delta}\varphi^j$, we can further assume $t_0^j = 0$. Finally, since the error $e^{it_n^j \Delta}\varphi^j - \varphi^j$ is small in L^2 and can be absorbed into the error w_n^J, we can assume $t_n^j \equiv 0$.

After this discussion, the condition we can impose in (8.2) is either $t_n^j \to \pm\infty$ or $t_n^j \equiv 0$ for each given j. We now define a nonlinear profile $V^j : I^j \times \mathbb{R}^2 \to \mathbb{C}$ associated with φ^j as follows:

- If $t_n^j \equiv 0$, we define V^j as the maximal lifespan solution with initial data φ^j. The corresponding nonlinear profile is the scaled version of V^j:

$$V_n^j = (\lambda_n^j)^{-1} V^j \left(\frac{t}{(\lambda_n^j)^2}, \frac{x}{\lambda_n^j} \right).$$

- If $t_n^j \to +\infty$, we define V_j as the maximal lifespan solution that scatters to $e^{it\Delta}\varphi^j$ as $t \to +\infty$. The corresponding nonlinear solution is given as

$$V_n^j = (\lambda_n^j)^{-1} V^j\Big(t_n^j + \frac{t}{(\lambda_n^j)^2}, \frac{x}{\lambda_n^j}\Big).$$

Note $V_n^j(0) = (\lambda_n^j)^{-1} V^j(t_n^j, \frac{x}{\lambda_n^j})$, we have for each j

$$\Big\| V_n^j(0) - (\lambda_n^j)^{-1} e^{it_n^j\Delta}\varphi^j\Big(\frac{\cdot}{\lambda_n^j}\Big)\Big\|_2 = \|V^j(t_n^j) - e^{it_n^j\Delta}\varphi^j\|_2$$
$$\to 0 \text{ as } n \to \infty.$$

Therefore for each $J \geq 1$

$$\Big\| u_n(0) - \sum_{j=1}^{J} V_n^j(0) - w_n^J \Big\|_2 \to 0 \text{ as } n \to \infty.$$

Moreover, from the mass decoupling condition, we have

$$\lim_{n\to\infty} \Big(M(u_n(0)) - \sum_{j=1}^{J} M(\varphi^j) - M(w_n^J) \Big) = 0,$$

$$\sum_{j=1}^{\infty} M(\varphi^j) \leq M_c.$$

This implies

$$\sup_j M(\varphi^j) \leq M_c.$$

We now discuss two cases:

Case 1: There exist an $\epsilon > 0$ such that

$$\sup_j M(\varphi^j) \leq M_c - \epsilon.$$

Case 2: $\sup_j M(\varphi^j) = M_c$.

We first focus on the Case 1. By the inductive assumption on the critical mass M_c and $M(V_n^j) = M(\varphi^j) \leq M_c - \epsilon$, we know V_n^j is global and

$$S(V_n^j) \leq L(M(\varphi^j)) \leq L(M_c - \epsilon) < \infty.$$

We will eventually derive a bound on u_n thus get a contradiction. To do so, we define

$$u_n^J(t) = \sum_{j=1}^{J} V_n^j(t) + e^{it\Delta} w_n^J$$

and prove u_n^J is an approximate solution of u_n for large n, J by using stability result. To achive this goal, we will prove the following three Claims.

Claim 8.4 (Boundedness of u_n^J).

$$\lim_{J\to\infty} \limsup_{n\to\infty} S(u_n^J) \lesssim M_c.$$

Claim 8.5 (Asymptotic agreement of the data).

$$\lim_{n\to\infty} M(u_n^J(0) - u_n(0)) = 0.$$

Claim 8.6 (Smallness in error term).

$$\lim_{J\to\infty} \limsup_{n\to\infty} \|(i\partial_t + \Delta)u_n^J - |u_n^J|^2 u_n^J\|_{L^{\frac{4}{3}}_{t,x}(\mathbb{R}\times\mathbb{R}^2)} = 0.$$

The direct consequence of the three claims together with stability is that u_n is global with $S(u_n) \lesssim M_c < \infty$ for sufficiently large n which is obviously a contradiction! Thus Case 1 can not happen.

We will postpone the proof the three claims to the end, now we deal with the second case:

$$\sup_j M(\phi^j) = M_c.$$

Note from previous discussion

$$\sum_{j=1}^{J^*} M(\phi^j) \le M_c.$$

We see $J^* = 1$ hence we can write

$$\begin{cases} u_n(0) = \lambda_n^{-1}(e^{it_n\Delta}\phi)(\frac{x}{\lambda_n}) + w_n, \\ M(\phi) = M_c, \; M(u_n) \nearrow M_c, \\ t_n \equiv 0 \text{ or } t_n \to \pm\infty. \end{cases} \tag{8.3}$$

Note from the decoupling of Mass,

$$\lim_{n\to\infty} [M(u_n(0)) - M(\phi) - M(w_n)] = 0,$$

we have $M(w_n) \to 0$, i.e, the error is small in L^2-space. Finally to achieve our goal we need to preclude the case when $t_n \to \pm\infty$. We argue by contradiction. Suppose $t_n \to \infty$, we estimate

$$\begin{aligned} S_{\ge 0}(e^{it\Delta}u_n(0)) &\lesssim S_{\ge 0}\left(e^{it\Delta}\lambda_n^{-1}(e^{it_n\Delta}\phi)\left(\frac{x}{\lambda_n}\right)\right) + S(e^{it\Delta}w_n) \\ &= S_{\ge 0}\left(\lambda_n^{-1}\left(e^{i(t_n+\frac{t}{\lambda_n^2})\Delta}\phi\right)\left(\frac{x}{\lambda_n}\right)\right) + S(e^{it\Delta}w_n) \\ &= S_{\ge 0}(e^{i(t+t_n)\Delta}\phi) + S_{\ge 0}(e^{it\Delta}w_n) \\ &= S_{\ge t_n}(e^{it\Delta}\phi) + S_{\ge 0}(e^{it\Delta}w_n) \to 0 \text{ as } n \to \infty. \end{aligned}$$

From the stability result, $\{u_n(t)\}$ is a good approximation of 0-*solution* on the time interval $[0, \infty)$, therefore

$$S_{\geq 0}(u_n(t)) \leq C < \infty$$

as n sufficiently large, which is again a contradiction. The case when $t_n \to -\infty$ can be treated similarly. Finally as $t_n \equiv 0$, we end up having

$$\begin{cases} u_n(0) = \lambda_n^{-1}\phi(\frac{x}{\lambda_n}) + w_n, \\ M(w_n) \to 0. \end{cases}$$

From this we know

$$M(w_n) = M(\lambda_n u_n(0, \lambda_n x) - \phi) \to 0.$$

Using the notation in the proposition we know $\{N_n^{-1} u_n(0, \frac{x}{N_n})\}$ is precompact in L^2_{rad}.

It remains to prove the three claims.

From the condition

$$\sum_{j=1}^{\infty} M(\varphi^j) \leq M_c$$

we know $\exists\ J_0$ such that when $j \geq J_0, M(\varphi^j) \leq \eta_0$. Here η_0 is the threshold for scattering of small solutions. Moreover for these j,

$$S(V^j) \lesssim M(V^j)^2 = M(\varphi^j)^2 \lesssim M(\varphi^j).$$

Those j for which $M(\varphi^j) > \eta_0$ is only finite, therefore $\forall 1 \leq j < \infty$

$$S(V^j) = S(V_n^j) \lesssim M(\varphi^j) \text{ uniformly in } j. \tag{8.4}$$

Now from triangle

$$\begin{aligned} S(u_n^J) &\lesssim S(\sum_{j=1}^{J} V_n^j) + S(e^{it\Delta} w_n^J) \\ &\lesssim \sum_{j=1}^{J} S(V_n^j) + C(J) \sum_{j \neq k} \int_{\mathbb{R}\times\mathbb{R}^2} |V_n^j||V_n^k|^3 dxdt + S(e^{it\Delta} w_n^J) \\ &\lesssim \sum_{j=1}^{J} M(\varphi^j) + C(J) \sum_{j \neq k} \int_{\mathbb{R}\times\mathbb{R}^2} |V_n^j||V_n^k|^3 dxdt + M_c. \end{aligned}$$

Therefore

$$\lim_{J\to\infty} \overline{\lim_{n\to\infty}} S(u_n^J) \lesssim M_c$$

provided the orthogonal result:

$$\forall j \neq k, \int_{\mathbb{R}\times\mathbb{R}^2} |V_n^j||V_n^k|^3 dxdt \to \text{ as } n \to \infty. \tag{8.5}$$

Indeed, recall $V_n^j = (\lambda_n^j)^{-1} V^j(t_n^j + \frac{t}{(\lambda_n^j)^2}, \frac{x}{\lambda_n^j})$. From density argument it suffices to show (8.5) by assuming V^j and V^k are $C_c^\infty(\mathbb{R} \times \mathbb{R}^2)$ functions. Undoing the scaling

$$\begin{aligned}
&\int_{\mathbb{R}\times\mathbb{R}^2} |V_n^j||V_n^k|^3 dxdt \\
&= \int_{\mathbb{R}\times\mathbb{R}^2} (\lambda_n^j)^{-1} \left| V^j\left(t_n^j + \frac{t}{(\lambda_n^j)^2}, \frac{x}{\lambda_n^j}\right)\right| (\lambda_n^k)^{-1} \left| V^k\left(t_n^k + \frac{t}{(\lambda_n^k)^2}, \frac{x}{\lambda_n^k}\right)\right|^3 dxdt \\
&= \int |V^j(s,y)| \frac{\lambda_n^j}{\lambda_n^k} \left| V^k\left(t_n^k - \frac{(\lambda_n^j)^2}{(\lambda_n^k)^2} t_n^j + \frac{(\lambda_n^j)^2}{(\lambda_n^k)^2} s, \frac{\lambda_n^j}{\lambda_n^k} y\right)\right|^3 dsdy.
\end{aligned}$$

Here, we use the change of variable:

$$t_n^j + \frac{t}{(\lambda_n^j)^2} = s, \frac{x}{\lambda_n^j} = y,$$
$$t = (s - t_n^j)(\lambda_n^j)^2, x = y\lambda_n^j.$$

By Hölder

$$\begin{aligned}
&\int_{\mathbb{R}\times\mathbb{R}^2} |V_n^j|^2 |V_n^k|^3 dxdt \\
&\leq \min\left(\|V^j\|_{\infty,\infty} \frac{\lambda_n^k}{\lambda_n^j} \|V^k\|_{3,3}^3, \quad \|V^j\|_{1,1} \frac{\lambda_n^j}{\lambda_n^k} \|V^k\|_{\infty,\infty}^3 \right) \\
&\to 0, \text{ as } n \to \infty.
\end{aligned}$$

If in the orthogonality condition

$$\lim_{n\to\infty} \frac{\lambda_n^j}{\lambda_n^k} + \frac{\lambda_n^k}{\lambda_n^j} < \infty.$$

Then we can assume, passing to a subsequence if necessary that

$$\begin{cases} \frac{\lambda_n^j}{\lambda_n^k} \to \lambda_0 \in (0, \infty), \\ \frac{1}{\lambda_n^j \lambda_n^k} |t_n^j (\lambda_n^j)^2 - t_n^k (\lambda_n^k)^2| \to \infty. \end{cases}$$

For n sufficiently large,

$$\int_{\mathbb{R}\times\mathbb{R}^2} |V_n^j||V_n^k|^3 dxdt \equiv 0$$

as the supports in t do not overlap.

This finishes the proof of (8.5), hence Claim 8.4.

Now we consider Claim 8.5.

Note

$$u_n(0) = \sum_{j=1}^J (\lambda_n^j)^{-1} (e^{it_n^j \Delta} \varphi^j) \left(\frac{x}{\lambda_n^j}\right) + w_n^J,$$

$$u_n^J(0) = \sum_{j=1}^J (\lambda_n^j)^{-1} V^j \left(t_n^j, \frac{x}{\lambda_n^j}\right) + w_n^J,$$

we have

$$\|u_n(0) - u_n^J(0)\|_2 \le \sum_{j=1}^{J} \|V^j(t_n^j) - e^{it_n^j\Delta}\varphi^j\|_2 \to 0$$

from the construction of V^j.

Proof of Claim 8.6.

Denote $F(z) = |z|^2 z$ as the nonlinear function, we have

$$|u_n^J|^2 u_n^J = F(\sum_{j=1}^{J} V_n^j + e^{it\Delta} w_n^J),$$

$$(i\partial_t + \Delta) u_n^J = \sum_{j=1}^{J} F(V_n^j).$$

Therefore u_n^J solves the approximate equation

$$(i\partial_t + \Delta) u_n^J = F(u_n^J) + e$$

with

$$\begin{aligned} e &= \sum_{j=1}^{J} F(V_n^j) - F(\sum_{j=1}^{J} V_n^j + e^{it\Delta} w_n^J) \\ &= \sum_{j=1}^{J} F(V_n^j) - F\Big(\sum_{j=1}^{J} V_n^j\Big) + F\Big(\sum_{j=1}^{J} V_n^j\Big) - F\Big(\sum_{j=1}^{J} V_n^j + e^{it\Delta} w_n^J\Big), \end{aligned}$$

$$|e| \le C(J) \sum_{j\neq k} V_n^j |V_n^k|^2 + |e^{it\Delta} w_n^J| \left(\left(\sum_{j=1}^{J} V_n^j \right)^2 + |e^{it\Delta} w_n^J|^2 \right)$$

$$\|e\|_{L^{\frac{4}{3}}_{t,x}} \le C(J) \sum_{j\neq k} \int |V_n^j| |V_n^k|^3 dx dt + \|e^{it\Delta} w_n^J\|_{L^4_{t,x}} \left(\left\| \sum_{j=1}^{J} V_n^j \right\|^2_{L^4_{t,x}} + \|e^{it\Delta} w_n^J\|^2_{L^4_{t,x}} \right).$$

Inserting the condition

$$\lim_{J\to\infty} \overline{\lim_{n\to\infty}} \|e^{it\Delta} w_n^J\|_{4,4} = 0,$$

we obtain

$$\lim_{J\to\infty} \overline{\lim_{n\to\infty}} \|e\|_{\frac{4}{3},\frac{4}{3}} = 0.$$

The last claim is proved. This finishes the proof of Proposition 8.3. □

Following the same argument, we can easily show:

Corollary 8.7. *Assume the radial scattering result fails for defocusing 2-D cubic NLS. Then there exists minimal mass blowup solution $u_c : I \times \mathbb{R}^2 \to \mathbb{C}$ in the following sense:*

$$\begin{cases} M(u_c) = M_c, \\ S_{\geq 0}(u_c) = S_{\leq 0}(u_c) = \infty. \end{cases}$$

Moreover, u_c is almost periodic modula scaling: there exist $N(t) : I \to \mathbb{R}^+$ such that $\{N(t)^{-1}u_c(t, \frac{x}{N(t)})\}_{t \in I}$ is precompact in L^2. Property like this is called almost periodic modulo scaling (APMS).

Proof. From the same analysis as in the previous proposition, the failure of the scattering result yields a critical mass and a minimizing sequence: $u_n : I_n \times \mathbb{R}^2 \to \mathcal{C}$ for which

$$\begin{cases} M(u_n) \nearrow M_c, \\ S_{\geq 0}(u_n) = S_{\leq 0}(u_n) \to \infty. \end{cases} \tag{8.6}$$

Applying the proposition, there exists $N_n > 0$ s.t. passing to a subsequence,

$$\begin{cases} N_n^{-1}u_n(0, \frac{x}{N_n}) \to \phi, \\ M(\phi) = M_c. \end{cases}$$

We now define u_c as the maximal lifespan solution with initial data ϕ. Obviously from mass conservation

$$M(u_c) = M_c.$$

Moreover, $S_{\geq 0}(u_c) = S_{\leq 0}(u_c) = \infty$ otherwise (8.6) together with stability result give a contradiction.

Fix this solution u_c and take any sequence $\{t_n\} \subset I$. From the assumption on u_c we know

$$\begin{cases} M(u_c) = M_c, \\ S_{\geq t_n}(u_c) = S_{\leq t_n}(u_c) = \infty. \end{cases}$$

Applying Palais-Smale condition, there exists $N(t_n) \in \mathbb{R}^+$ and a subsequence such that

$$N(t_n)^{-1}u\left(t_n, \frac{x}{N(t_n)}\right)$$

converges in L^2. Corollary is proved. □

Let's first get a sense of $N(t)$ by giving a couple of examples.

1. $u(t,x) = e^{it}R(x)$. $N(t) = 1$, soliton solution.
2. $u(t,x) = e^{it\Delta}u_0(x)$. No $N(t)$ is defined, dispersive solution does not have compactness.
3. $u(t,x) = (1-t)^{-1}e^{\frac{i|x|^2}{4(1-t)} - \frac{it}{1-t}}R(\frac{x}{1-t})$, $N(t) = (1-t)^{-1}$.

The case when $N(t) \sim 1$ corresponds to soliton solutions; $N(t) \to +\infty$ corresponds to wave collapse. We now prove several properties for $N(t)$.

8.1 Discussion on $N(t)$ and almost periodic modulo scaling solution

Proposition 8.8. *Let* $u(t) : I \times \mathbb{R}^2 \to \mathbb{C}$ *be a non-trivial maximal lifespan solution to NLS. Let* $N(t) : I \to \mathbb{R}^+$ *be such that* $\{N(t)^{-1}u(t, \frac{x}{N(t)})\}$ *is precompact in* L^2*. Then*

(1) *For any compact set* $J \subset I$*, there exists* $k > 0$

$$\frac{1}{k} \leq \inf_{t \in J} N(t) \leq \sup_{t \in J} N(t) \leq k.$$

(2) *Local constancy: There exists* $\delta = \delta(u) > 0,\ C = C(u) > 0$ *such that for any* $t_0 \in I, I_{t_0} = [-N(t_0)^{-2}\delta + t_0, t_0 + \delta N(t_0)^{-2}] \subset I$*, and*

$$\sup_{t \in I_{t_0}} N(t) \sim \inf_{t \in I_{t_0}} N(t).$$

Proof. We first prove (1) and argue by contradiction. Assume (1) fails, there exist $t_n \in J$ and $t_n \to t_0 \in J$ such that

$$N(t_n) \to \infty.$$

Consider the sequence $\{N(t_n)^{-1}u(t_n, \frac{x}{N(t_n)})\}$. On the one hand, passing to a subsequence

$$N(t_n)^{-1}u\left(t_n, \frac{x}{N(t_n)}\right) \to \varphi \Rightarrow M(\varphi) = M(u).$$

On the other hand, from the fact $u \in C(I; L^2)$, we know $u(t_n) \to u(t_0)$ in L^2 thus

$$N(t_n)^{-1}u\left(t_n, \frac{x}{N(t_n)}\right) \rightharpoonup 0 \text{ weakly in } L^2.$$

Therefore $\varphi \equiv 0$ and $M(u) = M(\varphi) = 0$. A contradiction! The lower bound can also be proved similarly.

We now look at (2) and prove $I_{t_0} \subset I$. Suppose this fails, then there exist $\delta_n \to 0$ and $t_n \in I$ for which $t_n + \delta_n N(t_n)^{-2} \notin I$.

We define scaled solution

$$u^{[t_n]}(t, x) = N(t_n)^{-1}u\left(t_n + \frac{t}{(N(t_n))^2}, \frac{x}{N(t_n)}\right).$$

Then $u^{[t_n]}$ is a sequence of maximal lifespan solution containing 0, but not δ_n. Consider now the sequence $u^{[t_n]}(0)$, passing to a subsequence if necessary,

$$u^{[t_n]}(0) \to \varphi \text{ in } L^2.$$

From the standard local theory, $\exists!$ solution $V(t, x) : [-T, T] \times \mathbb{R}^2 \to \mathbb{C}$ with initial data φ. From stability result as n sufficiently large, the lifespan of $u^{[t_n]}$ contains $[-T, T]$, which is a contradiction!

Finally we prove the local consistency. Assume not true, there exist a sequence of time pairs $\{t_n, t'_n\} \subset I$ for which

$$\begin{cases} |t'_n - t_n| N^2(t_n) \to 0 \text{ as } n \to \infty, \\ \frac{N(t'_n)}{N(t_n)} \to \infty. \end{cases}$$

Now let

$$s_n = (t'_n - t_n) N^2(t_n),$$
$$u^{[t_n]}(t,x) = N(t_n)^{-1} u\left(t_n + \frac{t}{N(t_n)^2}, \frac{x}{N(t_n)}\right).$$

Obviously

$$u^{[t_n]}(s_n, x) = N(t_n)^{-1} u\left(t'_n, \frac{x}{N(t_n)}\right)$$
$$= \left(\frac{N(t_n)}{N(t'_n)}\right)^{-1} N(t'_n)^{-1} u\left(t'_n, \frac{x}{N(t'_n)} \frac{N(t'_n)}{N(t_n)}\right).$$

On the one hand, passing to subsequence

$$u^{[t_n]}(0) \to \varphi \text{ in } L^2.$$

From the standard local theory and the stability result, $\exists[-T, T]$ such that the lifespan of $u^{[t_n]}$ contains $[-T, T]$ and the Strichartz norm of $u^{[t_n]}$ is uniformly bounded on $[-T, T]$. Therefore $u^{[t_n]}$ is continuous uniformly at $t = 0$ and

$$u^{[t_n]}(s_n) \to \varphi \text{ in } L^2.$$

On the other hand, from the condition $\frac{N(t'_n)}{N(t_n)} \to \infty$, we obtain

$$u^{[t_n]}(s_n) \rightharpoonup 0 \text{ weakly in } L^2.$$

Therefore $M(\varphi) = M(u) = 0$, a contradiction! □

Corollary 8.9 ($N(t)$ at blowup). *Let u be APMS on maximal interval I with frequency function $N(t)$. If $\sup I < \infty$ then $N(t) \gtrsim_u |\sup I - t|^{-\frac{1}{2}}$. In particular,*

$$N(t) \to \infty \text{ as } t \to \sup I. \tag{8.7}$$

If $\sup I = \infty$, then for any $t_0 \in I$, we have

$$N(t) \gtrsim_{t_0,u} \langle t - t_0 \rangle^{-\frac{1}{2}}. \tag{8.8}$$

The same statement holds for $\inf I$. *Here,* $\langle t \rangle = (1 + t^2)^{\frac{1}{2}}$.

Proof. If $\sup I < \infty$, then the fact that I contains $[t - \delta N(t)^{-1}, t + \delta N(t)^{-2}]$ implies $\sup I > t + \delta N(t)^{-2}$, we get (8.7). If $\sup I = \infty$, we discuss two cases, $t_0 \in [t - \delta N(t)^{-2}, t + \delta N(t)^{-2}]$ and $t_0 \notin [t - \delta N(t)^{-2}, t + \delta N(t)^{-2}]$. In the first case $N(t) \sim N(t_0) \sim 1$ so is (8.8). In the second case, we have $|t - t_0| \gtrsim N(t)^{-2}$, $\langle t - t_0 \rangle \gtrsim N(t)^{-2}$ thus (8.8). □

We will also rely on the following important property for $N(t)$:

Proposition 8.10 (Quasi-Continuous dependence of $N(t)$ on u). *Let $u_n : I_n \times \mathbb{R}^2 \to \mathbb{C}$ be a sequence of APMS solutions with frequency scale function $N_n(t)$ and compactness modulus function $C(\eta) : \mathbb{R}^+ \to \mathbb{R}^+$, independent of n. Let u_n converge locally uniformly to a non-zero solution u on the lifespan I, i.e. $\forall$ compact set $K \subset I$, for n sufficiently large, $K \subset I_n$ and*

$$\|u_n - u\|_{L_t^\infty L_x^2 \cap L_{t,x}^4 (K \times \mathbb{R}^2)} \to 0.$$

Then u is APMS with $N(t)$ and the same compactness modulus function $C(\eta)$. Moreover,

$$\liminf_{n\to\infty} N_n(t) \lesssim_{u,c} N(t) \lesssim \limsup_{n\to\infty} N_n(t) \ \forall t \in I.$$

The proposition can be proved following the similar idea as Proposition 8.8 for instance, we will leave the proof to the interested readers.

Finally, based on everything we know about $N(t)$ so far, we can reduce our consideration to more specific enemies, for which we have a control on $N(t)$.

Proposition 8.11 (Reduce further to enemies). *Assume the scattering result fails. Then there exists a minimal mass blowup solution $u(t,x)$ which is APMS on $[0,\infty)$ with $N(t) \leq 1$.*

Proof. Our previous analysis gives us the following: suppose the scattering result fails, then there exists critical mass $M_c > 0$ and critical solution $V(t,x)$ such that

$$\begin{cases} V : J \times \mathbb{R}^2 \to \mathbb{C}, J \text{ open }, 0 \in J, \\ M(V) = M_c, \\ S_{\geq 0}(V) = S_{\leq 0}(V) = \infty, \\ V \text{ is APMS with frequency function given by } N_V(t). \end{cases}$$

Now we will construct another solution u with the additional property that $N_u(t) \leq 1$.

Let $J_n \subset J$ be compact intervals approaching J. Since $N(t)$ is continuous and J_n is compact, we have

$$1 \lesssim_{J_n} N_V(t) \lesssim_{J_n} 1.$$

We can take $t_n \in J_n$ for which

$$N_V(t_n) \geq \frac{1}{2} N_V(t), \forall t \in J_n.$$

As $S_{J_n}(V) \to \infty$, without loss of generality, we can assume $S_{\geq t_n, t \in J_n}(V) \to \infty$ as $n \to \infty$.

Define

$$V^{[t_n]} = N_V(t_n)^{-1} V\Big(t_n + \frac{t}{N_V(t_n)^2}, \frac{x}{N_V(t_n)}\Big).$$

Then $V^{[t_n]}$ is simply the scaled copy of V on maximal life span $\tilde{I}_n$

$$V^{[t_n]} : \tilde{I}_n = (J - t_n)N_V(t_n)^2 \times \mathbb{R}^2 \to \mathbb{C},$$
$$N_{V^{[t_n]}}(t) = \frac{N_V(t_n + \frac{t}{N_V(t_n)^2})}{N_V(t_n)} \leq 2, \ \forall\, t \in \tilde{J}_n = (J_n - t_n)N(t_n)^2.$$

Note passing to subsequence,

$$V^{[t_n]}(0) = N_V(t_n)^{-1} V\left(t_n, \frac{x}{N_V(t_n)}\right) \to \varphi \in L^2(\mathbb{R}^2). \tag{8.9}$$

Let $w(t,x)$ be the forward maximal lifespan solution with initial data φ

$$w : [0, T_{\max}) \times \mathbb{R}^2 \to \mathbb{C}.$$

We will show w is APMS forward in time, $S_{\geq 0}(w) = \infty$ and the frequency scale function associated with w is uniformly bounded. Note from here, the construction of u (with $N_u(t) \leq 1$) simply follows from a scaling argument.

First we show $S_{\geq 0}(w) = \infty$. Suppose this is not true and $S_{\geq 0}(w) < \infty$, the classical local theory implies $T_{\max} = \infty$. From (8.9) and stability result we know

$$S_{\geq 0} V^{[t_n]} = S_{\geq t_n}(V) < \infty,$$

uniformly for sufficiently large n, which is a contradiction. Therefore $S_{\geq 0}(w) = \infty$.

Next we show w is APMS. From Proposition 8.10, it suffices to verify $V^{[t_n]}|_{\tilde{J}_n \cap (0,\infty)}$ converges locally uniformly to w. Let $t > 0$ and $t < T_{\max}$, we consider the compact interval $[0,t]$. Being a strong solution implies

$$S_{[0,t]} w < \infty.$$

Applying stability result again, we know for n sufficiently large

$$[0,t] \subset \tilde{J}_n(\text{ otherwise }, S_{\tilde{J}_n \cap [0,\infty)} V^{[t_n]} \lesssim 1, \text{ uniformly in } n)$$

and on $[0,t]$, $\|V^{[t_n]} - w\|_{L^4_{t,x} \cap L^\infty L^2([0,t])} \to 0$ as $n \to \infty$. Therefore on $[0, T_{\max})$, there exists a frequency scale function $N_w(t)$ such that

$$N_w(t) \lesssim \limsup_{n \to \infty} N_{V^{[t_n]}}(t) \lesssim 2, \forall t \in [0, T_{\max}).$$

Moreover, from the boundedness of $N_w(t)$, $T_{\max} = \infty$.

Now we obtain a solution w whose frequency scale function is $N_w(t) \leq C$ on $[0,\infty)$. Then for

$$u(t,x) = \frac{1}{C} w\left(\frac{t}{C^2}, \frac{x}{C}\right)$$

we can easily verify that $N_u(t) \leq 1$ on $[0,\infty)$. The Proposition is proved. □

Remark 8.12. *The solution u we obtain is only APMS in half time slab!*

Proposition 8.13 (Equivalent description of AMPS). *Let $u(t) : I \times \mathbb{R}^2 \to \mathbb{C}$ be APMS with frequency function $N(t)$. Then there exists a compactness modulus function $C(\eta) : \mathbb{R}^+ \to \mathbb{R}^+$ such that $\forall\ \eta > 0$,*

$$\int_{|x|\geq \frac{C(\eta)}{N(t)}} |u(t,x)|^2 < \eta,$$

$$\int_{|\xi|\geq C(\eta)N(t)} |\hat{u}(t,\xi)|^2 < \eta.$$

Proof. "$\Rightarrow$"

From the definition, we know

$$\left\{N(t)^{-1}u\left(t, \frac{x}{N(t)}\right)\right\}_{t\in I}$$

is precompact in $L^2(\mathbb{R}^2)$. Therefore it is totally bounded and can be covered by finite $\epsilon - web$. Take $\frac{\eta}{100}$ for example, there exist $f_1, f_2, \cdots, f_k \in L^2(\mathbb{R}^2)$ such that $\forall t \in I$, we can find f_j such that

$$\left\|N(t)^{-1}u\left(t, \frac{x}{N(t)}\right) - f_j(x)\right\|_{L^2(\mathbb{R}^2)} \leq \frac{\eta}{100}. \tag{8.10}$$

For this given $f_j(x)$,there exists $C_j(\eta)$ for which

$$\int_{|x|\geq C_j(\eta)} |f_j(x)|^2 dx \lesssim \frac{\eta}{100}. \tag{8.11}$$

Taking $C(\eta) = \max_{1\leq j\leq k} C_j(\eta)$ and using change of variable we get

$$\int_{|x|\geq \frac{C(\eta)}{N(t)}} |u(t,x)|^2 \leq \eta.$$

The localization in Fourier side can also be obtained by Parseval.

"$\Leftarrow$"

It is easy to see that the conclusion follows from the sufficient condition on compact set in L^2:

A bounded set $\mathcal{K}$ in $L^2(\mathbb{R}^n)$ is precompact if and only if

$$\int_{|x|\geq R} |f(x)|^2 dx \to 0, \quad \int_{|\xi|>R} |\hat{f}(\xi)|^2 d\xi \to 0$$

as $R \to \infty$ uniformly in $f \in \mathcal{K}$. □

Proposition 8.14 (Refined Duhamel). *Let $u(t,x)$ be APMS on I. Then $e^{-it\Delta}u(t) \rightharpoonup 0$ weakly in L^2 as $t \to \sup I$ and $t \to \inf I$. As a consequence*

$$\begin{aligned} u(t,x) &= w - \lim_{T\to\sup I} i\int_t^T e^{i(t-s)\Delta}F(u)(s)ds \\ &= w - \lim_{T\to\inf I} i\int_t^T e^{i(t-s)\Delta}F(u)(s)ds. \end{aligned} \tag{8.12}$$

Proof. From Duhamel formula, $\forall\, T \in I$,

$$u(t) = e^{i(t-T)\Delta}u(T) - i\int_T^t e^{i(t-s)\Delta}F(u(s))ds.$$

The formula (8.12) then follows if we can show

$$e^{-it\Delta}u(t) \rightharpoonup 0 \text{ weakly in } L^2(\mathbb{R}^2) \tag{8.13}$$

as $t \to \sup I$ and $t \to \inf I$. We only consider the case when $t \to \sup I$ as the other one is similar. We first assume $\sup I < \infty$. $\forall \varphi \in C_c^\infty(\mathbb{R}^2)$, we have

$$\begin{aligned} |\langle u(t), \varphi\rangle| &= \left|\int_{\mathbb{R}^2} u(t,x)\bar{\varphi}dx\right| \\ &\leq \|u(t)\|_{L^2(|x|\geq \frac{C(\eta)}{N(t)})}\|\varphi\|_2 + \|u\|_2\|\varphi\|_{L^2(|x|<\frac{C(\eta)}{N(t)})} \\ &\leq \eta^{\frac{1}{2}}\|\varphi\|_2 + M(u)^{\frac{1}{2}}\|\varphi\|_{L^2(|x|<\frac{C(\eta)}{N(t)})}. \end{aligned} \tag{8.14}$$

Let $t \to \sup I$, as $N(t) \to \infty$ as $t \to \sup I < \infty$ (Corollary 8.9), we know the second term goes to 0. Hence

$$|\langle u(t), \varphi\rangle| \lesssim \eta^{\frac{1}{2}} + o_t(1).$$

From the arbitrariness of η, we prove that $u(t) \rightharpoonup 0$, weakly in L^2. Finally as $\sup I < \infty$, and $t \to e^{it\Delta}$ is continuous in L^2, we establish (8.13).

Now we consider $\sup I = \infty$. It suffices to show $\forall \varphi \in C_c^\infty(\mathbb{R}^2)$

$$\lim_{t\to\infty} \langle u(t), e^{it\Delta}\varphi\rangle_{L^2} = 0. \tag{8.15}$$

Using Hölder and Corollary 8.9, we estimate

$$\begin{aligned} |\langle u(t), e^{it\Delta}\varphi\rangle| &\leq \|u(t,x)\|_2\|e^{it\Delta}\varphi\|_{L^2(|x|\leq\frac{C(\eta)}{N(t)})} + \eta^{\frac{1}{2}}\|\varphi\|_2 \\ &\lesssim M(u)^{\frac{1}{2}}\frac{C(\eta)}{N(t)}\|e^{it\Delta}\varphi\|_\infty + \eta^{\frac{1}{2}}\|\varphi\|_2 \\ &\lesssim M(u)^{\frac{1}{2}}C(\eta)\langle t - t_0\rangle^{\frac{1}{2}}|t|^{-1} + \eta^{\frac{1}{2}}\|\varphi\|_2. \end{aligned}$$

Take $t \to \infty$ the $\eta \to 0$ we obtain (8.15). □

We continue to introduce the preliminaries needed for the future analysis.

Our ultimate goal is to obtain the boundedness of scattering size $S(u) = \int_{J\times\mathbb{R}^2} |u(t,x)|^4 dtdx$ uniform in J for any finite mass solution. This can be done by proving the nonexistence of minimal mass blowup solution, we first get a relation between $S(u)$ and $N(t)$ if u is a minimal blowup solution which is AMPS. The intuition we should have is $u(t,x) \sim N(t)R(xN(t))$ thus

$$\|u\|^4_{L^4_{t,x}(J)} \sim \|N(t)N(t)^{-\frac{3}{4}}\|^4_{L^4_t} \sim \int_J N^2(t)dt.$$

Indeed we have

Lemma 8.15. *Let $u : I \times \mathbb{R}^2 \to \mathbb{C}$ be APMS with frequency parameter $N(t)$. Then for any subinterval $J \subset I$*

$$\int_J N(t)^2 dt \lesssim \int_J \int_{\mathbb{R}^2} |u(t,x)|^4 dx dt \lesssim 1 + \int_J N(t)^2 dt.$$

Proof. We first prove

$$\int_J N(t)^2 dt \lesssim \int_J \int_{\mathbb{R}^2} |u(t,x)|^4 dx dt.$$

As u is APMS, there exist $C \gg 1$ such that

$$1 \lesssim_u \int_{|x| \le \frac{C}{N(t)}} |u(t,x)|^2 dx \lesssim \frac{C}{N(t)} \left(\int_{\mathbb{R}^2} |u(t,x)|^4 dx \right)^{\frac{1}{2}}.$$

Square both sides and integrating over J we have

$$\int_J N(t)^2 dt \lesssim \int_J \int_{\mathbb{R}^2} |u(t,x)|^4 dx dt.$$

Next we show the second half.

Assume first that $\int_J N(t)^2 dt \le \delta$, choose $t_0 \in J$ such that $|J| N(t_0)^2 \le 2\delta$, using Duhamel and Strichartz we have

$$u(t) = e^{i(t-t_0)\Delta} u(t_0) - i \int_{t_0}^t e^{i(t-s)\Delta} |u|^2 u(s) ds,$$

$$\begin{aligned} \|u\|_{L^4_{t,x}(J)} &\le \|u_{>C(\eta)N(t_0)}(t_0)\|_2 + \|P_{<C(\eta)N(t_0)} e^{i(t-t_0)\Delta} u(t_0)\|_{L^4_{t,x}(J)} + \|u\|_{L^4_{t,x}(J)} \\ &\le \eta + (C(\eta)N(t_0))^{\frac{1}{2}} |J|^{\frac{1}{4}} \|u(t_0)\|_2 + \|u\|^4_{L^4_{t,x}(J)} \\ &\le \eta + C(\eta)\delta^{\frac{1}{4}} + \|u\|^4_{L^4_{t,x}(J)}. \end{aligned}$$

First take η small, then δ small, a continuation argument gives $\|u\|_{L^4_{t,x}(J)} \lesssim 1$. In the general case, partition J into subintervals $J = \bigcup I_j$ such that on each I_j, $\int_{I_j} N(t)^2 dt \le \delta$. Easy to see $\sharp\{I_j\} = \frac{1 + \int N(t)^2 dt}{\delta}$. On each time interval applying the same argument we have

$$\begin{aligned} \|u\|^4_{L^4_{t,x}} &= \sum \int \int_{I_j} |u(t,x)|^4 dx dt \\ &\lesssim \sharp\{I_j\} \lesssim 1 + \int_J N(t)^2 dt. \end{aligned}$$

□

Remark 8.16. *In d-dimensions, the corresponding statement is*

$$\int_J N(t)^2 dt \lesssim \int_J \int_{\mathbb{R}^d} |u(t,x)|^{\frac{2(d+2)}{d}} dx dt \lesssim 1 + \int_J N(t)^2 dt.$$

Here, $u(t,x)$ is AMPS in the sense $\{N(t)^{-\frac{d}{2}} u(t, \frac{x}{N(t)})\}_{t \in I}$ is precompact in $L^2(\mathbb{R}^d)$.

Lemma 8.17 (2nd version of local constancy). *Let u be APMS with $N(t)$. Suppose*

$$\int_{J\times\mathbb{R}^2} |u(t,x)|^4 dtdx \leq C.$$

Then $N(t_1) \sim_C N(t_2), \forall t_1, t_2 \in J$.

Proof. By partition the interval J, we can assume $C = \epsilon \ll 1$. Using Duhamel, we write

$$u(t_2) = e^{i(t_2-t_1)\Delta}u(t_1) - i\int_{t_1}^{t_2} e^{i(t_2-t)\Delta}|u|^2u(t)dt.$$

From the frequency localization, we know there exists η such that

$$\|P_{c(\eta)N(t_i)\leq C(\eta)N(t_i)}u(t_i)\|_2 \geq \frac{99}{100}M^{\frac{1}{2}}(u).$$

Denote $P^i_{med} = P_{c(\eta)N(t_i)\leq\cdot\leq C(\eta)N(t_i)}$. Using this lower bound, Duhamel formula and the Strichartz, we estimate

$$\begin{aligned}\|P^2_{med}u(t_1)\|_2 &\geq \|P^2_{med}u(t_2)\|_2 - C\|u\|^3_{4,4}\\ &\geq \frac{99}{100}M^{\frac{1}{2}}(u) - c\varepsilon^{\frac{3}{4}} \geq \frac{98}{100}M^{\frac{1}{2}}(u).\end{aligned}$$

Therefore the two frequencies must overlap, otherwise

$$\|u(t_1)\|_2 \geq 2\frac{98}{100}M^{\frac{1}{2}} \geq \frac{3}{2}M^{\frac{1}{2}}, \text{ a contradiction!}$$

The overlapping fact gives that

$$c(\eta)N(t_1) \leq C(\eta)N(t_2),\ c(\eta)N(t_2) \leq C(\eta)N(t_1).$$

Hence $N(t_1) \sim N(t_2)$ on J. □

Remark. In d dimensions,

$$\iint_{J\times\mathbb{R}^d} |u|^{\frac{2(d+2)}{d}} dxdt \leq C$$

implies $N(t_1) \sim N(t_2), \forall t_1, t_2 \in J$.

8.2 Estimates

In this subsection, we will introduce some preliminary estimates used to prove scattering, including the important Morawetz estimates. We start with the following bilinear Strichartz estimate, which is used to deal with the interactions between high-low frequencies. We state the result in d dimensions.

Lemma 8.18 (Bilinear Strichartz). *Let $d \geq 1$ and $M \leq N$, then*

$$\|[e^{it\Delta}P_Mf][e^{it\Delta}P_Ng\|_{L^2_{t,x}(\mathbb{R}\times\mathbb{R}^d)} \lesssim M^{\frac{d-1}{2}}N^{-\frac{1}{2}}\|f\|_2\|g\|_2. \tag{8.16}$$

When $d = 1$ we require $M \leq \frac{1}{4}N$, so $P_MP_N = 0$.

Proof. When $d = 2$

$$\begin{aligned} RHS(8.16) &\le \|e^{it\Delta}P_M f\|_{4,4}\|e^{it\Delta}P_N g\|_{4,4} \\ &\lesssim \|P_M f\|_2 \|P_N g\|_2. \end{aligned}$$

When $d \ge 3$

$$\begin{aligned} RHS(8.16) &\le \|e^{it\Delta}P_M f\|_{4,4}\|e^{it\Delta}P_N g\|_{4,4} \\ &\lesssim M^{-d(\frac14-\frac{d-1}{2d})}N^{-d(\frac14-\frac{d-1}{2d})}\|e^{it\Delta}P_M f\|_{4,\frac{2d}{d-1}}\|e^{it\Delta}P_N f\|_{4,\frac{2d}{d-1}} \\ &\lesssim M^{\frac{d-2}{4}}N^{\frac{d-2}{4}}\|f\|_2\|g\|_2. \end{aligned}$$

This settles the case if $M \sim N$ and $d \ge 2$. So in all dimensions, we can assume $M \le \frac14 N$.

By Parseval and duality

$$\begin{aligned} &\|e^{it\Delta}P_M f \cdot e^{it\Delta}P_N g\|_{2,2} \qquad (8.17)\\ &= \|\mathcal{F}_{t,x}\big(e^{it\Delta}P_M f e^{it\Delta}P_N g\big)(\tau,\xi)\|_{L^2_\tau L^2_\xi} \\ &= \sup_{\|F\|_{L^2_{\tau,\xi}}\le 1}\left\langle \int \delta(\tau-\tau'-|\xi-\eta|^2)\widehat{P_M f}(\xi-\eta)\delta(\tau'-|\eta|^2)\widehat{P_N g}(\eta)d\tau' d\eta, F(\tau,\xi)\right\rangle_{L^2_{\tau,\xi}} \\ &= \sup_{\|F\|_{L^2_{\tau,\xi}}\le 1}\int \widehat{P_M f}(\xi-\eta)\widehat{P_N g}(\eta)\bar F(|\xi-\eta|^2+|\eta|^2,\xi)d\xi d\eta \\ &= \sup_{\|F\|_{2,2}\le 1}\int \widehat{P_M f}(\xi)\widehat{P_N g}(\eta)\bar F(|\eta|^2+|\xi|^2,\eta+\xi)d\eta d\xi. \qquad (8.18) \end{aligned}$$

By orthogonal change of variable if necessary we may assume the first coordinate of η dominate, i.e,

$$\eta_1 - \xi_1 \sim N.$$

We now apply change of variable:

$$\begin{gathered} (\xi,\eta) \to (\omega,\beta,\zeta) \\ w = |\xi|^2+|\eta|^2,\ \beta = (\xi_2,\xi_3,\cdots,\xi_d),\ \zeta = \xi+\eta. \end{gathered}$$

Note that $|\beta| \lesssim M$ and the Jacobian $|\mathcal{J}| \sim N^{-1}$, we have

$$\begin{aligned} (8.18) &= \int \hat f_M(\xi)\hat g_N(\eta)\bar F(\omega,\zeta)\mathcal{J} d\omega d\zeta d\beta \\ &\lesssim \left(\int |\hat f_M(\xi)\hat g_N(\eta)|^2|\mathcal{J}|^2 d\omega d\zeta d\beta\right)^{\frac12}\left(\int |F(\omega,\zeta)|^2 1_{|\beta|\lesssim M} d\omega d\zeta d\beta\right)^{\frac12} \\ &\lesssim M^{\frac{d-1}{2}}\|F\|_{2,2}N^{-\frac12}\left(\int |\hat f_M(\xi)\hat g_N(\eta)|^2 d\xi d\eta\right)^{\frac12} \\ &\lesssim M^{\frac{d-1}{2}}N^{-\frac12}\|f\|_2\|g\|_2. \end{aligned}$$

□

Corollary 8.19 (Bilinear Strichartz II). *Let M, N be the same as in the last Lemma. Given any spacetime slab $I \times \mathbb{R}^d, t_0 \in I$ and functions u, v defined on $I \times \mathbb{R}^d$, we have*

$$\|P_{\geq N}uP_{\leq M}v\|_{L^2_{t,x}} \lesssim M^{\frac{d-1}{2}}N^{-\frac{1}{2}}(\|P_{\geq N}u(t_0)\|_2 + \|(i\partial_t + \Delta)u_{\geq N}\|_{L^{\frac{2(d+2)}{d+4}}_{t,x}})$$
$$\cdot(\|P_{\leq M}v(t_0)\|_2 + \|(i\partial_t + \Delta)v_{\leq N}\|_{L^{\frac{2(d+2)}{d+4}}_{t,x}}).$$

The very last estimate is a monotonicity formula known as Morawetz estimate. Such estimate was first introduced by Morawetz in Klein-Gordon equation [35]. In 1978, Lin-Strauss [31] adapted the estimate to Schrödinger setting. Morawetz estimate gives an a priori control on the spacetime norm of the solution and were proved to be effective in showing the scattering of the solution, especially in energy subcritical problem. Later, the estimate is also shown to play a very important role in critical problems. In the following, we will give a short discussion on the Morawetz estimate, and deduce the one that will be used in our problem.

Consider NLS in d-dimensions

$$iu_t + \Delta u = \mu|u|^p u.$$

It can be verified that

$$\begin{aligned}\frac{d^2}{dt^2}\int a(x)|u(t,x)|^2dx &= 2\frac{d}{dt}\int \nabla a(x)\mathrm{Im}(\bar{u}(t,x)\nabla u(t,x))dx \\ &= 2\int_{\mathbb{R}^d} \partial_{x_j}\partial_{x_k}a(x)\mathrm{Re}(\partial_j u\partial_k\bar{u})dx \\ &\quad + \frac{p\mu}{p+2}\int_{\mathbb{R}^d}|u(t,x)|^{p+2}\Delta a(x)dx \\ &\quad - \frac{1}{2}\int_{\mathbb{R}^d}|u(t,x)|^2\Delta^2 a(x)dx.\end{aligned}$$

Take $a(x) = |x|$. Easy to check in $d \geq 3$,

$$\nabla a(x) = \frac{x}{|x|},\ \Delta^2 a(x) < 0,\ \Delta a(x) = \frac{d-1}{|x|},\ \partial_j\partial_k a \text{ is positive definite,}$$
$$\left|\int \nabla a(x)\mathrm{Im}(\bar{u}\nabla u)dx\right| \lesssim \|u\|_2\|\nabla u\|_2.$$

Integrating over $\mathbb{R}$, we obtain for $\mu > 0$ and $d \geq 3$

$$\int_{\mathbb{R}}\int_{\mathbb{R}^d}\frac{|u(t,x)|^{p+2}}{|x|}dxdt \lesssim \|u\|_2\sup_{t\in\mathbb{R}}\|\nabla u(t)\|_2.$$

Remark 8.20. *The restriction that $d \geq 3$ is used to prove $\Delta^2 a(x) \leq 0$ in the* *distributional sense.*

$$a(x) = |x|, \ \nabla a(x) = \frac{x}{|x|}, \ \ \Delta a(x) = \frac{d-1}{|x|},$$
$$\nabla\left(\frac{1}{|x|}\right) = -\frac{x}{|x|^3}, \Delta\left(\frac{1}{|x|}\right) = \frac{d-3}{|x|^3}, \quad d>3,$$
$$-\Delta\left(\frac{1}{|x|}\right) = 4\pi\delta(x), \ d=3.$$

The classical Morawetz introduced above has a weight $\frac{1}{|x|}$ in it. It is more useful to treat the radial solutions. In 2003, I-team [10] discovered the interaction Morawetz which gives the global space-time control of solution with no weights. Next we introduce a simple version of the proof.

Take $3d$ cubic defocusing NLS as an example

$$iu_t + \Delta u = |u|^2 u.$$

Define

$$M(t) = \int_{\mathbb{R}^6} \frac{x_j - y_j}{|x-y|} \mathrm{Im}[\bar{u}(t,x)\bar{u}(t,y)\partial_{x_j}(u(t,x)u(t,y))]dxdy.$$

From simple Hölder inequality we have

$$|M(t)| \lesssim \|u\|_2^3 \|\nabla u\|_2.$$

On the other hand

$$\frac{d}{dt}M(t) = \int_{\mathbb{R}^6} \frac{x_j - y_j}{|x-y|}\frac{\partial}{\partial t}\mathrm{Im}(\bar{u}(t,x)\partial_{x_j}u(t,x))|u(t,y)|^2 dxdy \tag{8.19}$$
$$+ \int_{\mathbb{R}^6} \frac{x_j - y_j}{|x-y|}\mathrm{Im}(\bar{u}(t,x)\partial_{x_j}u(t,x))\frac{\partial}{\partial t}|u(t,y)|^2 dxdy \tag{8.20}$$

Note from the local momentum conservation,

$$\partial_t \mathrm{Im}(\bar{u}\partial_j u) = -2\partial_k \mathrm{Re}(\partial_k u \partial_j \bar{u}) + \frac{1}{2}\partial_j \Delta(|u|^2) - \frac{1}{2}\partial_j(|u|^4),$$

we have

$$(8.19) = I + II + III.$$

$$I = \int_{\mathbb{R}^6} \frac{x_j - y_j}{|x-y|}(-2)\partial_k \mathrm{Re}(\partial_k u \partial_j \bar{u})(t,x)|u|^2(t,y)dxdy$$
$$= 2\int_{\mathbb{R}^6}\left[\frac{\delta_{jk}}{|x-y|} - \frac{(x_j-y_j)(x_k-y_k)}{|x-y|^3}\right]\mathrm{Re}(\partial_k u \partial_j \bar{u})(t,x)|u|^2(t,y)dxdy$$
$$= 2\int_{\mathbb{R}^6}\left[\frac{|\nabla u(t,x)|^2}{|x-y|} - \frac{|(x-y)\cdot\nabla u(t,x)|^2}{|x-y|^3}\right]|u|^2(t,y)dxdy.$$

$$
\begin{aligned}
II &= \int_{\mathbb{R}^6} \frac{x_j - y_j}{|x-y|} \frac{1}{2} \partial_{x_j} \Delta |u|^2(t,x) |u(t,y)|^2 dxdy \\
&= -\frac{1}{2} \int_{\mathbb{R}^6} \frac{2}{|x-y|} \Delta |u|^2(t,x) |u|^2(t,y) dxdy \\
&= 4\pi \int_{\mathbb{R}^6} |u|^2(t,x) |u|^2(t,y) \delta(x,y) dxdy \\
&= 4\pi \int_{\mathbb{R}^6} |u(t,x)|^4 dx.
\end{aligned}
$$

$$
\begin{aligned}
III &= \int_{\mathbb{R}^6} \frac{x_j - y_j}{|x-y|} \left(\frac{1}{2}\right) \partial_{x_j} |u|^4(t,x) |u|^2(t,y) dxdy \\
&= \frac{1}{2} \int_{\mathbb{R}^6} \frac{2}{|x-y|} |u|^4(t,x) |u|^2(t,y) dxdy \geq 0.
\end{aligned}
$$

Now we write

$$
(8.20) = \int_{\mathbb{R}^6} \frac{x_j - y_j}{|x-y|} \mathrm{Im}(\bar{u} \partial_j u(t,x)) \partial_t |u|^2(t,y) dxdy.
$$

Note $\partial_t |u|^2 = 2\mathrm{Re}(u_t \bar{u}) = -2\mathrm{Im} \nabla \cdot (\bar{u} \nabla u)$, we continue to write

$$
(8.20) = 2 \int \left(\frac{(x_j - y_j)(x_k - y_k)}{|x-y|^3} - \frac{\delta_{jk}}{|x-y|} \right) \mathrm{Im}(\bar{u} \partial_j u)(t,x) \mathrm{Im}(\bar{u} \partial_k u)(t,y) dxdy.
$$

Denote $\vec{n} = \frac{x-y}{|x-y|}$,

$$
\begin{aligned}
(8.20) =& 2 \int_{\mathbb{R}^6} \frac{1}{|x-y|} \big(\mathrm{Im}(\bar{u} \nabla u)(x) \cdot \vec{n} \ \mathrm{Im}(\bar{u} \nabla u)(y) \cdot \vec{n} \\
&- \mathrm{Im}(\bar{u} \nabla u)(x) \mathrm{Im}(\bar{u} \nabla u)(y) \big) dxdy \\
=& 2 \int_{\mathbb{R}^6} \frac{1}{|x-y|} \mathrm{Im} \bar{u} \ (\nabla u - \nabla u \cdot \vec{n}\vec{n})(t,x) \mathrm{Im} \bar{u} \ (\nabla u - \nabla u \cdot \vec{n}\vec{n})(t,y) dxdy.
\end{aligned}
$$

Using Hölder inequality, one has

$$
\begin{aligned}
|8.20| \leq& \int_{\mathbb{R}^6} \frac{1}{|x-y|} |u(t,x)|^2 |\nabla u - \nabla u \cdot \vec{n} \ \vec{n}|^2(t,y) dxdy \\
&+ \int_{\mathbb{R}^6} \frac{1}{|x-y|} |u(t,y)|^2 |\nabla u - \nabla u \cdot \vec{n} \ \vec{n}|^2(t,x) dxdy.
\end{aligned}
$$

An easy change of variable gives $I + (8.20) \geq 0$. Finally integrating t over I, we have

$$
\int\int_{I \times \mathbb{R}^3} |u(t,x)|^4 dtdx \lesssim \|u\|_2^3 \sup_{t \in I} \|\nabla u\|_2.
$$

Exercise. Prove in *d-dimensions* ($d \geq 3$)

$$\int_{I\times\mathbb{R}^d} (-\Delta^2|x-y|)|u(t,x)|^2|u(t,y)|^2 dxdydt \lesssim \|u\|_2^3 \sup_{t\in I} \|u(t)\|_{\dot{H}^1}.$$

Next we explain an application of the interaction Morawetz inequality in proving scattering for subcritical NLS.

Theorem 8.21. *Consider*

$$\begin{cases} iu_t + \Delta u = |u|^2 u, \\ u(0) \in H^1(\mathbb{R}^3). \end{cases}$$

Then u exists globally and scatters.

Proof. We only sketch the steps of proof.

Step 1: Global existence was obtained before by usual contraction argument and energy conservation.

Step 2: Scattering is reduced to showing

$$\|u\|_{L^5_{t,x}(\mathbb{R}\times\mathbb{R}^3)} \leq C(\|u_0\|_{H^1}). \tag{8.21}$$

Indeed, if (8.21) holds, then the Strichartz norm

$$\|u\|_{S^1(\mathbb{R}\times\mathbb{R}^3)} := \|\langle\nabla\rangle u\|_{\infty,2} + \|\langle\nabla\rangle u\|_{2,6}$$

is finite. Then $u_\pm = u_0 - i\int_0^{\pm\infty} e^{i(t-s)\Delta}|u|^2 u(s)ds \in H^1$.

Step 3: To prove (8.21), we use Morawetz inequality which implies

$$\int\int_{\mathbb{R}\times\mathbb{R}^3} |u(t,x)|^4 dtdx \leq C(E,M).$$

Partition time interval

$$\mathbb{R} = \bigcup_{j=1}^{J} I_j$$

such that

$$\int_{I_j\times\mathbb{R}^3} |u(t,x)|^4 \sim \eta.$$

Applying Strichartz on $I_j = [t_j, t_{j+1})$ we have

$$\begin{aligned} \|u\|_{5,5} &\lesssim \|u(t_j)\|_{H^1} + \||\nabla|^{\frac{1}{2}}(|u|^2 u)\|_{\frac{10}{7},\frac{10}{7}} \\ &\lesssim 1 + \||\nabla|^{\frac{1}{2}} u\|_{\frac{10}{3},\frac{10}{3}} \|u\|_{5,5}^2 \\ &\lesssim 1 + \|u\|_{5,5}\|u\|_{S^1(I_j)}\|u\|_{4,4}^{\frac{2}{3}}\|u\|_{10,10}^{\frac{1}{3}} \\ &\lesssim 1 + \eta^{\frac{2}{3}}\|u\|_{5,5} \\ \Rightarrow \|u\|_{L^5_{t,x}(I_j)} &\lesssim 1. \end{aligned}$$

Summing bounds on I_j gives the global $L^5_{t,x}$-*norm* control. □

The scattering will rely on the Morawetz estimate in high dimensions $d \geq 3$. However, Morawetz estimate require solutions have at least $\dot{H}^{\frac{1}{4}}$ regularity thus is not directly useful. A first step is to establish a superficial connection between Morawetz estimate and $N(t)$.

Heuristically, our solution is

$$u(t,x) = N(t)^{\frac{d}{2}} R(xN(t)).$$

So in $d = 3$,

$$\int_{J\times\mathbb{R}^3} |u(t,x)|^4 dtdx \sim \int_J N(t)^{\frac{3}{2}\times 4} N(t)^{-\frac{3}{4}\times 4} dt = \int_J N(t)^3 dt.$$

One can also check in $d \geq 4$,

$$\int_{J\times\mathbb{R}^{2d}} (-\Delta^2|x-y|)|u(t,x)|^2|u(t,y)|^2 dxdydt \sim \int_J N(t)^3 dt \sim \int_J N(t)^3 dt.$$

In lower dimensions $d = 1, 2$, the proof of Morawetz estimate is tricky, we will not discuss here. To complete the whole picture, we will go back to 3-*dimensions*.

Lemma 8.22 (Discritize $N(t)$). *Let u be APMS with frequency scale $N(t)$. Suppose $J = \bigcup J_k$ with $S_I(u) \sim \epsilon$. Let $N(J_k) = \sup_{t\in J_k} N(t)$. Then*

$$\sum_k N(J_k) \sim \int_J N(t)^3 dt.$$

We will skip the proof. The intuition we should have here is

1. local constancy on each J_k,
2. J_k is more or less a characterize interval with the length $N(t_k)^{-2}$.

Therefore

$$\begin{aligned}\int_J N(t)^3 dt &= \sum \int_{J_k} N(t)^3 dt \\ &\sim \sum N^3(t_k)|J_k| \sim \sum N(t_k).\end{aligned}$$

Before proceeding, we summarize the preliminary work we did so far and we will state them in the general dimensions

1. If scattering results for *d-dimensional* mass critical NLS

$$iu_t + \Delta u = |u|^{\frac{4}{d}} u$$

fails for radial data, then there exists minimal mass blowup solution which is APMS with frequency scale $N(t)$.
2. Using scaling argument, we can further reduce our enemy to the case : u is APMS with $N(t) \leq 1$ on $[0,\infty)$. (Note we don't care what's happening for $t \leq 0$.)

3. Refined Duhamel formula

$$u(t) = w - \lim_{T\to\infty} i \int_t^T e^{i(t-s)\Delta} |u|^{\frac{4}{d}} u(s) ds.$$

4. Morawetz inequality: for $d \geq 3$,

$$\int\int_{J\times\mathbb{R}^{2d}} -\Delta^2(|x-y|)|u(t,x)|^2|u(t,y)|^2 dxdydt \leq C\|u\|_2^3 \sup_{t\in J}\|u(t)\|_{\dot{H}^1}.$$

In particular, in $d = 3$, we have

$$\int_{J\times\mathbb{R}^3} |u(t,x)|^4 dtdx \leq C\|u\|_2^3 \sup_{t\in J}\|u(t)\|_{\dot{H}^1}.$$

Bilinear Strichartz: Let $d \geq 1, M \leq \frac{1}{4}N$. Then

$$\|P_{\geq N}uP_{\leq M}V\|_{L^2_{t,x}(J\times\mathbb{R}^d)} \lesssim \frac{M^{\frac{d-1}{2}}}{N^{\frac{1}{2}}}\|P_{\geq N}u\|_{S^0_*(J\times\mathbb{R}^2)}\|P_{\leq M}V\|_{S^0_*(J\times\mathbb{R}^2)}.$$

Here, $\|u\|_{S^0_*(J\times\mathbb{R}^2)} = \|u(t_0)\|_2 + \|(i\partial_t+\Delta)u\|_{L_{t,x}^{\frac{2(d+2)}{d+4}}}$. (RMK: $d \geq 2$, the estimate is trivially true for $M \gtrsim N$.)

5. Local constancy of $N(t)$: $\exists\delta(u)$ such that $N(t_1) \sim N(t_2)$ for $t_1, t_2 \in [t_0 - \delta N^{-2}(t_0), t_0 + \delta N^{-2}(t_0)]$.
6. $N(t)$ VS Strichartz norm

$$\int_J N(t)^2 dt \lesssim \int_{J\times\mathbb{R}^d} |u(t,x)|^{\frac{2(d+2)}{d}} dtdx \lesssim 1 + \int_J N(t)^2 dt.$$

We are ready to give the proof of the scattering result. We will present the case when $d = 3$ and when the data is spherically symmetric.

9 Proof of the scattering result

Let $d = 3$ and we consider the scattering conjecture for radial solutions. Here we follow the proof in [13].

Suppose the conjecture fails, the discussion from previous section show that there exist an radial, APMS solution $u(t,x)$ on $[0,\infty)$ with $N(t) \leq 1$. We distinguish two cases:

Case 1: $\int_0^\infty N(t)^3 dt = K < \infty$.

Case 2: $\int_0^\infty N(t)^3 dt = \infty$.

In Case 1, as $\int_0^\infty N(t)^3 dt$ scales the same with Morawetz inequality (which is $\dot{H}^{\frac{1}{4}}$ in scaling), the boundedness indicates the solution has higher regularity. We will set up an iteration scheme to prove $H^{1+\epsilon}$ regularity and get a contradiction from there.

In Case 2, we will get a contradiction from a frequency localized Morawetz inequality.

We first consider Case 1: $\int_0^\infty N(t)^3 dt = K < \infty$.

To get higher regularity, we need to prove high frequency has decay. To this end, we will establish an inductive estimate on high frequency in certain Strichartz norm.

Lemma 9.1. *Let $J \subset [0,\infty)$ be compact and $J = \bigcup J_k$ such that $S_{J_k}(u) \sim \epsilon, 0 < s < 1 + \frac{4}{3}$. Then*

$$\|u_{>N}\|_{2,6} \leq \|u_{>N}(t_0)\|_2 + \sum_{M \leq \eta N} \left(\frac{M}{N}\right)^s \|u_{>M}\|_{2,6}$$
$$+\eta^{\frac{4}{3}} \|u_{>\eta N}\|_{2,6} + \frac{C_0 K^{\frac{1}{2}}}{(\eta N)^{\frac{1}{2}}} \sup_{J_k} \|u_{>\eta N}\|_{S^0_*(J_k \times \mathbb{R}^3)}.$$

Here the $(2,6)$ norm is on $J \times \mathbb{R}^3$, t_0 can be chosen as any point in J.

Proof. Due to the local constancy of $N(t)$ on J_k, we will consider $N(t)$ as piece wise constant:

$$N(t) \equiv N(J_k), \ \forall\, t \in J_k.$$

From Strichartz estimate, we only need to estimate $\|P_{>N}F(u)\|_{2,\frac{6}{5}}$ with $F(u) = |u|^{\frac{4}{3}}u$. We write

$$\begin{aligned}
F(u) &= F(u_{\leq \eta N}) + O(|u|^{\frac{4}{3}})u_{>\eta N} \\
&= F(u_{\leq \eta N}) + O(|u_{>C_0 N(t)}|^{\frac{4}{3}} + |u_{\leq C_0 N(t)}|^{\frac{4}{3}})u_{>\eta N} \\
&= F(u_{\leq \eta N}) + O(|u_{>C_0 N(t)}|^{\frac{4}{3}} + \chi_{> \frac{C_0}{N(t)}} |u_{\leq C_0 N(t)}|^{\frac{4}{3}})u_{>\eta N} \\
&\quad + O(\chi_{\leq \frac{C_0}{N(t)}} |u_{\leq C_0 N(t)}|^{\frac{4}{3}})u_{>\eta N} \\
&:= A + O(B)u_{>\eta N} + O(C)u_{>\eta N}.
\end{aligned}$$

We estimate the contribution from every term

$$\begin{aligned}
\|B\|_{2,\frac{6}{5}} &\leq \|u_{>\eta N}\|_{2,6}(\|u_{>C_0 N(t)}\|^{\frac{4}{3}}_{\infty,2} + \|\chi_{> \frac{C_0}{N(t)}} u_{\leq C_0 N(t)}\|^{\frac{4}{3}}_{\infty,2}) \\
&\leq \eta^{\frac{4}{3}} \|u_{>\eta N}\|_{2,6}, \\
\|P_{>N}A\|_{2,6} &\leq \|P_{>N}F(u_{\leq \eta N})\|_{2,6} \leq N^{-s}\||\nabla|^s F(u_{<\eta N})\|_{2,\frac{6}{5}} \\
&\leq N^{-s}\||\nabla|^s u_{\leq \eta N}\|_{2,6}\|u\|^{\frac{4}{3}}_{\infty,2} \\
&\lesssim \sum_{M \leq \eta N} \left(\frac{M}{N}\right)^s \|u_{>M}\|_{2,6}.
\end{aligned}$$

These two terms will be treated inductively. Term C is not, instead, it will provide the true improvement from the bilinear Strichartz. On each J_k

$$\begin{aligned}
&\|\chi_{\leq \frac{C_0}{N(t)}} |u_{\leq C_0 N(t)}|^{\frac{4}{3}} u_{>\eta N}\|_{2,\frac{6}{5}(J_k)} \\
&\leq \|u_{\leq C_0 N(t)} u_{>\eta N}\|_{2,2} \|u_{<C_0 N(t)}\|_{\infty,2}^{\frac{1}{3}} \|\chi_{\leq \frac{C_0}{N(t)}}\|_{\infty,6} \\
&\leq \frac{C_0 N(J_k)}{(\eta N)^{\frac{1}{2}}} \left(\frac{C_0}{N(J_k)}\right)^{\frac{1}{2}} \|P_{>\eta N} u\|_{S_*^0(J_k)} \\
&\leq C_0^{\frac{3}{2}} \left(\frac{N(J_k)}{\eta N}\right)^{\frac{1}{2}} \|P_{>\eta N} u\|_{S_*^0(J_k)}.
\end{aligned}$$

Summing each J_k together, we get

$$\begin{aligned}
\|C\|_{2,\frac{6}{5}(J)}^2 &= \sum_k \|C\|_{2,\frac{6}{5}(J_k)}^2 \\
&\leq \sum_k C_0^3 \frac{N(J_k)}{\eta N} \|P_{>\eta N} u\|_{S_*^0(J_k)}^2 \\
&\leq \frac{C_0^3}{\eta N} \left(\sum_k N(J_k)\right) \sup_k \|u_{>\eta N}\|_{S_*^0(J_k)}^2, \\
\|C\|_{2,\frac{6}{5}} &\leq \frac{C_0^{\frac{3}{2}}}{(\eta N)^{\frac{1}{2}}} K^{\frac{1}{2}} \sup_k \|u_{>\eta N}\|_{S_*^0(J_k)}.
\end{aligned}$$

Collecting the estimates for each piece gives rise to the right estimate. □

Based on this lemma, we have the following bound

Theorem 9.2. *Let $J \subset [0,\infty)$ be compact and $\int_J N(t)^3 dt \leq K < \infty$. Then*

$$\|u_{>N}\|_{2,6} \leq C_1 \left[\left(\frac{K}{N}\right)^{\frac{1}{2}} + \sigma_J(N)\right]. \tag{9.1}$$

Moreover, for $\frac{1}{2} \leq s_0 < 1 + \frac{4}{3}$,

$$\|u_{>N}\|_{L_t^2 L_x^6(J)} \lesssim \frac{K^{\frac{1}{2}}}{N^{s_0}} \|u\|_{L_t^\infty \dot{H}^{s_0 - \frac{1}{2}}([0,\infty))} + \sigma_J(N). \tag{9.2}$$

Here, for dyadic number M, $\sigma_J(M)$ is the frequency envelop that majorizes $\inf_{t\in J} \|u_{>M}(t)\|_2$: let $M = 2^l$,

$$\sigma_J(M) = \sum_{j\in\mathbb{Z}} 2^{-\frac{|j-l|}{3}} \inf_{t\in J} \|u_{>2^j}(t)\|_2.$$

Remark 9.3. *It's easy to verify that for $M \geq N$*

$$\sigma_J(M) \leq \sigma_J(N) \leq \left(\frac{M}{N}\right)^{\frac{1}{3}} \sigma_J(M).$$

Proof of Theorem 9.2.

Proof. We first partition $J = \bigcup J_k$ such that $S_{J_k}(u) \sim \epsilon$, on each J_k,

$$\begin{aligned}\|u\|_{S^0_*(J_k)} &\leq \|u\|_{\infty,2} + \||u|^{\frac{4}{3}}u\|_{\frac{10}{7},\frac{10}{7}} \\ &\lesssim 1 + S_{J_k}(u) \lesssim 1.\end{aligned}$$

Plugging this to Lemma 9.1 we obtain

$$\begin{aligned}\|u_{>N}\|_{2,6} \leq \sigma_J(N) + C_0 \Bigg[\sum_{M \leq \eta N} \frac{M}{N} \|u_{>M}\|_{2,6} \\ + \eta^{\frac{4}{3}} \|u_{>\eta N}\|_{2,6} + \frac{K^{\frac{1}{2}}}{(\eta N)^{\frac{1}{2}}} \Bigg].\end{aligned} \tag{9.3}$$

We will inductively prove the bound (9.1).

First, as J is compact interval, $\|u\|_{2,6} \leq C(J)$. Hence (9.1) holds for very small N.

Next, plugging the inductive assumption (9.1) into (9.3), we have

$$\begin{aligned}\|u_{>N}\|_{2,6} &\lesssim \sigma_J(N) + C_0 \Bigg[\sum_{M \leq \eta N} \frac{M}{N} \left(C_1 \sigma_J(M) + C_1 \left(\frac{K}{M}\right)^{\frac{1}{2}} \right) \\ &\quad + \eta^{\frac{4}{3}} \left(C_1 \sigma_J(\eta_N) + C_1 \left(\frac{K}{N\eta}\right)^{\frac{1}{2}} \right) \Bigg] + C_0 \frac{K^{\frac{1}{2}}}{(\eta N)^{\frac{1}{2}}} \\ &\leq \sigma_J(N)(1 + C_0 C_1 \eta^{\frac{2}{3}} + C_0 C_1 \eta) \\ &\quad + C_0 \left(\frac{K}{\eta N}\right)^{\frac{1}{2}} + C_0 C_1 \eta \left(\frac{K}{N}\right)^{\frac{1}{2}} + C_0 C_1 \eta^{\frac{1}{3}} \left(\frac{K}{N}\right)^{\frac{1}{2}} \\ &\leq C_1 \left[\sigma_J(N) + \left(\frac{K}{N}\right)^{\frac{1}{2}} \right].\end{aligned}$$

If C_1 is taken sufficiently large.

To see (9.2), we modify the above argument a little bit. On each J_k

$$\begin{aligned}\|u_{>\eta N}\|_{S^0_*(J_k \times \mathbb{R}^3)} &\leq \|u_{>\eta N}\|_{\infty,2} + \|P_{>\eta N}(|u|^{\frac{4}{3}}u)\|_{2,\frac{6}{5}} \\ &\leq (N\eta)^{-(s_0 - \frac{1}{2})} \|u\|_{L_t^\infty \dot{H}^{s_0 - \frac{1}{2}}} \\ &\quad + (\eta N)^{-(s_0 - \frac{1}{2})} \||\nabla|^{s_0 - \frac{1}{2}} u\|_{\infty,2} \|u\|_{2,6} \|u\|_{\infty,2}^{\frac{1}{3}} \\ &\lesssim (\eta N)^{-(s_0 - \frac{1}{2})} \|u\|_{L_t^\infty \dot{H}^{s_0 - \frac{1}{2}}}.\end{aligned}$$

Therefore, the estimate from Lemma 9.1 implies

$$\begin{aligned}\|u_{>N}\|_{2,6} \lesssim \sigma_J(N) + \sum_{M \leq \eta N} \frac{M}{N} \|u_{>M}\|_{2,6} + \eta^{\frac{4}{3}} \|u_{>\eta N}\|_{2,6} \\ + C_0 \frac{K^{\frac{1}{2}}}{(\eta N)^{s_0}} \|u\|_{L_t^\infty \dot{H}^{s_0 - \frac{1}{2}}}.\end{aligned}$$

Repeating the same argument above yields (9.2). □

Remark 9.4. *The above argument actually gives*

$$\|P_{>N}(|u|^{\frac{4}{3}}u)\|_{2,\frac{6}{5}} \lesssim \frac{K^{\frac{1}{2}}}{N^{s_0}}\|u\|_{L_t^\infty \dot{H}^{s_0-\frac{1}{2}}([0,\infty))}.$$

Finally we use this crucial result to prove higher regularity of u.

Theorem 9.5. *Let u be APMS with $N(t) \le 1$ on $[0,\infty)$. If $\int_0^\infty N(t)^3 dt = K < \infty$, then $u \in L_t^\infty \dot{H}^s([0,\infty))$ with $0 \le s < 1 + \frac{4}{3}$.*

Proof. Without loss of generality, we assume $N(t)$ is piecewise constant. Note

$$\sum N(J_k) = \int_0^\infty N(t)^3 dt = K < \infty, \text{ thus } N(J_k) \to 0, \text{ as } k \to \infty,$$

we have for each fixed N,

$$\|u_{>N}(t)\|_2 \to 0 \text{ as } t \to \infty.$$

Applying dominated convergence, $\sigma_{J_T}(N) \to 0$ for $J_T = [0,T)$ as $T \to \infty$. Applying (9.2) first on each J_T then take $T \to \infty$, we have

$$\begin{cases} \|u_{>N}\|_{2,6(\mathbb{R}^+)} \lesssim \frac{K^{\frac{1}{2}}}{N^{s_0}}\|u\|_{L_t^\infty \dot{H}^{s_0-\frac{1}{2}}(\mathbb{R}^+)}, \\ \|P_{>N}F(u)\|_{2,\frac{6}{5}(\mathbb{R}^+)} \lesssim \frac{K^{\frac{1}{2}}}{N^{s_0}}\|u\|_{L_t^\infty \dot{H}^{s_0-\frac{1}{2}}(\mathbb{R}^+)}, \end{cases} \quad \frac{1}{2} \le s_0 < 1 + \frac{4}{3}.$$

Applying refined Strichartz, $\forall t \ge 0$

$$\begin{aligned} \|u_{>N}(t)\|_2 &\lesssim \|P_{>N}F(u)\|_{2,\frac{6}{5}(\mathbb{R}^+)} \\ &\lesssim \frac{K^{\frac{1}{2}}}{N^{s_0}}\|u\|_{L_t^\infty \dot{H}^{s_0-\frac{1}{2}}}. \end{aligned}$$

Starting with $s_0 = \frac{1}{2}$, applying this estimate finitely many times yields that

$$u \in L_t^\infty H^s(\mathbb{R}^+),\ 0 \le s < 1 + \frac{4}{3}.$$ □

Finally we get a contradiction from this result.

Theorem 9.6. *There is no APMS solution on $[0,\infty)$ with $N(t) \le 1$, $\int_0^\infty N(t)^3 dt = K < \infty$ and $u \in L_t^\infty H^{1+\epsilon}(\mathbb{R}^+)$.*

Proof. Interpolating between $u \in L_t^\infty H^{1+\epsilon}(\mathbb{R}^+)$ with frequency localization we have for some $C(\eta)$

$$\|\nabla P_{>C(\eta)N(t)}u(t)\|_2 \le \eta, \forall \eta > 0.$$

We now compute the $\dot{H}^1$-norm of $u(t)$:

$$\begin{aligned} \|u(t)\|_{\dot{H}^1} &\le \|\nabla P_{\le C(\eta)N(t)}u(t)\|_2 + \|\nabla P_{>C(\eta)N(t)}u(t)\|_2 \\ &\lesssim C(\eta)N(t) + \eta. \end{aligned}$$

From Gagliardo-Nireberg inequality,

$$E \lesssim (C(\eta)N(t) + \eta)^2, \forall \eta > 0, \forall t \in \mathbb{R}^+.$$

First taking $t \to \infty$, $(N(t) \to 0)$, then taking $\eta \to 0$, we have

$$E = 0 \Rightarrow u \equiv 0.$$

A contradiction! □

We now deal with Case 2:

$u(t)$ is APMS on $[0, \infty)$ with $N(t) \leq 1$ and $\int_0^\infty N(t)^3 dt = \infty$. The way to kill this enemy is through interaction Morawetz inequality. The strategy is described as follows: Let K be a sufficiently large number (we will eventually get a contradiction by taking $K \to \infty$). From the compactness and $N(t) \leq 1$, we know (take $\eta = \frac{1}{1000}$ for example)

$$\begin{aligned} 1 &\lesssim \int_{|x| \leq \frac{C(\eta)}{N(t)}} |P_{\leq K} u(t,x)|^2 dx \\ &\lesssim \left(\int_{|x| \leq \frac{C(\eta)}{N(t)}} |P_{\leq K} u(t,x)|^4 dx \right)^{\frac{1}{2}} \left(\frac{C(\eta)}{N(t)} \right)^{\frac{3}{2}}. \end{aligned}$$

Integrating over $J \subset [0, \infty)$, we obtain

$$\int_J N(t)^3 \lesssim \int_{J \times \mathbb{R}^3} |P_{\leq K} u(t,x)|^4 dt dx. \tag{9.4}$$

If we cheat and assume $P_{\leq K} u$ is a solution to NLS. Then we are done, as from Morawetz inequality,

$$\int_{J \times \mathbb{R}^3} |P_{\leq K} u(t,x)|^4 dt dx \leq \|u\|_2^3 \sup_{t \in J} \|\nabla P_{\leq K} u(t)\|_2 \lesssim K.$$

Letting $J \to [0, \infty)$ and noting $\int_0^\infty N(t)^3 dt = \infty$, we have a contradiction.

The true case is much more involved simply because $u_{\leq K}$ is not solution.

Let $J \subset [0, \infty)$ be compact and on J

$$\int_J N(t)^3 dt = K, K \geq 1.$$

The contradiction will be seen if we can show, $\exists\ \eta \ll 1$ such that

$$\int_{J \times \mathbb{R}^3} |P_{\leq K} u(t,x)|^4 dt dx \leq \eta K, \text{ for all } K \geq 1.$$

Theorem 9.7. *There exist $\eta \ll 1$ and $K_0(\eta)$ s.t for all $K \geq K_0(\eta)$ and J for which $\int_J N(t)^3 dt = K$, we have*

$$\int_{J \times \mathbb{R}^3} |P_{\leq K} u(t,x)|^4 dt dx \leq \eta K.$$

Proof. Define

$$M(t) = \int \frac{x_j - y_j}{|x-y|} \mathrm{Im}[\bar{u}_{\leq K}(t,x)\partial_j u_{\leq K}(t,x)] |u_{\leq K}(t,y)|^2 dx dy.$$

First from trivial Hölder inequality, we know

$$\begin{aligned} M(t) &\leq \|u\|_2^3 \|\nabla u_{\leq K}\|_2 \\ &\lesssim \|\nabla P_{\leq C(\eta)} u\|_2 + \|\nabla P_{C(\eta)\leq \cdot \leq K} u\|_2 \\ &\lesssim C(\eta) + K\|u_{\geq C(\eta)}\|_2 \\ &\leq \frac{1}{2}\eta K \text{ for } K \gg C(\eta). \end{aligned}$$

Now we compute $\frac{d}{dt}M(t)$. The goal is to show

$$\int_J \frac{d}{dt} M(t) \geq \int_J \int_{\mathbb{R}^3} |P_{\leq K} u|^4 dx dt - \frac{1}{2}\eta K. \tag{9.5}$$

To see (9.5), we start with the following identity: let w be such that $iw_t + \Delta w = f$. Then

$$\begin{aligned} \partial_t \mathrm{Im}(\bar{w}\partial_j w) = &-2\partial_k \mathrm{Re}(\partial_k w \partial_j \bar{w}) + \frac{1}{2}\partial_j \Delta(|w|^2) \\ &+ \mathrm{Re}(\bar{f}\partial_j w - \bar{w}\partial_j f) \end{aligned}$$

Replace w by $u_{\leq K}$, f by

$$P_{\leq K} F(u) = F(u_{\leq K}) + P_{\leq K}F(u) - F(u_{\leq K}),$$

we have

$$\begin{aligned} \partial_t \mathrm{Im}(\bar{u}_{\leq K}\partial_j u_{\leq K}) = &-2\partial_k \mathrm{Re}(\partial_k u_{\leq K} \partial_j \bar{u}_{\leq K}) + \frac{1}{2}\partial_j \Delta(|u_{\leq K}|^2) \\ &+ \mathrm{Re}(\bar{F}(u_{\leq K})\partial_j u_{\leq K} - \bar{u}_{\leq K}\partial_j F(u_{\leq K})) \\ &+ \mathrm{Re}([P_{\leq K}\bar{F}(u) - \bar{F}(u_{\leq K})]\partial_j u_{\leq K} \\ &- \bar{u}_{\leq K}\partial_j [P_{\leq K}F(u) - F(u_{\leq K})]). \end{aligned} \tag{9.6}$$

Similarly, it is easy to find out

$$\begin{aligned} \partial_t |u_{\leq K}|^2 &= 2\mathrm{Re}(\bar{u}_{\leq K}\partial_t u_{\leq K}) \\ &= -2\nabla \cdot \mathrm{Im}(\bar{u}_{\leq K}\nabla u_{\leq K}) + \mathrm{Im}(P_{\leq K}F(u)\bar{u}_{\leq K}). \end{aligned} \tag{9.7}$$

Plugging (9.6), (9.7) into time derivative of $M(t)$, from the previous analysis on Morawetz estimate we know the first three of RHS (9.6) together with the first of RHS (9.7) will give the lower bound $\int |P_{\leq K}u(t,x)|^4 dt dx$. We are left with showing the contribution from the error terms is no bigger than ηK, i.e:

$$\begin{aligned} &\bigg| \int_{J\times\mathbb{R}^6} \frac{x_j - y_j}{|x-y|} \Big\{ [P_{\leq K}F(u) - F(u_{\leq K})]\partial_j \bar{u}_{\leq K} \\ &-\bar{u}_{\leq K}\partial_j \big[P_{\leq K}F(u) - F(u_{\leq K})\big] \Big\}(t,x)|u_{\leq K}(t,y)|^2 dx dy dt \bigg| \leq \frac{1}{100}\eta K. \end{aligned} \tag{9.8}$$

$$\left| \int_{J\times\mathbb{R}^6} \frac{x_j - y_j}{|x-y|} \mathrm{Im}(\bar{u}_{\leq K}\partial_j u_{\leq K})(t,x) \mathrm{Im}(P_{\leq K}F(u)\bar{u}_{\leq K}) dx dy dt \right| \leq \frac{1}{100}\eta K. \tag{9.9}$$

To get (9.8), (9.9), we need space time control on u. We will embed such estimates Proposition 9.8 and Corollary 9.9 into the proof.

The starting point is an improvement of the Theorem 9.2, where we proved

$$\|u_{>N}\|_{2,6(J)} \lesssim \left(\frac{K}{N}\right)^{\frac{1}{2}} + \sigma_J(N). \tag{9.10}$$

We will show a finer version of this estimate as follows:

Proposition 9.8.

$$\|u_{>N}\|_{2,6(J)} \leq C_1 \tilde{\sigma}_J(N)^{\frac{1}{2}} \left(\left(\frac{K}{N}\right)^{\frac{1}{2}} + 1\right). \tag{9.11}$$

Here, $\tilde{\sigma}_J(N)$ is the frequency envelop that majorizes $\|u_{>N}\|_{\infty,2(J)}$.

Proof. From Lemma 9.1, we have

$$\|u_{>N}\|_{2,6(J)} \lesssim \tilde{\sigma}_J(N) + \sum_{M \leq \eta N} \frac{M}{N} \|u_{>M}\|_{2,6} + \eta^{\frac{4}{3}} \|u_{>\eta N}\|_{2,6}$$
$$+ \frac{C_0 K^{\frac{1}{2}}}{(\eta N)^{\frac{1}{2}}} \sup_{J_k} \|u_{>\eta N}\|_{S^0_*(J_k \times \mathbb{R}^3)}.$$

Here J_k is subinterval for which $J = \bigcup J_k$ and $S_{J_k}(u) \sim \epsilon$. Note

$$\begin{aligned}
\|u_{>\eta N}\|_{S^0_*(J_k)} &\leq \tilde{\sigma}_J(\eta_N) + \|P_{>\eta N} F(u)\|_{2,\frac{6}{5}(J_k)} \\
&\leq \tilde{\sigma}_J(\eta_N) + \|P_{>\eta N} F(u_{<\tilde{\sigma}_J(N)N})\|_{2,\frac{6}{5}} + \|u^{\frac{4}{3}} u_{>\tilde{\sigma}_J(N)N}\|_{2,\frac{6}{5}} \\
&\lesssim \tilde{\sigma}_J(\eta N) + (\eta N)^{-1} \tilde{\sigma}_J(N) N + \tilde{\sigma}_J(N \tilde{\sigma}_J(N)) \\
&\lesssim \left(\frac{1}{\tilde{\sigma}_J(N)}\right)^{\frac{1}{3}} \tilde{\sigma}_J(N) + \eta^{-\frac{1}{3}} \tilde{\sigma}_J(N) + \eta^{-1} \tilde{\sigma}_J(N) \\
&\lesssim (\tilde{\sigma}_J(N))^{\frac{1}{2}} \eta^{-1}.
\end{aligned}$$

Hence

$$\|u_{>N}\|_{2,6} \lesssim \tilde{\sigma}_J(N) + \sum_{M \leq \eta N} \frac{M}{N} \|u_{>M}\|_{2,6} + \eta^{\frac{4}{3}} \|u_{>\eta N}\|_{2,6}$$
$$+ \frac{C_0 K^{\frac{1}{2}}}{(\eta N)^{\frac{1}{2}}} \eta^{-1} (\tilde{\sigma}_J(N))^{\frac{1}{2}}. \tag{9.12}$$

First, from (9.10) we know (9.11) for sufficiently small N. Now assume (9.11) for all frequencies $\le \eta N$, we have

$$\begin{aligned}
\|u_{>N}\|_{2,6} &\le \tilde{\sigma}_J(N) + \sum_{M\le\eta N} \frac{M}{N} C_1 \tilde{\sigma}_J(M)^{\frac{1}{2}} \left(\left(\frac{K}{M}\right)^{\frac{1}{2}} + 1\right) \\
&\quad + \eta^{\frac{4}{3}} C_1 \tilde{\sigma}_J(\eta N)^{\frac{1}{2}} \left(\left(\frac{K}{\eta N}\right)^{\frac{1}{2}} + 1\right) + C_0 \eta^{-\frac{3}{2}} \left(\frac{K}{N}\right)^{\frac{1}{2}} (\tilde{\sigma}_J(N))^{\frac{1}{2}} \\
&\le \tilde{\sigma}_J(N)^{\frac{1}{2}} \left[1 + C_1 \sum_{M\le\eta N} \frac{M}{N}\left(\frac{N}{M}\right)^{\frac{1}{6}} + C_1 \eta^{\frac{4}{3}} \eta^{-\frac{1}{6}}\right] \\
&\quad + \tilde{\sigma}_J(N)^{\frac{1}{2}} \left[C_1 \sum_{M\le\eta N} \left(\frac{M}{N}\right)^{1-\frac{1}{6}} \left(\frac{K}{M}\right)^{\frac{1}{2}}\right. \\
&\quad \left. + \eta^{\frac{4}{3}} \eta^{-\frac{1}{6}} \eta^{-\frac{1}{2}} \left(\frac{K}{N}\right)^{\frac{1}{2}} C_1 + C_0 \eta^{-\frac{3}{2}} \left(\frac{K}{N}\right)^{\frac{1}{2}}\right] \\
&\le C_1 \tilde{\sigma}_J(N)^{\frac{1}{2}} \left[1 + \left(\frac{K}{N}\right)^{\frac{1}{2}}\right].
\end{aligned}$$

If we chose C_1 large enough. □

As a corollary of this proposition, we have

Corollary 9.9. *For* $\frac{1}{2} < s \le 1$,

$$\begin{gathered}
\| |\nabla|^s P_{\le K} u\|_{2,6} \le o(K^s) (i.e. \tfrac{o(K^s)}{K^s} \to 0 \text{ as } K \to \infty), \\
\| |\nabla|^s P_{\le K} F(u)\|_{2,\frac{6}{5}} \lesssim K^s, \\
\|P_{\le K}(F(u)) - F(u_{\le K})\|_{2,\frac{6}{5}} \lesssim 1.
\end{gathered}$$

Proof. Using dyadic decomposition, we can write

$$\begin{aligned}
&\| |\nabla|^s P_{\le K} u\|_{2,6} \\
&\le \sum_{N\le K} N^s \|P_{\ge N} u\|_{2,6} \\
&\lesssim \sum_{N\le K\cdot\rho(K)} N^s \left(\frac{K}{N}\right)^{\frac{1}{2}} + \left(\sum_{K\ge N > K\cdot\rho(K)} N^s \left(\frac{K}{N}\right)^{\frac{1}{2}}\right) \Big(\tilde{\sigma}_J(K\cdot\rho(K))\Big)^{\frac{1}{2}}.
\end{aligned}$$

Here $\rho(K) \to 0$ as $K \to \infty$, $K\rho(K) \to \infty$ as $K \to \infty$, and we have used the monotone decreaseness of $\tilde{\sigma}(N)$. A quick computation gives

$$\begin{aligned}
\| |\nabla|^s P_{\le K} u\|_{2,6} &\lesssim \rho(K)^{s-\frac{1}{2}} K^s + \big(\tilde{\sigma}_J(K\rho(K))\big)^{\frac{1}{2}} K^s \\
&= o(K^s).
\end{aligned}$$

Now we prove the second one. Using the following estimate

$$\begin{aligned}
P_N F(u) &= P_N F(u_{\le N}) + O(u_{>N} |u|^{\frac{4}{3}}), \\
\|P_N F(u_{\le N})\|_{2,\frac{6}{5}} &\le N^{-1} \|\nabla P_{\le N}\|_{2,6} \|u\|_{\infty,2}^{\frac{4}{3}} \lesssim 1,
\end{aligned}$$

we have

$$\begin{aligned}
\||\nabla|^s P_{\le K} F(u)\|_{2,\frac{6}{5}} &\le \sum_{N\le K} N^s \|P_N F(u)\|_{2,\frac{6}{5}} \\
&\le \sum_{N\le K} N^s (1 + \|u_{>N}\|_{2,6}) \\
&\lesssim K^s + \sum_{N\le K} N^s (\frac{K}{N})^{\frac{1}{2}} \lesssim K^s.
\end{aligned}$$

This settles the second one. Finally, we prove the last one, write

$$\begin{aligned}
P_{\le K} F(u) - F(u_{\le K}) &= P_{\le K} F(u_{\le K}) - F(u_{\le K}) + P_{\le K}(F(u) - F(u_{\le K})) \\
&= P_{>K} F(u_{\le K}) + P_{\le K}(u_{>K} O(|u|^{\frac{4}{3}})).
\end{aligned}$$

A simple Hölder yields

$$\begin{aligned}
\|P_{\le K} F(u) - F(u_{\le K})\|_{2,\frac{6}{5}} &\lesssim \|P_{>K} F(u_{\le K})\|_{2,\frac{6}{5}} + \|u_{>K}\|_{2,6} \\
&\lesssim K^{-1} \|\nabla F(u_{\le K})\|_{2,\frac{6}{5}} + 1 \\
&\lesssim 1.
\end{aligned}$$

Corollary 9.9 is then proved. □

Now let's go back to controlling (9.8), (9.9) with these estimates. We first estimate (9.8), the contribution from the first integrand can be estimated as follows:

$$\begin{aligned}
&\int_{J\times\mathbb{R}^6} \frac{x_j - y_j}{|x-y|} [P_{\le K} F(u) - F(u_{\le K})] \partial_j \bar{u}_{\le K}(x,t) |u_{\le K}(t,y)|^2 dxdydt \\
&\lesssim \|u_{\le K}\|_{\infty,2}^2 \|\nabla u_{\le K}\|_{2,6} \|P_{\le K} F(u) - F(u_{\le K})\|_{2,\frac{6}{5}} \\
&\lesssim o(K).
\end{aligned}$$

For the second part, we can use integration by parts to transfer ∇, when ∇ hits $\bar{u}_{\le K}$, it is the same with the first, we are then left to consider for $G = P_{\le K} F(u) - F(u_{\le K})$,

$$\begin{aligned}
&\int \frac{1}{|x-y|} [\bar{u}_{\le K} G(x)] |u_{\le K}(t,y)|^2 dxdydt \\
&\lesssim \|\Delta^{-1}(\bar{u}_{\le K} G) |u_{\le K}|^2\|_{1,1} \\
&\le \|u_{\le K}\|_{8,\frac{36}{13}}^2 \|\Delta^{-1}(\bar{u}_{\le K} G)\|_{\frac{4}{3},\frac{18}{5}} \\
&\lesssim \|\nabla^{\frac{1}{6}} u_{\le K}\|_{8,\frac{12}{5}}^2 \|\bar{u}_{\le K} G\|_{\frac{4}{3},\frac{18}{17}} \\
&\lesssim \||\nabla|^{\frac{1}{6}} u_{\le K}\|_{8,\frac{12}{5}}^2 \|u_{\le K}\|_{4,9} \|G\|_{2,\frac{6}{5}} \\
&\lesssim \||\nabla|^{\frac{1}{6}} u_{\le K}\|_{8,\frac{12}{5}}^2 \||\nabla|^{\frac{2}{3}} u_{\le K}\|_{4,3}. \qquad (9.13)
\end{aligned}$$

Interpolating between

$$\|\nabla u_{\leq K}\|_{2,6} \leq o(K),$$
$$\|u_{\leq K}\|_{\infty,2} \lesssim 1,$$

we can bound

$$\||\nabla|^{\frac{1}{6}} u_{\leq K}\|_{8,\frac{12}{5}} \lesssim o(K^{\frac{1}{6}}),$$
$$\||\nabla|^{\frac{2}{3}} u_{\leq K}\|_{4,3} \lesssim o(K^{\frac{2}{3}}).$$

Plugging it back in (9.13), we see (9.13) $\lesssim o(K)$, (9.8) is proved.

Now we consider the estimate for (9.9), we write

$$\operatorname{Im}(P_{\leq K}F(u)\bar{u}_{\leq K}) = \operatorname{Im}(F(u)\bar{u}_{\leq K}) - \operatorname{Im}(P_{>K}F(u)\bar{u}_{\leq K}). \tag{9.14}$$

The contribution from the second term in (9.9) using (9.14) can be controlled as follows

$$\begin{aligned}
&\left|\int_{J\times\mathbb{R}^6} \frac{x_j - y_j}{|x-y|}\operatorname{Im}(\bar{u}_{\leq K}\partial_j u_{\leq K})(t,x)\operatorname{Im}(P_{>K}F(u)\bar{u}_{\leq K})dxdydt\right| \\
&\leq \left|\int_{J\times\mathbb{R}^6} \frac{x_j - y_j}{|x-y|}\operatorname{Im}(\bar{u}_{\leq K}\partial_j u_{\leq K})(t,x)\operatorname{Im}\left(\nabla\cdot\frac{\nabla}{\Delta}P_{>K}F(u)\bar{u}_{\leq K}\right)dtdxdy\right| \\
&\lesssim \|u_{\leq K}\nabla u_{\leq K}\|_{\infty,1}\left\|\frac{\nabla}{\Delta}P_{>K}F(u)\cdot\nabla u_{\leq K}\right\|_{1,1} && (9.15)\\
&\quad + \left\|\Delta^{-1}\left[\operatorname{Im}\left(\frac{\nabla}{\Delta}P_{>K}F(u)\bar{u}_{\leq K}\right)\right]\cdot\operatorname{Im}(\bar{u}_{\leq K}\nabla u_{\leq K})\right\|_{1,1}. && (9.16)
\end{aligned}$$

Using estimate for $P_{>K}F(u)$:

$$\begin{aligned}
\|P_{>K}F(u)\|_{2,\frac{6}{5}} &\lesssim \|P_{>K}F(u_{\leq K})\|_{2,\frac{6}{5}} + \|u^{\frac{4}{3}}u_{>K}\|_{2,\frac{6}{5}} \\
&\lesssim K^{-1}\|\nabla u_{\leq K}\|_{2,6} + \|u_{>K}\|_{2,6} \\
&\lesssim 1
\end{aligned}$$

and Bernstein, Hölder, we have

$$\begin{aligned}
(9.15) &\lesssim K\cdot K^{-1}\|P_{>K}F(u)\|_{2,\frac{6}{5}}\|\nabla u_{\leq K}\|_{2,6} \\
&\leq o(K).
\end{aligned}$$

(9.16) can be treated almost identical with (9.13):

$$\begin{aligned}
(9.16) &\lesssim \|\bar{u}_{\leq K}\nabla u_{\leq K}\|_{4,\frac{18}{13}}\left\|\Delta^{-1}\left(\bar{u}_{\leq K}\frac{\nabla}{\Delta}P_{>K}F(u)\right)\right\|_{\frac{4}{3},\frac{18}{5}} \\
&\lesssim K\|u_{\leq K}\|^2_{8,\frac{36}{13}}\|u_{\leq K}\|_{4,9}\||\nabla|^{-1}P_{>K}F(u)\|_{2,\frac{6}{5}} \\
&\lesssim \||\nabla|^{\frac{1}{6}}u_{\leq K}\|^2_{8,\frac{12}{5}}\||\nabla|^{\frac{2}{3}}u_{\leq K}\|_{4,3}\|P_{>K}F(u)\|_{2,\frac{6}{5}} \\
&\lesssim o(K).
\end{aligned}$$

Therefore the second half in (9.9) has the estimate

$$\left|\int_{J\times\mathbb{R}^6} \frac{x_j - y_j}{|x-y|}\mathrm{Im}(\bar{u}_{\leq K}\partial_j u_{\leq K})(t,x)\mathrm{Im}(P_{>K}F(u)\bar{u}_{\leq K})dxdydt\right| \leq o(K).$$

The contribution from the first term $\mathrm{Im}(F(u)\bar{u}_{\leq K})$ to (9.9) is discussed as follows. Note

$$\begin{aligned}\mathrm{Im}(F(u)\bar{u}_{\leq K}) &= \mathrm{Im}(F(u)\bar{u}_{>K}) = \mathrm{Im}(F(u_{\leq K})\bar{u}_{>K}) + O(|u|^{\frac{4}{3}}u_{>K}\bar{u}_{>K}) \\ &= \mathrm{Im}\Big(F(u_{\leq K})\nabla\cdot\nabla\Delta^{-1}\bar{u}_{>K}\Big) + O(|u|^{\frac{4}{3}}u_{>K}\bar{u}_{>K}).\end{aligned}$$

From Hölder, integration by parts,

$$\begin{aligned}&\left|\int_{J\times\mathbb{R}^6} \frac{x_j - y_j}{|x-y|}\mathrm{Im}(\bar{u}_{\leq K}\partial_j u_{\leq K})(t,x)\mathrm{Im}(F(u)\bar{u}_{\leq K})dxdydt\right| \\ &\lesssim \|u_{\leq K}\nabla u_{\leq K}\|_{\infty,1}\||u|^{\frac{4}{3}}u_{>K}\bar{u}_{>K}\|_{1,1} \\ &\quad + \|u_{\leq K}\nabla u_{\leq K}\|_{\infty,1}\|\nabla F(u_{\leq K})\nabla\Delta^{-1}\bar{u}_{>K}\|_{1,1} \\ &\quad + \|\bar{u}_{\leq K}\nabla u_{\leq K}\Delta^{-1}(F(u_{\leq K})\nabla\Delta^{-1}u_{>K})\|_{1,1} \\ &\lesssim K\|u_{>K}\|_{2,6}^2 + k\|\nabla u_{\leq K}\|_{2,6}\||\nabla|^{-1}u_{>K}\|_{2,6} \\ &\quad + \|\bar{u}_{\leq K}\nabla u_{\leq K}\|_{\infty,\frac{9}{8}}\|\Delta^{-1}(F(u_{\leq K})\nabla^{-1}u_{>K})\|_{1,9} \\ &\lesssim o(K) + \|u_{\leq K}\|_{\infty,2}\|\nabla u_{\leq K}\|_{\infty,(\frac{1}{2}-\frac{1}{9})^{-1}}\|F(u_{\leq K})\nabla^{-1}u_{>K}\|_{1,\frac{9}{7}} \\ &\leq o(K) + \|\nabla^{\frac{4}{3}}u_{\leq K}\|_{\infty,2}\|F(u_{\leq K})\|_{2,\frac{18}{11}}\|\nabla^{-1}u_{>K}\|_{2,6} \\ &\leq o(K) + K^{\frac{4}{3}}\|\nabla^{\frac{2}{3}}F(u_{\leq K})\|_{2,\frac{6}{5}}K^{-1}\|u_{>K}\|_{2,6} \\ &\leq o(K).\end{aligned}$$

Collecting all the estimates together, we proved (9.9), (9.10). The contradiction is then established, so is the non-existence of Case 2. □

10 Appendix: Marcinkiewicz interpolation theorem

Unlike the Riesz interpolation theorem, the Marcinkiewicz interpolation deals with more general operators.

Definition 10.1. *Assume operator T is an operator between two spaces of functions. T is called quasilinear if*

$$\begin{aligned}|T(f+g)(x)| &\leq |Tf(x)| + |Tg(x)|, \\ |T(cf)(x)| &\leq |c||Tf(x)|, \ \forall \text{ complex number } c.\end{aligned}$$

Another remarkable improvement for Marcinkiewicz interpolation is that it can deal with the operator only having weak boundedness at the end point. We start with the weak L^p space.

Let $(X, d\mu)$ be a measure space, for $f \in L^p(X)$, we easily see that

$$\begin{aligned}\{x : |f(x)| > \lambda\} &= \int_X \chi_{\{x:|f(x)|>\lambda\}} dx \\ &\leq \int_X \frac{|f(x)|^p}{\lambda^p} dx \\ &\lesssim \lambda^{-p} \|f\|_p^p.\end{aligned}$$

Therefore

$$\sup_{\lambda>0} \lambda m\{x; |f(x)| > \lambda\}^{\frac{1}{p}} \lesssim \|f\|_p.$$

This motivates us to define

Definition 10.2 (Weak L^p space)**.** *Let $(X, d\mu)$ be a measure space. Let $1 \leq p < \infty$. Define the weak L^p space*

$$L^{p,\infty} = \{f;\ \|f\|_{p,\infty} = \sup_{\lambda>0} \lambda m\{x : |f(x)| > \lambda\}^{\frac{1}{p}} < \infty\}.$$

When $p = \infty$, we formally define it as the normal L^∞ space.

Several remarks about this definition.

Remark 10.3. 1. *For $1 \leq p < \infty$, $L^p(X) \subset L^{p,\infty}(X)$ and $L^p \neq L^{p,\infty}$. Indeed, one can easily check that*

$$|x|^{-\frac{n}{p}} \notin L^p(\mathbb{R}^n, dx).$$

On the other hand,

$$\begin{aligned}\lambda m\{x : |f(x)| > \lambda\}^{\frac{1}{p}} &= \lambda m\{x : |x| < \lambda^{-\frac{p}{n}}\}^{\frac{1}{p}} \\ &= \lambda(\lambda^{-p})^{\frac{1}{p}} = 1.\end{aligned}$$

That is, $|x|^{-\frac{n}{p}} \in L^{p,\infty}(\mathbb{R}^n, dx)$.

2. *$L^{p,\infty}$ is a quasi-normed space because*

$$\|f + g\|_{p,\infty} \leq 2(\|f\|_{p,\infty} + \|g\|_{p,\infty}).$$

3. *For $1 \leq q < \infty$, One can also discuss the general Lorentz space $L^{p,q}(X, d\mu)$ whose quasinorm is defined by*

$$\|f\|_{p,q} = \|\lambda m\{x \in X; |f(x)| > \lambda\}^{\frac{1}{p}}\|_{L^q(\frac{d\lambda}{\lambda})}.$$

It is easy to verify that for $p = q$, $L^{p,q} = L^p$. For $1 \leq q_1 < q_2 \leq \infty$, the following embedding hold

$$L^{p,q_1} \subset L^{p,q_2}.$$

We have enough preparation to introduce

Theorem 10.4 (Marcinkiewicz interpolation, diagonal version). *Let $(X, d\mu)$, $(Y, d\nu)$ be two measure spaces. Let $1 \le p_0 < p < p_1 \le \infty$ and*

$$T: \ L^{p_i}(X, d\mu) \to L^{p_i}(Y; d\nu), \ i = 0, 1$$

be weakly L^{p_i} bounded in the sense that

$$\|Tf\|_{p_0,\infty} := \sup_{\lambda>0} \lambda \ mes\{|Tf(x)| > \lambda\}^{\frac{1}{p_0}} \le C_0 \|f\|_{p_0},$$

$$\|Tf\|_{p_1,\infty} := \sup_{\lambda>0} \lambda \ mes\{|Tf(x)| > \lambda\}^{\frac{1}{p_1}} \le C_1 \|f\|_{p_1}.$$

Then T is strongly L^p bounded,

$$\|Tf\|_p \lesssim C_p \|f\|_p.$$

Proof. We first discuss the first case $p_1 < \infty$. The computation starts from the layer cake representation. For any $f \in L^p$,

$$\begin{aligned}
\|f\|_p^p &= \int_{\mathbb{R}^n} |f(x)|^p dx \\
&= \frac{1}{p} \int_{\mathbb{R}^n} \int_0^{|f(x)|} \lambda^{p-1} d\lambda dx \\
&= \frac{1}{p} \int_0^\infty \int_{x:|f(x)|>\lambda} dx \lambda^{p-1} d\lambda \\
&= \frac{1}{p} \int_0^\infty \lambda^{p-1} m\{x : |f(x)| > \lambda\} d\lambda.
\end{aligned}$$

We can use this to estimate the L^p norm of Tf. To this end, we first decompose

$$f(x) = f_\lambda(x) + f^\lambda(x); \ f_\lambda(x) = f(x)\chi_{|f(x)|\le\lambda}, \ f^\lambda(x) = f(x)\chi_{|f(x)|>\lambda}.$$

Obviously for $f \in L^p$, $f_\lambda \in L^{p_1}$, $f^\lambda \in L^{p_0}$. Second, since T is sublinear,

$$|Tf(x)| \le |Tf_\lambda(x)| + |Tf^\lambda(x)|.$$

Hence

$$m\{x : |Tf(x)| > \lambda\} \le m\{x : |Tf_\lambda(x)| > \lambda/2\} + m\{x : |Tf^\lambda(x)| > \lambda/2\}.$$

We now continue the estimate by using the weak L^{p_i} boundedness of T:

$$\begin{aligned}
\|Tf\|_p^p &= \frac{1}{p} \int_0^\infty \lambda^{p-1} m\{x : |Tf(x)| > \lambda\} d\lambda \\
&\lesssim_p \int_0^\infty \lambda^{p-1} m\{x : |Tf_\lambda(x)| > \lambda/2\} d\lambda \\
&\quad + \int_0^\infty \lambda^{p-1} m\{x : |Tf^\lambda(x)| > \lambda/2\} d\lambda \\
&\lesssim_p \int_0^\infty \lambda^{p-p_1-1} \|f_\lambda\|_{p_1}^{p_1} d\lambda + \int_0^\infty \lambda^{p-p_0-1} \|f^\lambda\|_{p_0}^{p_0} d\lambda.
\end{aligned}$$

As

$$\begin{aligned}\int_0^\infty \lambda^{p-p_1-1}\|f_\lambda\|_{p_1}^{p_1}d\lambda &= \int_0^\infty \lambda^{p-p_1-1}\int_{\{x:|f(x)|<\lambda\}}|f(x)|^{p_1}dxd\lambda\\ &\le \int_0^\infty |f(x)|^{p_1}\int_{|f(x)|}^\infty \lambda^{p-p_1-1}d\lambda dx\\ &\lesssim \int_0^\infty |f(x)|^{p_1}|f(x)|^{p-p_1}dx\\ &= \|f\|_p^p,\end{aligned}$$

similarly one can get

$$\int_0^\infty \lambda^{p-p_0-1}\|f^\lambda\|_{p_0}^{p_0}d\lambda \lesssim \|f\|_p^p.$$

Combining the two pieces together we finally have

$$\|Tf\|_p^p \lesssim \|f\|_p^p.$$

In the case $p_1=\infty$, we notice that

$$\|f_\lambda\|_\infty \le \lambda$$

and

$$\|Tf_\lambda\|_\infty \le C_1\|f_\lambda\|_\infty \le C_1\lambda.$$

Instead of considering f, we consider $\frac{f}{2C_1}$. From

$$\left|T\left(\frac{f}{2C_1}\right)(x)\right| \le \frac{1}{2C_1}|Tf(x)|,$$

we know

$$\begin{aligned}m\left\{x:\left|T\left(\frac{f}{2C_1}\right)(x)\right|>\lambda\right\} &\le m\{x:1/(2C_1)|Tf(x)|>\lambda\}\\ &= m\{x:|Tf(x)|>2C_1\lambda\}\\ &\le m\{x:|Tf^\lambda(x)|>C_1\lambda\}+m\{x:|Tf_\lambda(x)|>C_1\lambda\}\\ &= m\{x:|Tf^\lambda(x)|>C_1\lambda\}.\end{aligned}$$

Plug it in we have

$$\begin{aligned}\left\|T\left(\frac{f}{2C_1}\right)\right\|_p^p &= \int_0^\infty \lambda^{p-1}m\left\{x:\left|T\left(\frac{f}{2C_1}\right)(x)\right|>\lambda\right\}d\lambda\\ &\le \int_0^\infty \lambda^{p-1}m\{x:|Tf^\lambda(x)|>C_1\lambda\}d\lambda\\ &\le \int_0^\infty \lambda^{p-1}\frac{\|f^\lambda\|_{p_0}^{p_0}}{(C_1\lambda)^{p_0}}d\lambda\end{aligned}$$

$$\lesssim \int_0^\infty \lambda^{p-p_0-1}\|f^\lambda\|_{p_0}^{p_0} d\lambda$$
$$\lesssim \|f\|_p^p.$$

Renormalizing this estimate yields the desired result. □

The full version of Marcinkiewicz interpolation involves more general index but can be proved similarly. Here we only record the detailed statement and skip the proof.

Theorem 10.5 (Marcinkiewicz interpolation, general version). *Let $1 \le p_i \le q_i \le \infty$, $i = 0, 1$ and $q_0 \ne q_1$. Let $1 < p \le q < \infty$ and there exist $\theta \in (0,1)$ such that*

$$\frac{1}{p} = \frac{1-\theta}{p_0} + \frac{\theta}{p_1},$$
$$\frac{1}{q} = \frac{1-\theta}{q_0} + \frac{\theta}{q_1}.$$

Suppose T is quasilinear operator of weak type (p_i, q_i), i. e:

$$\|Tf\|_{q_i,\infty} \le C_i f \|f\|_{p_i}.$$

Then T is bounded from L^p to L^q with the bound depending on C_i and θ.

Next we give an example of Marcinkiewicz interpolation theorem.

Theorem 10.6 (Hardy-Littlewood-Paley Theorem). *For any $1 < p \le 2$, there exists c_p such that*

$$\left(\int_{\mathbb{R}^n} |\hat{f}(x)|^p |x|^{(p-2)n} dx\right)^{\frac{1}{p}} \le c_p \left(\int_{\mathbb{R}^n} |f(x)|^p dx\right)^{\frac{1}{p}}. \tag{10.1}$$

For any $2 \le q < \infty$, there exists c_q such that

$$\left(\int_{\mathbb{R}^n} |\hat{f}(x)|^q dx\right)^{\frac{1}{q}} \le c_q \left(\int_{\mathbb{R}^n} |f(x)|^q |x|^{(q-2)n} dx\right)^{\frac{1}{q}}. \tag{10.2}$$

Remark 10.7. *We first note that the power of x is suggested by the scaling. Indeed, under the scaling transformation*

$$f(x) \to f(\lambda x),$$

$$LHS(10.1) = \|\hat{f}|x|^{\frac{s}{p}}\|_p \to \|\lambda^{-n}\hat{f}(\lambda^{-1}x)|\lambda^{-1}x|^{\frac{s}{p}}\|_p \lambda^{\frac{s}{p}} = \lambda^{-n+\frac{s}{p}+\frac{n}{p}}\|\hat{f}|x|^{\frac{s}{p}}\|_p,$$
$$RHS(10.1) \sim \|f\|_p \to \lambda^{-\frac{n}{p}}\|f\|_p.$$

Matching the exponent in λ we see $s = (p-2)n$.

We now give the proof of this theorem.

Proof. Note the $RHS(10.1) = \|\hat{f}|x|^n\|_{L^p(|x|^{-2n}dx)}$, we define two measure spaces

$$(X,\mu) = (\mathbb{R}^n, dx)\,(Y,\nu) = (\mathbb{R}^n, |x|^{-2n}dx)$$

and the operator

$$Tf(x) = |x|^n\hat{f}(x).$$

Then for $f \in L^2(\mathbb{R}^n)$, from Parseval inequality $\|\hat{f}\|_2 = \|f\|_2$, we immediately get

$$\|Tf\|_{L^2(Y,d\nu)} \leq \|f\|_{L^2(X,d\mu)}.$$

Now we prove T is weak $(1,1)$ type.

For any $\lambda > 0$, as the set

$$\begin{aligned}\{x \in Y;\ |Tf(x)| > \lambda\} &= \{x \in Y;\ |x|^n|\hat{f}(x)| > \lambda\}\\ &\subset \{x \in Y;\ |x|^n\|f\|_1 > \lambda\}\\ &\subset \left\{x \in Y; |x|^n > \frac{\lambda}{\|f\|_1}\right\}\\ &= \int_Y \chi_{|x|>\left(\frac{\lambda}{\|f\|_1}\right)^{\frac{1}{n}}} d\nu\\ &= \int_{\mathbb{R}^n} |x|^{-2n}\chi_{|x|>\left(\frac{\lambda}{\|f\|_1}\right)^{\frac{1}{n}}} dx\\ &\lesssim \frac{\|f\|_1}{\lambda},\end{aligned}$$

taking supremum in λ we get

$$\sup_{\lambda>0}\lambda\{x \in Y;\ |Tf(x)| > \lambda\} \lesssim \|f\|_1.$$

Therefore T is weak $(1,1)$ type operator. Interpolating with L^2 boundedness of T yields the estimate (10.1).

(10.2) is the dual estimate of (10.1). We therefore prove the theorem. □

Exercise.

1. Let $\phi(x) \geq 0$ satisfy

$$m\{x:\ \phi(x) > t\} \lesssim t^{-1}.$$

Show there exist a constant $c > 0$ such that

$$\left(\int_{\mathbb{R}^n}\left|\frac{\hat{f}(x)}{\phi(x)}\right|^p \phi^2(x)dx\right)^{\frac{1}{p}} \leq c\|f\|_p,\ \forall 1 < p \leq 2.$$

2. Give a counter example showing the condition $q_0 \neq q_1$ in the Marcinkiewicz interpolation is unremovable. (Hint: Consider

$$Tf(x) = \frac{1}{\sqrt{x}}\int_0^1 f(t)dt.\ )$$

3. Try to prove the general version of Marcinkiewicz interpolation theorem.

References

[1] J. Bergh and J. Lofstrom, *Interpolation Spaces, An Introduction*, Springer-Verlag, 1976.

[2] P. Begout and A. Vargas, *Mass concentration phenomena for the L^2-critical nonlinear Schrödinger equation*, Trans. Amer. Math. Soc. **359** (2007), 5257–5282.

[3] J. Bourgain, *Global well-posedness of defocusing 3D critical NLS in the radial case*, J. Amer. Math. Soc **12** (1999), 145–171.

[4] J. Bourgain, *New Global Well-posedness Results for Non-linear Schrödinger Equations*, AMS Publications, 1999.

[5] R. Carles and S. Keraani, *On the role of quadratic oscillations in nonlinear Schrödinger equation II. The L^2-critical case*, Trans. Amer. Math. Soc. **359** (2007), 33–62.

[6] T. Cazenave and F. B. Weissler, *Some remarks on the nonlinear Schrödinger equation in the critical case*, Nonlinear semigroups, Partial Differential Equations and Attractors, Lecture Notes in Math. **1394** (1989), 18–29.

[7] T. Cazenave and F. B. Weissler, *Critical nonlinear Schrödinger Equation*, Non. Anal. TMA **14** (1990), 807–836.

[8] M. Christ, J. Colliander and T. Tao, *Ill-posedness for nonlinear Schrödinger and wave equations*, to appear Annales IHP. math.AP/0311048.

[9] M. Christ and A. Kiselev, *Maximal operators associated to filtrations*, J. Funct. Anal. **179** (2001), 409–425.

[10] J. Colliander, M. Keel, G. Staffilani, H. Takaoka and T. Tao, *Global existence and scattering for rough solutions of a nonlinear Schrödinger equation on $\mathbb{R}^3$*, Comm. Pure Appl. Math. **57** (2004), 987–1014.

[11] J. Colliander, M. Keel, G. Staffilani, H. Takaoka and T. Tao, *Global well-posedness and scattering for the energy-critical nonlinear Schrödinger equation in $\mathbb{R}^3$*, Annals of Math. **167** (2007), 767–865.

[12] P. Constantin and J. C. Saut, *Local smoothing properties of dispersive equations*, J. Amer. Math. Soc. **1** (1988), 413–439.

[13] B. Dodson, *Global well-posedness and scattering for the defocusing L^2-critical nonlinear Schrödinger equation when $d \geq 3$*, J. Amer. Soc, **25**(2012), 429–463.

[14] B. Dodson, *Global well-posedness and scattering for the defocusing L^2-critical nonlinear Schrödinger equation when $d \geq 2$*, Preprint, math.AP/1006.1375.

[15] B. Dodson, *Global well-posedness and scattering for the defocusing L^2-critical nonlinear Schrödinger equation when $d \geq 1$*, Preprint, math.AP/1010.0040.

[16] B. Dodson, *Global well-posedness and scattering for the mass critical nonlinear Schrödinger equation with mass below the mass of the ground state*, Preprint, math.AP/1104.1114.

[17] D. Foschi, *Inhomogeneous Strichartz estimates*, J. Hyperbolic Differ. Equ. **2** (2005), 1–24.

[18] J. Ginibre and G. Velo, *Scattering theory in the energy space for a class of nonlinear Schrödinger equations*, J. Math. Pure. Appl. **64** (1985), 363–401.

[19] P. Gérard, *Description du défaut de compacité de l'injection de Sobolev*, ESAIM Control Optim. Calc. Var. **3** (1998), 213–233.

[20] T. Hmidi and S. Keraani, *Blowup theory for the critical nonlinear Schrödinger equations revisited*, Int. Math. Res. Not. **46** (2005), 2815–2828.

[21] M. Keel and T. Tao, *Endpoint Strichartz estimates*, Amer. J. Math. **120** (1998), 955–980.

[22] C. Kenig and F. Merle, *Global well-posedness, scattering and blow-up for the energy-critical, focusing nonlinear Schrödinger equation in the radial case*, Invent. Math. **166** (2006), 645–675.

[23] S. Keraani, *On the defect of compactness for the Strichartz estimates for the Schrödinger equations*, J. Diff. Eq. **175** (2001), 353–392.

[24] S. Keraani, *On the blow-up phenomenon of the critical nonlinear Schrödinger equations*. J. Diff. Eq. **235**, (2006), 171–192.

[25] R. Killip and M. Visan, *Nonlinear Schrödinger Equation at Critical Regularity*, Clay Mathematics Proceedings. **17**, 2013.

[26] R. Killip and M. Visan, *The focusing energy-critical nonlinear Schrödinger equation in dimensions five and higher*, Amer. J. Math. **132** (2010), 361–424.

[27] R. Killip, T. Tao and M. Visan, *The cubic Schrödinger equation with radial data*, J. Eur. Math. Soc. **11** (2009), 1203–1258.

[28] R. Killip, M. Visan and X. Zhang, *The mass-critical nonlinear Schrödinger equation with radial data in dimentions three and higher,* Analysis and PDE **1** (2008), 229–266.

[29] M. Keel, T. Tao, *Endpoint Strichartz estimates*, Amer. Math. J. **120** (1998), 955–980.

[30] J. Bourgain and D. Li, *On an endpoint Kato-Ponce inequality*, Differential Integral Equations **27** (2014), 1037–1072.

[31] J. Lin and W. Strauss, *Decay and scattering of solutions of a nonlinear Schrödinger equation*, J. Funct. Anal. **30** (1978), 245–263.

[32] F. Merle and Y. Tsutsumi, *L^2-concentration of blow-up solutions for the nonlinear Schrödinger equation with the critical power nonlinearity*, J. Diff. Eq. **84** (1990), 205–214.

[33] F. Merle, *Determination of blow-up solutions with minimal mass for nonlinear Schrödinger equations with critical power*, Duke Math. J. **69** (1993), 427–454.

[34] F. Merle and L. Vega, *Compactness at blow-up time for L^2 solutions of the crtical nonlinear Schrödinger equation in 2D,* Int. Math. Res. Not. **8** (1998), 399–425.

[35] C. Morawetz, *Time decay for the nonlinear Klein-Gordon equation*, Proc. Roy. Soc. A **306** (1968), 291–296.

[36] P. Sjölin, *Regularity of solutions to the Schrödinger equation*, Duke Math. J. **55** (1987), 699–715.

[37] E. M. Stein, *Singular Integrals and Differentiability Properties of Functions*, Princeton University Press, 1970.

[38] E. M. Stein, *Harmonic Analysis*, Princeton University Press, 1993.

[39] T. Tao, *Nonlinear dispersive equations: local and global analysis*, CBMS regional conference series in mathematics, July 2006.

[40] T. Tao, *Spherically averaged endpoint Strichartz estimates for the two-dimensional Schrdinger equation*, Comm. PDE **25** (2000), 1471–1485.

[41] T. Tao, *A sharp bilinear restrictions estimate for paraboloids*, Geom. Funct.

Anal. **13** (2003), 1359–1384.

[42] T. Tao, *Global well-posedness and scattering for the higher-dimensional energy-critical non-linear Schrödinger equation for radial data*, New York Journal of Mathematics **11** (2005), 57–80.

[43] T. Tao, M. Visan and X. Zhang, *The nonlinear Schrödinger equation with combined power type nonlinearities,* Comm. PDE **32** (2007), 1281–1343.

[44] T. Tao, M. Visan and X. Zhang, *Minimal-mass blowup solutions of the mass-critical NLS*, Forum Math. **20** (2008), 881–919.

[45] T. Tao, M. Visan and X. Zhang, *Global well-posedness and scattering for the mass-critical defocusing NLS with spherical symmetry in higher dimensions*, Duke Math. J. **140** (2007), 165–202.

[46] L. Vega, *Schrödinger equations: pointwise convergence to the initial data*, Proc. Amer. Math. Soc **102** (1988), 874–878.

[47] M. Visan, *The defocusing energy-critical nonlinear Schrödinger equation in high dimensions*, Duke Math. J. **138** (2007), 281–374.

[48] M. Weinstein, *Nonlinear Schrödinger equations and sharp interpolation estimates,* Comm. Math. Phys. **87** (1983), 567–576.